T0360795

Deterministic and Stochastic Topics in Computational Finance

Deterministic and Stochastic Topics in
Computational Finance

Ovidiu Calin

Princeton University

 **World Scientific**

NEW JERSEY · LONDON · SINGAPORE · BEIJING · SHANGHAI · HONG KONG · TAIPEI · CHENNAI · TOKYO

Published by

World Scientific Publishing Co. Pte. Ltd.

5 Toh Tuck Link, Singapore 596224

USA office: 27 Warren Street, Suite 401-402, Hackensack, NJ 07601

UK office: 57 Shelton Street, Covent Garden, London WC2H 9HE

British Library Cataloguing-in-Publication Data
A catalogue record for this book is available from the British Library.

DETERMINISTIC AND STOCHASTIC TOPICS IN COMPUTATIONAL FINANCE

ISBN 978-981-3203-07-5
ISBN 978-981-3203-08-2 (pbk)

Printed in Singapore

Preface

The present book invites the reader to a journey through the mathematical models of financial markets wonderland. This type of approach is customarily called *computational finance*, to both emphasize the theoretically computational aspect and make the distinction from the *empirical finance*, which uses much less sophisticated mathematical methods to perform its analysis. The attributes *deterministic* and *stochastic* present in the title suggest both types of models approached in the book. The first type is just an approximation of the former, and is applicable when special conditions are satisfied (e.g. just for the short run or in the case when the noise that impacts the market parameters is neglected).

It is worth to remark that none of the models presented in this book is an ultimate correct model of the markets. Contrary to the laws of physical world, which stay unchanged for ever, the models that govern financial markets proved not to be time invariant; this continuous change of the market behavior makes all models we know so far to be incorrect to a larger or a smaller extent. Furthermore, probably no *right* model will be ever found, since the dynamics of the markets are driven by the humans psychology itself, whose nature seems to be of little mathematical nature. Nevertheless, the best we can do, given these environment conditions, is to cast the market behavior into mathematical formulas as faithfully as we can. Consequently, the book presents mathematical models starting from relative simple ones, such as Black-Scholes, and heading to very complex, but extremely popular these days, such as stochastic volatility models or stochastic returns models.

This book had grown from the lecture notes written for a one-semester course on topics on computational finance for senior and graduate students. The prerequisites contain an introduction to stochastic calculus course (covered, for instance, by the book [15]), as well as the usual Calculus sequence courses. The book is addressed to undergraduate and graduate students in Masters in Finance programs as well as to individuals who will be using this material to become more efficient in their practical applications.

The book presents continuous-time models for financial markets and their closed form solutions. A distinguished emphasis is put on the fact that there are two main and actually equivalent ways of pricing options: using stochastic calculus by computing expectations and solving parabolic partial differential equations. There are several books covering topics of mathematical finance, some using the former technique, while others taking advantage of the latter. The equivalence between these two pricing techniques is given by a central result, called the Feynmann-Kac formula, which provides the solution of a parabolic partial differential equation as an expectation of a certain stochastic process. The present book takes equally advantage of both techniques, the reader being exposed to a variety of examples treated from both stochastic and PDEs points of view.

The book is written in a very clear and simple way, and maybe less rigorous sometimes, especially when it comes to the mathematical regularity conditions, which were always assumed satisfied. These rigor sacrifices were made with the hope of increasing the book audience among non-mathematicians, including Business and Economics students with a limited mathematical background. The appendices contain most of the mathematical prerequisites to make the reader follow the text, without the need of an extra text. For a more mathematically oriented approach the reader is referred to the books of Etheridge [31], Gulisashvili [37] and Kijima [49]. For the heat kernels method beyond pricing derivatives the reader is referred to the book of Avramidi [4]. For the less mathematically sophisticated reader, we suggest the reference textbook of Hull [42].

In this book the reader has the advantage to be able to actually compute a large number of prices for European, American or Asian derivatives. The key feature of this textbook is the large number of detailed examples and proposed problems, most of them solved in the last chapter, from which will benefit both the beginner as well as the advanced student. A flow chart indicating the possible order the reader can follow can be found at the end of this preface.

This book had grown from a series of lectures and courses given by the author at Princeton University (USA), Eastern Michigan University (USA), Kuwait University (Kuwait) and Fu-Jen University (Taiwan). The student body was very variate. I had math, statistics, computer science, economics and business majors. At the initial stage, several students read the first draft of these notes and provided valuable feedback, supplying a list of corrections, which might not be exhaustive. Finding any typos or making comments regarding the present material are welcome.

The text is divided into five parts. Part I, *Introduction*, introduces the reader into the world of determinism and stochasticity, including some statistical methods for regression. Part II, *Interest Rates and Bonds*, presents several stochastic models for the spot rate and prices bonds in these cases. Part III, *Risk-Neutral Valuation Pricing*, uses the method of expectations to price the most important European derivatives; it also contains material regarding martingale measures and the risk-neutral measure. Part IV, *PDE Approach*, introduces the reader into the Black-Scholes analysis of plain vanilla options, as well as Asian and American options. In the case of American options, where no exact formulas exist, we provided some approximations near the boundary, and in the case of perpetuities we provided closed form solutions. Part V, *Stochastic Volatility and Return Models*, is that part of the book which deals with the most up-to-date developments in the financial field. These include stochastic volatility models, such as Heston, GARCH and AR(1). The last chapter is dedicated to VAR process used to price options in the case when the underlying asset has a stochastic rate of return.

Last, but not least, we notify the reader that this book is not a recipe book on teaching how to make money on the stock market. On the contrary, the present book teaches how to keep "what you already have" rather than "making some extra more". The former deals with sophisticated mathematical models that remove risk, called hedging, while the latter is just speculation, based on trading market irregularities. The reader interested in "speculation" might want to check with some materials involving pattern recognition, wavelets and Fourier transforms, machine learning, as well as other statistical approaches geared towards finding a "signal" or an "alpha" in the market on which the trader can bid. For an introductory textbook in using statistical analysis to the study of financial data the reader is referred to Carmona [20]. For a method to identify arbitrage opportunities in the market by means of *cointegration*, the reader may consult Tsay [63].

Heartfelt thanks go to the reviewers who made numerous comments and observations contributing to the quality of this book, and whose time is very much appreciated. I am specially indebt to Thomas Cosimano who pointed me out the use of OU processes in the study of stochastic rate of return models. Finally, I would like to express my gratitude to World Scientific Publishing team, especially to Rok-Ting Tan for her excellent job done under pressing time constraints.

Princeton, NJ, USA, 2016 Ovidiu Calin

Chapters Diagram

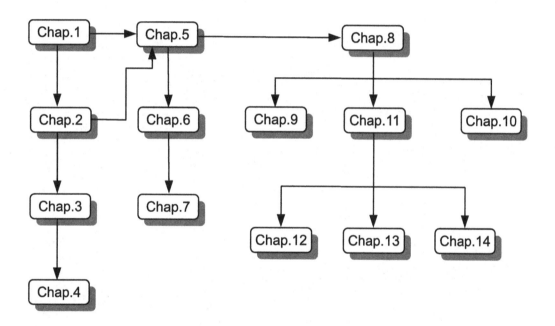

List of Notations and Symbols

$(\Omega, \mathcal{F}, P)$	Probability space
Ω	Sample space
$\mathcal{F}$	σ-field
X	Random variable
X_t	Stochastic process
$\mathcal{F}_t$	Filtration
$\dfrac{dW_t}{dt}$	White noise
W_t, B_t	Brownian motion
$\Delta W_t, \Delta B_t$	Jumps of the Brownian motion during time interval Δt
dW_t, dB_t	Infinitesimal jumps of the Brownian motion
$\mathbb{E}(X_t)$	The mean of X_t
$\mathbb{E}^x(X_t)$	Expectation of X_t, given $X_0 = x$
$\mathbb{E}[X_t|X_0 = x]$	Expectation of X_t, given $X_0 = x$
$\mathbb{E}(X_t|\mathcal{F}_s)$	Conditional expectation of X_t, given $\mathcal{F}_s$
$Var(X_t)$	Variation of X_t
$F_X(x)$	Probability distribution function of X
$p_X(x)$	Probability density function of X
$p(t; x, y)$	Transition density function
$\mathbb{E}[\,\cdot\,]$	Expectation operator
$Var(X)$	Variance of the random variable X
$cov(X, Y)$	Covariance of X and Y
$corr(X, Y)$	Correlation of X and Y
$\Gamma(\,\cdot\,)$	Gamma function
$B(\,\cdot\,,\cdot\,)$	Beta function
$\mathcal{N}(\mu, \sigma^2)$	Normally distributed with mean μ and variance σ^2

N_t	Poisson process
M_t	Compensated Poisson process
S_t	Stock price
μ	Drift rate
σ	Volatility, standard deviation
σ_{imp}	Implied volatility
r	Risk-free interest rate
δ	Continuous dividend rate
r_t	Spot rate
$c(t)$	European call price
$C(t), C_A(t)$	American call price
$p(t)$	European put price
$P(t), P_A(t)$	American put price
ν_t, v_t	Stochastic volatility
$b(t), b(t,v)$	Free-boundary
$P(t,T)$	Bond price
$R(t,T)$	Yields curve
$f(t,T)$	Instantaneous forward rate
V_T	Derivative payoff
V_t	Derivative value at t
$\tau = T - t$	Time to maturity
Π_t	Portfolio value
$\partial_{x_k}, \dfrac{\partial}{\partial x_k}$	Partial derivative with respect to x_k
$\mathbb{R}^n$	n-dimensional Euclidean space
$\|x\|$	Euclidean norm $(= \sqrt{x_1^2 + \cdots + x_n^2})$
$\mathbb{I}_A, 1_A, \chi_A$	The characteristic function of A
$\|f\|_{L^2}$	The L^2-norm $(= \sqrt{\int_a^b f(t)^2\, dt})$
$L^2[0,T]$	Squared integrable functions on $[0,T]$
$C^2(\mathbb{R}^n)$	Functions twice differentiable with second derivative continuous
$C_0^2(\mathbb{R}^n)$	Functions with compact support of class C^2.

Contents

Part I

Introduction

Chapter 1

Determinism or Stochasticity?

1.1 Determinism and Semi-determinism

Any system (physical, economic, financial, etc.) can be described by a certain number of parameters. In order to know and understand the system, one observes or measures these parameters that drive the dynamics of the system. We shall assume we are able to measure all parameters with precise accuracy (even if in some cases this cannot be performed due to some uncertainty principle). They will describe the system as coordinates of a point in a parameter space. The number of parameters denotes the dimension of the space which models the configuration of the system.

For instance, a moving particle in the plane is characterized by its position and velocity, (x, v), so the parameter space is 6-dimensional, while the balance amount in a certificate of deposit account is characterized by a single number, so the system is 1-dimensional. If the system evolves in time, its parameters are functions of time, and the system can be represented by a curve in the aforementioned space. One of the most important problems is to predict and describe the trajectories of the analyzed systems. These can be of three types, and they are discussed briefly in the following.

A system is called *determinist* if both future and past states of the system are uniquely determined by the present state of the system. Most systems studied by classical mechanics are determinist. For instance, if the present position, x_0, and velocity, v_0, of a car are known at time t_0, then integrating the equations of motion, $x''(t) = a$, with constant acceleration a, one can find both future and past positions and velocities of the car, according to the formulas

$$v(t) = v_0 + a(t - t_0)$$
$$x(t) = x_0 + v_0(t - t_0) + \frac{1}{2}a(t - t_0)^2,$$

where t is the time parameter. For $t > t_0$ we obtain the future states of the car, while for $t < t_0$ we obtain its past. The future and the past play symmetric roles for this deterministic problem.

A similar determinist process occurs in finance when observing the value of money. An amount K of money at time t_0 worths $Ke^{r(t-t_0)}$ at time $t > t_0$ and $Ke^{-r(t-t_0)}$ at time $t < t_0$, provided the interest rate r is constant. This shows that the present state of the system, K, determines uniquely the future and past states of the system.

The determinist systems are described by the theory of ordinary differential equations, see Arnold [3].

A system is called *semi-determinist* if only the future (or the past) states of the system are uniquely determined by the present state of the system, while the past (the future) is not. Heat propagation, smoke evolution, and ink diffusion are semi-determinist processes. They can be predicted in the future but cannot be traced back in time. The explanation of this phenomenon is given in Chapter 8 and it is based on the formula provided by Proposition 8.1.2, which solves a forward heat equation

$$\frac{\partial u}{\partial \tau} - \frac{\partial^2 u}{\partial x^2} = 0$$
$$u(0, x) = f(x),$$

with the solution given by the convolution between the fundamental solution and the initial temperature

$$u(\tau, x) = \int_{\mathbb{R}} G(\tau, y - x) f(y)\, dy, \qquad \tau > 0,$$

where

$$G(\tau, x) = \frac{1}{\sqrt{4\pi\tau}} e^{-\frac{x^2}{4\tau}}, \qquad \tau > 0. \tag{1.1.1}$$

The expression $p(y, x; \tau) = G(\tau, y - x)$ represents the transition probability of a particle starting at x at time $t = 0$ and arriving at y at time $t = \tau$. Since the initial condition acts at any point y as an initial density of heat, then the expression $G(\tau, y - x) f(y)\, \Delta y$ represents the heat transferred from the interval $[y, y + \Delta y]$ to the point x within time τ. To find the heat at the point x, we need to take into account the effects of all the above contributions, which is obtained by the summation $\sum G(\tau, y-x)f(y)\,\Delta y$, which after taking the limit $\Delta y \to dy$, yields the aforementioned integral.

The semi-determinist processes are of utmost importance for finance. The unpredictable movements of stocks are modeled by Brownian motions, whose transition probabilities are of the form (1.1.1). Similar processes are used to model particles of ink or smoke, which diffuse in a liquid or atmosphere. The heat or diffusion equation used in finance takes the form of a backward heat-type equation, called the Black-Scholes equation. The past states of securities based on the underlying stocks can be predicted, while their future states cannot. The impossibility of foreseeing the future price of securities written on stocks leads to methods of computing the present price from future prices, which are provided in the contract at the expiration time. This is the reason why for pricing, let's say an American option, the financial analyst has to work backwards through a tree, which models the multiple possible movements of the stock price.

The semi-determinist systems are mathematically described both by the theory of partial differential equations and stochastic calculus. This is why there are two main direct approaches to derivatives pricing in finance.

1.2 How to Measure

All exact sciences are based on observing and measuring some variables, and the more accurate this is done the more we can infer about the system. To measure a variable means to assign a real number to it at a certain time, when the measurement occurs. In order to determine the state of a system one needs to measure all the parameters which defines the system. Then, using the measurements already done, one would like to predict future states of the system in the most accurate possible way.

In finance we measure interest rate, market volatility, stock prices, or different security prices (bonds, futures, options, etc.). When measuring the interest rate or volatility, we assign a percent, or a number between 0 and 1; when measuring a stock or a security price, we assign a certain number of dollars to it, etc.

The measuring process involves two ingredients:

(i) the time t when the measurement occurs;

(ii) the information available to the one who measures at time t, denoted by $\mathcal{F}_t$.

If S_t denotes the price of a stock at time t, and we would like to measure today the tomorrow's price of the stock, S_{t+1}, then we cannot do it, because the information available today about the market, $\mathcal{F}_t$, is insufficient to determine the price of the stock in the future. In this case, the variable S_{t+1} is not measurable at time t.

Another example involves measuring the *historical volatility*, which is the stock's volatility computed from the market data available until time t; this is measurable with respect to the information $\mathcal{F}_t$.

A grasp on the information $\mathcal{F}_t$ is provided by knowing all the stocks, interest rates and the security prices until and at time t. The information $\mathcal{F}_t$ consists in the set of all possible events, which already happened until time t, or they look like might happen. The events which already happened are sure events in the market (they occurred with 100% probability); other events have certain probabilities of happening. They all influence the price of stocks. For instance, a possible strike for the workers of a certain airline has an influence in the stock market, even if it is not a sure event. The higher the odds of the strike, the larger the impact on the stock market.

Hence, each element of the information set $\mathcal{F}_t$ has associated a probability. To set the notations, we write: if $A \in \mathcal{F}_t$ is an event, then $P(A)$ is the probability of A, i.e. a number between 0 and 1, describing the odds of the event A.

The complement of an event, or the union and intersection of two events in $\mathcal{F}_t$ belongs also to $\mathcal{F}_t$; this provides a structure of σ-*algebra* (or σ-*field*), see Appendix B.1.

1.3   An Uncertainty Principle

The mathematical modeling of financial markets is governed by some sort of uncertainty principle. Any model depends on a certain number of parameters. The more complex the model is, the more parameters are involved; for instance the Heston model encounters more parameters than the classical Black-Scholes model. These parameters are usually estimated from the historical data using statistical analysis. However, even if a model is more complex and might look theoretically appealing it is not always the case that it is also useful in practice.

Consider the case of a model that has only few parameters. Even if in this case the estimation of parameters can be sharp, the price payed for this is the incapacity of the model to grasp the complexity of the real world markets, leading to an under-fit.

On the other hand, consider a model with lots of parameters that looks to accommodate faithfully even the smallest details of the market dynamics. Even if the model looks theoretically robust, when it comes to parameters estimation the model might become useless. This is because the parameters cannot be well-estimated. For instance, if the maximum likelihood method is used, the parameters are estimated by the critical points of the log-likelihood function, which is a multiple-variable function. The set of critical points might have a complicated structure; or, the maximum of the surface defined by the log-likelihood function might be so flat that an estimation would be a difficult task to perform; or, there might be multiple maxima.

In conclusion, the uncertainty principle consists in the fact that we cannot determine accurately both the model parameters and the market state; if we construct a complex model meant to capture market details, then the parameters estimation becomes difficult; if the model has a small number of parameters that can be estimated well, then the model might be too simplistic and not able to fully describe the market. Hence, adopting a proper financial model is a tradeoff between complexity and the number of involved parameters. Most financial models involve 2, 3 or 4 parameters, as we shall see in the next chapters.

1.4   Market Completeness

Derivatives valuation is based on the opportunity of being able to hedge the risk away. This means, that entering a financial contract, whose price is affected by a random source of uncertainty, we can always offset the risk by entering other trading product(s), such as buying(selling) the underlying stock or other derivative(s). This type of argument is used when developing the Black-Scholes equation. Since we do not question the possibility of forming a risk-less portfolio starting from a given contract, it means that we assume the market is rich enough in financial instruments that can be used to offset the risk. These type of markets are called *complete*.

In the case when not all sources of risk can be hedged away, the market is called *incomplete*. This is the case of derivative valuation in environments such as diffusions with jumps (market shocks), stochastic volatility (such as Heston, GARCH or AR models), or stochastic rate of return (VAR model). Despite of the fact that a risk-less portfolio cannot be constructed in these cases, there is a way around it to allow for a pricing method. This consists in introducing an extra term, λ, subtracted from the real world drift rate, μ, called *cumulative price of risk*, which is associated with all the noise in the market. Sometimes, in order to obtain a closed form expression for the derivative price, we assume $\mu\lambda = r$, which is equivalent with assuming that in the economy all risk can be hedged. We have done this when pricing bonds and swaps in Chapter 8.

1.5   Market Efficiency

Thinking of markets as markets of information, we are facing the problem whether the market prices today summarize all the available information about future values, i.e. whether the markets are *information efficient*. In order to be able to obtain a fair price for securities, to assume that the asset price, S_t, is $\mathcal{F}_t$-measurable, to be able to trade portfolios and apply the *no arbitrage principle*, there is need to assume the so called *efficient market hypothesis*. This states that prices reflect all publicly available information and also adjust instantly to reflect the news in the market. All the pricing done in this book assumes that the efficient market hypothesis holds.

It is worth the remark that whenever the efficient market hypothesis holds trading rules should not work. In an informationally efficient market the fundamental analysis performed by professional managers has not much decision power in picking stocks over an amateur, see Cochrane [22]. Therefore, market efficiency is a natural consequence of market competition. It was a theme developed by Fama in late 1960s and early 1970s, see [52], and it has become the organizing principle for over 30 years of empirical work in financial economics.

1.6 Stopping Times

This section will gently introduce decisions driven by market information, which are needed later. Consider the market modeled by a probability space $(\Omega, \mathcal{F}, P)$. This means that Ω is the set of all possible states the market might take, while $\mathcal{F}$ stands for the total information provided by all events in the market. The probability P measures the occurrence chance of the events in $\mathcal{F}$. The market information until time t, denoted by $\mathcal{F}_t$, provides an ascending sequence of σ-fields describing the information flow

$$\mathcal{F}_s \subset \mathcal{F}_t \subset \mathcal{F}, \qquad \forall s < t.$$

The object $(\mathcal{F}_t)_{t \geq 0}$ is called a *filtration* and will play an important role in taking market decisions. Assume that the decision to buy or sell an asset before or at time t is determined by the information $\mathcal{F}_t$ available at time t. Consequently, this decision time can be modeled by a random variable $\tau : \Omega \to [0, \infty]$, which satisfies

$$\{\omega; \tau(\omega) \leq t\} \in \mathcal{F}_t, \qquad \forall t \geq 0.$$

This means that given the information set $\mathcal{F}_t$, we know whether the event $\{\omega; \tau(\omega) \leq t\}$ had occurred or not. We note that the possibility $\tau = \infty$ is also included, since the decision to keep the asset for ever (or never buy it) counts also as a possible event. A random variable τ with the previous properties is called a *stopping time*.

Exercise 1.6.1 *Let $\mathcal{F}_t$ be the information available until time t regarding the evolution of a stock. Assume the price of the stock at time $t = 0$ is \$20 per share. Which of the following decisions are stopping times, and why:*

(a) *Sell the stock when it reaches for the first time the price of $60 per share;*

(b) *Buy the stock when it reaches for the first time the price of $10 per share;*

(c) *Sell the stock at the end of the year;*

(d) *Sell the stock either when it reaches for the first time $80 or at the end of the year.*

(e) *Keep the stock either until the initial investment doubles or until the end of the year;*

(f) *Sell the stock when it reaches the maximum level it will ever be;*

(g) *Wait until the stock is $30 and sell after one week;*

(h) *Buy the stock when it reaches the lowest level it will ever be.*

1.7 Martingales and Submartingales

Why do people buy stocks? Because they expect their value to increase over time. If S_t denotes the value of the stock at time t, then the expected value of the stock at some time u in the future (i.e. $t < u$) is given by the conditional expectation $\mathbb{E}[S_u|\mathcal{F}_t]$. We may regard of this expression as the predicted value of the stock at time u, given the market information $\mathcal{F}_t$. The fact that the stock value is expected to increase can be written as

$$\mathbb{E}[S_u|\mathcal{F}_t] \geq S_t, \qquad t < u, \tag{1.7.2}$$

i.e. the future predicted value exceeds the observed value of the stock at time t. We also notice that S_t is $\mathcal{F}_t$-measurable and has a value that does not explode in time (it is integrable). A random variable with a behavior similar with the one of S_t is called an $\mathcal{F}_t$-*submartingale.*

If in relation (1.7.2) the inequality is replaced by identity, i.e.

$$\mathbb{E}[S_u|\mathcal{F}_t] = S_t, \qquad t < u, \tag{1.7.3}$$

then S_t is an $\mathcal{F}_t$-*martingale*. In this case the future predicted values are equal with the last observation made. More properties of these processes can be found in the Appendix B.5

1.8 Optional Stopping Theorem

From the financial point of view, this theorem says that if you buy a "fair" asset at some initial time and adopt a strategy of deciding when to sell it, then the expected price at the selling time is just the initial price. Hence, one cannot make money by buying and selling an asset whose price is a martingale. Fortunately, the price of a stock is not a martingale (it is actually a submartingale), and people can still expect to make money buying and selling stocks.

Theorem 1.8.1 (Optional Stopping Theorem) *Let $(M_t)_{t\geq 0}$ be a right continuous $\mathcal{F}_t$-martingale and τ be a stopping time with respect to $\mathcal{F}_t$ such that τ is bounded, i.e. there is an $N < \infty$ such that $\tau \leq N$. Then $\mathbb{E}[M_\tau] = \mathbb{E}[M_0]$. If M_t is an $\mathcal{F}_t$-submartingale, then $\mathbb{E}[M_\tau] \geq \mathbb{E}[M_0]$.*

The previous theorem is a special case of the more general Optional Stopping Theorem of Doob:

Theorem 1.8.2 *Let M_t be a right continuous martingale and σ, τ be two bounded stopping times, with $\sigma \leq \tau$. Then M_σ, M_τ are integrable and*

$$\mathbb{E}[M_\tau | \mathcal{F}_\sigma] = M_\sigma \quad a.s.$$

In particular, taking expectations, we have

$$\mathbb{E}[M_\tau] = \mathbb{E}[M_\sigma] \quad a.s.$$

In the case when M_t is a submartingale then $\mathbb{E}[M_\tau] \geq \mathbb{E}[M_\sigma]$ almost sure.

Doob's theorem has the following financial interpretation: if one buys an asset and has two different strategies of deciding when to sell the asset, the expected price at selling in both cases is the same. More precisely, the selling price at the time of the first strategy is the expected selling price at the time of the second strategy, given the market information at the time of the first selling strategy. For proofs and details on these theorems the reader may consult Çinlar [21] or Shiryaev [62] .

1.9   Can Random be Deterministic?

Coin tossing or rolling dies are very common examples of random experiments. But how "random" are such processes? Are they really completely random? The previous questions infer that might be quite possible that some deterministic experiments to be actually disguised as random phenomena. This section will briefly cover this peculiar topic.

In probability theory we characterize the final state of a flipped coin, not by a single possible outcome, but by two outcomes, with equal chances of occurrence, called a probability distribution. The only possible outcomes for the coin tossing experiment are heads (H) and tails (T). For a fair (flat, unweighed) coin the odds of getting either H or T are $1/2$. Consequently, this approach implies that coin flipping is not a determinist process. At least, this is what anyone would expect.

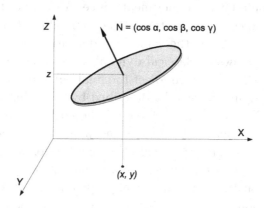

Figure 1.1: *The parameterization of a tossed coin.*

Now, let's take a closer look at the experiment. Naturally tossed coins obey the laws of classical mechanics, which are actually deterministic; hence the future states of the coin depend solely on the initial conditions. In order to model this, we consider a fixed frame in $\mathbb{R}^3$ with orthonormal axes Ox, Oy and Oz, which will be used to describe the motion of the coin. A coin moving in $\mathbb{R}^3$ space can be considered as a round piece of a planar region. It is known that a plane is completely determined by a given point and its normal direction. Using this fact, the coin can be completely described by the following 6 parameters, see Fig.1.1:

• the coordinates of the coin center, (x, y, z); the first two are the coordinates of the projection onto the table (XY-plane), while z is the vertical height of the center;

• the angles made by the normal to the coin with the axis Ox, Oy and Oz, denoted by (α, β, γ).

Hence, the state of the coin is a point in the following 6-dimensional space

$$(x, y, z, \alpha, \beta, \gamma) \in \mathbb{R} \times \mathbb{R} \times [0, \infty) \times [0, 2\pi) \times [0, 2\pi) \times [-\pi/2, \pi/2].$$

This is the parameter space on which the coin dynamics evolves. The motion of the flipped coin corresponds to a continuous curve

$$t \to (x(t), y(t), z(t), \alpha(t), \beta(t), \gamma(t)), \qquad 0 \leq t \leq \tau$$

in the aforementioned 6-dimensional space, with τ denoting landing time. At time τ the motion stops, and the final state can be observed, or measured. At landing we have $z(\tau) = 0$ and $\gamma(\tau) \in \{-\pi, \pi\}$. It is the value of $\gamma(\tau)$ that produces the outcome for the experiment, since this defines the orientation of the final state of the coin. For instance, the final value $\gamma(\tau) = \pi$ corresponds to H and $\gamma(\tau) = -\pi$ to T. Is the value $\gamma(\tau)$ deterministic?

According to the laws of classical dynamics, the value of $\gamma(\tau)$ is certainly deterministic. But the procedure to get the final outcome is not that straightforward. The coin dynamics can be described by a system of ordinary differential equations. The only acting force on the coin is gravity; the spinning coin has also kinetic energy of rotation and angular momentum of rotation relative to one of the diameters. Neglecting the friction with the air, texture of the surface, moisture, etc., and assuming that there is no precession (implied by the fact that the rotation axis is perpendicular on the normal to the coin), Keller [47] showed that the final outcome is completely determined by the initial upward velocity v_z and the rate of spin. A more complete analysis, involving also precession, can be found in Diaconis et al.[25].

For finding the landing time, τ, it suffices to find the positive solution of the equation $z(\tau) = 0$, where the vertical height is given by

$$z(t) = z_0 + v_z t - \frac{1}{2} g t^2,$$

where z_0 is the launching height, v_z is the vertical component of the initial launching velocity, and g denotes the gravitational acceleration. Even if the equations of motion of the flipping coin cannot be integrated in the general case, the best one can hope is to find conservations laws, such as the conservation of energy and momentum. The final state $\gamma(\tau)$ is explicitly computable from the initial conditions, at least theoretically.

However, $\gamma(\tau)$ is highly sensitive to small changes in the initial conditions, especially if the rotation rate is large. Since when throwing a coin, we do not have enough control or precision on our hand, this will drastically affect the final outcome. In fact, not the movement of the coin is unpredictable, but our own hand control. The sensitivity of the solution has a lower order of magnitude than our biological sensitivity and locomotor control. In theory, if one can measure precisely the initial data, then the final state of the coin can be predicted with certainty.

A phenomenon with "small changes in, big changes out" is called *chaotic*. One can actually construct pseudorandom numbers by feeding chaotic behavior into a computer.

If that much trouble is found in the analysis of such a simplistic experiment, we can imagine the difficulty faced in the case of interpreting stock markets, which are driven by millions of parameters. This is the main reason why stochastic calculus, which makes important simplifying assumptions, is more useful than a real deterministic analysis. However, in the day when the computation power will increase tremendously due to the use of quantum computers, the deterministic analysis will be used at its full capacity. This will change completely the way stochastic processes are analyzed, and real life simulations of stock markets might then become a possibility.

1.10 Change of Time Scale

From the point of view of time scale, a financial problem can be analyzed under two aspects: the short-range view, where the time is measured in seconds or minutes, and the long-range view, where the time is measured in years. The time scale used can emphasize one aspect over the other.

The problem analyzed from the short-range point of view is a stochastic problem, as unexpected changes may occur at any time. If the effect of these changes average out over time, then the long-range view leads to a deterministic problem. In this case, the amount of money (which otherwise is quantified by the cent) is considered as a smooth quantity that can be described by differential equations.

This duality is not new; it has been used in Physics where probability tools are used to describe behavior of particles at the small scale, while the description of objects at the large scale requires the use of classical differential techniques.

We shall describe in the following a couple of financial models that are probabilistic in the short-range, but become deterministic in the long-range.

1. Consider a credit card account, M_t, that holds an amount of M_0 dollars at time $t = 0$. We assume that amounts of \$1 are withdrawn from the account at random times, without replacement. If $S_1, S_2, S_3, \cdots$ denote the withdrawal times, then the account balance changes as

$$
\begin{aligned}
M_{S_k} &= M_0 - k, \qquad k \geq 1 \\
M_t &= M_{S_k}, \quad \text{for } S_k \leq t < S_{k+1},
\end{aligned}
$$

see Fig.1.2.

It makes sense to consider the time periods elapsed between two withdrawals, $T_k = S_k - S_{k-1}$, to be independent and exponentially distributed with the same mean, $\lambda > 0$. Under these hypothesis the balance at time t can be described by the stochastic process

$$
M_t = M_0 - N_t,
$$

where N_t is a Poisson process with constant rate λ, which is the withdrawal rate, see Appendix C.1. This mathematical model is consistent with the following set of conditions:

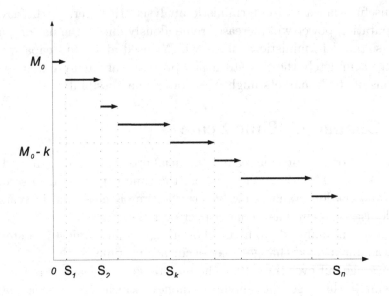

Figure 1.2: *The dynamics of the account from the short-range point of view.*

(a) If $0 < t_1 < t_2 < t_3$, then the amount of money withdrawn from the account during the time interval $[t_1, t_2]$ is independent of the amount withdrawn during the time interval $[t_2, t_3]$.

(b) The probability that exactly k dollars are withdrawn during the time interval $[t, t + \Delta t]$ is

$$P(M_t - M_{t \mid \Delta t} = k) = P(N_{t+\Delta t} - N_t = k) = \frac{(\lambda \Delta t)^k}{k!} e^{-\lambda \Delta t}.$$

As a consequence of (b) we have:

(c) The probability that exactly \$1 is withdrawn from the account during the time interval $[t, t + \Delta t]$ is

$$P(M_t - M_{t+\Delta t} = 1) = P(N_{t+\Delta t} - N_t = 1) = \lambda \Delta t + o(\Delta t),$$

where $o(\Delta t)$ possesses the property that

$$\lim_{\Delta t \to 0} \frac{o(\Delta t)}{\Delta t} = 0. \tag{1.10.4}$$

(d) The probability that more than \$1 is withdrawn from the account during the time interval $[t, t + \Delta t]$ is $o(\Delta t)$.

We note that under this model the average amount of money in the bank account decreases at the constant rate λ

$$\mathbb{E}[M_t] = M_0 - \lambda t,$$

and it is expected to deplete at the time $t = M_0/\lambda$.

All these considerations apply for the short-range point of view. We'll show next that the aforementioned probabilistic model yields a deterministic model for the long-range time scale.

Consider a large time instance $t > 0$ (by "large" we mean that t is measured in years rather than in seconds) and we shall determine a formula for the account balance at time t, denoted by $M(t)$. We start by dividing the interval $[0, t]$ into n equal subintervals of length $\Delta t = \dfrac{t}{n}$, determined by the division

$$0 = t_0 < t_1 < \cdots < t_n = t,$$

with $t_j = \dfrac{jt}{n}$. The number n is considered large enough such that the previous probabilistic analysis applies on each of the intervals $[t_k, t_{k+1}]$. Then the probability that no money is spent during the first time interval $[0, t_1]$ is $1 - \lambda\Delta t + o(\Delta t)$. This comes from condition (d) and the formula for the probability of the complementary event. In general, the probability of the event that no money is withdrawn during the time interval $[t_k, t_{k+1}]$ is $1 - \lambda\Delta t + o(\Delta t)$. Since these events are independent (see condition (a) above) then the probability that no money is withdrawn during the time interval $[0, t]$ is given by

$$
\begin{aligned}
P(M_0 - M(t) = 0) &= P\left(\bigcap_{k=0}^{n-1} \{M_{t_k} - M_{t_{k+1}} = 0\}\right) \\
&= \prod_{k=0}^{n-1} P(M_{t_k} - M_{t_{k+1}} = 0) \\
&= (1 - \lambda\Delta t)^n + o(\Delta t) \\
&= \left(1 - \frac{\lambda t}{n}\right)^n + o(\Delta t)
\end{aligned}
$$

Since the right side does not depend on n, taking $n \to \infty$, we obtain

$$P(M_0 - M(t) = 0) = e^{-\lambda t}. \tag{1.10.5}$$

The equation (1.10.5) can be interpreted by stating that the proportion of the initial balance M_0 that has not been spent during the time interval $[0, t]$ is $e^{-\lambda t}$. This implies the following deterministic formula for the remaining money in the account at time t

$$M(t) = M_0 e^{-\lambda t}.$$

We note that in the long-range scale case the balance decreases exponentially with the decay rate λ, which is the rate of withdraw. The associated deterministic dynamics is obtained by calculating the variation of the balance during

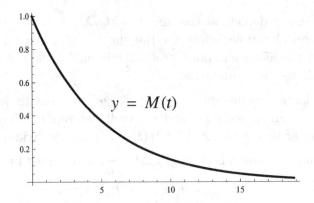

Figure 1.3: *The dynamics of the account from the long-range point of view.*

the time interval Δt as

$$
\begin{aligned}
\Delta M(t) &= M(t + \Delta t) - M(t) = M_0 e^{-\lambda t}(e^{-\lambda \Delta t} - 1) \\
&= -\lambda M_0 \Delta t + o(\Delta t),
\end{aligned}
$$

where we used the asymptotic formula

$$
e^{-\lambda t} = 1 - \lambda \Delta t + o(\Delta t). \tag{1.10.6}
$$

Dividing by Δt, taking $\Delta t \to 0$, using the definition of the derivative and formula (1.10.4) we get

$$
\frac{dM(t)}{dt} = -\lambda M(t).
$$

This represents the balance dynamics under the long-range point of view, see Fig.1.3.

2. We shall revise the previous model under the more general assumption that the rate of withdrawals is not constant in time. In this case the problem can be modeled using a non-homogeneous Poisson process, N_t, which satisfies

$$
P(N_b - N_a = k) = \frac{1}{k!}\Lambda(a, b)^k e^{-\Lambda(a,b)}, \quad k = 0, 1, 2, \cdots,
$$

where

$$
\Lambda(a, b) = \int_a^b \lambda(u)\, du
$$

and $\lambda(u)$ denotes the density of the withdrawal rate. Since $M_t = M_0 - N_t$, the chance that during the time interval $[t, t + \Delta t]$ exactly k dollars are withdrawn is

$$
P(M_t - M_{t+\Delta t} = k) = P(N_{t+\Delta t} - N_t = k) = \frac{1}{k!}\Lambda(t, t + \Delta t)^k e^{-\Lambda(t,t+\Delta t)}.
$$

Since for Δt small

$$\Lambda(t, t + \Delta t) = \int_t^{t+\Delta t} \lambda(u)\,du = \lambda(t)\Delta t + o(\Delta t),$$

using (1.10.6) we can write

$$\begin{aligned}
P(M_t - M_{t+\Delta t} = 1) &= e^{-\Lambda(t,t+\Delta t)}\Lambda(t, t + \Delta t) \\
&= (1 - \Lambda(t, t + \Delta t))\Lambda(t, t + \Delta t) + o(\Delta t) \\
&= \lambda(t)\Delta t - \lambda(t)^2 \Delta t^2 + o(\Delta t) \\
&= \lambda(t)\Delta t + o(\Delta t).
\end{aligned}$$

Consequently, the complementary event will have the probability

$$P(M_t - M_{t+\Delta t} = 0) = 1 - \lambda(t)\Delta t + o(\Delta t). \tag{1.10.7}$$

As before, we shall use formula (1.10.7) to change the time scale from local to global. Let $t > 0$ and consider the division $0 = t_0 < t_1 < \cdots < t_n = t$, with $t_j = j\Delta t = jt/n$. Using the independence property

$$\begin{aligned}
P(M_0 - M(t) = 0) &= P\left(\bigcap_{k=0}^{n-1} \{M_{t_k} - M_{t_{k+1}} = 0\} \right) \\
&= \prod_{k=0}^{n-1} \left(1 - \lambda(t_k)\Delta t\right) + o(\Delta t) \\
&= e^{\ln \prod_{k=0}^{n-1} \left(1 - \lambda(t_k)\Delta t\right)} + o(\Delta t) \\
&= e^{\sum_{k=0}^{n-1} \ln \left(1 - \lambda(t_k)t/n\right)} + o(\Delta t), \tag{1.10.8}
\end{aligned}$$

where we used that $e^{x+o(h)} = e^x + o(h)$. Using the asymptotic formula

$$\ln(1 - x) = -x + o(x)$$

we obtain that for n large we have

$$\sum_{k=0}^{n-1} \ln\left(1 - \lambda(t_k)\frac{t}{n}\right) = -\sum_{k=0}^{n-1} \lambda(t_k)\frac{t}{n},$$

which for $n \to \infty$ tends to the integral

$$-\int_0^t \lambda(u)\,du = -\Lambda(0, t).$$

Hence, for $n \to \infty$, formula (1.10.8) becomes

$$P(M_0 - M(t) = 0) = e^{-\Lambda(0,t)}.$$

This states that a fraction equal to $e^{-\Lambda(0,t)}$ of the initial balance will remain in the account after time t. Therefore, the balance at time t is given by

$$M(t) = M_0 e^{-\Lambda(0,t)} = M_0 e^{-\int_0^t \lambda(u)\,du}.$$

The deterministic dynamics can be described by computing a change in the account balance during the time step Δt:

$$
\begin{aligned}
M(t+\Delta t) - M(t) &= M_0 e^{-\int_0^{t+\Delta t} \lambda(u)\,du} - M_0 e^{-\int_0^t \lambda(u)\,du} \\
&= M_0 e^{-\int_0^t \lambda(u)\,du} e^{-\int_t^{t+\Delta t} \lambda(u)\,du} - M_0 e^{-\int_0^t \lambda(u)\,du} \\
&= M(t)\left(e^{-\int_t^{t+\Delta t} \lambda(u)\,du} - 1\right) \\
&= M(t)\left(e^{-\lambda(t)\Delta t + o(\Delta t)} - 1\right) \\
&= M(t)\left(1 - \lambda(t)\Delta t - 1 + o(\Delta t)\right) \\
&= -\lambda(t)M(t)\Delta t + o(\Delta t),
\end{aligned}
$$

which leads to the differential equation

$$M'(t) = -\lambda(t)M(t).$$

Exercise 1.10.1 *The initial amount in a bank account is M_0. Deposits of $1 are executed at random times at a constant rate r.*

(a) Analyze the problem from the short-range point of view.

(b) Find the long-range dynamics for the balance in the account.

Exercise 1.10.2 *Consider a bank account with the following properties:*

(i) Deposits of $1 are executed at random times at the constant rate r.

(ii) Amounts of $1 are randomly withdrawn at the constant rate λ.

(iii) The deposits and withdraws are modeled by independent Poisson processes, N_t and $\widetilde{N}_t$.

Then

(a) Write the short-range formula for the account balance, M_t, in terms of N_t and $\widetilde{N}_t$.

(b) Write the long-range dynamics equation for the account balance at time t.

Chapter 2

Calibration to the Market

Regression is used when calibrating financial models to the market, i.e. finding the values of the model parameters for which the model best fits the market observations with respect to a certain goodness of fit measure. Regression can be either deterministic, or stochastic. The former looks for a best fit of the market data by a curve, while the latter approximates the data by a stochastic process, which depends on some parameters. Regression provides a procedure by which these parameters can be inferred such that the model approximates the true data, minimizing a certain error.

2.1   Deterministic Regression

Assume n measurements have been made at times $t_1, \cdots, t_n$, with the results $y_1, \cdots, y_n$. The deterministic regression consists in finding a curve which best fits the data, see Fig.2.1 a. In most cases the curve has to be chosen from a family of parameterized curves $\psi_\xi(t)$, and the problem becomes the one of finding the value of the parameter $\xi = (\xi_1, \cdots, \xi_k) \in \mathbb{R}^k$ for which a certain goodness of fit is minimized.

The procedure of finding the best fit parameter in the case of least squares regression has the following geometric interpretation. We start by noting that the assignment

$$\Sigma: \quad \xi \to \big(\psi_\xi(t_1), \cdots, \psi_\xi(t_n)\big) \in \mathbb{R}^n$$

is a k-hypersurface in $\mathbb{R}^n$. The observations vector $(y_1, \cdots, y_n)$ can be considered as the coordinates of a point P in $\mathbb{R}^n$. If Q is an arbitrary point on the hypersurface Σ given by coordinates $\psi_\xi(t_j)$, the Euclidean distance between the points P and Q is given by

$$dist(P, Q) = \sqrt{\sum_{j=1}^{n} \big(\psi_\xi(t_j) - y_j\big)^2}.$$

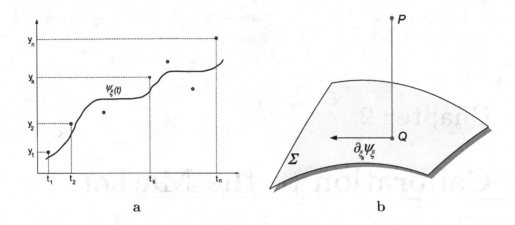

Figure 2.1: **a.** *The regression curve ψ_ξ;* **b.** *The shortest distance from P to the surface Σ occurs for PQ orthogonal to Σ.*

The minimum of this distance is achieved when Q is the orthogonal projection of P onto Σ, see Fig.2.1 b. This is equivalent with minimizing the *sum of the squared errors*

$$F(\xi) = \sum_{j=1}^{n} \left(\psi_\xi(t_j) - y_j \right)^2,$$

with respect to the parameter ξ.

Note that the partial derivatives can be written as

$$
\begin{aligned}
\partial_{\xi_k} F(\xi) &= 2 \sum_{j=1}^{n} \left(\psi_\xi(t_j) - y_j \right) \partial_{\xi_k} \psi_\xi(t_j) \\
&= 2 \langle \overrightarrow{PQ}, \partial_{\xi_k} \psi_\xi \rangle,
\end{aligned}
$$

where $\overrightarrow{PQ}$ denotes the vector that starts at P and ends at Q. Using that $\partial_{\xi_k} \psi_\xi$ is a tangent vector to the hypersurface Σ at Q, it follows that the *normal equation*

$$\partial_{\xi_k} F(\xi) = 0, \qquad k = 1, \cdots, n$$

is equivalent with the fact that PQ is normal to the tangent plane $T_Q \Sigma$.

To conclude, the best least squares curve (regression curve), ψ_ξ, is the one for which the parameter ξ satisfies the equations

$$\boxed{\sum_{j=1}^{n} \left(\psi_\xi(t_j) - y_j \right) \partial_{\xi_k} \psi_\xi(t_j) = 0, \qquad k = 1, \cdots, n.} \qquad (2.1.1)$$

Sometimes these equations might not have a solution, a fact that translates as saying that the model ψ_ξ is not a good candidate for fitting the observations. Other times the solution exists, and might be even unique, but there is no explicit expression available for it. In these cases the use of numerical methods is the only way to find the parameter.

If the observations point P is close enough to the regression surface ψ_ξ, then standard geometric properties state the existence and uniqueness of the orthogonal projection Q, which corresponds to the existence of the best fitting model. Here, an important role is played by choosing the correct model ψ_ξ, which might be done such that it avoids the existence of a secondary orthogonal projection.

If P is not close enough to the regression surface ψ_ξ, in order for PQ to be of minimum length, some convexity condition needs to be satisfied. To see this, we compute the Hessian of F, and show that it is positive definite

$$
\begin{aligned}
\partial_{\xi_j}\partial_{\xi_k}F(\xi) &= 2\langle\overrightarrow{PQ},\partial_{\xi_j}\partial_{\xi_k}\psi_\xi\rangle + 2\langle\partial_{\xi_j}\psi_\xi,\partial_{\xi_k}\psi_\xi\rangle \\
&= 2\langle\overrightarrow{PQ},\partial_{\xi_j}\partial_{\xi_k}\psi_\xi\rangle + 2g_{jk}(\xi),
\end{aligned}
$$

where $g_{jk}(\xi)$ denotes the Riemannian metric on the hypersurface Σ. It is well known that $g_{jk}(\xi)$ is positive definite. If the surface is convex, then the matrix $\partial_{\xi_j}\partial_{\xi_k}\psi_\xi$ is also positive definite, and if assume that P is on the correct side of the surface, then the first term is positive definite. This sums up to a positive definite behavior for the Hessian of F.

In the following we shall present a few classical examples of regression that have explicit solutions.

Example 2.1.1 (Linear regression) In this case the family of curves $\psi_\xi(t) = mt + b$ depends on two parameters $\xi = (m, b)$. This forms a 2-dimensional surface Σ, which actually is a plane. The sum of squared errors becomes in this case

$$
F(m, b) = \sum_{j=1}^{n}(mt_j + b - y_j)^2.
$$

Since $\partial_m\psi_\xi = (t_1, \cdots, t_n)$ and $\partial_b\psi_\xi = (1, \cdots, 1)$, then the regression equations (2.1.1) become the following linear system in m and b

$$
\sum_{j=1}^{n}(mt_j + b - y_j)t_j = 0
$$

$$
\sum_{j=1}^{n}(mt_j + b - y_j) = 0.
$$

The unique solution is given by

$$m^* = \frac{n \sum t_j y_j - \sum t_j \sum y_j}{n \sum t_j^2 - (\sum t_j)^2}, \quad b^* = \frac{\sum t_j^2 \sum y_j - \sum t_j \sum t_j y_j}{n \sum t_j^2 - (\sum t_j)^2}, \quad (2.1.2)$$

and the regression line has the equation $\psi(t) = m^* t + b^*$.

Exercise 2.1.2 (Quadratic regression) *Consider n observations (t_1, y_1), $\cdots$, (t_n, y_n). Use the method of least squares to find formulas for the regression parameters a, b and c for the quadratic model $\psi_{a,b,c}(t) = at^2 + bt + c$ which best fits the data.*

Exercise 2.1.3 *Consider the observations (t_j, y_j) given by the following table*

t_j	y_j
0.14	2.44
0.60	1.80
1.30	1.30
2.16	1.22
2.60	1.30
3.00	1.42
3.60	2.00
4.00	2.60
4.37	3.80
4.42	4.05
4.58	4.80

(a) *Use either the closed formulas developed in Exercise 2.1.2 or the Excel Solver feature to find the regression quadratic model $\psi_{a,b,c}(t) = at^2 + bt + c$.*
(b) *Plot the table data and the regression quadratic model.*

Exercise 2.1.4 (Exponential regression) *Consider the 2-parameters family of curves $\psi_\xi(t) = be^{rt}$, with $\xi = (b, r)$. Find the regression parameters formulas for this model.*

Exercise 2.1.5 (Logarithmic regression) *Find the regression parameters formulas for the logarithmic model*

$$\psi_{a,b}(t) = a + b \ln t.$$

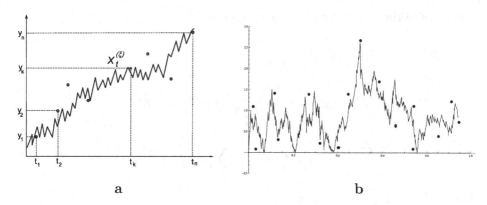

Figure 2.2: **a.** *The process* $X_t^{(\xi)}$ *best fits data* (t_j, y_j); **b.** *The process* $X_t = \sigma|W_t|$ *best fits data* (t_j, y_j).

2.2 Stochastic Regression

Let $X_t^{(\xi)}$ be a family of stochastic processes depending on the parameter vector $\xi = (\xi_1, \cdots, \xi_k)$, and consider n observations (t_j, y_j), $1 \leq j \leq n$. The stochastic regression consists in finding the value of the parameter ξ for which the process $X_t^{(\xi)}$ best fits the data, see Fig.2.2 a. This means to find ξ such that the expectation of the sum of errors

$$F(\xi) = \sum_{j=1}^{n} \mathbb{E}\big[(X_{t_j}^{(\xi)} - y_j)^2\big]$$

reaches its minimum value. The estimated parameter ξ minimizes the error between the values predicted by the model $X_{t_j}^{(\xi)}$ and data observations, y_j. In order to further express the variational equations, we denote by $p_\xi(x, t)$ the probability density function of $X_t^{(\xi)}$, and briefly recall its main properties:

1) $p_\xi(x, t) \geq 0$

2) $\displaystyle\int p_\xi(x, t)\, dx = 1$

3) $P(a < X_t^{(\xi)} < b) = \displaystyle\int_a^b p_\xi(x, t)\, dx.$

Using the definition of expectation

$$\mathbb{E}[X_{t_j}^{(\xi)}] = \int x p_\xi(x, t_j)\, dx,$$

we can write the goodness of fit in terms of p_ξ

$$
\begin{aligned}
F(\xi) &= \sum_{j=1}^{n} \mathbb{E}\big[(X_{t_j}^{(\xi)} - y_j)^2\big] \\
&= \sum_{j=1}^{n} \mathbb{E}[(X_{t_j}^{(\xi)})^2] - 2\sum_{j=1}^{n} y_j \mathbb{E}[X_{t_j}^{(\xi)}] + \|y\|^2 \\
&= \sum_{j=1}^{n} \int x^2 p_\xi(x, t_j)\, dx - 2\sum_{j=1}^{n} y_j \int x p_\xi(x, t_j)\, dx + \|y\|^2,
\end{aligned}
$$

where $\| \ \|$ stands for the Euclidean norm

$$
\|y\| = \sqrt{y_1^2 + \cdots + y_n^2}.
$$

It is not hard to see that the variational equations $\partial_{\xi_r} F(\xi) = 0$ can be written as

$$
\boxed{\sum_{j=1}^{n} \int (x^2 - 2y_j x)\partial_{\xi_r} p_\xi(x, t_j)\, dx = 0, \qquad r = 1, \cdots, k.} \tag{2.2.3}
$$

There is no general explicit formula for the solution ξ. In the following we shall consider the regression for a few particular families of stochastic processes, and try to obtain explicit formulas for the best fit parameter ξ in these cases.

Example 2.2.1 Let $\varphi_\xi : \mathbb{R} \to \mathbb{R}$, with $\xi = (\xi_1, \cdots, \xi_k)$, be a family of continuous functions, and consider the family of stochastic processes $X_t^{(\xi)} = \varphi_\xi(W_t)$, where W_t is a one-dimensional Brownian motion, see Appendix C.2. Denote by

$$
\phi_t(x) = \frac{1}{\sqrt{2\pi t}} e^{-\frac{x^2}{2t}}, \qquad t > 0
$$

the probability density of W_t. Then the expectation of the sum of squares errors can be written in terms of ϕ_t as follows

$$
\begin{aligned}
F(\xi) &= \sum_{j=1}^{n} \mathbb{E}[(X_{t_j}^{(\xi)})^2] - 2\sum_{j=1}^{n} \mathbb{E}[X_{t_j}^{(\xi)}] + \|y\|^2 \\
&= \sum_{j=1}^{n} \int \varphi_\xi^2(x)\phi_{t_j}(x)\, dx - 2\sum_{j=1}^{n} y_j \int \varphi_\xi(x)\phi_{t_j}(x)\, dx + \|y\|^2.
\end{aligned}
$$

Hence, the variational equations $\partial_{\xi_r} F(\xi) = 0$ take the following form

$$
\boxed{\sum_{j=1}^{n} \int (\varphi_\xi(x) - y_j)\partial_{\xi_r} \varphi_\xi(x)\phi_{t_j}(x)\, dx = 0, \qquad r = 1, \cdots, k.} \tag{2.2.4}
$$

Even this system is still too complex to be solved explicitly in ξ. The next examples deal with some particular forms of the function φ_ξ.

Example 2.2.2 Consider the 1-parameter family of functions $\varphi_\xi(x) = \sigma|x|$, with $\xi = \sigma > 0$. The corresponding stochastic process is the reflected Brownian motion $X_t = \sigma|W_t|$. This process would be a good candidate if $y_j \geq 0$ and y_j does not display any upward trend, see Fig. 2.2 b.
The expectation of the sum of squares errors can be evaluated as

$$
\begin{aligned}
F(\sigma) &= \sum_{j=1}^{n} \mathbb{E}[(\sigma|W_{t_j}| - y_j)^2] \\
&= \sum_{j=1}^{n} [\sigma^2 \mathbb{E}[W_{t_j}^2] - 2\sigma y_j \mathbb{E}[|W_{t_j}|] + \|y\|^2] \\
&= \sigma^2 \sum_{j=1}^{n} t_j - 2\sigma \sum_{j=1}^{n} y_j \sqrt{\frac{2t_j}{\pi}} + \|y\|^2,
\end{aligned}
$$

which is quadratic in the parameter σ. The minimum of $F(\sigma)$ is obtained for the value

$$
\sigma^* = \sqrt{\frac{2}{\pi}} \frac{\sum \sqrt{t_j} y_j}{\sum t_j}.
$$

We note that we have used in the previous computation the formula for the expectation of $|W_t|$, see Exercise 2.2.3.

Exercise 2.2.3 *The process $X_t = |W_t|$ is called a Brownian motion reflected at the origin. Show that $\mathbb{E}[X_t] = \sqrt{2t/\pi}$.*

Exercise 2.2.4 *Find the regression parameters formulas for the stochastic model*

$$
X_t = \sigma|W_t| + b.
$$

Example 2.2.5 Consider the 2-parameter family stochastic process $X_t = e^{\sigma W_t + \mu}$. This would be a good choice in the case when the data y_j is positive and also displays an upward trend. The associated goodness of fit function is

$$
\begin{aligned}
F(\mu, \sigma) &= \sum_{j=1}^{n} \mathbb{E}[(e^{\sigma W_{t_j} + \mu} - y_j)^2] \\
&= \sum_{j=1}^{n} \mathbb{E}[e^{2\sigma W_{t_j} + 2\mu} - 2y_j e^{\sigma W_{t_j} + \mu} + y_j^2] \\
&= e^{2\mu} \sum_{j=1}^{n} e^{2\sigma^2 t_j} - 2e^{\mu} \sum_{j=1}^{n} y_j e^{\frac{1}{2}\sigma^2 t_j} + \|y\|^2 \\
&= A(\sigma)x^2 + B(\sigma)x + C,
\end{aligned}
$$

where

$$A(\sigma) = \sum_{j=1}^{n} e^{2\sigma^2 t_j}, \quad B(\sigma) = -2\sum_{j=1}^{n} y_j e^{\frac{1}{2}\sigma^2 t_j},$$

$$C = \|y\|^2, \quad x = e^{\mu} > 0.$$

Since F is quadratic in x, its minimum is reached for

$$x = -\frac{B(\sigma)}{2A(\sigma)}, \tag{2.2.5}$$

which can be written equivalently as

$$\mu = \ln\left[\frac{\sum y_j e^{\frac{1}{2}\sigma^2 t_j}}{\sum e^{2\sigma^2 t_j}}\right]. \tag{2.2.6}$$

The variational equation $\partial_\sigma F(\mu,\sigma) = 0$ is equivalent to

$$x = -\frac{B'(\sigma)}{A'(\sigma)}. \tag{2.2.7}$$

Equating relations (2.2.5) and (2.2.7) implies the following equation satisfied by the critical value σ

$$\frac{B(\sigma)}{2A(\sigma)} = \frac{B'(\sigma)}{A'(\sigma)}.$$

This can be written explicitly as

$$\frac{\sum e^{2\sigma^2 t_j}}{\sum t_j e^{2\sigma^2 t_j}} = \frac{2\sum y_j e^{\frac{1}{2}\sigma^2 t_j}}{\sum t_j y_j e^{\frac{1}{2}\sigma^2 t_j}}. \tag{2.2.8}$$

The critical point (μ^*, σ^*) of $F(\mu,\sigma)$ is the solution of the system of equations (2.2.6)-(2.2.8). The regression model is given by $X_t = e^{\sigma^* W_t + \mu^*}$.

Exercise 2.2.6 (a) *Use Ito's formula, see Appendix C.6, to get*

$$d(W_t^4) = 4W_t^3 \, dW_t + 6W_t^2 \, dt;$$

(b) *Integrate in (a) and then take the expectation to obtain* $\mathbb{E}[W_t^4] = 3t^2$;

(c) *Find the stochastic regression parameter σ for the stochastic model*

$$X_t = \sigma W_t^2.$$

Exercise 2.2.7 *Find the stochastic regression parameter formulas for the model* $X_t = \mu t + \sigma W_t^2$.

Remark 2.2.8 It is worth noting that the method of least squares provides different answers if applied for $X_t^{(\xi)}$ and $Y_t^{(\xi)} = f(X_t^{(\xi)})$, in the sense that the goodness of fit functions

$$F(\xi) \;=\; \sum_{j=1}^{n} \mathbb{E}[(X_{t_j}^{(\xi)} - x_j)^2]$$

$$G(\xi) \;=\; \sum_{j=1}^{n} \mathbb{E}[(f(X_{t_j}^{(\xi)}) - f(x_j))^2]$$

have distinct minima points ξ, for nonlinear functions f. In the virtue of this remark, the regression model $X_t = e^{\mu t + \sigma W_t}$ with observations (t_j, x_j) provides a different result than applying the method of least squares to its logarithm $Y_t = \ln X_t = \mu t + \sigma W_t$ with observations (t_j, y_j), $y_j = \ln x_j$.

2.3 Calibration

Financial models, either deterministic or stochastic, depend on one or more parameters ξ. These parameters have to be determinated from the available market data, such as stock prices or actively traded securities prices y_j observed at time instances t_j. The procedure by which one infers the parameters values from market prices is called *calibration* of the model to the market.

Example 2.3.1 Consider the constant yield deterministic model for the price of a zero-coupon bond at time t, paying \$1 at the expiration time T

$$P(t, T) = e^{-r(T-t)}.$$

In order to estimate the value of the rate r it suffices to observe the market prices of n bonds with time to expiration $\tau_j = T - t_j$

$$y_j = P(t_j, T), \qquad j = 1, \cdots, n.$$

The rate r has to be estimated such that the error between the observed and predicted bond prices is the smallest possible. In order to do this, form the goodness of fit function

$$F(r) = \sum_{j=1}^{n} \left(P(t_j, T) - y_j \right)^2 = \sum_{j=1}^{n} \left(e^{-r\tau_j} - y_j \right)^2$$

and set the normal equation $F'(r) = 0$, which does not have an explicit solution. In order to further simplifications, substitute $u = e^{-r}$, and consider the value of u for which

$$G(u) = \sum_{j=1}^{n} \left(u^{\tau_j} - y_j \right)^2$$

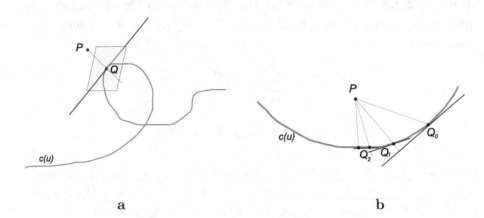

a   **b**

Figure 2.3: **a.** *PQ is normal to the curve $c(u)$;* **b.** *The iterative method providing a sequence (Q_n) of approximations of the orthogonal projection Q.*

is optimized. The equation $G'(u) = 0$ is equivalent to

$$\sum_{j=1}^{n} \tau_j u^{\tau_j - 1} (u^{\tau_j} - y_j) = 0. \tag{2.3.9}$$

This equation still does not have an explicit solution and consequently, numerical methods have to be employed for finding an estimation of the solution. Most numerical methods ask for an initial guess, which is of foremost importance for the convergence of the numerical procedure. One way to do this is to consider the solution of each of the equations $e^{-r(T-t_j)} = y_j$, given by $r_j = -\frac{\ln y_j}{T - t_j}$, and then consider their average

$$\bar{r} = \frac{1}{n} \sum_{j=1}^{n} r_j$$

as an initial guess. This will correspond to $\bar{u} = e^{-\bar{r}}$ as an initial start for the equation (2.3.9).

In the following we shall describe its geometrical significance. Relation (2.3.9) can be written equivalently as

$$\langle \dot{c}(u), \vec{p}(u) \rangle = 0,$$

where $\dot{c}(u) = (\tau_1 u^{\tau_1 - 1}, \cdots, \tau_n u^{\tau_n - 1})$ is the velocity vector to the n-dimensional curve $c(u) = (u^{\tau_1}, \cdots, u^{\tau_n})$, which traces the prices of n bonds as a function of $u = e^{-r}$. The vector $\vec{p}(u)$ denotes the position vector of the curve $c(u)$ with respect to the observations point $P = (y_1, \cdots, y_n)$. Condition (2.3.9) represents the orthogonality relation between the vector $\vec{p}(u)$ and the curve $c(u)$. It is equivalent with stating that the normal plane to the curve at the point $c(u)$ passes through the point $P = (y_j)$, see Fig. 2.3 a.

The normal projection onto the curve $c(u)$ can be obtained by an iterative procedure, which is an alteration of Newton's tangents method. This will be presented in the following. Let $Q_0 = c(\bar{u})$ be the initial guess. Consider the tangent line to $c(u)$ at Q_0, called ℓ_0. Let Q_1 be the point where the normal line from P to ℓ_0 intersects the curve $c(u)$. Substitute Q_1 in the role of Q_0 and construct the point Q_2. Inductively, we construct a sequence of points Q_n on the curve $c(u)$. Its limit, $Q = \lim_{n\to\infty} Q_n$, has the property that PQ is normal on the curve $c(u)$. We skip the proof, but the reader can see that this seems clear at least for the case depicted in Fig.2.3 b.

Example 2.3.2 The price of a bond which pays \$1 at maturity $T = 10$ years is given by $P(t,T) = e^{-r(T-t)}$. We shall use the method of least squares to estimate the parameter r from the following market data:

t_j	$P(t_j, T)$
1	0.04
2	0.07
4	0.28
6	0.64
8	0.92

The goodness of fit function in this case is given by

$$F(r) = (e^{-9r} - 0.04)^2 + (e^{-8r} - 0.07)^2 + (e^{-6r} - 0.280)^2$$
$$+ (e^{-4r} - 0.64)^2 + (e^{-2r} - 0.92)^2, \qquad 0 < r < 1.$$

Its graph is convex and is given by Fig.2.4 a. The minimum is realized for the value $r^* = 0.18561$, so the best fitting curve has the expression $P(t, 10) = e^{-0.18561(10-t)}$, see Fig.2.4 b.

Example 2.3.3 The cash value in the case of stochastic spot rates

$$r_t = r + \sigma W_t$$

is given by the model

$$M_t = M_0 e^{rt + \sigma Z_t},$$

where $Z_t = \int_0^t W_s\, ds$ is the integrated Brownian motion, see Appendix C.2. The parameters r and σ can be obtained by calibration. Consider the market

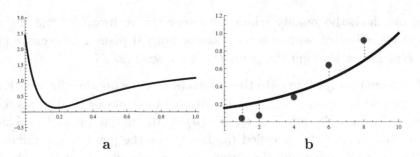

Figure 2.4: **a.** *The graph of the goodness of fit function $F(r)$;* **b.** *The exponential model and the data plot.*

observations $y_j = M_{t_j}$, $j = 1, \ldots, n$. The goodness of fit function can be computed as

$$
\begin{aligned}
F(r, \sigma) &= \sum_{j=1}^{n} \mathbb{E}[(M_{t_j} - y_j)^2] \\
&= M_0^2 \sum_{j=1}^{n} e^{2rt_j} e^{\frac{2}{3}\sigma^2 t_j^3} - 2M_0 \sum_{j=1}^{n} y_j e^{rt_j + \frac{1}{6}\sigma^2 t_j^3} + \|y\|^2.
\end{aligned}
$$

It is not hard to see that the normal equations

$$
\partial_r F = 0, \qquad \partial_\sigma F = 0
$$

can be written explicitly in the form

$$
\sum_{j=1}^{n} e^{2rt_j} e^{\frac{2}{3}\sigma^2 t_j^3} t_j = \sum_{j=1}^{n} e^{rt_j} e^{\frac{1}{6}\sigma^2 t_j^3} y_j t_j
$$

$$
2\sum_{j=1}^{n} e^{2rt_j} e^{\frac{2}{3}\sigma^2 t_j^3} t_j^3 = \sum_{j=1}^{n} e^{rt_j} e^{\frac{1}{6}\sigma^2 t_j^3} y_j t_j^3.
$$

Again, there is no explicit formula for the solution of the previous system. Numerical methods need to be used.

Example 2.3.4 A popular model for the stock price S_t is given by the geometrical Brownian motion

$$
S_t = S_0 e^{(\mu - \sigma^2/2)t + \sigma W_t}. \tag{2.3.10}
$$

This model depends on two parameters: μ, the return, and σ, the stock volatility. The calibration to the market of the model consists in evaluating the parameters μ and σ such that the model is the most consistent with the market data observations $y_j = S_{t_j}$, $j = 1, \ldots, n$, in the sense of the least squares error.

Let $r = \mu - \sigma^2/2$. Then the goodness of fit function is

$$F(r, \sigma) = \sum_{j=1}^{n} \mathbb{E}[(S_{t_j} - y_j)^2]$$

$$= S_0^2 \sum_{j=1}^{n} e^{2(r+\sigma^2)t_j} - 2S_0 \sum_{j=1}^{n} y_j e^{(r+\frac{1}{2}\sigma^2)t_j} + \|y\|^2,$$

where we used the formula $\mathbb{E}[e^{cW_t}] = e^{\frac{1}{2}c^2 t}$, $c \in \mathbb{R}$. A straightforward computation shows that the normal equations

$$\partial_r F = 0, \qquad \partial_\sigma F = 0$$

after some algebraic manipulations become

$$S_0 \sum_{j=1}^{n} t_j e^{2(r+\sigma^2)t_j} = \sum_{j=1}^{n} y_j t_j e^{(r+\frac{1}{2}\sigma^2)t_j} \qquad (2.3.11)$$

$$2\sigma S_0 \sum_{j=1}^{n} t_j e^{2(r+\sigma^2)t_j} = \sigma \sum_{j=1}^{n} y_j t_j e^{(r+\frac{1}{2}\sigma^2)t_j}. \qquad (2.3.12)$$

If $\sigma \neq 0$, then dividing the second equation by σ and subtracting from the first one yields

$$S_0 \sum_{j=1}^{n} t_j e^{2(r+\sigma^2)t_j} = 0,$$

which does not have any real solutions. Hence $\sigma = 0$. The parameter r can be found from equation (2.3.11) where we set $\sigma = 0$. It turns out that the model (2.3.10), which best fits the data in the least squares sense, is in fact *deterministic*, given by $S_t = S_0 e^{r^* t}$, where r^* is the solution of equation (2.3.11). Hence the need for alternative methods to the least squares method, which provide nontrivial stochastic regression models.

2.4 Alternatives to the Method of Least Squares

The method of least squares uses the minimization of the Euclidean distance to estimate parameters. There are also other measures whose optimization leads to parameter estimations. We shall encounter some of the most convenient in the following.

2.4.1 The Maximum Likelihood Method

Let $P(A\,|\,\xi)$ be the probability of the event A, given the parameter ξ. Then consider the observed data $\{y_1, \cdots, y_n\}$ at time instances $\{t_1, \cdots, t_n\}$, respectively, and try to match the stochastic process $X_t^{(\xi)}$, which depends on the parameter ξ, to the given data. Since we do not expect this to be a perfect fit, we need to determine the value of the model parameter ξ, which makes most likely to observe $X_{t_j}^{(\xi)}$ in a proximity of y_j at time t_j, see Fig.2.5. In order to do this, consider the probability

$$P\big(X_{t_j}^{(\xi)} \in (y_j, y_j + dy)\,|\,\xi\big) = p_\xi(y_i, t_i)dy, \qquad 1 \le j \le n, \qquad (2.4.13)$$

and define the *likelihood function* of parameter ξ, given observations (t_j, y_j), by

$$L(\xi; y_j, t_j) = \prod_{j=1}^{n} p_\xi(y_i, t_i).$$

We notice that the previous expression involves a product, which is compatible with the independence of observations. The *maximum likelihood method* consists in optimizing the likelihood function $L(\xi; y_j, t_j)$, maximizing the product of probabilities (2.4.13). The parameter ξ is obtained as a solution of the equation

$$\frac{\partial L(\xi; y_j, t_j)}{\partial \xi} = 0.$$

However, it is computationally more convenient to optimize the log-likelihood function

$$G(\xi; y_j, t_j) = \ln L(\xi; y_j, t_j) = \sum_{j=1}^{n} \ln p_\xi(y_j, t_j).$$

This is based on the fact that L and G have the same critical points ξ. These points satisfy the equation

$$\sum_{j=1}^{n} \frac{\partial_\xi p_\xi(y_j, t_j)}{p_\xi(y_j, t_j)} = 0.$$

Example 2.4.1 Consider observations $\{(t_j, y_j)\}$, $1 \le j \le n$, and the model $X_t = \mu t + \sigma W_t$, where W_t is a Brownian motion. The model parameters in this case are $\xi = (\mu, \sigma)$. Using that $W_t \sim \mathcal{N}(0, t)$, the associated likelihood function becomes

$$L(\mu, \sigma; y_j, t_j) = \prod_{j=1}^{n} \frac{1}{\sqrt{2\pi\sigma^2 t_j}} e^{-\frac{(y_j - \mu t_j)^2}{2\sigma^2 t_j}},$$

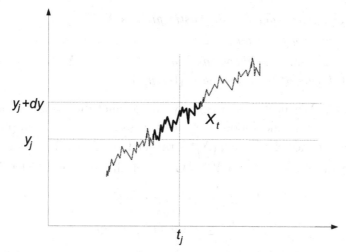

Figure 2.5: *The process X_t is in a proximity of y_j at time t_j.*

and the log-likelihood function is

$$G(\mu, \sigma; y_j, t_j) = -\ln(2\pi)^{n/2} - n \ln \sigma - \frac{1}{2} \sum_{j=1}^{n} \ln t_j - \frac{1}{2\sigma^2} \sum \frac{(y_j - \mu t_j)^2}{t_j}.$$

Since

$$\partial_\mu G(\mu, \sigma; y_j, t_j) = \frac{1}{\sigma^2} \left(\sum_{j=1}^{n} y_j - \mu \sum_{j=1}^{n} t_j \right)$$

$$\partial_\sigma G(\mu, \sigma; y_j, t_j) = \frac{1}{\sigma} \left(-n + \frac{1}{\sigma^2} \sum_{j=1}^{n} \frac{(y_j - \mu t_j)^2}{t_j} \right),$$

the model parameters are given by

$$\mu^* = \frac{\sum_{j=1}^{n} y_j}{\sum_{j=1}^{n} t_j}$$

$$\sigma^{*2} = \frac{1}{n} \sum_{j=1}^{n} \frac{(y_j - \mu t_j)^2}{t_j}.$$

It is worth noting that the model parameters obtained by the least squares method are different than the ones obtained previously, where the latter are estimated by the deterministic model

$$\mu = \frac{\sum_{j=1}^{n} y_j}{\sum_{j=1}^{n} t_j^2}$$

$$\sigma^2 = 0.$$

Exercise 2.4.2 *Consider the stochastic process $X_t = \sigma|W_t|$.*
(a) Find the density function of X_t;
(b) Use the maximum likelihood method for the model X_t to estimate the parameter σ in terms of the given data $\{(t_j, y_j)\}$.

Remark 2.4.3 We note that contrary to the method of least squares, the method of maximum likelihood provides the same solution if applied for the more convenient model $Y_t^{(\xi)} = f(X_t^{(\xi)})$ rather than $X_t^{(\xi)}$. The observations x_j for $X_t^{(\xi)}$ become $y_j = f(x_j)$ for $Y_t^{(\xi)}$. Consider the probability densities

$$P(X_t^{(\xi)} \in dx) = p(t, x; \xi)\, dx$$

$$P(Y_t^{(\xi)} \in dy) = q(t, y; \xi)\, dy,$$

related by $p(t, x; \xi) = q(t, y; \xi) f'(x)$. The likelihood functions of $X_t^{(\xi)}$ and $Y_t^{(\xi)}$ are given by

$$L^p(t_j, x_j; \xi) = \prod_{j=1}^{n} p(t_j, x_j; \xi),$$

$$L^q(t_j, y_j; \xi) = \prod_{j=1}^{n} q(t_j, y_j; \xi).$$

A computation shows that

$$
\begin{aligned}
\partial_\xi \ln L^p(t_j, x_j; \xi) &= \partial_\xi \sum_{j=1}^{n} \ln p(t_j, x_j; \xi) \\
&= \partial_\xi \sum [\ln q(t_j, y_j; \xi) + \ln f'(x_j)] \\
&= \partial_\xi \ln L^q(t_j, y_j; \xi).
\end{aligned}
$$

Hence, the likelihood functions $L^p(t_j, x_j; \xi)$ and $L^q(t_j, y_j; \xi)$ have the same critical points ξ. In the next exercise this method is applied for $f(x) = \ln(x)$.

Exercise 2.4.4 *Consider the observations $\{(t_j, y_j)\}$ and the geometric Brownian motion model $X_t = e^{\mu t + \sigma W_t}$. Find the maximum likelihood estimation of the parameters μ and σ.*

2.4.2   The Maximum Entropy Method

Consider the model $X_t^{(\xi)}$ and the given data (t_j, y_j). Another measure that can be optimized is the *entropy*

$$H(\xi; y_j, t_j) = -\sum_{j=1}^{n} p_\xi(y_j, t_j) \ln p_\xi(y_j, t_j). \qquad (2.4.14)$$

The log-likelihood $\ln p_\xi(y_j, t_j)$ represents the information that the model $X_t^{(\xi)}$ is in the proximity of y_j at time t_j for the model parameter value ξ. This method intends to find the value of ξ for which the average information associated with the data (t_j, y_j) is maximum. It can also be regarded as a method of finding the parameter ξ by setting maximum ignorance level. The normal equation $\partial_{\xi_k} H(\xi; y_j, t_j) = 0$ can be written as

$$\sum_{j=1}^{n} (\ln p_\xi(y_j, t_j) + 1) \partial_{\xi_k} p_\xi(y_j, t_j) = 0. \tag{2.4.15}$$

In general, this equation does not have an explicit solution.

2.4.3 The Kullback-Leibler Relative Entropy

Assume that data (t_j, y_j) is obtained from n observation of a stochastic process Y_t, i.e.

$$y_j = Y_{t_j}, \qquad 1 \le j \le n.$$

Let $q(y, t)$ denote the probability density associated with the process Y_t. We need to find the parameter ξ such that the model $X_t^{(\xi)}$ best fits the stochastic process Y_t at the given observation times. For this we define the *Kullback-Leibler relative entropy*

$$D_{KL}(q\|p)(\xi) = \sum_{j=1}^{n} q(y_j, t_j) \ln \frac{q(y_j, t_j)}{p_\xi(y_j, t_j)}, \tag{2.4.16}$$

which is a measure of inefficiency of assuming data distributed according to p_ξ, the probability density of $X_t^{(\xi)}$, when in fact it is distributed as q. The value of the parameter ξ for which the goodness of fit between the two distributions is reached satisfies the equation

$$\sum_{j=1}^{n} q(y_j, t_j) \partial_{\xi_k} p_\xi(y_j, t_j) / p_\xi(y_j, t_j) = 0. \tag{2.4.17}$$

2.4.4 The Cross Entropy

Again, assume that data (t_j, y_j) are obtained from n observation of a stochastic process Y_t, whose probability density is $q(y, t)$. In order to find the parameter ξ such that the model $X_t^{(\xi)}$ best fits the stochastic process Y_t at the given observation times we may consider the *cross entropy*

$$S(q, p)(\xi) = -\sum_{j=1}^{n} q(y_j, t_j) \ln p_\xi(y_j, t_j), \tag{2.4.18}$$

which is an information measure regarded as an error metric between the two probability densities. The optimality equation satisfied by ξ is the same as equation (2.4.17).

It is not clear in general which method provides the best result. It always depends on the model involved. And for a given model some methods might provide better estimations for the parameter ξ than other methods.

For more details on information measures and their properties and relation to information geometry the reader is referred to the monograph [17].

Part II

Interest Rates and Bonds

Chapter 3

Modeling Stochastic Rates

Elementary Calculus provides powerful methods for modeling phenomena from the real world. However, the real world is imperfect, due to multiple perturbation effects, and in order to study it properly, one needs to employ methods of Stochastic Calculus.

3.1 Deterministic versus Stochastic Calculus

In this section we shall consider a simple finance problem and solve it in both frames of deterministic and stochastic calculus and then compare the results.

The deterministic model Consider the amount of money $M(t)$ at time t invested in a bank account that pays interest at a constant rate r. The differential equation which models this problem is

$$dM(t) = rM(t)dt, \qquad (3.1.1)$$

i.e. the instantaneous relative rate of return $\dfrac{dM(t)}{M(t)}$ is proportional with the time interval dt. Given the initial investment $M(0) = M_0$, the account balance at time t is given by $M(t) = M_0 e^{rt}$, which is the solution of (3.1.1).

The stochastic model In the real world the interest rate r is not constant. It may be assumed constant only for a small amount of time, such as one day or one week.[1] The interest rate changes unpredictably over time. This can be modeled in a few different ways. For instance, for the sake of simplicity, we

---

[1] At the time this book was written, the interest rates in US were frozen to a very low level for several years; however, in a normal market the rates fluctuate randomly.

may assume that the interest rate at time t is given by the diffusion process $r_t = r + \sigma W_t$, where $\sigma > 0$ is a constant that controls the volatility of the rate, and W_t is a Brownian motion process. The process r_t represents a diffusion that starts at $r_0 = r$, with constant mean, $\mathbb{E}[r_t] = r$, and variance proportional to the time elapsed, $Var[r_t] = \sigma^2 t$. With this change in the model, the account balance at time t becomes a stochastic process M_t that satisfies the following stochastic differential equation

$$dM_t = (r + \sigma W_t)M_t dt, \qquad t \geq 0. \tag{3.1.2}$$

In order to solve this equation, we write it as $dM_t - r_t M_t dt = 0$ and multiply by the integrating factor $e^{-\int_0^t r_s \, ds}$. We can check that

$$d\left(e^{-\int_0^t r_s \, ds}\right) = -e^{-\int_0^t r_s \, ds} r_t dt$$

$$dM_t \, d\left(e^{-\int_0^t r_s \, ds}\right) = 0,$$

since $dt^2 = dt \, dW_t = 0$. Using the product rule, the equation becomes exact

$$d\left(M_t \, e^{-\int_0^t r_s \, ds}\right) = 0.$$

Integrating yields the solution

$$M_t = M_0 e^{\int_0^t r_s \, ds} = M_0 e^{\int_0^t (r + \sigma W_s) \, ds}$$
$$= M_0 e^{rt + \sigma Z_t},$$

where $Z_t = \int_0^t W_s \, ds$ is the integrated Brownian motion process, see Appendix C.2. Since Z_t is normally distributed, $Z_t \sim \mathcal{N}(0, t^3/3)$, then its moment generating function is $m(\sigma) = \mathbb{E}[e^{\sigma Z_t}] = e^{\sigma^2 t^3/6}$. This implies

$$\mathbb{E}[e^{\sigma Z_t}] = e^{\sigma^2 t^3/6}$$
$$Var[e^{\sigma Z_t}] = m(2\sigma) - m(\sigma) = e^{\sigma^2 t^3/3}(e^{\sigma^2 t^3/3} - 1).$$

The solution $M_t = M_0 e^{rt + \sigma Z_t}$ is log-normally distributed, with the mean and variance

$$\mathbb{E}[M_t] = M_0 e^{rt} \mathbb{E}[e^{\sigma Z_t}] = M_0 e^{rt + \sigma^2 t^3/6}$$
$$Var[M_t] = M_0^2 e^{2rt} Var[e^{\sigma Z_t}] = M_0^2 e^{2rt + \sigma^2 t^3/3}(e^{\sigma^2 t^3/3} - 1).$$

Conclusion We shall make a few interesting remarks. If $M(t)$ and M_t represent the balance at time t in the cases when r is constant and r_t is stochastic, respectively, then

$$\mathbb{E}[M_t] = M_0 e^{rt} e^{\sigma^2 t^3/6} > M_0 e^{rt} = M(t).$$

This means that we expect to have more money in the account in the latter case rather than in the former. Similarly, a bank can expect to make more money when lending at a stochastic interest rate than at a constant interest rate. This inequality is due to the convexity of the exponential function. If $X_t = rt + \sigma Z_t$, then Jensen's inequality, see Appendix B.7, yields

$$\mathbb{E}[e^{X_t}] \geq e^{\mathbb{E}[X_t]} = e^{rt}.$$

3.2 Langevin's Equation

We shall consider another stochastic extension of the equation (3.1.1). We denote now by M_t the wealth of a bank at time t, which is invested at rate r. At the same time, we shall allow for continuously random deposits and withdrawals, which can be modeled by an unpredictable term, given by $\alpha \, dW_t$, with α constant. This makes sense if a large number of customers execute transactions at any time. The obtained equation

$$\boxed{dM_t = rM_t dt + \alpha dW_t, \quad t \geq 0} \tag{3.2.3}$$

is called *Langevin's equation*.

We shall solve equation (3.2.3) as a linear stochastic equation. Multiplying by the integrating factor e^{-rt} yields

$$d(e^{-rt} M_t) = \alpha e^{-rt} dW_t.$$

Integrating we obtain

$$e^{-rt} M_t = M_0 + \alpha \int_0^t e^{-rs} \, dW_s.$$

Hence the solution is

$$\boxed{M_t = M_0 e^{rt} + \alpha \int_0^t e^{r(t-s)} \, dW_s.} \tag{3.2.4}$$

This is called the *Ornstein-Uhlenbeck process*. Since the last term is a Wiener integral, which is normally distributed, we have that M_t is Gaussian with the mean

$$\mathbb{E}[M_t] = M_0 e^{rt} + \mathbb{E}[\alpha \int_0^t e^{r(t-s)} \, dW_s] = M_0 e^{rt}$$

and variance

$$Var[M_t] = Var\left[\alpha \int_0^t e^{r(t-s)} \, dW_s\right] = \frac{\alpha^2}{2r}(e^{2rt} - 1).$$

It is worth noting that the expected balance is equal to $M_0 e^{rt}$, i.e. is the balance in a world where r is constant. The variance for t small is approximately equal to $\alpha^2 t$, which is the variance of αW_t.

If the constant α is replaced by a random function $\alpha(t, W_t)$, the equation (3.2.3) becomes

$$dM_t = rM_t dt + \alpha(t, W_t) dW_t, \quad t \geq 0.$$

Using a similar argument we arrive at the following solution:

$$\boxed{M_t = M_0 e^{rt} + \int_0^t e^{r(t-s)} \alpha(t, W_t) \, dW_s.}\tag{3.2.5}$$

In general, this process is not Gaussian. Its mean and variance are given by

$$\mathbb{E}[M_t] = M_0 e^{rt}$$
$$Var[M_t] = \int_0^t e^{2r(t-s)} \mathbb{E}[\alpha^2(t, W_t)] \, ds.$$

The integral in equation (3.2.5) can be computed explicitly in just a few cases. For instance, if $\alpha(t, W_t) = e^{\sqrt{2r} W_t}$, using the following formula given in [15], p. 163

$$\int_0^t e^{-\frac{\lambda^2 s}{2} + \lambda W_s} \, dW_s = \frac{1}{\lambda} \left(e^{-\frac{\lambda^2 t}{2} + \lambda W_t} - 1 \right),$$

with $\lambda = \sqrt{2r}$, we can work out (3.2.5) in an explicit form

$$\begin{aligned}
M_t &= M_0 e^{rt} + \int_0^t e^{r(t-s)} e^{\sqrt{2r} W_t} \, dW_s \\
&= M_0 e^{rt} + e^{rt} \int_0^t e^{-rs} e^{\sqrt{2r} W_t} \, dW_s \\
&= M_0 e^{rt} + e^{rt} \frac{1}{\sqrt{2r}} \left(e^{-rs + \sqrt{2r} W_t} - 1 \right) \\
&= M_0 e^{rt} + \frac{1}{\sqrt{2r}} \left(e^{\sqrt{2r} W_t} - e^{rt} \right).
\end{aligned}$$

3.3 Equilibrium Models

Let r_t denote the spot rate at time t. This is the rate at which one can invest for the shortest period of time.[2] For the sake of simplicity, we assume the interest rate r_t satisfies an equation that involves one source of uncertainty.[3]

---

[2] This is also called the short-time rate or the instantaneous rate.
[3] This means the stochastic differential equation involves only one Brownian motion.

If the spot rates were differentiable, the drift term affected by the random fluctuations of the market would be written as

$$\frac{dr_t}{dt} = m(r_t) + \text{``noise''}.$$

This can be formalized as in the following stochastic differential equation

$$\boxed{dr_t = m(r_t)dt + \sigma(r_t)dW_t.} \tag{3.3.6}$$

In this model the drift rate and volatility of the spot rate r_t do not depend explicitly on the time t. There are several classical choices for $m(r_t)$ and $\sigma(r_t)$ that will be discussed in the following sections. It is worth noting that in the case of equilibrium models the spot rate r_t determines the term structure and the bond price. We shall deal with this subject in more detail in Chapter 4.

3.3.1 Rendleman and Bartter Model

The model introduced in 1986 by Rendleman and Bartter [59] assumes that the short-time rate satisfies the process

$$\boxed{dr_t = \mu r_t dt + \sigma r_t dW_t.}$$

The growth rate μ and the volatility σ are considered constants. This equation describes a geometric Brownian motion and its solution is given by

$$r_t = r_0 e^{(\mu - \frac{\sigma^2}{2})t + \sigma W_t}.$$

The distribution of r_t is log-normal, with

$$\ln \frac{r_t}{r_0} \sim N\left(\left(\mu - \frac{\sigma^2}{2}\right)t, \sigma^2 t\right).$$

Its mean and variance are given by

$$\begin{aligned}
\mathbb{E}[r_t] &= r_0 e^{(\mu - \frac{\sigma^2}{2})t}\mathbb{E}[e^{\sigma W_t}] = r_0 e^{(\mu - \frac{\sigma^2}{2})t}e^{\sigma^2 t/2} = r_0 e^{\mu t}. \\
Var[r_t] &= r_0^2 e^{2(\mu - \frac{\sigma^2}{2})t}Var[e^{\sigma W_t}] = r_0^2 e^{2(\mu - \frac{\sigma^2}{2})t}e^{\sigma^2 t}(e^{\sigma^2 t} - 1) \\
&= r_0^2 e^{2\mu t}(e^{\sigma^2 t} - 1).
\end{aligned}$$

One of the disadvantages of this model is that in the long run the rate expectation $\mathbb{E}[r_t]$ becomes unboundedly large. This is the reason why the next two models incorporate the *mean reverting* phenomenon of interest rates. This means that in the long run the rate converges towards an average level. These models are more realistic and are based on economic arguments.

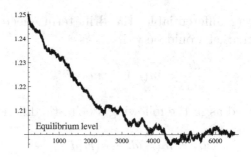

Figure 3.1: *A simulation of* $dr_t = a(b - r_t)dt + \sigma dW_t$, *with* $r_0 = 1.25$, $a = 3$, $\sigma = 1\%$, $b = 1.2$.

3.3.2 Vasicek Model

Vasicek's assumption is that the short-term interest rates should satisfy the mean reverting stochastic differential equation

$$dr_t = a(b - r_t)dt + \sigma dW_t, \qquad (3.3.7)$$

with a, b, σ positive constants, see Vasicek [66].

If assume the spot rates were deterministic, then taking $\sigma = 0$, we obtain the ordinary differential equation

$$dr_t = a(b - r_t)dt,$$

with the solution

$$r_t = b + (r_0 - b)e^{-at}.$$

This implies that the rate r_t is pulled towards level b at the rate a. This means that, if $r_0 > b$, then r_t is decreasing towards b, and if $r_0 < b$, then r_t is increasing towards the horizontal asymptote b. The term σdW_t in Vasicek's model adds some "white noise" to the process. In the following we shall find an explicit formula for the spot rate r_t in the stochastic case.

Proposition 3.3.1 *The solution of the equation* (3.3.7) *is given by*

$$r_t = b + (r_0 - b)e^{-at} + \sigma e^{-at} \int_0^t e^{as} \, dW_s. \qquad (3.3.8)$$

The process r_t is Gaussian with mean and variance

$$\mathbb{E}[r_t] = b + (r_0 - b)e^{-at};$$
$$Var[r_t] = \frac{\sigma^2}{2a}(1 - e^{-2at}).$$

Proof: Multiplying equation (3.3.7) by the integrating factor e^{at} yields

$$d\left(e^{at}r_t\right) = abe^{at}\,dt + \sigma e^{at}\,dW_t.$$

Integrating between 0 and t and dividing by e^{at} we get

$$\begin{aligned}
r_t &= r_0 e^{-at} + be^{-at}(e^{at} - 1) + \sigma e^{-at}\int_0^t e^{as}\,dW_s \\
&= b + (r_0 - b)e^{-at} + \sigma e^{-at}\int_0^t e^{as}\,dW_s.
\end{aligned}$$

Since the spot rate r_t is the sum between the deterministic function $r_0 e^{-at} + be^{-at}(e^{at} - 1)$ and a multiple of a Wiener integral, it follows that r_t is Gaussian (see Proposition C.6.3, Appendix C.6), with

$$\begin{aligned}
\mathbb{E}[r_t] &= b + (r_0 - b)e^{-at} \\
Var(r_t) &= Var\left[\sigma e^{-at}\int_0^t e^{as}\,dW_s\right] = \sigma^2 e^{-2at}\int_0^t e^{2as}\,ds \\
&= \frac{\sigma^2}{2a}(1 - e^{-2at}).
\end{aligned}$$

∎

The following consequence explains the name of *mean reverting rate*.

Remark 3.3.2 Since $\lim_{t\to\infty}\mathbb{E}[r_t] = \lim_{t\to\infty}(b + (r_0 - b)e^{-at}) = b$, the process is mean reverting, i.e. the spot rate r_t tends to b as $t \to \infty$. This limit is approached exponentially fast. The variance tends in the long run to $\frac{\sigma^2}{2a}$, so the random market fluctuations interfering with the mean reversion are of magnitude $\frac{\sigma}{\sqrt{2a}}$, so that weak mean reverting processes have large volatility.

Since r_t is normally distributed, the Vasicek model has been criticized for allowing negative interest rates[4] and unbounded large rates. However, if one considers r_t as the real interest rate, as given by the difference between the spot rate and the inflation rate, then in this case it may become negative. See Fig.3.1 for a simulation of the short-term interest rates for the Vasicek model.

---

[4]Starting with January 2015 the Swiss Bank start charging negative interest rates to banks and corporations; negative interest rates occured also in Japan, see "Below zero", *The Economist*, Nov. 14, 1998, p.81.

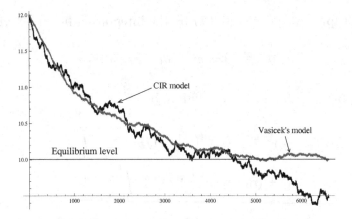

Figure 3.2: *Comparison between a simulation in CIR and Vasicek models, with parameter values $a = 3$, $\sigma = 15\%$, $r_0 = 12$, $b = 10$. Note that the CIR process tends to be more volatile than Vasicek's.*

Exercise 3.3.3 *Let $0 \leq s < t$. Find the following expectations*

(a) $\mathbb{E}[W_t \int_0^s W_u e^{au} \, du]$.

(b) $\mathbb{E}[\int_0^t W_u e^{au} \, du \int_0^s W_v e^{av} \, dv]$.

Exercise 3.3.4 (a) *Find the probability that r_t is negative.*
(b) *What happens with this probability when $t \to \infty$?*
(c) *Find the rate of change of this probability.*
(d) *Compute $cov(r_s, r_t)$.*

3.3.3   Calibration of Vasicek's Model

Assume the spot rate has been observed at n time instances, $t_1, t_2, \cdots, t_n$, with the readings $r_1, r_2, \cdots, r_n$, respectively. We are interested in the estimation of the model parameters a, b and σ for a Vasicek's model that best fits the observed data. We shall accomplish this by stochastic regression, trying two of the methods introduced in Chapter 2.

The Method of Least Squares

For the sake of simplicity we shall use the notation $\tau_j = e^{-at_j}$. In this case the model parameters are $\xi = (a, b, \sigma)$ and the stochastic model is $X_t^{(\xi)} = r_t$. The predicted values of the spot rates at time instances t_j are

$$\hat{r}_j = b + (r - b)\tau_j + \sigma\tau_j \int_0^t e^{as} \, dW_s, \quad 1 \leq j \leq n.$$

Then the goodness of fit function given as the sum of squares error is given by

$$
\begin{aligned}
F(a,b,\sigma) &= \sum_{j=1}^{n} \mathbb{E}[(\widehat{r}_j - r_j)^2] \\
&= \sum_{j=1}^{n} \mathbb{E}\Big[\Big(\underbrace{b - r_j + (r-b)\tau_j}_{=A_j} + \underbrace{\sigma\tau_j \int_0^t e^{as}\,dW_s}_{=B_j}\Big)^2\Big] \\
&= \sum_{j=1}^{n} A_j^2 + 2\sum_{j=1}^{n} A_j \underbrace{\mathbb{E}[B_j]}_{=0} + \sum_{j=1}^{n} \mathbb{E}[B_j^2] \\
&= \sum_{j=1}^{n} A_j^2 + \frac{\sigma^2}{2a}\Big(n - \sum_{j=1}^{n} \tau_j^2\Big).
\end{aligned}
$$

The variational equation $\dfrac{\partial F}{\partial \sigma} = 0$ becomes

$$
\sum_{j=1}^{n} \tau_j^2 = n. \tag{3.3.9}
$$

The left side is an increasing function of a, given by $f(a) = \sum_{j=1}^{n} e^{2at_j}$, so the equation (C.4.6) has at most a solution. Since we can easily verify that $f(0) = n$, it follows that $a = 0$ is the only solution. Hence, the Vasicek model has a vanishing parameter and therefore it becomes uninteresting. The conclusion is that the least squares method is not appropriate for this stochastic regression.

The Method of Maximum Likelihood

Since r_t has a Gaussian distribution, its probability density evaluated at (r_j, t_j) is

$$
p_{a,b,\sigma}(r_j, t_j) = \sqrt{\frac{a}{\pi}} \frac{1}{\sigma\sqrt{1-\tau_j^2}} \exp\Big(-\frac{a(r_j - b - (r_0 - b)\tau_j)^2}{\sigma^2(1-\tau_j^2)}\Big),
$$

and the resulting likelihood function becomes

$$
\begin{aligned}
L(a,b,\sigma; r_j, t_j) &= \prod_{j=1}^{n} p_{a,b,\sigma}(r_j, t_j) \\
&= \Big(\frac{a}{\pi}\Big)^{n/2} \frac{\sigma^{-n}}{\prod_{j=1}^{n}(1-\tau_j^2)^{1/2}} \\
&\quad \exp\Big(-\frac{a}{\sigma^2}\sum_{j=1}^{n} \frac{(r_j - b - (r_0 - b)\tau_j)^2}{1-\tau_j^2}\Big),
\end{aligned}
$$

which yields the log-likelihood function

$$
\begin{aligned}
G(a, b, \sigma; r_j, t_j) &= \ln L(a, b, \sigma; r_j, t_j) \\
&= \frac{n}{2}(\ln a - \ln \pi) - n \ln \sigma - \frac{1}{2}\sum_{j=1}^{n}\ln(1 - \tau_j^2) \\
&\quad - \frac{a}{\sigma^2}\sum_{j=1}^{n}\frac{(r_j - b - (r_0 - b)\tau_j)^2}{1 - \tau_j^2}.
\end{aligned}
$$

Multiplying the equation $\dfrac{\partial G}{\partial \sigma} = 0$ by σ^3 we obtain a linear equation with solution

$$
\sigma^2 = \frac{2a}{n}\sum_{j=1}^{n}\frac{(r_j - b - (r_0 - b)\tau_j)^2}{1 - \tau_j^2}. \tag{3.3.10}
$$

This relation provides the parameter σ is terms of parameters b and a, unknown yet. The equation $\dfrac{\partial G}{\partial b} = 0$ is a linear equation in b with solution

$$
b = \frac{\sum_{j=1}^{n}\frac{r_0\tau_j - r_j}{1 + \tau_j}}{\sum_{j=1}^{n}\frac{\tau_j - 1}{\tau_j + 1}}. \tag{3.3.11}
$$

This provides the value of parameter b in terms of parameter a. We shall find next an equation whose solution is the parameter a.

Using $\partial \tau_j / \partial a = -t_j \tau_j$, elementary differentiation rules provide

$$
\begin{aligned}
\frac{\partial G}{\partial a} &= \frac{n}{2a} - \sum_{j=1}^{n}\frac{1}{1 - \tau_j^2}\left(\tau_j^2 t_j + (r_j - b - (r_0 - b)\tau_j)^2\right) \\
&\quad - \frac{a}{\sigma^2}\sum_{j=1}^{n}\frac{\partial}{\partial a}\left\{\frac{(r_j - b - (r_0 - b)\tau_j)^2}{1 - \tau_j^2}\right\}.
\end{aligned}
$$

The derivative in the above summation can be calculated using logarithmic differentiation

$$
\begin{aligned}
\frac{\partial}{\partial a}\left\{\frac{(r_j - b - (r_0 - b)\tau_j)^2}{1 - \tau_j^2}\right\} &= 2\frac{(r_j - b - (r_0 - b)\tau_j)^2}{1 - \tau_j^2} \\
&\quad \left(\frac{(r_0 - b)\tau_j t_j}{r_j - b - (r_0 - b)\tau_j} - \frac{\tau_j^2 t_j}{1 - \tau_j^2}\right).
\end{aligned}
$$

The equation $\dfrac{\partial G}{\partial a} = 0$ becomes a nonlinear equation solely in the unknown a

$$\frac{n}{2a} = \frac{2a}{\sigma^2} \sum_{j=1}^{n} \frac{(r_j - b - (r_0 - b)\tau_j)^2}{1 - \tau_j^2} \left(\frac{(r_0 - b)\tau_j t_j}{r_j - b - (r_0 - b)\tau_j} - \frac{\tau_j^2 t_j}{1 - \tau_j^2} \right)$$

$$+ \sum_{j=1}^{n} \frac{1}{1 - \tau_j^2} \left(\tau_j^2 t_j + (r_j - b - (r_0 - b)\tau_j)^2 \right). \tag{3.3.12}$$

There is no closed form solution for the equation (3.3.12). Solving it by numerical methods provides the value of the parameter a. Substituting in (3.3.11) yields the value of parameter b. Then formula (3.3.10) gives the value for the parameter σ.

3.3.4 Cox-Ingersoll-Ross Model

The Cox-Ingersoll-Ross (CIR) model assumes that the spot rates verify the stochastic equation

$$\boxed{dr_t = a(b - r_t)dt + \sigma\sqrt{r_t}\, dW_t,} \tag{3.3.13}$$

with a, b, σ positive constants, see [23]. Two main advantages of this model are:

- the process exhibits mean reversion.
- it is not possible for the interest rates to become negative.

A process that satisfies equation (3.3.13) is called a *CIR process*. In mathematics this is known under the name of Feller process, and its transition density is given by

$$\boxed{p_t(y_0, y) = \frac{2a}{\sigma^2(e^{at} - 1)} \left(\frac{y}{y_0} \right)^{\nu/2} e^{a(1+\nu/2)t - \frac{2a(y_0 + e^{at}y)}{\sigma^2(e^{at} - 1)}} I_\nu\left(\frac{4a\sqrt{e^{at}y_0 y}}{\sigma^2(e^{at} - 1)} \right),}$$

where $\nu = \dfrac{2ab}{\sigma^2} - 1$, and

$$I_\nu(z) = \sum_{n=0}^{\infty} \frac{1}{n!\,\Gamma(n + \nu + 1)} \left(\frac{z}{2} \right)^{\nu + 2n}$$

is the Bessel function of index ν. The development of the aforementioned formula can be found, for instance, in Gulisashvili [37].

In the following we shall compute its first two moments starting from the equation (3.3.13).

Multiplying by e^{at} yields the exact equation

$$d(e^{at} r_t) = abe^{at}\, dt + \sigma e^{at} \sqrt{r_t}\, dW_t.$$

Integrating between 0 and t we obtain

$$r_t = (r_0 - b)e^{-at} + b + \sigma e^{-at} \int_0^t e^{au} \sqrt{r_u}\, dW_u.$$

Using that an Ito integral has zero expectation, we get the following formula for the mean of the spot rate

$$\mathbb{E}[r_t] = \mu_t = (r_0 - b)e^{-at} + b. \tag{3.3.14}$$

Squaring the expression

$$r_t = \mu_t + \sigma e^{-at} \int_0^t e^{au} \sqrt{r_u}\, dW_u,$$

and taking the expectation leads to

$$
\begin{aligned}
\mathbb{E}[r_t^2] - \mu_t^2 &= \sigma^2 e^{-2at} \mathbb{E}\left[\left(\int_0^t e^{au} \sqrt{r_u}\, dW_u\right)^2\right] \\
&= \sigma^2 e^{-2at} \int_0^t e^{2au} \mathbb{E}[r_u]\, du \\
&= \sigma^2 e^{-2at} \int_0^t e^{2au}\left(b + (r_0 - b)e^{-au}\right) du \\
&= \frac{\sigma^2}{a} e^{-2at}\left[\frac{b}{2}(e^{at} - 1)(e^{at} + 1) + (r_0 - b)(e^{at} - 1)\right] \\
&= \frac{\sigma^2}{a} e^{-2at}(e^{at} - 1)\left(r_0 + \frac{b}{2}(e^{at} - 1)\right).
\end{aligned}
$$

Hence, the variance is

$$Var(r_t) = \frac{\sigma^2}{a} e^{-2at}(e^{at} - 1)\left(r_0 + \frac{b}{2}(e^{at} - 1)\right).$$

A computation shows that

$$\lim_{t\to\infty} \mathbb{E}[r_t] = b, \qquad \lim_{t\to\infty} Var(r_t) = \frac{b\sigma^2}{2a},$$

i.e. the process is mean reverting.

    The calibration to the market of a CIR process follows roughly the same idea as in the case of Vasicek model, but it is much more elaborate and we shall not attempt it here.

Exercise 3.3.5 *Let* $\mu_k(t) = \mathbb{E}[r_t^k]$ *be the kth moment of* r_t. *Find a formula for the moment* $\mu_3(t)$, *where* r_t *satisfies the CIR process (3.3.13).*

3.4 No-arbitrage Models

In the following models the drift rate is a function of time, which is chosen such that the model is consistent with the term structure. For these models the term structure is an input to the model. The detailed study of the term structure is done in Chapter 4.

3.4.1 Ho and Lee Model

The first no-arbitrage model was proposed in 1986 by Ho and Lee [40]. The model was presented initially in the form of a binomial tree. The continuous time-limit of this model is

$$\boxed{dr_t = \theta(t)dt + \sigma dW_t.}$$
(3.4.15)

In this model $\theta(t)$ is the average direction in which r_t moves and it is considered independent of r_t, while σ is the standard deviation of the short rate. The solution process is Gaussian and is given by

$$r_t = r_0 + \int_0^t \theta(s)\, ds + \sigma W_t.$$

If $f(0,t)$ denotes the forward rate at time t as seen at time 0, it will be shown in section 4.5.1 that $\theta(t) = \partial_t f(0,t) + \sigma^2 t$. Using that $r_0 = f(0,0)$ is the initial spot rate, the solution of (3.4.15) becomes

$$r_t = f(0,t) + \frac{1}{2}\sigma^2 t^2 + \sigma W_t.$$

This formula provides the spot rate in terms of the forward rate.

Exercise 3.4.1 *Let r_t be the rate given by the Ho-Lee model.*

(a) *Prove that $r_t = r_s + \int_s^t \theta(u)\, du + \sigma W_{t-s}$, for any $0 < s < t$;*

(b) *Show the following integral average formula*

$$\frac{1}{t-s}\int_s^t \theta(u)\, du = \mathbb{E}[r_t|\mathcal{F}_t] - r_s,$$

where $\mathbb{E}[\cdot|\mathcal{F}_t]$ denotes the conditional expectation with respect to the information $\mathcal{F}_t$ generated by $\{W_s; s \le t\}$.

(c) *Prove that $\theta(s)$ is the instantaneously average velocity of spot rates, i.e.*

$$\theta(s) = \lim_{t \to s} \frac{\mathbb{E}[r_t|\mathcal{F}_s] - r_s}{t-s}.$$

3.4.2   Hull and White Model

The model proposed by Hull and White [41] is an extension of the Ho and Lee model that incorporates mean reversion

$$\boxed{dr_t = (\theta(t) - ar_t)dt + \sigma dW_t,}$$ (3.4.16)

with a and σ constants. We can solve the equation multiplying by the integrating factor e^{at}

$$d(e^{at}r_t) = \theta(t)e^{at}dt + \sigma e^{at}dW_t.$$

Integrating between 0 and t yields

$$r_t = r_0e^{-at} + e^{-at}\int_0^t \theta(s)e^{as}\,ds + \sigma e^{-at}\int_0^t e^{as}\,dW_s.$$ (3.4.17)

Since the first two terms are deterministic and the last is a Wiener integral, the process r_t is Gaussian.

The function $\theta(t)$ can be calculated from the term structure (see section 4.5.2) as

$$\theta(t) = \partial_t f(0,t) + af(0,t) + \frac{\sigma^2}{2a}(1 - e^{-2at}).$$

Then

$$
\begin{aligned}
\int_0^t \theta(s)e^{as}\,ds &= \int_0^t \partial_s f(0,s)e^{as}\,ds + a\int_0^t f(0,s)e^{as}\,ds \\
&\quad + \frac{\sigma^2}{2a}\int_0^t (1 - e^{-2as})e^{as}\,ds \\
&= f(0,t)e^{at} - r_0 + \frac{\sigma^2}{a^2}\big(\cosh(at) - 1\big),
\end{aligned}
$$

where we used that $f(0,0) = r_0$. The deterministic part of r_t becomes

$$r_0e^{-at} + e^{-at}\int_0^t \theta(s)e^{as}\,ds = f(0,t) + \frac{\sigma^2}{a^2}e^{-at}\big(\cosh(at) - 1\big).$$

An algebraic manipulation shows that

$$e^{-at}\big(\cosh(at) - 1\big) = \frac{1}{2}(1 - e^{at})^2.$$

Substituting into (3.4.17) yields

$$\boxed{r_t = f(0,t) + \frac{\sigma^2}{2a^2}(1 - e^{at})^2 + \sigma e^{-at}\int_0^t e^{as}\,dW_s.}$$

For a better understanding of the solution behavior, we note that the mean and variance are

$$
\begin{aligned}
\mathbb{E}[r_t] &= f(0,t) + \frac{\sigma^2}{2a^2}(1 - e^{at})^2 \\
Var(r_t) &= \sigma^2 e^{-2at} Var\left[\int_0^t e^{as}\, dW_s\right] = \sigma^2 e^{-2at} \int_0^t e^{2as}\, ds \\
&= \frac{\sigma^2}{2a}(1 - e^{-2at}),
\end{aligned}
$$

and hence, in the long run the random fluctuations are of order of magnitude $\sigma/\sqrt{2a}$.

Exercise 3.4.2 *Consider a model that describes the dynamics of the spot rate r_t by*

$$
dr_t = \left(\mu'(t) + \lambda(\mu(t) - r_t)\right)dt + \sigma(t)\, dW_t.
$$

(a) *Verify that the solution is given by*

$$
r_t = r_0 e^{-\lambda t} + \mu(t) - \mu(0)e^{-\lambda t} + \int_0^t e^{-\lambda(t-u)}\sigma(u)\, dW_u.
$$

(b) *Show that r_t is normally distributed, with mean and variance*

$$
\begin{aligned}
\mathbb{E}[r_t] &= r_0 e^{-\lambda t} + \mu(t) - \mu(0)e^{-\lambda t} \\
Var(r_t) &= \int_0^t e^{-2\lambda(t-u)}\sigma(u)^2\, du.
\end{aligned}
$$

3.5 Nonstationary Models

These models assume both θ and σ as functions of time. In the following we shall discuss two models with this property.

3.5.1 Black, Derman and Toy Model

The binomial tree of Black, Derman and Toy [10] is equivalent with the following continuous model of short-time rate

$$
\boxed{d(\ln r_t) = \left[\theta(t) + \frac{\sigma'(t)}{\sigma(t)}\ln r_t\right]dt + \sigma(t)dW_t.} \tag{3.5.18}
$$

Making the substitution $u_t = \ln r_t$, we obtain a linear equation in u_t

$$
du_t = \left[\theta(t) + \frac{\sigma'(t)}{\sigma(t)}u_t\right]dt + \sigma(t)dW_t.
$$

The equation can be written equivalently as

$$\frac{\sigma(t)du_t - d\sigma(t)\,u_t}{\sigma^2(t)} = \frac{\theta(t)}{\sigma(t)}dt + dW_t,$$

which after using the quotient rule becomes

$$d\left[\frac{u_t}{\sigma(t)}\right] = \frac{\theta(t)}{\sigma(t)}dt + dW_t.$$

Integrating and solving for u_t leads to

$$u_t = \frac{u_0}{\sigma(0)}\sigma(t) + \sigma(t)\int_0^t \frac{\theta(s)}{\sigma(s)}ds + \sigma(t)W_t.$$

This implies that u_t is Gaussian and hence $r_t = e^{u_t}$ is log-normal for each t. Using $u_0 = \ln r_0$ and

$$e^{\frac{u_0}{\sigma(0)}\sigma(t)} = e^{\frac{\sigma(t)}{\sigma(0)}\ln r_0} = r_0^{\frac{\sigma(t)}{\sigma(0)}},$$

we obtain the following explicit formula for the spot rate

$$\boxed{r_t = r_0^{\frac{\sigma(t)}{\sigma(0)}}\, e^{\sigma(t)\int_0^t \frac{\theta(s)}{\sigma(s)}ds}\, e^{\sigma(t)W_t}.} \tag{3.5.19}$$

Since $\sigma(t)W_t$ is normally distributed with mean 0 and variance $\sigma^2(t)t$, the log-normal variable $e^{\sigma(t)W_t}$ has

$$\mathbb{E}[e^{\sigma(t)W_t}] = e^{\sigma^2(t)t/2}$$
$$Var[e^{\sigma(t)W_t}] = e^{\sigma(t)^2 t}(e^{\sigma(t)^2 t} - 1).$$

Hence

$$\mathbb{E}[r_t] = r_0^{\frac{\sigma(t)}{\sigma(0)}}\, e^{\sigma(t)\int_0^t \frac{\theta(s)}{\sigma(s)}ds}\, e^{\sigma^2(t)t/2}$$
$$Var[r_t] = r_0^{\frac{2\sigma(t)}{\sigma(0)}}\, e^{2\sigma(t)\int_0^t \frac{\theta(s)}{\sigma(s)}ds}\, e^{\sigma(t)^2 t}(e^{\sigma(t)^2 t} - 1).$$

Exercise 3.5.1 (a) *Solve the Black, Derman and Toy model in the case when σ is constant.*
(b) *Show that in this case the spot rate r_t is log-normally distributed.*
(c) *Find the mean and the variance of r_t.*

Exercise 3.5.2 *The following model was introduced by Black and Karasinski in 1991, see [11]:*

$$d(\ln r_t) = \left[\theta(t) - a(t)\ln r_t\right]dt + \sigma(t)dW_t.$$

Find the explicit formula for r_t.

Chapter 4

Bonds, Forward Rates and Yield Curves

This chapter deals with the following three financial equivalent instruments: bonds, forward rates and yield curves. Knowing any one of these provides full information on the other two. This makes possible to approach the interest rate market from multiple perspectives. The second part of the chapter presents several bond prices computations in the case when the spot rate is stochastic.

4.1 Bonds

A *bond* is a contract between a buyer and a financial institution (bank, government, etc.) by which the financial institution agrees to pay a certain principal to the buyer at a determined time T in the future, plus some periodic coupon payments during the life time of the contract. If there are no coupons to be payed, the bond is called *a zero coupon bond* or a *discount bond*. Its price at any time $0 \leq t \leq T$ will be denoted by $P(t, T)$.

Consider first the case of deterministic rates:

(i) If spot rates are constant, $r_t = r$, then the price at time t of a discount bond that pays off \$1 at time T is given by

$$P(t, T) = e^{-r(T-t)}. \tag{4.1.1}$$

(ii) If spot interest rates depend deterministic on time, $r_t = r(t)$, the bond formula becomes

$$P(t, T) = e^{-\int_t^T r(s)\, ds}. \tag{4.1.2}$$

This formula can be obtained as a consequence of (4.1.1). Dividing the time interval $[t, T]$ into n equal subintervals of length $\Delta s = (T - t)/n$

$$t = s_0 < s_1 < \cdots < s_{n-1} < s_n = T,$$

the value of the bond at time s_{k-1} is obtaining by discounting the value of the bond at time s_k

$$
\begin{aligned}
P(s_n, T) &= P(T, T) = 1 \\
P(s_{n-1}, T) &= e^{-r(s_{n-1})\Delta s} P(s_n, T) \\
P(s_{n-2}, T) &= e^{-r(s_{n-2})\Delta s} P(s_{n-1}, T) \\
\cdots &= \cdots \cdots \cdots \\
P(s_0, T) &= e^{-r(s_0)\Delta s} P(s_1, T).
\end{aligned}
$$

Multiplying on columns yields

$$P(t, T) = P(s_0, T) = e^{-\sum_{k=0}^{n-1} r(s_k)\Delta s}.$$

The formula (4.1.2) is obtained by taking the limit $n \to \infty$.

However, in general, the spot rates r_t are stochastic. In this case, the formula $e^{-\int_t^T r_s \, ds}$ is a random variable, and the price of the bond in this case can be calculated using the conditional expectation as of time t

$$P(t, T) = \mathbb{E}[e^{-\int_t^T r_s \, ds} | \mathcal{F}_t], \tag{4.1.3}$$

where $\mathcal{F}_t$ denotes the information available in the market at time t.

The bond price can be seen as the graph of a surface in $\mathbb{R}^3$ given by

$$(t, T) \to \Psi(t, T) = \big(t, T, P(t, T)\big),$$

where $0 \le t \le T$, see Fig. 4.1. The surface satisfies the boundary condition

$$P(t, T)\big|_{t=T} = 1, \qquad T \ge 0.$$

It is worth noting that the coordinate curves $T \to P(t, T)$ are deterministic while $t \to P(t, T)$ are stochastic. This makes possible to describe the bond surface by providing the stochastic model for the bond price, for instance, as

$$
\begin{aligned}
dP(t, T) &= r_t P(t, T) dt + \nu(t, T) P(t, T) \, dW_t \tag{4.1.4} \\
P(T, T) &= 1,
\end{aligned}
$$

where r_t is the spot rate and $\nu(t, T)$ is the bond volatility at time t (and might also depend on the entire price history until time t). Since the bond pays off with certainty \$1 at time T, the volatility must decline to zero at maturity, i.e. $\nu(T, T) = 0$. The differential $dP(t, T)$ is taken with respect to t and for avoiding confusions, it is denoted sometimes by $d_h P(t, T)$.

Equation (4.1.4) represents a one-factor model for the bond price, since the market uncertainty has only one noise source, dW_t. We shall stick with this simple model for the rest of the chapter.

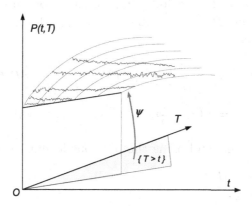

Figure 4.1: *The stochastic surface* $\Psi(t,T) = \big(t, T, P(t,T)\big)$.

4.2   Yield

If the interest rates are constant, $r_t = r$, the bond price formula $P(t,T) = e^{-r(T-t)}$ enables to retrieve the rate as $r = -\ln P(t,T)/(T-t)$. If the interest rates are deterministic, $r_t = r(t)$, then the similar formula

$$\bar{r} = \frac{1}{T-t} \int_t^T r(s)\, ds$$

provides the mean interest rate.

In the case of stochastic interest rates, a similar rate of distinguished importance can be defined as in the following. The *yield* of the bond is defined as

$$R(t,T) = -\ln P(t,T)/(T-t), \qquad t \leq T. \tag{4.2.5}$$

This is equivalent with the inverted formula

$$P(t,T) = e^{-R(t,T)(T-t)}.$$

The assignment

$$(t,T) \to R(t,T)$$

defines the *term structure surface*. The coordinate curves

$$T \to R(t,T)$$

are deterministic and smooth, while

$$t \to R(t,T)$$

are stochastic processes. In the following we shall compute the law of this process. Assuming that $P(t, T)$ satisfies equation (4.1.4), Ito's formula (see Appendix C.6) provides

$$
\begin{aligned}
d \ln P(t, T) &= \frac{1}{P(t, T)} dP(t, T) - \frac{1}{2 P(t, T)^2} dP(t, T)^2 \qquad (4.2.6) \\
&= \left(r_t - \frac{1}{2} \nu(t, T)^2 \right) dt + \nu(t, T) \, dW_t.
\end{aligned}
$$

Applying the product rule and using the previous formula we have

$$
\begin{aligned}
dR(t, T) &= d \left(\frac{\ln P(t, T)}{t - T} \right) \\
&= d \left(\frac{1}{t - T} \right) \ln P(t, T) + \frac{1}{t - T} d \ln P(t, T) \\
&= \frac{1}{T - t} \left(R(t, T) - (r_t - \nu(t, T)^2 / 2) \right) dt - \frac{1}{T - t} \nu(t, T) \, dW_t.
\end{aligned}
$$

Hence, the yield in terms of the spot rate and bond price volatility is given by

$$
dR(t, T) = \frac{1}{T - t} \left(R(t, T) - r_t + \nu(t, T)^2 / 2 \right) dt - \frac{1}{T - t} \nu(t, T) \, dW_t. \quad (4.2.7)
$$

The spot rate, r_t, can be obtained from the yield as in the following. Consider a small interval of time Δt. Then a discount bond with maturity $T = t + \Delta t$ has the price given by

$$
P(t, t + \Delta t) = e^{-R(t, t + \Delta t) \Delta t}.
$$

Comparing with the asymptotic relation

$$
P(t, t + \Delta t) = e^{-r_t \Delta t}, \qquad \Delta t \to 0,
$$

we can retrieve the spot rate from the yield via the formula $r_t = R(t, t)$. Moreover,

$$
\begin{aligned}
r_t &= \lim_{\Delta t \to 0} R(t, t + \Delta t) = - \lim_{\Delta t \to 0} \frac{\ln P(t, t + \Delta t)}{\Delta t} \\
&= - \lim_{\Delta t \to 0} \frac{\ln P(t, t + \Delta t) - \ln P(t, t)}{\Delta t} \\
&= - \frac{\partial}{\partial T} \ln P(t, T) \Big|_{T = t},
\end{aligned}
$$

where we used the condition $P(t, t) = 1$. This shows how can the spot rate be retrieved from the bond price. However, in general, knowing the spot rate r_t is not enough to recover $P(t, T)$. In order to be able to do this, we need forward rates.

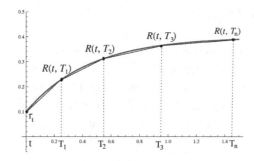

Figure 4.2: *The construction of the yield curve by interpolation.*

Formula (4.2.5) shows that when bond prices increase, yields decrease lowering the rates for mortgages and other loans. The market price for bonds can be driven up by bond buying programs, such as the one employed by the European Central Bank for the period March 2015 – September 2016. The expected effect of this program is a stimulation of the European economy.

4.3   Bootstrap Method

The construction of the yield curve starting from bond prices is called the *bootstrap method*. Consider n discount bonds with time to maturity $\tau_j = T_j - t$ and face values V_j, $1 \leq j \leq n$. Their prices, P_j, can be directly observed from the bond market, as in the next table:

Bond price	Face value	Maturity
P_1	V_1	T_1
P_2	V_2	T_2
...	...	...
P_n	V_n	T_n

The bond price formula $P_j = V_j e^{-R(t,T_j)(T_j - t)}$ implies

$$R(t, T_j) = \frac{1}{T_j - t} \ln \frac{V_j}{P_j}, \quad 1 \leq j \leq n.$$

The yield curve $R(t, T)$ can be constructed by interpolation from the previous values, starting with the initial value $R(t, t) = r_t$, which is the spot rate at time t, see Fig.4.2.

4.4   Forward Rates

This section is dedicated to the stochastic time evolution of the forward curve. Let $0 \leq t < T_1 < T_2$, and denote by $f(t, T_1, T_2)$ the *forward rate* as seen at time t for the time period between T_1 and T_2. This means that if one enters a contract at time t that pays \$1 at time T_2, then its price at time T_1 is obtained discounting by the forward rate between T_1 and T_2 as

$$V = e^{-f(t,T_1,T_2)(T_2-T_1)}.$$

Now, consider a discount bond that pays off the amount V at time T_1. Its price at time t is

$$P(t, T_1)V = P(t, T_1)e^{-f(t,T_1,T_2)(T_2-T_1)}. \tag{4.4.8}$$

In order to avoid arbitrage opportunities, this price has to be equal to the value of a bond at time t that pays of \$1 at maturity T_2, so

$$P(t, T_2) = P(t, T_1)V. \tag{4.4.9}$$

Equating relations (4.4.8)-(4.4.9) and solving for the forward rate, implies

$$f(t, T_1, T_2) = -\frac{\ln P(t, T_2) - \ln P(t, T_1)}{T_2 - T_1}. \tag{4.4.10}$$

The process followed by $f(t, T_1, T_2)$ can be computed using Ito's formulas similar to (4.2.7)

$$d \ln P(t, T_1) = \left(r_t - \frac{1}{2}\nu(t, T_1)^2\right)dt + \nu(t, T_1)\,dW_t$$

$$d \ln P(t, T_2) = \left(r_t - \frac{1}{2}\nu(t, T_2)^2\right)dt + \nu(t, T_2)\,dW_t$$

and then taking the differential in (4.4.10) we obtain

$$df(t, T_1, T_2) = \frac{\nu(t, T_2)^2 - \nu(t, T_1)^2}{2(T_2 - T_1)}dt - \frac{\nu(t, T_2) - \nu(t, T_1)}{T_2 - T_1}dW_t. \tag{4.4.11}$$

It is worthy to note that $f(t, T_1, T_2)$ depends only on the bonds volatilities.

The *instantaneous forward rate of borrowing* is defined by

$$f(t, T) = \lim_{\Delta t \to 0} f(t, T, T + \Delta t) = -\frac{\partial}{\partial T}\ln P(t, T), \tag{4.4.12}$$

where we made use of (4.4.10). $f(t, T)$ represents the rate at time T as seen from time t. For $t = T$ this becomes the spot rate at time t

$$r_t = f(t, t). \tag{4.4.13}$$

Exercise 4.4.1 *Let $0 < t < T_1 < T_2 < T_3$. Show that the forward rate $f(t, T_1, T_3)$ can be written as the weighted average of partial forwards rates as follows:*

$$f(t, T_1, T_3) = \frac{T_3 - T_2}{T_3 - T_1} f(t, T_2, T_3) + \frac{T_2 - T_1}{T_3 - T_1} f(t, T_1, T_2).$$

State and prove a generalization.

The forward rate can be retrieved from instantaneous forward rates as an integral average as in the following.

Proposition 4.4.2 *Let $t < a < b$. Then*

$$f(t, a, b) = \frac{1}{b - a} \int_a^b f(t, \tau) \, d\tau.$$

Proof: Consider the equidistant division $a = T_0 < T_1 < \cdots < T_{n-1} < T_n = b$, with width $\Delta \tau$. A generalization of Exercise 4.4.1 to n intervals writes as

$$f(t, a, b) = \frac{1}{b - a} \sum_{j=0}^{n-1} f(t, T_j, T_{j+1})(T_{j+1} - T_j)$$

$$= \frac{1}{b - a} \sum_{j=0}^{n-1} f(t, T_j, T_j + \Delta \tau) \Delta \tau.$$

Taking $\Delta \tau \to 0$ yields

$$f(t, a, b) = \frac{1}{b - a} \lim_{n \to \infty} \sum_{j=0}^{n-1} f(t, T_j, T_j + \Delta \tau) \Delta \tau$$

$$= \frac{1}{b - a} \int_a^b f(t, \tau, \tau) \, d\tau = \frac{1}{b - a} \int_a^b f(t, \tau) \, d\tau.$$

∎

Exercise 4.4.3 (Forward contract) *Enter a contract at time t to pay K at time T_1 in order to receive \$1 at a later time T_2. Show the following equivalent formulas for the payment:*

(a) $K = e^{-f(t, T_1, T_2)(T_2 - T_1)}$;

(b) $K = e^{-\int_t^T f(t, s) \, ds}$;

(c) $K = P(t, T_2)/T(t, T_1)$.

Exercise 4.4.4 (Coupon bearing bond) *Let $0 < t < T_1 < \cdots < T_n < T$. Consider a contract that pays off the cash ammounts $c_1, \cdots, c_n$ at times T_1, $\cdots$, T_n, and a final payment of \$1 at time T. Show that the price of the contract at time t is*

$$C(t) = P(t, T) + \sum_{i=1}^{n} c_i P(t, T_i).$$

4.5 Single-Factor HJM Models

The process followed by the instantaneous forward rate $f(t, T)$ can be obtained from (4.4.11) making $T_1 = T$, $T_2 = T + \Delta t$ and taking $\Delta t \to 0$. The resulting process is

$$df(t, T) = \nu(t, T)\frac{\partial}{\partial T}\nu(t, T)dt - \frac{\partial}{\partial T}\nu(t, T)\, dW_t. \tag{4.5.14}$$

This corresponds to a one-factor model, which is fully described by the bond volatility $\nu(t, T)$. Following Health, Jarrow and Morton [38] we write

$$df(t, T) = m(t, T)dt - s(t, T)\, dW_t.$$

Assuming $\nu(t, t) = 0$, the following relation between the drift and volatility holds

$$m(t, T) = s(t, T) \int_t^T s(t, u)\, du.$$

The forward rate $f(t, T)$ can be written in terms of the initial forward rate as

$$f(t, T) = f(0, T) + \int_0^t m(\tau, T)\, d\tau - \int_0^t s(\tau, T)\, dW_\tau. \tag{4.5.15}$$

The process for the spot rate The aforementioned one-factor (HJM) model for the forward rate provides a model for the spot rate. Making $T = t$ in (4.5.15) and using (4.4.13) yields

$$r_t = f(0, t) + \int_0^t m(\tau, t)\, d\tau - \int_0^t s(\tau, t)\, dW_\tau, \tag{4.5.16}$$

where

$$m(\tau, t) = \nu(\tau, t)\frac{\partial}{\partial t}\nu(\tau, t) \tag{4.5.17}$$

$$s(\tau, t) = \frac{\partial}{\partial t}\nu(\tau, t). \tag{4.5.18}$$

In the case when the volatility $\nu(t, T)$ is independent of both the interest rate r_t and its history, then the spot rate is normally distributed with the mean and variance given by

$$\mathbb{E}[r_t] = f(0, t) + \int_0^t m(\tau, t)\, d\tau$$

$$Var(r_t) = \int_0^t s(\tau, t)^2\, d\tau.$$

In the following we shall encounter a few familiar cases for the one-factor (HJM) model.

4.5.1   Ho-Lee Model

Let $\nu(t, T) = -\sigma(T - t)$, with σ constant. Then (4.5.17)-(4.5.18) provide

$$m(\tau, t) = \sigma^2(t - \tau)$$
$$s(\tau, t) = -\sigma.$$

Substituting in (4.5.16) we obtain the following formula for the spot rate

$$r_t = f(0, t) + \frac{1}{2}\sigma^2 t^2 + \sigma W_t. \tag{4.5.19}$$

The associated stochastic differential equation is given by

$$dr_t = (\partial_t f(0, t) + \sigma^2 t)dt + \sigma\, dW_t.$$

Substituting $\theta(t) = \partial_t f(0, t) + \sigma^2 t$ we arrive at the Ho-Lee model

$$dr_t = \theta(t)dt + \sigma\, dW_t. \tag{4.5.20}$$

Hence, the Ho-Lee model is a one-factor (HJM) model.

Now we shall work out the computation in the reverse way, i.e. we assume that r_t satisfies the Ho-Lee model (4.5.20) and we show that the drift function $\theta(t)$ depends on the yield curve $f(0, t)$. Solving the equation (4.5.20) we obtain

$$r_t = r_0 + \Theta(t) + \sigma W_t,$$

where $\Theta(t) = \int_0^t \theta(s)\, ds$. Integrating again yields

$$\int_0^t r_s\, ds = r_0 t + \int_0^t \Theta(s)\, ds + \sigma Z_t,$$

where $Z_t = \int_0^t W_s\,ds$. Next we compute the bond price using formula (4.1.3)

$$
\begin{aligned}
P(0,t) &= \mathbb{E}\left[e^{-\int_0^t r_s\,ds}\middle|\mathcal{F}_0\right] \\
&= e^{-r_0 t - \int_0^t \Theta(s)\,ds}\mathbb{E}\left[e^{-\sigma Z_t}\middle|\mathcal{F}_0\right] \\
&= e^{-r_0 t - \int_0^t \Theta(s)\,ds + \sigma^2 t^3/6},
\end{aligned}
$$

where we took out the deterministic part and used that $\sigma Z_t \sim \mathcal{N}(0, \sigma^2 t^3/3)$. Formula (4.4.12) provides the yield curve from the bond price

$$
\begin{aligned}
f(0,t) &= -\frac{\partial}{\partial t}\ln P(0,t) \\
&= -\frac{\partial}{\partial t}\left(-r_0 t - \int_0^t \Theta(s)\,ds + \frac{1}{6}\sigma^2 t^3\right) \\
&= r_0 + \Theta(t) - \frac{1}{2}\sigma^2 t^2.
\end{aligned}
$$

Solving for $\Theta(t)$ and then differentiating in

$$
\Theta(t) = f(0,t) - r_0 + \frac{1}{2}\sigma^2 t^2
$$

yields

$$
\theta(t) = \partial_t f(0,t) + \sigma^2 t,
$$

which provides the drift, $\theta(t)$, in terms of the slope of the yield curve. The initial forward curve $f(0,t)$ is an input to the model, in the sense that the function $\theta(t)$ can be read from the forward curve available at time $t = 0$.

4.5.2   Hull and White Model

Assume the volatility is $v(t,T) = -\frac{\sigma}{a}(1 - e^{-a(T-t)})$. Then we can infer from the equations (4.5.17)-(4.5.18) that

$$
\begin{aligned}
m(\tau,t) &= \frac{\sigma^2}{a}\left(e^{-a(t-\tau)} - e^{-2a(t-\tau)}\right) \\
s(\tau,t) &= -\sigma e^{-a(t-\tau)}.
\end{aligned}
$$

An elementary computation provides

$$
\int_0^t m(\tau,t)\,d\tau = \frac{\sigma^2}{a^2}\left(\frac{1}{2}(1 + e^{-2at}) - e^{-at}\right).
$$

Substituting in (4.5.16), we obtain

$$r_t = f(0,t) + \frac{\sigma^2}{a^2}\left(\frac{1}{2}(1 + e^{-2at}) - e^{-at}\right) + \sigma e^{-at}\int_0^t e^{a\tau}\,dW_\tau. \tag{4.5.21}$$

Comparing with the solution of the Hull-White model (3.4.16), which is given by

$$r_t = r_0 e^{-at} + e^{-at}\int_0^t \theta(s)e^{as}\,ds + \sigma e^{-at}\int_0^t e^{a\tau}\,dW_\tau, \tag{4.5.22}$$

we obtain the following integral equation in $\theta(t)$

$$r_0 + \int_0^t \theta(s)e^{as}\,ds = e^{at}\left[f(0,t) + \frac{\sigma^2}{a^2}\left(\frac{1}{2}(1 + e^{-2at}) - e^{-at}\right)\right].$$

Solving for $\theta(t)$ by differentiation, we obtain

$$\theta(t) = \partial_t f(0,t) + af(0,t) + \frac{\sigma^2}{2a}(1 - e^{-2at}).$$

This shows that the initial forward curve $f(0,t)$ is an input for the Hull and White model. We also note that taking $a \to 0$ in the Hull and White model recovers Ho-Lee model.

In the following we shall work out the computation in the reversed way. Assume that r_t satisfies equation (3.4.16)

$$dr_t = (\theta(t) - ar_t)dt + \sigma dW_t,$$

with the solution given by the Gaussian process (3.4.17)

$$\begin{aligned}
r_t &= r_0 e^{-at} + e^{-at}\int_0^t \theta(s)e^{as}\,ds + \sigma e^{-at}\int_0^t e^{as}\,dW_s \\
&= r_0 e^{-at} + e^{-at}\alpha(t) + \sigma e^{-at}\beta_t,
\end{aligned}$$

where

$$\alpha(t) = \int_0^t \theta(s)e^{as}\,ds, \qquad \beta_t = \int_0^t e^{as}\,dW_s.$$

Integrating between 0 and t

$$\int_0^t r_s\,ds = \frac{r_0(e^{-at} - 1)}{a} + \int_0^t e^{-as}\alpha(s)\,ds + \sigma\int_0^t e^{-as}\beta_s\,ds$$

and using the bond price formula (4.1.3) we have

$$\begin{aligned}
P(0,t) &= \mathbb{E}\left[e^{-\int_0^t r_s\,ds}\Big|\mathcal{F}_0\right] \\
&= e^{\frac{r_0(1 - e^{-at})}{a}}e^{-\int_0^t e^{-as}\alpha(s)\,ds}\,\mathbb{E}\left[e^{-\sigma\int_0^t e^{-as}\beta_s\,ds}\Big|\mathcal{F}_0\right] \\
&= e^{\frac{r_0(1 - e^{-at})}{a}}e^{-\int_0^t e^{-as}\alpha(s)\,ds}e^{\frac{\sigma^2}{2a^2}\left(t + \frac{1 - e^{-2at}}{2a} + \frac{2}{a}(1 - e^{-at})\right)},
\end{aligned}$$

where we took out the measurable functions, dropped the independent condition, and used Exercise 4.5.1, part (d).

$$\frac{\sigma^2}{2a^2}\left(t + \frac{1 - e^{-2at}}{2a} + \frac{2}{a}(1 - e^{-at})\right)$$

Exercise 4.5.1 *Consider the process* $X_t = \int_0^t e^{-as}\beta_s\,ds$. *Show that:*
(a) $\mathbb{E}[X_t] = 0$.

(b) $\mathbb{E}[X_t\beta_t] = \dfrac{1}{a^2}\left(\dfrac{e^{at} + e^{-at}}{2} - 1\right)$.

(c) $\mathbb{E}[X_t^2] = \dfrac{1}{2a^2}\left(t + \dfrac{1 - e^{-2at}}{2a} + \dfrac{2}{a}(1 - e^{-at})\right)$.

(d) $\mathbb{E}[e^{\sigma X_t}] = e^{\frac{\sigma^2}{2a^2}\left(t + \frac{1 - e^{-2at}}{2a} + \frac{2}{a}(1 - e^{-at})\right)}$.

The yield curve can be obtained from the bond price as

$$
\begin{aligned}
f(0,t) &= -\frac{\partial}{\partial t}\ln P(0,t)\\
&= \frac{\partial}{\partial t}\left(\frac{r_0}{a}(e^{-at} - 1) + \int_0^t e^{-as}\alpha(s)\,ds\right.\\
&\quad \left. + \frac{\sigma^2}{2a^2}\left(-t + \frac{e^{-2at} - 1}{2a} + \frac{2}{a}(e^{-at} - 1)\right)\right)\\
&= -r_0 e^{-at} + e^{-at}\alpha(t) + \frac{\sigma^2}{2a^2}(-1 - e^{-2at}\quad 2e^{-at})\\
&= e^{-at}\left(-r_0 - \frac{\sigma^2}{a^2}\right) + e^{-at}\alpha(t) - \frac{\sigma^2}{2a^2} - \frac{\sigma^2}{2a^2}e^{-2at}. \quad (4.5.23)
\end{aligned}
$$

Solving for $\alpha(t)$,

$$\alpha(t) = e^{at}f(0,t) + r_0 + \frac{\sigma^2}{2a^2} + \frac{\sigma^2}{2a^2}e^{at} + \frac{\sigma^2}{2a^2}e^{-at},$$

and then differentiating, we obtain

$$\theta(t)e^{at} = ae^{at}f(0,t) + e^{at}\frac{\partial}{\partial t}f(0,t) + \frac{\sigma^2}{2a^2}ae^{at} - \frac{\sigma^2}{2a^2}ae^{-at}.$$

Dividing by e^{at} we obtain $\theta(t)$ in terms of the term structure

$$\theta(t) = \frac{\partial}{\partial t}f(0,t) + af(0,t) + \frac{\sigma^2}{2a}(1 - e^{-2at}). \quad (4.5.24)$$

Exercise 4.5.2 *What happens if we allow $a \to 0$ in relation (4.5.24)? What do you obtain?*

Exercise 4.5.3 *(a) Let $y(t) = f(0,t)$. Show that $y(t)$ satisfies the following linear differential equation*

$$y' + ay = \theta(t) - \frac{\sigma^2}{2a}(1 - e^{-2at})$$

$$y(0) = r_0.$$

(b) Verify that the solution is given by formula (4.5.23).

Exercise 4.5.4 *Find the forward curve $f(t,T)$ in the case of*
(a) Ho-Lee model;
(b) Hull and White model.

4.5.3 Vasicek Model

Considering $\nu(t,T) = \frac{\sigma}{a}e^{-a(T-t)}$, then equations (4.5.17)-(4.5.18) imply

$$m(\tau,t) = -\frac{\sigma^2}{a}e^{-2a(t-\tau)}$$

$$s(\tau,t) = -\sigma e^{-a(t-\tau)}.$$

Substituting in (4.5.16) and performing the integration, we obtain

$$r_t = f(0,t) - \frac{\sigma^2}{4a}(1 - e^{-2at}) + \sigma e^{-at}\int_0^t e^{a\tau}\,dW_\tau.$$

Comparing with the Ornstain-Uhlenbeck process

$$r_t = b + (r_0 - b)e^{-at} + \sigma e^{-at}\int_0^t e^{a\tau}\,dW_\tau,$$

which is a solution of the Vasicek's model (3.3.7), we infer

$$f(0,t) = b + (r_0 - b)e^{-at} + \frac{\sigma^2}{4a}(1 - e^{-2at}).$$

This shows that in the case of the Vasicek's model the initial forward curve is an output of the model.

Exercise 4.5.5 *Express the parameters a, b, and σ in terms of the initial derivatives of the initial forward curve $\partial_t f(0,t)|_{t=0}$, $\partial_t^2 f(0,t)|_{t=0}$, and $\partial_t^3 f(0,t)|_{t=0}$. Why is this method limited in practical applications?*

Exercise 4.5.6 *Let $f(0,t) = 0.5 - 0.2e^{-0.15t}$. Find the spot rate r_t and evaluate $\mathbb{E}[r_t|\mathcal{F}_0]$ in the following cases:*
(a) Ho-Lee model;
(b) Hull and White model.

4.6   Relation Formulas

In this section we shall deal with the one-to-one relations among bond prices $P(t,T)$, instantaneous forward rates $f(t,T)$ and yield curves $R(t,T)$.

Proposition 4.6.1 *The price of a discount bond paying off \$1 at time T can be evaluated as*

$$P(t,T) = e^{-\int_t^T f(t,s)\,ds}. \tag{4.6.25}$$

Proof: Integrating between t and T in the definition formula for instantaneously forward rates (4.4.12)

$$f(t,\tau) = -\frac{\partial}{\partial\tau}\ln P(t,\tau),$$

we obtain

$$\int_t^T f(t,\tau)\,d\tau = \ln P(t,T) - \ln P(t,t) = \ln P(t,T),$$

where we used $P(t,t) = 1$. Solving for $P(t,T)$ we arrive at formula (4.6.25).

■

Proposition 4.6.2 *The relations between yield curves and instantaneous forward rates are given by*

$$R(t,T) \quad = \quad \frac{1}{T-t}\int_t^T f(t,s)\,ds \tag{4.6.26}$$

$$f(t,T) \quad = \quad R(t,T) + (T-t)\frac{\partial}{\partial T}R(t,T). \tag{4.6.27}$$

Proof: Writing the bond price in terms of the yield

$$P(t,T) = e^{-R(t,T)(T-t)}$$

and comparing with equation (4.6.25) implies

$$R(t,T)(T-t) = \int_t^T f(t,s)\,ds, \tag{4.6.28}$$

which implies the former relation. Then differentiating with respect to T in (4.6.28) and using product rule and Fundamental Theorem of Calculus, we obtain the latter relation.

■

It is worth noting that if $R(t,T)$ increases with maturity T, then the latter formula implies $f(t,T) > R(t,T)$. Taking $t = T$, we obtain the relation with the spot rate

$$R(t,t) = r_t = f(t,t).$$

Exercise 4.6.3 *Find the stochastic differential equation of $R(t,T)$ with respect to t.*

Exercise 4.6.4 *Consider the forward rate satisfying the model*

$$df(t,T) = \alpha dt + \sigma dW_t, \qquad t \le T,$$

with α and σ constants. Given the initial forward curve

$$f(0,T) = 0.4 - 0.15e^{-0.2T}.$$

(a) Find the yield curve $R(t,T)$;
(b) Find the bond price $P(t,T)$.

Exercise 4.6.5 *Assume the spot rate is given by $r_t = r + \sigma W_t$, with r and σ constants.*
(a) Find the instantaneous forward rate $f(t,T)$;
(b) Find the yield curves $R(t,T)$.

The next few sections deal with models that have the special feature that there is a closed formula price for the associated discount bond. This is based on the ability of evaluating explicitly the conditional expectation $\mathbb{E}[e^{-\int_t^T r_s\,ds}|\mathcal{F}_t]$. One main ingredients in doing this is the formula

$$\mathbb{E}[e^{\int_t^T \varphi(s)\,dW_s}|\mathcal{F}_t] = e^{\frac{1}{2}\int_t^T \varphi^2(s)\,ds},$$

which reduces a conditional expectation to an exponential. This follows from the fact that the Wiener integral $\int_t^T \varphi(s)\,dW_s$ is normally distributed with mean zero and variance $\int_t^T \varphi^2(s)\,ds$, and is independent of the information in the market at time t, see Proposition C.6.3, Appendix C.6, for details.

4.7 A Simple Spot Rate Model

We shall start by considering a simplistic example, which despite of the fact that does not have any market applicability, will serve our later computational purposes. Assume the spot rate r_t satisfies the stochastic equation

$$dr_t = \sigma dW_t,$$

with $\sigma > 0$, constant. This corresponds to a spot rate which difusses starting from an initial rate r as

$$r_t = r + \sigma W_t, \qquad t \ge 0.$$

We shall show that the price of the associated zero-coupon bond that pays off $1 at time T is given by the formula

$$P(t,T) = e^{-r_t(T-t)+\frac{1}{6}\sigma^2(T-t)^3}.$$

Let $0 < t < T$ be fixed. We start by writing future spot rates in terms of the present spot rates. This can be achieved by solving the stochastic differential equation, and obtaining for any $t < s < T$

$$r_s = r_t + \sigma(W_s - W_t).$$

Integrating over the remaining life of the bond yields

$$\int_t^T r_s \, ds = r_t(T - s) + \sigma \int_t^T (W_s - W_t) \, ds.$$

Then taking the exponential, we obtain the price of the bond at time t

$$
\begin{aligned}
P(t,T) &= \mathbb{E}\left[e^{-\int_t^T r_s \, ds}\,\Big|\,\mathcal{F}_t\right] \\
&= e^{-r_t(T-t)}\mathbb{E}\left[e^{-\sigma \int_t^T (W_s-W_t) \, ds}\,\Big|\,\mathcal{F}_t\right] \\
&= e^{-r_t(T-t)}\mathbb{E}\left[e^{-\sigma \int_t^T (W_s-W_t) \, ds}\right] \\
&= e^{-r_t(T-t)}\mathbb{E}\left[e^{-\sigma \int_0^{T-t} W_s \, ds}\right] \\
&= e^{-r_t(T-t)}e^{\frac{\sigma^2}{2}\frac{(T-t)^3}{3}}.
\end{aligned}
$$

In the second identity we took the $\mathcal{F}_t$-measurable part out of the expectation, while in the third identity we dropped the condition since $W_s - W_t$ is independent of the information set $\mathcal{F}_t$ for any $t < s$. The fourth identity invoked the stationarity of the Brownian motion. The last identity follows from the fact that $-\sigma \int_0^{T-t} W_s \, ds$ is normally distributed with mean 0 and variance $\frac{\sigma^2(T-t)^3}{3}$, and from the expression of the moment generating function of a normal distribution.

It is worth noting that for this model the price of the bond, $P(t,T)$, depends only the spot rate at time t, r_t, and the time to maturity $T-t$. This is not a realistic model since it states that if today's spot rate is known, then this would suffice for computing the bond price; this would not take into consideration all the future market movements between today and maturity, which would definitely have an effect on the bond price.

Term Structure The yield curves are given by

$$
\begin{aligned}
R(t,T) &= -\frac{1}{T-t}\ln P(t,T) = r_t - \frac{1}{6}\sigma^2(T-t)^2 \\
&= r - \frac{1}{6}\sigma^2(T-t)^2 + \sigma W_t.
\end{aligned}
$$

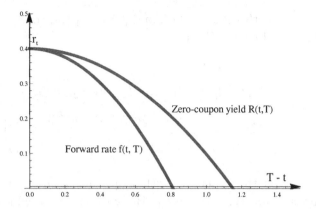

Figure 4.3: *The term structure situation for $r_t = r + \sigma W_t$.*

Then the instantaneous forward rate becomes

$$f(t,T) = R(t,T) + (T-t)\frac{\partial}{\partial T}R(t,T)$$

$$= r_t - \frac{1}{2}\sigma^2(T-t)^2 = r - \frac{1}{2}\sigma^2(T-t)^2 + \sigma W_t. \quad (4.7.29)$$

The forward rates lay below the yield curves, and the term structure is downward sloping, see Fig.4.3. One major disadvantage of this model is that the rates become negative for $T-t$ large.

Exercise 4.7.1 *Asume the spot rates exhibit positive jumps of size $\sigma > 0$ and satisfy the stochastic equation*

$$dr_t = \sigma dN_t,$$

where N_s is a Poisson process with rate λ.
(a) Find the price of the associated zero-coupon bond that pays off \$1 at time T;
(b) Compute the yield curve $R(t,T)$;
(c) Determine the formula for the instantaneous forward rates.

4.8   Bond Price for Ho-Lee Model

In the case of Ho-Lee model the initial term structure is an input to the model, and the explicit formula for the spot rate is given by relation (4.5.19)

$$r_t = f(0,t) + \frac{1}{2}\sigma^2 t^2 + \sigma W_t. \quad (4.8.30)$$

We shall show that for this model the bond price depends on the spot rate r_t and the initial forward curve. We have

$$
\begin{aligned}
P(t,T) &= \mathbb{E}\big[e^{-\int_t^T r_s\, ds}\big|\mathcal{F}_t\big] \\
&= \mathbb{E}\big[e^{-\int_t^T f(0,s)\, ds - \frac{1}{2}\sigma^2 \int_t^T s^2\, ds - \sigma \int_t^T W_s\, ds}\big|\mathcal{F}_t\big] \\
&= \mathbb{E}\big[e^{-(T-t)f(0,t,T) - \frac{1}{6}\sigma^2 (T^3 - t^3) - \sigma \int_t^T W_s\, ds}\big|\mathcal{F}_t\big] \\
&= e^{-(T-t)f(0,t,T) - \frac{1}{6}\sigma^2 (T^3 - t^3)} \mathbb{E}\big[e^{-\sigma \int_t^T W_s\, ds}\big|\mathcal{F}_t\big]. \quad (4.8.31)
\end{aligned}
$$

We shall evaluate next the expectation that appears in the previous relation. First we create the jump $W_s - W_s$ in the Brownian motion, which is independent of information $\mathcal{F}_t$

$$
\sigma \int_t^T W_s\, ds = \sigma \int_t^T (W_s - W_t)\, ds + \sigma W_t(T - t).
$$

Using the stationarity of Brownian motion, taking the measurable factors out and dropping the independent condition, we have

$$
\begin{aligned}
\mathbb{E}\big[e^{-\sigma \int_t^T W_s\, ds}\big|\mathcal{F}_t\big] &= \mathbb{E}\big[e^{-\sigma \int_t^T (W_s - W_t)\, ds - \sigma W_t(T-t)}\big|\mathcal{F}_t\big] \\
&= e^{-\sigma W_t(T-t)} \mathbb{E}\big[e^{-\sigma \int_t^T (W_s - W_t)\, ds}\big|\mathcal{F}_t\big] \\
&= e^{-\sigma W_t(T-t)} \mathbb{E}\big[e^{-\sigma \int_t^T (W_s - W_t)\, ds}\big] \\
&= e^{-\sigma W_t(T-t)} \mathbb{E}\big[e^{-\sigma \int_0^{T-t} W_u\, du}\big] \\
&= e^{-\sigma W_t(T-t)} e^{\frac{1}{6}\sigma^2 (T-t)^3}. \quad (4.8.32)
\end{aligned}
$$

Note that in the last identity we used that $\int_0^{T-t} W_u\, du \sim \mathcal{N}\big(0, \frac{(T-t)^3}{3}\big)$. Since from (4.8.30) we can solve for the Brownian motion in terms of the spot rate as

$$
\sigma W_t = r_t - f(0,t) - \frac{1}{2}\sigma^2 t^2,
$$

then (4.8.32) becomes

$$
\mathbb{E}\big[e^{-\sigma \int_t^T W_s\, ds}\big|\mathcal{F}_t\big] = e^{-r_t(T-t)} e^{f(0,t)(T-t) + \frac{1}{2}\sigma^2 t^2 (T-t)} e^{\frac{1}{6}\sigma^2 (T-t)^3}.
$$

Substituting into (4.8.31), then after doing the algebra we obtain

$$
\boxed{P(t,T) = e^{-r_t(T-t)} A(t,T),} \quad (4.8.33)
$$

with

$$
A(t,T) = \exp[-f(0,t,T)(T-t) + f(0,t)(T-t) - \frac{1}{2}\sigma^2 t(T-t)^2]. \quad (4.8.34)
$$

The associated yield curve is

$$R(t,T) = -\frac{\ln P(t,T)}{T-t}$$

$$= r_t + f(0,t,T) - f(0,t) + \frac{1}{2}\sigma^2 t(T-t).$$

Interest Term Structure The bond formula (4.8.33) combined with relation $P(t,T) = e^{-R(t,T)(T-t)}$ imply the following formula for the yield curve

$$R(t,T) = r_t + f(0,t,T) - f(0,t) + \frac{1}{2}\sigma^2 t(T-t).$$

From here, the forward rates can be obtained using formula (4.6.27)

$$\begin{aligned} f(t,T) &= R(t,T) + (T-t)\frac{\partial}{\partial T}R(t,T) \\ &= r_t + f(0,t,T) - f(0,t) + \sigma^2 t(T-t) \\ &\quad + (T-t)\partial_T f(0,t,T). \end{aligned} \tag{4.8.35}$$

It is worthy to note that the bond formula can be written in terms of the initial bond prices. Using

$$f(0,t) = -\frac{\partial \ln P(0,t)}{\partial t}$$

$$f(0,t,T) = -\frac{1}{T-t}[\ln P(0,T) - \ln P(0,t)],$$

formula (4.8.34) becomes

$$A(t,T) = \exp\left[\ln\frac{P(0,T)}{P(0,t)} - (T-t)\frac{\partial \ln P(0,t)}{\partial t} - \frac{1}{2}\sigma^2 t(T-t)^2\right].$$

4.9 Bond Price for Vasicek's Model

The next result regarding bond pricing is due to Vasicek [66]. Its initial proof was based on partial differential equations, while here we provide an approach solely based on expectations.

Proposition 4.9.1 *Assume the spot rate r_t satisfies the model*

$$dr_t = a(b - r_t)dt + \sigma dW_t. \tag{4.9.36}$$

Then the price of the zero-coupon bond that pays \$1 at time T is given by

$$P(t,T) = A(t,T)e^{-B(t,T)r_t}, \tag{4.9.37}$$

where

$$B(t,T) = \frac{1 - e^{-a(T-t)}}{a}$$

$$A(t,T) = e^{\frac{(B(t,T)-T+t)(a^2 b - \sigma^2/2)}{a^2} - \frac{\sigma^2 B(t,T)^2}{4a}}.$$

Proof: The computation follows the idea of section 4.7. Integrating equation (4.9.36) between t and T, and then taking the exponential, we arrive at

$$e^{-\int_t^T r_s \, ds} = e^{\frac{1}{a}(r_T - r_t) - \frac{\sigma}{a}(W_T - W_t) - b(T-t)}. \tag{4.9.38}$$

The bond price is computed by taking the conditional expectation as of time t in relation (4.9.37). In order to do this we need to express the terms in the exponent of the right side more conveniently.

We start with the computation of the difference $(r_T - r_t)$. Multiplying equation (4.9.36) by e^{as} leads to the exact equation

$$d(e^{as} r_s) = abe^{as} ds + \sigma e^{as} dW_s,$$

which after integration between t and T yields

$$e^{aT} r_T - e^{at} r_t = b(e^{aT} - e^{at}) + \sigma \int_t^T e^{as} \, dW_s.$$

Solving for r_T, then subtracting r_t and dividing by a we obtain

$$\frac{1}{a}(r_T - r_t) = -r_t B(t,T) + bB(t,T) + \frac{\sigma}{a} e^{-aT} \int_t^T e^{as} \, dW_s. \tag{4.9.39}$$

The second exponent in the right side of (4.9.38) can be written as an integral

$$\frac{\sigma}{a}(W_T - W_t) = \frac{\sigma}{a} \int_t^T dW_s. \tag{4.9.40}$$

Substituting (4.9.39) and (4.9.40) into (4.9.38) we obtain

$$e^{-\int_t^T r_s \, ds} = e^{-r_t B(t,T)} e^{bB(t,T) - b(T-t)} e^{\frac{\sigma}{a} \int_t^T (e^{as-aT} - 1) \, dW_s}. \tag{4.9.41}$$

Taking the measurable part out, then dropping the independent condition in the conditional expectation, the price of the discount bond at time t becomes

$$\begin{aligned}
P(t,T) &= \mathbb{E}[e^{-\int_t^T r_s \, ds} | \mathcal{F}_t] \\
&= e^{-r_t B(t,T)} e^{bB(t,T) - b(T-t)} \mathbb{E}[e^{\frac{\sigma}{a} \int_t^T (e^{as-aT} - 1) \, dW_s}] \\
&= e^{-r_t B(t,T)} A(t,T),
\end{aligned}$$

where we denoted

$$A(t,T) = \mathbb{E}[e^{\frac{\sigma}{a}\int_t^T (e^{as-aT}-1)\,dW_s}]. \tag{4.9.42}$$

As a Wiener integral, $\int_t^T (e^{as-aT} - 1)\,dW_s$ is normally distributed with zero mean and variance given by

$$\int_t^T (e^{as-aT} - 1)^2\,ds \;=\; -\frac{a}{2}B(t,T)^2 - B(t,T) + (T - t).$$

Then the log-normal variable $e^{\frac{\sigma}{a}\int_t^T (e^{as-aT}-1)\,dW_s}$ has the mean

$$\mathbb{E}[e^{\frac{\sigma}{a}\int_t^T (e^{as-aT}-1)\,dW_s}] = e^{\frac{1}{2}\frac{\sigma^2}{a^2}[-\frac{a}{2}B(t,T)^2 - B(t,T) + (T-t)]}.$$

Substituting in (4.9.42) and performing the following algebraic computations

$$bB(t,T) - b(T - t) + \frac{1}{2}\frac{\sigma^2}{a^2}[-\frac{a}{2}B(t,T)^2 - B(t,T) + (T - t)]$$

$$= -\frac{\sigma^2}{4a}B(t,T)^2 + (T - t)\left(\frac{\sigma^2}{2a^2} - b\right) + \left(b - \frac{\sigma^2}{2a^2}\right)B(t,T)$$

$$= -\frac{\sigma^2}{4a}B(t,T)^2 + \frac{1}{a^2}(a^2 b - \sigma^2/2)\big(B(t,T) - T + t\big),$$

we obtain

$$A(t,T) = e^{\frac{(B(t,T)-T+t)(a^2 b - \sigma^2/2)}{a^2} - \frac{\sigma^2 B(t,T)^2}{4a}}.$$

It is worth noting that $P(t,T)$ depends on the time to maturity, $T - t$, and the spot rate r_t. ∎

Exercise 4.9.2 *Find the price of an infinitely lived bond in the case when spot rates satisfy Vasicek's model.*

The term structure Let $R(t,T)$ be the continuously compounded interest rate at time t for a term of $T - t$. Using the formula for the bond price $P(t,T) = e^{-R(t,T)(T-t)}$ we get

$$R(t,T) = -\frac{1}{T - t}\ln P(t,T).$$

Using formula for the bond price (4.9.37) yields

$$R(t,T) = -\frac{1}{T - t}\ln A(t,T) + \frac{1}{T - t}P(t,T)r_t.$$

A few possible shapes of the term structure $R(t,T)$ in the case of Vasicek's model are given in Fig. 4.4.

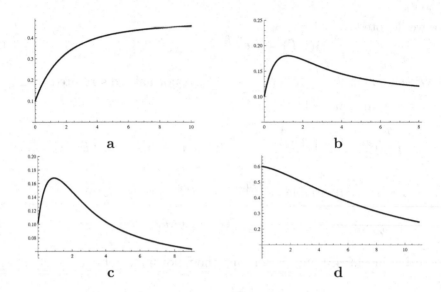

Figure 4.4: *The yield on zero-coupon bond versus maturity time $T - t$ in the case of Vasicek's model:* **a.** $a = 1$, $b = 0.5$, $\sigma = 0.6$, $r_t = 0.1$; **b.** $a = 1$, $b = 0.5$, $\sigma = 0.8$, $r_t = 0.1$; **c.** $a = 1$, $b = 0.5$, $\sigma = 0.97$, $r_t = 0.1$; **d.** $a = 0.3$, $b = 0.5$, $\sigma = 0.3$, $r_t = 0.r$.

4.10 Bond Price for CIR's Model

In the case when r_t satisfy the Cox, Ingersoll, and Ross (CIR) model (3.3.13), the zero-coupon bond price has a similar form as in the case of Vasicek's model

$$P(t,T) = A(t,T)e^{-B(t,T)r_t}.$$

The functions $A(t,T)$ and $B(t,T)$ are given in this case by

$$B(t,T) = \frac{2(e^{\gamma(T-t)} - 1)}{(\gamma + a)(e^{\gamma(T-t)} - 1) + 2\gamma}$$

$$A(t,T) = \left(\frac{2\gamma e^{(a+\gamma)(T-t)/2}}{(\gamma + a)(e^{\gamma(T-t)} - 1) + 2\gamma} \right)^{2ab/\sigma^2},$$

where $\gamma = (a^2 + 2\sigma^2)^{1/2}$. For details the reader can consult Cox, Ingersoll, and Ross [23].

Exercise 4.10.1 *Consider the spot rate r_t satisfying the Black, Derman and Toy model (3.5.18).*

(a) Find the price of the associated discount bond;

(b) Match the function parameter $\theta(t)$ to the forward curve $f(0, t)$.

4.11 Mean Reverting Model with Jumps

We shall study a model for the short-term interest rate r_t, which is mean reverting and also incorporates jumps. Let N_t be the Poisson process of constant rate λ and $M_t = N_t - \lambda t$ denote the compensated Poisson process, which is a martingale with respect to the information set $\mathcal{F}_t = \sigma\{N_s; s \leq t\}$, induced by N_t, see Appendix C.1.

Consider the following model for the spot rate

$$dr_t = a(b - r_t)dt + \sigma dM_t, \tag{4.11.43}$$

with a, b, σ positive constants. It is worth noting the similarity with the Vasicek's model, which is obtained by replacing the process dM_t by dW_t, where W_t is a one-dimensional Brownian motion.

Making abstraction of the uncertainty source σdM_t, this implies that the rate r_t is pulled towards level b at the rate a. This means that, if for instance $r_0 > b$, then r_t is decreasing towards b. The term σdM_t adds jumps of size σ to the process. The fact that these jumps have equal size is a limitation of the model, but will certainly help in finding a trackable formula.

The stochastic differential equation (4.11.43) can be solved explicitly using the method of integrating factor, see Øksendal [57]. Multiplying by e^{at} we get an exact equation; integrating between 0 and t yields the following closed form formula for the spot rate

$$r_t = b + (r_0 - b)e^{-at} + \sigma e^{-at} \int_0^t e^{as} dM_s. \tag{4.11.44}$$

Since the integral with respect to dM_t is a martingale, the expectation of the spot rate is

$$\mathbb{E}[r_t] = b + (r_0 - b)e^{-at}.$$

In the long run the mean $\mathbb{E}[r_t]$ tends to b, which shows the mean reversion. Using the formula (see [15], p.134, Exercise 5.8.7)

$$\mathbb{E}\left[\left(\int_0^t f(s) dM_s\right)^2\right] = \lambda \int_0^t f(s)^2 ds$$

the variance of the spot rate can be computed as follows

$$
\begin{aligned}
Var(r_t) &= \sigma^2 e^{-2at} Var\left(\int_0^t e^{as} dM_s\right) = \sigma^2 e^{-2at} \mathbb{E}\left[\left(\int_0^t e^{as} dM_s\right)^2\right] \\
&= \frac{\lambda \sigma^2}{2a}(1 - e^{-2at}).
\end{aligned}
$$

It is worth noting that in the long run the variance tends to the constant $\frac{\lambda \sigma^2}{2a}$, which is proportional with the jumps frequency λ.

The spot rate can also be written explicitly in terms of the waiting times S_k of the Poisson process N_t, see Appendix C.1. First, we note that

$$
\int_0^t e^{as} \, dM_s = \int_0^t e^{as} \, dN_s - \lambda \int_0^t e^{as} \, ds
$$

$$
= \sum_{k=1}^{N_t} e^{aS_k} - \frac{\lambda}{a}(e^{at} - 1).
$$

Then substituting in formula (4.11.44), we obtain

$$
r_t = b - \frac{\lambda\sigma}{a} + e^{-at}\left(r_0 - b + \frac{\lambda\sigma}{a}\right) + \sigma e^{-at} \sum_{k=1}^{N_t} e^{aS_k}.
$$

In the following we shall find the value of the associated zero-coupon bond.

Let $\mathcal{F}_t$ be the σ-algebra generated by the random variables $\{N_s; s \leq t\}$. This contains all information about jumps occurred until time t. This information is used in the computation of the bond value

$$
P(t, T) = \mathbb{E}\left[e^{-\int_t^T r_s \, ds}|\mathcal{F}_t\right].
$$

Proposition 4.11.1 *Assume the spot rate r_t satisfies the mean reverting model with jumps*

$$
dr_t = a(b - r_t)dt + \sigma dM_t.
$$

Then the price of the zero-coupon bond that pays off $1 at time T is given by

$$
P(t, T) = A(t, T)e^{-B(t,T)r_t},
$$

where

$$
B(t, T) = \frac{1 - e^{-a(T-t)}}{a}
$$

$$
A(t, T) = \exp\{(b + \frac{\lambda\sigma}{a})B(t, T)
$$

$$
+ [\lambda(1 - \frac{\sigma}{a}) - b](T - t) - \lambda e^{-\sigma a} \int_0^{T-t} e^{\frac{\sigma}{a}e^{-ax}} \, dx\}.
$$

Proof: The first part of this computation mimics closely the one done in the case of Vasicek's model. Integrating the stochastic differential equation between t and T yields

$$
-a \int_t^T r_s \, ds = r_T - r_t - \sigma(M_T - M_t) - ab(T - t).
$$

Taking the exponential we obtain

$$e^{-\int_t^T r_s\,ds} = e^{\frac{1}{a}(r_T - r_t) - \frac{\sigma}{a}(M_T - M_t) - b(T-t)}. \tag{4.11.45}$$

Multiplying in the stochastic differential equation by e^{as} yields the exact equation

$$d(e^{as}r_s) = abe^{as}ds + \sigma e^{as}dM_s.$$

Integrate between t and T to get

$$e^{aT}r_T - e^{at}r_t = b(e^{aT} - e^{at}) + \sigma\int_t^T e^{as}\,dM_s.$$

Solving for r_T and subtracting r_t yields

$$r_T - r_t = -r_t(1 - e^{a(T-t)}) + b(1 - e^{a(T-t)}) + \sigma e^{-aT}\int_t^T e^{as}\,dM_s.$$

Dividing by a and using the notation for $B(t,T)$ yields

$$\frac{1}{a}(r_T - r_t) = -r_t B(t,T) + bB(t,T) + \frac{\sigma}{a}e^{-aT}\int_t^T e^{as}\,dM_s.$$

Substituting into (4.11.45) and using that $M_T - M_t = \int_t^T dM_s$, we get

$$e^{-\int_t^T r_s\,ds} = e^{-r_t B(t,T)}e^{bB(t,T) - b(T-t)}e^{\frac{\sigma}{a}\int_t^T (e^{as-aT} - 1)\,dM_s}. \tag{4.11.46}$$

Taking the predictable part out and dropping the independent condition, the price of the zero-coupon bond at time t becomes

$$\begin{aligned}
P(t,T) &= \mathbb{E}\left[e^{-\int_t^T r_s\,ds}\big|\mathcal{F}_t\right] = e^{-r_t B(t,T)}e^{bB(t,T) - b(T-t)}\mathbb{E}\left[e^{\frac{\sigma}{a}\int_t^T (e^{as-aT} - 1)\,dM_s}\big|\mathcal{F}_t\right] \\
&= e^{-r_t B(t,T)}e^{bB(t,T) - b(T-t)}\mathbb{E}\left[e^{\frac{\sigma}{a}\int_t^T (e^{as-aT} - 1)\,dM_s}\big|\mathcal{F}_t\right] \\
&= e^{-r_t B(t,T)}A(t,T),
\end{aligned}$$

where

$$A(t,T) = e^{bB(t,T) - b(T-t)}\mathbb{E}\left[e^{\frac{\sigma}{a}\int_t^T (e^{as-aT} - 1)\,dM_s}\big|\mathcal{F}_t\right]. \tag{4.11.47}$$

We shall compute in the following the right side expectation. From Bertoin [9], p. 8, the exponential process

$$e^{\int_0^t u(s)\,dN_s + \lambda\int_0^t (1 - e^{u(s)})\,ds}$$

is an $\mathcal{F}_t$-martingale, with $\mathcal{F}_t = \sigma(N_u; u \leq t)$. This implies

$$\mathbb{E}\left[e^{\int_t^T u(s)\,dN_s}\big|\mathcal{F}_t\right] = e^{-\lambda\int_t^T (1 - e^{u(s)})\,ds}.$$

Using $dM_t = dN_t - \lambda dt$ yields

$$\mathbb{E}\left[e^{\int_t^T u(s)\,dM_s}|\mathcal{F}_t\right] = \mathbb{E}\left[e^{\int_t^T u(s)\,dN_s}|\mathcal{F}_t\right]e^{-\lambda\int_t^T u(s)\,ds} = e^{-\lambda\int_t^T (1+u(s)-e^{u(s)})\,ds}.$$

Let $u(s) = \frac{\sigma}{a}(e^{a(s-T)}-1)$ and substitute in (4.11.47); then after changing the variable of integration we obtain

$$
\begin{aligned}
A(t,T) &= e^{bB(t,T)-b(T-t)}\mathbb{E}\left[e^{\frac{\sigma}{a}\int_t^T (e^{as-aT}-1)\,dM_s}|\mathcal{F}_t\right]\\
&= e^{bB(t,T)-b(T-t)}e^{\lambda\int_0^{T-t}[1+\frac{\sigma}{a}(e^{-ax}-1)-e^{\frac{\sigma}{a}(e^{-ax}-1)}]\,dx}\\
&= e^{bB(t,T)-b(T-t)}e^{\lambda(T-t)+\frac{\lambda\sigma}{a}\left(B(t,T)-(T-t)\right)-\lambda\int_0^{T-t}e^{\frac{\sigma}{a}(e^{-ax}-1)}\,dx}\\
&= e^{(b+\frac{\lambda\sigma}{a})B(t,T)}e^{[\lambda(1-\frac{\sigma}{a})-b](T-t)}e^{-\lambda\int_0^{T-t}e^{\frac{\sigma}{a}(e^{-ax}-1)}\,dx}\\
&= e^{(b+\frac{\lambda\sigma}{a})B(t,T)}e^{[\lambda(1-\frac{\sigma}{a})-b](T-t)}e^{-\lambda e^{-\sigma a}\int_0^T {}^t e^{\frac{\sigma}{a}e^{-ax}}\,dx}\\
&= \exp\{(b+\frac{\lambda\sigma}{a})B(t,T)\\
&\quad +[\lambda(1-\frac{\sigma}{a})-b](T-t)-\lambda e^{-\sigma a}\int_0^{T-t}e^{\frac{\sigma}{a}e^{-ax}}\,dx\}. \quad (4.11.48)
\end{aligned}
$$

■

Evaluation using Special Functions The integral in the last term of formula (4.11.48) cannot be computed explicitly; for its evaluation we need to use some special functions.

The solution of the initial value problem

$$
\begin{aligned}
f'(x) &= \frac{e^x}{x}, \quad x > 0\\
\lim_{x\searrow 0} f(x) &= -\infty
\end{aligned}
$$

is the *exponential integral function* $f(x) = Ei(x)$, $x > 0$. This is a special function that can be evaluated numerically in MATHEMATICA by calling the function ExpIntegralEi[x]. For instance, for any $0 < \alpha < \beta$

$$\int_\alpha^\beta \frac{e^t}{t}\,dt = Ei(\beta) - Ei(\alpha).$$

The reader can find more details regarding the exponential integral function in Abramovitz and Stegun [1]. The last integral in the expression of $A(t,T)$ can be evaluated using this special function. Substituting $t = \frac{\sigma}{a}e^{-ax}$ we have

$$
\begin{aligned}
\int_0^{T-t} e^{\frac{\sigma}{a}e^{-ax}}\,dx &= \frac{1}{a}\int_{\frac{\sigma}{a}e^{-a(T-t)}}^{\frac{\sigma}{a}} \frac{e^t}{t}\,dt\\
&= \frac{1}{a}\left[Ei\left(\frac{\sigma}{a}\right) - Ei\left(\frac{\sigma}{a}e^{-a(T-t)}\right)\right].
\end{aligned}
$$

4.12 A Model with Pure Jumps

Consider the spot rate r_t satisfying the stochastic differential equation

$$dr_t = \sigma dM_t, \tag{4.12.49}$$

where σ is a positive constant denoting the volatility of the rate. This model is obtained when the rate at which r_t is pulled toward b is $a = 0$, so there is no mean reverting effect. This type of behavior can be noticed during a short period of time in a highly volatile market; in this case the behavior of r_t is mostly influenced by jumps.

The solution of (4.12.49) is

$$r_t = r_0 + \sigma M_t = r_0 - \lambda \sigma t + \sigma N_t,$$

which is an $\mathcal{F}_t$-martingale. The rate r_t has jumps of size σ that occur at the waiting times S_n, which have a gamma distribution with parameters $\alpha = n$ and $\beta = 1/\lambda$.

We shall compute next the value $P(t, T)$ at time t of a zero-coupon bond that pays the amount of \$1 at maturity T. This is given by the conditional expectation

$$P(t, T) = \mathbb{E}[e^{-\int_t^T r_s\, ds} | \mathcal{F}_t].$$

Integrating between t and s in equation (4.12.49) yields

$$r_s = r_t + \sigma(M_s - M_t), \qquad t < s < T.$$

And then

$$\int_t^T r_s\, ds = r_t(T - t) + \sigma \int_t^T (M_s - M_t)\, ds.$$

Taking out the predictable part and dropping the independent condition yields

$$
\begin{aligned}
P(t, T) &= \mathbb{E}[e^{-\int_t^T r_s\, ds} | \mathcal{F}_t] \\
&= e^{-r_t(T-t)} \mathbb{E}[e^{-\sigma \int_t^T (M_s - M_t)\, ds} | \mathcal{F}_t] \\
&= e^{-r_t(T-t)} \mathbb{E}[e^{-\sigma \int_0^{T-t} M_\tau\, d\tau}] \\
&= e^{-r_t(T-t)} \mathbb{E}[e^{-\sigma \int_0^{T-t} (N_\tau - \lambda\tau)\, d\tau}] \\
&= e^{-r_t(T-t) + \frac{1}{2}\sigma\lambda(T-t)^2} \mathbb{E}[e^{-\sigma \int_0^{T-t} N_\tau\, d\tau}]. \tag{4.12.50}
\end{aligned}
$$

We need to work out the expectation. Using integration by parts,

$$-\sigma \int_0^T N_t \, dt \;=\; -\sigma\left(TN_T - \int_0^T t \, dN_t\right)$$

$$=\; -\sigma\left(T \int_0^T dN_t - \int_0^T t \, dN_t\right)$$

$$=\; -\sigma \int_0^T (T-t) \, dN_t,$$

so

$$e^{-\sigma \int_0^T N_t \, dt} = e^{-\sigma \int_0^T (T-t) \, dN_t}.$$

Using the formula

$$\mathbb{E}\left[e^{\int_0^T u(t) \, dN_t}\right] = e^{-\lambda \int_0^T (1 - e^{u(t)}) \, dt}$$

we have

$$\mathbb{E}\left[e^{-\sigma \int_0^T N_t \, dt}\right] \;=\; \mathbb{E}\left[e^{-\sigma \int_0^T (T-t) \, dN_t}\right] = e^{-\lambda \int_0^T (1 - e^{-\sigma(T-t)}) \, dt}$$

$$=\; e^{-\lambda T} e^{-\frac{\lambda}{\sigma}(e^{-\sigma T} - 1)}$$

$$=\; e^{-\lambda\left(T + \frac{1}{\sigma}(e^{-\sigma T} - 1)\right)}.$$

Replacing T by $T - t$ and t by τ yields

$$\mathbb{E}\left[e^{-\sigma \int_0^{T-t} N_\tau \, d\tau}\right] = e^{-\lambda\left(T - t + \frac{1}{\sigma}(e^{-\sigma(T-t)} - 1)\right)}.$$

Substituting in (4.12.50) yields the formula for the bond price

$$P(t,T) \;=\; e^{-r_t(T-t) + \frac{1}{2}\sigma\lambda(T-t)^2} e^{-\lambda\left(T - t + \frac{1}{\sigma}(e^{-\sigma(T-t)} - 1)\right)}$$

$$=\; \exp\left\{-(\lambda + r_t)(T - t) + \frac{1}{2}\sigma\lambda(T - t)^2 - \frac{\lambda}{\sigma}(e^{-\sigma(T-t)} - 1)\right\}.$$

Proposition 4.12.1 *Assume the spot rate r_t satisfies $dr_t = \sigma dM_t$, with $\sigma > 0$ constant. Then the price of the zero-coupon bond that pays off \$1 at time T is given by*

$$P(t,T) = \exp\left\{-(\lambda + r_t)(T - t) + \frac{1}{2}\sigma\lambda(T - t)^2 - \frac{\lambda}{\sigma}(e^{-\sigma(T-t)} - 1)\right\}.$$

The yield curve is given by

$$R(t,T) \;=\; -\frac{1}{T-t} \ln P(t,T)$$

$$=\; r_t + \lambda - \frac{\lambda\sigma}{2}(T - t) + \frac{\lambda}{\sigma(T-t)}(e^{-\sigma(T-t)} - 1).$$

Exercise 4.12.2 *Find the instantaneous forward rate $F(t,T)$ for the model (4.12.49).*

Exercise 4.12.3 *(a) Prove the following formula*

$$d[(1+\sigma)^{N_t}] = \sigma(1+\sigma)^{N_t}.$$

(b) Use part (a) to find a solution of

$$dX_t = \sigma X_t \, dN_t$$
$$X_0 = 1.$$

Exercise 4.12.4 *Assume the spot rate satisfies the equation $dr_t = \sigma r_t dM_t$, where M_t is the compensated Poisson process. Using Exercise 4.12.3 find a solution of the form $r_t = r_0 e^{\phi(t)}(1+\sigma)^{N_t}$, where N_t denotes the Poisson process.*

Part III

Risk-Neutral Valuation Pricing

Chapter 5

Modeling Stock Prices

The price of a stock can be modeled by a continuous stochastic process which is the sum between a predictable and an unpredictable part. However, this type of model does not take into account market crashes. If those are to be taken into consideration, the stock price needs to contain a third component that models unexpected jumps. We shall also discuss the time when a stock reaches a given barrier, correlation of two stocks, the running maximum distribution of a stock and stock averages such as arithmetic, geometric and harmonic.

5.1 Constant Drift and Volatility Model

Let S_t be the price of a stock at time t. If $\mathcal{F}_t$ denotes the market information at time t, then S_t is a continuous process that is $\mathcal{F}_t$-measurable. The rate of return on the stock during the time interval Δt measures the percentage increase in the stock price between instances t and $t + \Delta t$ and it is given by $\dfrac{S_{t+\Delta t} - S_t}{S_t}$. When Δt is infinitesimally small, we obtain the instantaneous rate of return

$$\frac{dS_t}{S_t} = \lim_{\Delta t \to 0} \frac{S_{t+\Delta t} - S_t}{S_t}.$$

In the absence of market crashes, this can be modeled as the sum of two components:

- the deterministic part, μdt, due to the drift;
- the noisy part, σdW_t, due to unexpected market news and fluctuations.

Adding these parts together, yields

$$\frac{dS_t}{S_t} = \mu dt + \sigma dW_t,$$

which leads to the stochastic differential equation

$$\boxed{dS_t = \mu S_t dt + \sigma S_t dW_t.} \tag{5.1.1}$$

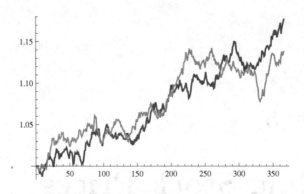

Figure 5.1:   *Two distinct simulations for the stochastic equation* $dS_t = 0.15S_t dt + 0.07S_t dW_t$, *with* $S_0 = 1$.

The parameters μ and σ are positive constants, which represent the *drift rate* and *volatility* of the stock. Since equation (5.1.1) is linear in S_t, standard results of existence and uniqueness yield a unique solution, provided an initial condition S_0 is given. The solution can be found using the method of variation of parameters. Considering a solution of the form $S_t = e^{\alpha(t)} e^{\sigma W_t}$, an application of Ito's formula leads to

$$
\begin{aligned}
dS_t &= de^{\alpha(t)} e^{\sigma W_t} + e^{\alpha(t)} de^{\sigma W_t} \\
&= \alpha'(t) e^{\alpha(t)} e^{\sigma W_t} dt + e^{\alpha(t)} \left(\sigma e^{\sigma W_t} dW_t + \frac{1}{2}\sigma^2 e^{\sigma W_t} dt \right) \\
&= \left(\alpha'(t) + \frac{1}{2}\sigma^2 \right) S_t dt + \sigma S_t dW_t.
\end{aligned}
$$

Comparing with (5.1.1) and equating the coefficients of dt implies

$$
\alpha'(t) + \frac{1}{2}\sigma^2 = \mu,
$$

and hence $\alpha(t) = (\mu - \frac{1}{2}\sigma^2)t + \alpha_0$. Concluding, the solution of (5.1.1) becomes

$$
\boxed{S_t = S_0 e^{(\mu - \frac{\sigma^2}{2})t + \sigma W_t},}
\tag{5.1.2}
$$

where S_0 denotes the price of the stock at time $t = 0$. It is worth noting that the stock price is $\mathcal{F}_t$-measurable, positive, and it has a log-normal distribution, with the mean and variance given by

$$
\begin{aligned}
\mathbb{E}[S_t] &= S_0 e^{\mu t} \tag{5.1.3} \\
Var[S_t] &= S_0^2 e^{2\mu t}(e^{\sigma^2 t} - 1). \tag{5.1.4}
\end{aligned}
$$

Two simulations of the stock price S_t can be seen in Fig.5.1. This model is also known under the name of Samuelson's model, see [61]. It was introduced in 1965 and preceded the work of Black, Scholes and Merton, which was done in early 1970s.

Exercise 5.1.1 *Let $\mathcal{F}_u$ be the information set at time u. Find $\mathbb{E}[S_t|\mathcal{F}_u]$ and $Var[S_t|\mathcal{F}_u]$ for $u \leq t$.[1] How these formulas become in the case $s = t$?*

Exercise 5.1.2 *Find the stochastic process followed by $\ln S_t$. What are the values of $\mathbb{E}[\ln(S_t)]$ and $Var[\ln(S_t)]$?*

Exercise 5.1.3 *Show that the stock S_t is a submartingale with respect to the information set $\mathcal{F}_t$.*

Exercise 5.1.4 *Find the stochastic differential equations associated with the following processes*

(a) $\dfrac{1}{S_t}$ (b) S_t^n (c) $(S_t - 1)^2$.

Exercise 5.1.5 (a) *Show that $\mathbb{E}[S_t^2] = S_0^2 e^{(2\mu+\sigma^2)t}$.*
(b) *Find a similar formula for $\mathbb{E}[S_t^n]$, with n positive integer.*

Exercise 5.1.6 (a) *Find the expectation $\mathbb{E}[S_t W_t]$.*
(b) *Find the correlation function $\rho_t = cor(S_t, W_t)$. What happens for t large?*

5.2   Correlation of Two Stocks

Consider two stock prices

$$dS_1 = \mu_1 S_1 dt + \sigma_1 S_1 dW_1(t) \qquad\qquad (5.2.5)$$
$$dS_2 = \mu_2 S_2 dt + \sigma_2 S_2 dW_2(t) \qquad\qquad (5.2.6)$$

driven by two correlated news terms with the correlation coefficient

$$\rho = cor(dW_1, dW_2).$$

The following result shows that in general the stock prices are positively correlated, and the stock volatilities play the determinant role in describing this:

---

[1]The conditional variance is defined by $Var(X|\mathcal{F}) = \mathbb{E}[X^2|\mathcal{F}] - \mathbb{E}[X|\mathcal{F}]^2$.

Proposition 5.2.1 *The correlation coefficient of stock prices S_1 and S_2 defined by (5.2.5)-(5.2.6) is*

$$cor(S_1, S_2) = \frac{e^{\sigma_1 \sigma_2 \rho t} - 1}{(e^{\sigma_1^2 t} - 1)^{1/2}(e^{\sigma_2^2 t} - 1)^{1/2}} > 0. \qquad (5.2.7)$$

In particular, if $\sigma_1 = \sigma_2$, then $cor(S_1, S_2) = \rho$.

Proof: Using Ito's lemma and formulas (5.2.5)-(5.2.6) we find the equation satisfied by $S_1 S_2$

$$
\begin{aligned}
d(S_1 S_2) &= S_1 dS_2 + S_2 dS_1 + dS_1 dS_2 \\
&= (\mu_1 + \mu_2 + \rho \sigma_1 \sigma_2) S_1 S_2 dt + \sigma_1 S_1 S_2 dW_1(t) + \sigma_2 S_1 S_2 dW_2(t).
\end{aligned}
$$

Integrate between 0 and t we obtain an integral stochastic equation

$$
\begin{aligned}
S_1(t) S_2(t) &= S_1(0) S_2(0) + (\mu_1 + \mu_2 + \rho \sigma_1 \sigma_2) \int_0^t S_1(u) S_2(u)\, du \\
&\quad + \sigma_1 \int_0^t S_1(u) S_2(u)\, dW_1(u) + \sigma_2 \int_0^t S_1(u) S_2(u)\, dW_2(u).
\end{aligned}
$$

Taking expectations on both sides and using that Wiener integrals have zero mean, we obtain an integral equation

$$\mathbb{E}[S_1(t) S_2(t)] = S_1(0) S_2(0) + (\mu_1 + \mu_2 + \rho \sigma_1 \sigma_2) \int_0^t \mathbb{E}[S_1(u) S_2(u)]\, du,$$

which can be transformed into a differential equation for $\Phi(t) = \mathbb{E}[S_1(t) S_2(t)]$:

$$\frac{d}{dt} \Phi(t) = (\mu_1 + \mu_2 + \rho \sigma_1 \sigma_2) \Phi(t), \qquad \Phi(0) = S_1(0) S_2(0).$$

The solution is

$$\Phi(t) = \mathbb{E}[S_1(t) S_2(t)] = S_1(0) S_2(0) e^{(\mu_1 + \mu_2 + \rho \sigma_1 \sigma_2)t}.$$

Now we can compute the covariance

$$
\begin{aligned}
cov(S_1, S_2) &= \mathbb{E}[S_1 S_2] - \mathbb{E}[S_1]\mathbb{E}[S_2] \\
&= S_1(0) S_2(0) e^{(\mu_1 + \mu_2 + \rho \sigma_1 \sigma_2)t} - S_1(0) e^{\mu_1 t} S_2(0) e^{\mu_2 t} \\
&= S_1(0) S_2(0) e^{(\mu_1 + \mu_2)t} \left(e^{\sigma_1 \sigma_2 \rho t} - 1 \right).
\end{aligned}
$$

Using the definition of correlation, after simplifications we obtain

$$cor(S_1, S_2) = \frac{cov(S_1, S_2)}{\sqrt{Var(S_1)} \sqrt{Var(S_2)}} = \frac{e^{\sigma_1 \sigma_2 \rho t} - 1}{(e^{\sigma_1^2 t} - 1)^{1/2}(e^{\sigma_2^2 t} - 1)^{1/2}}.$$

∎

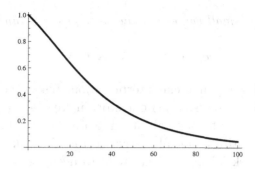

Figure 5.2: *The correlation function* $f(t) = \dfrac{e^{\sigma_1\sigma_2 t}-1}{(e^{\sigma_1^2 t}-1)^{1/2}(e^{\sigma_2^2 t}-1)^{1/2}}$ *in the case* $\sigma_1 = 0.15$, $\sigma_2 = 0.40$, $\rho = 1$.

Remark 5.2.2 (*a*) If the stocks have equal volatilities, $\sigma_1 = \sigma_2 = \sigma$, then

$$cor(S_1, S_2) = \frac{e^{\rho\sigma^2 t} - 1}{e^{\sigma^2 t} - 1}.$$

(*b*) If let $t \to 0$ in (5.2.7) we obtain the short time correlation of the stocks equal to the correlations of the noise terms

$$cor(S_1, S_2) = \rho.$$

(*c*) The stock prices correlation gets weaker as t increases:

$$cor(S_1, S_2) \to 0 \quad \text{as} \quad t \to \infty.$$

This follows from the asymptotic correspondence

$$\frac{e^{\sigma_1\sigma_2\rho t} - 1}{(e^{\sigma_1^2 t} - 1)^{1/2}(e^{\sigma_2^2 t} - 1)^{1/2}} \sim \frac{e^{\sigma_1\sigma_2\rho t}}{e^{\frac{\sigma_1^2+\sigma_2^2}{2}t}} = e^{-\frac{(\sigma_1-\sigma_2)^2}{2}t}e^{-\sigma_1\sigma_2(1-\rho)t} \to 0, \quad t \to \infty.$$

It follows that in the long run any two stocks tend to become uncorrelated, see Fig.5.2.

Corollary 5.2.3 (*i*) *Consider two stock prices driven by the same news term,* dW_t

$$\begin{aligned}
dS_1 &= \mu_1 S_1 dt + \sigma_1 S_1 dW_t & (5.2.8)\\
dS_2 &= \mu_2 S_2 dt + \sigma_2 S_2 dW_t. & (5.2.9)
\end{aligned}$$

Then their correlation is given by

$$cor(S_1, S_2) = \frac{e^{\sigma_1\sigma_2 t} - 1}{(e^{\sigma_1^2 t} - 1)^{1/2}(e^{\sigma_2^2 t} - 1)^{1/2}}.$$

(ii) In particular, for small values of t the stock prices S_1 and S_2 are positively strongly correlated

$$cor(S_1, S_2) \to 1 \quad as \ t \to 0.$$

This fact has the following financial interpretation. If some stocks are driven by the same market fluctuations, when one stock increases, then the other tends also to increase, at least for a small amount of time. In the case when some bad news affect an entire financial market sector, the risk becomes systemic, and hence if one stock fails, all others tend to decrease as well, leading to a strain on that financial market sector.

Exercise 5.2.4 *Consider two correlated Brownian motions, W_1, W_2, satisfying $\rho = cor(dW_1, dW_2)$. Show that*

$$\mathbb{E}[dW_1 \, dW_2] = \rho \, dt.$$

Exercise 5.2.5 *If X and Y are random variables and $\alpha, \beta \in \mathbb{R}$, show that*

$$cor(\alpha X, \beta Y) = \begin{cases} cor(X, Y), & if \ \alpha\beta > 0 \\ -cor(X, Y), & if \ \alpha\beta < 0. \end{cases}$$

Exercise 5.2.6 *Find the following*
(a) $cov\big(dS_1(t), dS_2(t)\big)$;
(b) $cor\big(dS_1(t), dS_2(t)\big)$.

Exercise 5.2.7 *Show that for any $\sigma_1, \sigma_2 > 0$, $\rho \in [-1, 1]$, and $t \geq 0$ the following inequality holds*

$$(e^{\sigma_1 \sigma_2 \rho t} - 1)^2 \leq (e^{\sigma_1^2 t} - 1)(e^{\sigma_2^2 t} - 1).$$

5.3   When Does a Stock Hit a Given Barrier?

Let $b > S_0$ and consider the first instance of time when the stock reaches the barrier b, see Fig.5.3, which symbolically can be written as

$$T_b = \inf\{t > 0; S_t \geq b\}.$$

Less rigorously, in practice the reader can think of T_b as the *minimum* (rather than *infimum*) of the positive time values t for which $S_t = b$. The random variable T_b is a stopping time, in the sense that the event $\{T_b \leq t\}$ belongs to $\mathcal{F}_t$, i.e. having given the market information at time t, we can decide whether the stock has reached the level b. Equivalently, T_b is the time needed for the stock S_t to hit the level b for the first time.

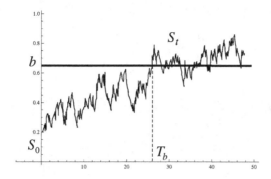

Figure 5.3: *The stopping time T_b when the stock S_t reaches the level b for the first time.*

In the following we shall compute the moment generating function for T_b, find its expectation and derive the formula for its probability law.

The moment generating function can be obtained by a manipulation involving martingales and the Optional Stopping Theorem (Theorem 1.8.1), as in the following.

Let $c > 0$ be a constant. Since $M_t = e^{-\frac{1}{2}c^2 t + cW_t}$ is an $\mathcal{F}_t$-martingale, see Appendix C.3, and T_b is a stopping time, the Optional Stopping Theorem provides

$$\mathbb{E}[M_{T_b}] = \mathbb{E}[M_0] = 1. \qquad (5.3.10)$$

Let $m = \mu - \sigma^2/2$ and assume that $m > 0$ for the rest of this section. Solving for W_t from the stock price expression

$$S_t = S_0 e^{mt + \sigma W_t}$$

we get

$$W_t = \frac{1}{\sigma}\left(\ln \frac{S_t}{S_0} - mt\right).$$

Substituting it into the expression of the martingale M_t, we obtain

$$M_t = \left(\frac{S_t}{S_0}\right)^{\frac{c}{\sigma}} e^{-(\frac{1}{2}c^2 + \frac{mc}{\sigma})t}.$$

Using that $S_{T_b} = b$, then taking the expectation yields

$$\mathbb{E}[M_{T_b}] = \left(\frac{b}{S_0}\right)^{\frac{c}{\sigma}} \mathbb{E}\left[e^{-(\frac{1}{2}c^2 + \frac{mc}{\sigma})T_b}\right].$$

Then equation (5.3.10) can be written as

$$\mathbb{E}\left[e^{-(\frac{1}{2}c^2 + \frac{mc}{\sigma})T_b}\right] = \left(\frac{S_0}{b}\right)^{\frac{c}{\sigma}}.$$

Substitute

$$c = -\frac{m}{\sigma} + \sqrt{2s + \frac{m^2}{\sigma^2}}, \qquad s > 0$$

and obtain

$$\mathbb{E}\left[e^{-sT_b}\right] = \left(\frac{S_0}{b}\right)^{-\frac{m}{\sigma^2} + \frac{1}{\sigma}\sqrt{2s+m^2/\sigma^2}} = \left(\frac{S_0}{b}\right)^{h(s)}, \qquad (5.3.11)$$

with

$$h(s) = \left(\frac{1}{2} - \frac{\mu}{\sigma^2}\right) + \frac{1}{\sigma}\sqrt{2s + \left(\frac{\mu}{\sigma} - \frac{\sigma}{2}\right)^2}.$$

Exercise 5.3.1 *Verify that the function $h(s)$ has the following properties:*
(a) $h(0) = 0$;
(b) $h(\mu) = 1$;
(c) $h'(0) = (\mu - \sigma^2/2)^{-1}$;
(d) $h''(0) = -\sigma^2(\mu - \sigma^2/2)^{-3}$.

Exercise 5.3.2 *Show that* $\mathbb{E}[e^{-\mu T_b}] = \dfrac{S_0}{b}$.

Exercise 5.3.3 *Prove that*

(a) $\quad -\dfrac{d}{ds}\mathbb{E}[e^{-sT_b}]\Big|_{s=0} = \mathbb{E}[T_b]$;

(b) $\quad \dfrac{d^2}{ds^2}\mathbb{E}[e^{-sT_b}]\Big|_{s=0} = \mathbb{E}[T_b^2]$.

Proposition 5.3.4 *The mean of the hitting time T_b is given by*

$$\mathbb{E}[T_b] = \frac{1}{\mu - \sigma^2/2} \ln\left(\frac{b}{S_0}\right).$$

Proof: Differentiating in (5.3.11) we have

$$\frac{d}{ds}\mathbb{E}[e^{-sT_b}] = \left(\frac{S_0}{b}\right)^{h(s)} \ln\frac{S_0}{b} h'(s)$$

and then taking the value at $s = 0$, we get

$$\frac{d}{ds}\mathbb{E}[e^{-sT_b}]\Big|_{s=0} = \left(\frac{S_0}{b}\right)^{h(0)} \ln\frac{S_0}{b} h'(0)$$

$$= -\ln\left(\frac{b}{S_0}\right)\frac{1}{\mu - \sigma^2/2}.$$

Exercise 5.3.3 implies the desired result. ∎

Exercise 5.3.5 *Use Exercise 5.3.3 to find*

(a) $\mathbb{E}[T_b^2]$;

(b) $Var(T_b)$.

The probability density of T_b Denote by $p_b(t)$ the probability density of the random variable T_b. We note that the domain of $p_b(t)$ is $[0, \infty)$. Then

$$\mathbb{E}\left[e^{-sT_b}\right] = \int_0^\infty e^{-st} p_b(t)\, dt$$

is the Laplace transform of $p_b(t)$, see Appendix A.2. Hence (5.3.11) becomes

$$\mathcal{L}(p_b(t))(s) = \left(\frac{S_0}{b}\right)^{h(s)},$$

where $\mathcal{L}$ denotes the Laplace transform. The probability density can be retrieved as an inverse Laplace transform

$$p_b(t) = \left(\mathcal{L}^{-1}\left(\frac{S_0}{b}\right)^{h(s)}\right)(t).$$

This inverse Laplace transform can actually be computed explicitly. Since

$$\left(\frac{S_0}{b}\right)^{h(s)} = e^{h(s)\ln\frac{S_0}{b}} = \left(\frac{S_0}{b}\right)^{\frac{1}{2}-\frac{\mu}{\sigma^2}} e^{-\frac{1}{\sigma}\ln\frac{b}{S_0}\sqrt{2s+(\frac{\mu}{\sigma}-\frac{\sigma}{2})^2}},$$

using the formula

$$\mathcal{L}^{-1}\left(e^{-c\sqrt{2s+n^2}}\right)(t) = \frac{ce^{-cn}}{\sqrt{2\pi}t^{3/2}} e^{-\frac{1}{2t}(c-nt)^2}$$

with

$$c = \frac{1}{\sigma}\ln\frac{b}{S_0}, \qquad n = \frac{\mu}{\sigma} - \frac{\sigma}{2},$$

an algebraic computation provides

$$\left(\mathcal{L}^{-1}\left(\frac{S_0}{b}\right)^{h(s)}\right)(t) = \frac{\ln\frac{b}{S_0}}{\sigma\sqrt{2\pi}t^{3/2}} e^{-\frac{1}{2t}[\frac{1}{\sigma}\ln\frac{b}{S_0}-(\frac{\mu}{\sigma}-\frac{\sigma}{2})t]^2},$$

and hence the probability density of T_b is given by the *inverse Gaussian distribution*, see Fig.5.4

$$\boxed{p_b(t) = \frac{\ln\frac{b}{S_0}}{\sigma\sqrt{2\pi}t^{3/2}} e^{-\frac{1}{2t}[\frac{1}{\sigma}\ln\frac{b}{S_0}-(\frac{\mu}{\sigma}-\frac{\sigma}{2})t]^2}, \qquad t > 0.} \qquad (5.3.12)$$

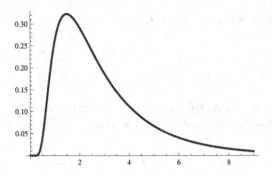

Figure 5.4: *The probability density of the stopping time T_b is skewed to the right.*

Exercise 5.3.6 *The density function of a random variable X having an inverse Gaussian distribution with parameters $\lambda > 0$ and $m > 0$ is given by*

$$f_X(x; m, \lambda) = \frac{\sqrt{\lambda}}{\sqrt{2\pi}x^{3/2}} e^{-\frac{\lambda(x-m)^2}{2m^2 x}}, \qquad x > 0.$$

(a) Show that the expressions for the parameters m and λ for the distribution (5.3.12) are

$$m = \frac{1}{\mu - \sigma^2/2} \ln \frac{b}{S_0}, \qquad \lambda = \left(\frac{1}{\sigma} \ln \frac{b}{S_0}\right)^2.$$

(b) Verify that

$$x = m\left[\left(1 + \frac{9m^2}{4\lambda^2}\right)^{1/2} - \frac{3m}{2\lambda}\right]$$

is the solution of the equation $\frac{\partial \ln f_X(x; \mu, \lambda)}{\partial x} = 0$.

(c) Find the mode (the value of t with the highest probability density) of the distribution (5.3.12).

Exercise 5.3.7 *Consider the doubling time of a stock*

$$T_2 = \inf\{t > 0; S_t \geq 2S_0\}.$$

(a) Find $\mathbb{E}[T_2]$ and $Var(T_2)$. Do these values depend of S_0?

(b) The expected return of a stock and its volatility are $\mu = 0.15$ and $\sigma = 0.20$. Find the expected time when the stock doubles its value.

Exercise 5.3.8 *Let $\bar{S}_t = \max_{u \leq t} S_u$ and $\underline{S}_t = \min_{u \leq t} S_u$ be the running maximum and minimum of the stock.*

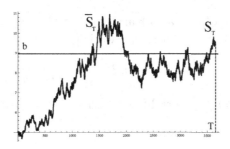

Figure 5.5: *The running maximum on the stock.*

(a) *Show that the events $\{\bar{S}_t \geq b\}$ and $\{T_b \leq t\}$ are the same.*

(b) *Use (a) to obtain the distribution function of $\bar{S}_t$;*

(c) *Use a similar argument to find the distribution function of $\underline{S}_t$.*

Exercise 5.3.9 *Let $b > S_0$. Without computing the integrals, describe the probability that:*

(a) *The stock S_t reaches the level b, before time T;*

(b) *The stock S_t reaches level b, after time T;*

(c) *The stock S_t reaches level b, before time T_2 and after time T_1.*

5.4 Probability to Hit a Barrier Before T

We shall deal next with the probability of a stock reaching a certain barrier before a given time T. This is represented by

$$P(T_b < T) = \int_0^T p_b(t)\,dt,$$

with $p_b(t)$ given in (5.3.12). Even if there is no explicit formula for this integral, it is remarkable that it can be reduced to an expression written in terms of the standard normal distribution function, $N(x) = \frac{1}{\sqrt{2\pi}}\int_{-\infty}^x e^{-\frac{z^2}{2}}\,dz$, which is well tabulated and widely used in the literature.

Since the event that the stock S_t reaches the barrier a before time T is the same as the event $\{\bar{S}_T \geq b\}$, see Exercise 5.3.8, then $P(T_b < T) = P(\bar{S}_T \geq b)$, and hence, it suffices to study the probability of the latter event. The study of the running maximum of the stock, $\bar{S}_T$, is related to the behavior of the running maximum of a Brownian motion with drift. This is related to a result describing the probability of a Brownian motion W_t that hits the line $y = \alpha + \gamma t$

in the time interval $[0, T]$ (see also Karatzas and Shreve [46], p.265). It is easy to see that the relation between those two is given by

$$P\Big(W_t \geq \alpha + \gamma t, \text{ for some } t \in [0, T]\Big) = P\Big(\max_{0 \leq t \leq T} (W_t - \gamma t) \geq \alpha\Big).$$

The next result computes the right side term.

Lemma 5.4.1 *Let* $\alpha > 0$. *Then*

$$P\Big(\max_{0 \leq t \leq T} (W_t - \gamma t) \geq \alpha\Big) = 1 - N\Big(\gamma \sqrt{T} + \frac{\alpha}{\sqrt{T}}\Big) + e^{-2\alpha\gamma} N\Big(\gamma \sqrt{T} - \frac{\alpha}{\sqrt{T}}\Big),$$

where $N(x) = \frac{1}{\sqrt{2\pi}} \int_{-\infty}^{x} e^{-\frac{z^2}{2}} dz$.

Proof: We shall do the proof for $\gamma < 0$. The case $\gamma > 0$ is similar. Consider the Brownian motion with drift, $X_t = W_t - \gamma t$, and let τ denote the hitting time

$$\tau = \inf\{t > 0; X_t \geq \alpha\}.$$

Since the event $\{\max_{0 \leq t \leq T} (W_t - \gamma t) \geq \alpha\}$ occurs if and only if $\tau \leq T$, using Exercise 5.4.7 (c) we have

$$P\Big(\max_{0 \leq t \leq T} (W_t - \gamma t) \geq \alpha\Big) = P(\tau \leq T) = \int_0^T \frac{\alpha}{\sqrt{2\pi}\tau^{3/2}} e^{-\frac{(\alpha+\gamma\tau)^2}{2\tau}} d\tau.$$

Hence, it suffices to show that

$$\int_0^T \frac{\alpha}{\sqrt{2\pi}\tau^{3/2}} e^{-\frac{(\alpha+\gamma\tau)^2}{2\tau}} d\tau = 1 - N\Big(\gamma \sqrt{T} + \frac{\alpha}{\sqrt{T}}\Big) + e^{-2\alpha\gamma} N\Big(\gamma \sqrt{T} - \frac{\alpha}{\sqrt{T}}\Big).$$
$$(5.4.13)$$

Even if a direct computation of the integral exists, we shall employ for the sake of simplicity a shorter method, which serves as a verification of formula (5.4.13). Consider the functions

$$f(T) = 1 - N\Big(\gamma \sqrt{T} + \frac{\alpha}{\sqrt{T}}\Big) + e^{-2\alpha\gamma} N\Big(\gamma \sqrt{T} - \frac{\alpha}{\sqrt{T}}\Big)$$

$$g(T) = \int_0^T \frac{\alpha}{\sqrt{2\pi}\tau^{3/2}} e^{-\frac{(\alpha+\gamma\tau)^2}{2\tau}} d\tau.$$

If we show that $f(0) = g(0)$ and $f'(T) = g'(T)$ for any $T \geq 0$, then it follows that $f(T) = g(T)$.

Using the properties of the cummulative distribution function $N(\cdot)$, we have

$$\begin{aligned} f(0) &= 1 - \lim_{T \to 0+} N\Big(\gamma \sqrt{T} + \frac{\alpha}{\sqrt{T}}\Big) + e^{-2\alpha\gamma} \lim_{T \to 0+} N\Big(\gamma \sqrt{T} - \frac{\alpha}{\sqrt{T}}\Big) \\ &= 1 - 1 + 0 = 0, \\ g(0) &= 0. \end{aligned}$$

The Fundamental Theorem of Calculus yields

$$g'(T) = \frac{\alpha}{\sqrt{2\pi}T^{3/2}}e^{-\frac{(\alpha+\gamma T)^2}{2T}}. \tag{5.4.14}$$

Using $N'(x) = \frac{1}{\sqrt{2\pi}}e^{-\frac{x^2}{2}}$, applying the chain rule we have

$$
\begin{aligned}
f'(T) &= -\frac{1}{\sqrt{2\pi}}e^{-\frac{1}{2}(\gamma\sqrt{T}+\alpha/\sqrt{T})^2}\left(\frac{\gamma}{2\sqrt{T}} - \frac{\alpha}{2T^{3/2}}\right) \\
&\quad + e^{-2\alpha\gamma}\frac{1}{\sqrt{2\pi}}e^{-\frac{1}{2}(\gamma\sqrt{T}-\alpha/\sqrt{T})^2}\left(\frac{\gamma}{2\sqrt{T}} + \frac{\alpha}{2T^{3/2}}\right) \\
&= -\frac{1}{\sqrt{2\pi}}e^{-\frac{(\alpha+\gamma T)^2}{2T}}\left(\frac{\gamma}{2\sqrt{T}} - \frac{\alpha}{2T^{3/2}}\right) \\
&\quad + \frac{1}{\sqrt{2\pi}}e^{-\frac{(\alpha+\gamma T)^2}{2T}}\left(\frac{\gamma}{2\sqrt{T}} + \frac{\alpha}{2T^{3/2}}\right) \\
&= \frac{\alpha}{\sqrt{2\pi}T^{3/2}}e^{-\frac{(\alpha+\gamma T)^2}{2T}}. \tag{5.4.15}
\end{aligned}
$$

Comparing equations (5.4.14) and (5.4.15) yields $f'(T) = g'(T)$. Hence $f(T) = g(T)$ for any $T \geq 0$. ∎

The following consequence provides the probability that a Brownian motion will ever intersect a given line:

Corollary 5.4.2 *We have*

$$P\left(W_t \geq \alpha + \gamma t, \text{ for some } t \geq 0\right) = \begin{cases} 1, & \text{if } \gamma \leq 0 \\ e^{-2\alpha\gamma}, & \text{if } \gamma > 0. \end{cases}$$

Proof: Take the limit $T \to \infty$ in Lemma 5.4.1 and use that $N(-\infty) = 0$ and $N(+\infty) = 1$. ∎

We come back now to the study of the probability of the event $\{\bar{S}_T \geq b\}$. The following notations, used in pricing lookback options, will be useful:

$$d_6 = \frac{\ln\frac{S_0}{b} + (\mu - \frac{\sigma^2}{2})T}{\sigma\sqrt{T}} \tag{5.4.16}$$

$$d_8 = \frac{\ln\frac{b}{S_0} + (\mu - \frac{\sigma^2}{2})T}{\sigma\sqrt{T}}. \tag{5.4.17}$$

The next result shows the connection between the running maximum on the stock, $\bar{S}_T$ and the expressions d_6, d_8.

Proposition 5.4.3 *Let $\bar{S}_T = \max\limits_{0 \le t \le T} S_t$, and let $b > 0$, see Fig. 5.5. Then*

$$P(\bar{S}_T \ge b) = N(d_6) + \left(\frac{b}{S_0}\right)^{\frac{2\mu}{\sigma^2}-1} N(-d_8). \tag{5.4.18}$$

Proof: The ideea of the proof is to reduce the problem from the running maximum of the stock to the running maximum of a Brownian motion with drift and then apply the previous lemma.

Let $m = \mu - \dfrac{\sigma^2}{2}$. Substituting $\gamma = -\dfrac{m}{\sigma}$ and $\alpha = \dfrac{1}{\sigma} \ln \dfrac{b}{S_0}$ in Lemma 5.4.1 we have

$$P(\bar{S}_T \ge b) \;=\; P\left(S_0 e^{\max\limits_{t \le T}(mt + \sigma W_t)} \ge b\right) = P\left(\max_{t \le T}\left(\frac{m}{\sigma}t + W_t\right) \ge \frac{1}{\sigma} \ln \frac{b}{S_0}\right)$$

$$= P(W_t - \gamma t \ge \alpha)$$

$$= 1 - N\left(\gamma\sqrt{T} + \frac{\alpha}{\sqrt{T}}\right) + e^{-2\alpha\gamma} N\left(\gamma\sqrt{T} - \frac{\alpha}{\sqrt{T}}\right). \tag{5.4.19}$$

Since

$$\gamma\sqrt{T} + \frac{\alpha}{\sqrt{T}} = \frac{\gamma T + \alpha}{\sqrt{T}} = \frac{-mT + \ln \frac{b}{S_0}}{\sigma\sqrt{T}} = -\frac{\ln \frac{S_0}{b} + (\mu - \frac{\sigma^2}{2})T}{\sigma\sqrt{T}} = -d_6;$$

$$\gamma\sqrt{T} - \frac{\alpha}{\sqrt{T}} = \frac{\gamma T - \alpha}{\sqrt{T}} = \frac{-mT - \ln \frac{b}{S_0}}{\sigma\sqrt{T}} = -\frac{\ln \frac{b}{S_0} + (\mu - \frac{\sigma^2}{2})T}{\sigma\sqrt{T}} = -d_8;$$

$$e^{-2\alpha\gamma} = e^{\frac{2m}{\sigma^2} \ln \frac{b}{S_0}} = \left(\frac{b}{S_0}\right)^{\frac{2m}{\sigma^2}} = \left(\frac{b}{S_0}\right)^{\frac{2\mu}{\sigma^2}-1},$$

substituting in (5.4.19) yields

$$P(\bar{S}_T \ge b) \;=\; 1 - N(-d_6) + \left(\frac{b}{S_0}\right)^{\frac{2\mu}{\sigma^2}-1} N(-d_8)$$

$$= N(d_6) + \left(\frac{b}{S_0}\right)^{\frac{2\mu}{\sigma^2}-1} N(-d_8).$$

∎

Corollary 5.4.4 *Let $b > 0$. The probability that the stock S_t will ever reach the barrier b is*

$$P(\sup_{t \ge 0} S_t \ge b) = \begin{cases} \left(\dfrac{b}{S_0}\right)^{\frac{2\mu}{\sigma^2}-1}, & \text{if } \mu < \frac{\sigma^2}{2} \\[2ex] 1, & \text{if } \mu \ge \frac{\sigma^2}{2}. \end{cases} \tag{5.4.20}$$

Proof: (*i*) Let $\mu > \dfrac{\sigma^2}{2}$. Since

$$P(\sup_{t \geq 0} S_t \geq b) = \lim_{T \to \infty} P(\bar{S}_T \geq b) = P(\bar{S}_\infty \geq b),$$

and $\lim_{T \to \infty} d_6 = +\infty$, $\lim_{T \to \infty} d_8 = +\infty$, using Proposition 5.4.3 we have

$$P(\bar{S}_\infty \geq b) = N(+\infty) + \left(\frac{b}{S_0}\right)^{\frac{2\mu}{\sigma^2}-1} N(-\infty) = 1.$$

(*ii*) Let $\mu < \dfrac{\sigma^2}{2}$. Since $\lim_{T \to \infty} d_6 = -\infty$, $\lim_{T \to \infty} d_8 = -\infty$, Proposition 5.4.3 implies

$$P(\bar{S}_\infty \geq b) = N(-\infty) + \left(\frac{b}{S_0}\right)^{\frac{2\mu}{\sigma^2}-1} N(+\infty) = \left(\frac{b}{S_0}\right)^{\frac{2\mu}{\sigma^2}-1}.$$

$\blacksquare$

To conclude this section, the probability of a stock reaching a certain barrier b, $b > S_0$, before a given time T is given by

$$P(T_b < T) = N(d_6) + \left(\frac{b}{S_0}\right)^{\frac{2\mu}{\sigma^2}-1} N(-d_8). \tag{5.4.21}$$

Exercise 5.4.5 *Let $F(b) = P(\bar{S}_T < b)$ be the probability function of $\bar{S}_T$ and $f(b) = F'(b)$ be its probability density.*
(*a*) *Find an expression for $F(b)$;*
(*b*) *Find the probability density $f(b)$.*

Exercise 5.4.6 (*a*) *Let $c > 0$ and $T > 0$. Using a similar proof as in Proposition 5.4.3 show that if $A(t)$ is a continuous function on $[0, +\infty)$, with $A(0) = 0$, then we have*

$$\int_0^T \frac{cA'(t)}{\sqrt{2\pi}A^2(T)} e^{-\frac{1}{2}\left(A(t)+\frac{c/2}{A(t)}\right)^2} dt = 1 - N\left(A(T) + \frac{c/2}{A(T)}\right) + e^{-c}N\left(A(T) - \frac{c/2}{A(T)}\right);$$

(*b*) *Use part (a) to find the value of the integral*

$$\int_0^T \frac{1}{t^2\sqrt{2\pi}} e^{-\frac{1}{2}\left(t+\frac{1}{t}\right)^2} dt.$$

Exercise 5.4.7 *Let $X_t = \sigma W_t + \mu t$ be a Brownian motion with drift, $\mu > 0$, and consider the stopping time*

$$\tau = \inf\{t > 0; X_t \geq a\},$$

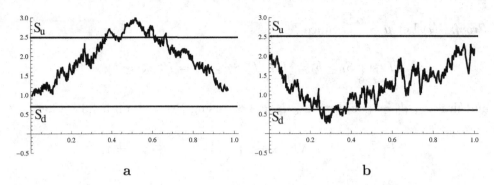

Figure 5.6: **a.** *The stock reaches the upper barrier S_u first;* **b.** *The stock reaches the lower barrier S_d first.*

with $\alpha > 0$.

(a) Apply the Optional Stopping Theorem to the martingale $M_t = e^{cW_t - \frac{1}{2}c^2 t}$ to show that

$$\mathbb{E}\left[e^{-(c\mu + \frac{1}{2}c^2)\tau}\right] = e^{-c\alpha}.$$

(b) Use part (a) to prove $\mathbb{E}[e^{-s\tau}] = e^{\frac{1}{\sigma^2}(\mu - \sqrt{2s\sigma^2 + \mu^2})\alpha}$, $s > 0$.

(c) Using the inverse Laplace transform, show that the probability density of τ is given by

$$p(t) = \frac{\alpha}{\sigma\sqrt{2\pi}t^{3/2}} e^{-\frac{1}{2t\sigma^2}(\alpha - \mu t)^2}, \qquad t > 0.$$

Exercise 5.4.8 *Let $X_t = \sigma W_t + \mu t$, with $\mu < 0$. Using a method similar with the one described in Exercise 5.4.7, prove that:*

(a) $\mathbb{E}[e^{-s\tau}] = e^{\frac{1}{\sigma^2}(\mu + \sqrt{2s\sigma^2 + \mu^2})\alpha}$, $s > 0$.

(b) The probability density of τ is given by

$$p(t) = \frac{\alpha}{\sigma\sqrt{2\pi}t^{3/2}} e^{-\frac{1}{2t\sigma^2}(\alpha + \mu t)^2}, \qquad t > 0.$$

5.5   Multiple Barriers

This section studies the case of the double barrier and deals with the probability that the stock price reaches a certain barrier before another barrier, see Fig.5.6 a,b. We still work under the assumption that the stock prices are log-normally distributed and that the condition $\mu > \sigma^2/2$ holds.

Theorem 5.5.1 *Let S_u and S_d be fixed, such that $S_d < S_0 < S_u$.*

(*i*) *The probability that the stock price S_t hits the upper value S_u before the lower value S_d is*

$$p = \frac{d^\gamma - 1}{d^\gamma - u^\gamma};$$

(*ii*) *The probability that the stock price S_t hits the lower value S_d before the upper value S_u is*

$$q = 1 - p = \frac{1 - u^\gamma}{d^\gamma - u^\gamma},$$

where $S_u/S_0 = u$, $S_d/S_0 = d$, and $\gamma = 1 - 2\mu/\sigma^2$.

Proof: Consider the Brownian motion with drift $X_t = mt + W_t$. Then Exercise 5.5.4 provides

$$P(X_t \text{ goes up to } \alpha \text{ before down to } -\beta) = \frac{e^{2m\beta} - 1}{e^{2m\beta} - e^{-2m\alpha}}. \qquad (5.5.22)$$

Choosing the following values for the parameters

$$m = \frac{\mu}{\sigma} - \frac{\sigma}{2} > 0, \quad \alpha = \frac{\ln u}{\sigma}, \quad \beta = -\frac{\ln d}{\sigma},$$

we have the sequence of identities

$$
\begin{aligned}
&\quad P(X_t \text{ goes up to } \alpha \text{ before down to } -\beta)\\
&= P(\sigma X_t \text{ goes up to } \sigma\alpha \text{ before down to } -\sigma\beta)\\
&= P(S_0 e^{\sigma X_t} \text{ goes up to } S_0 e^{\sigma\alpha} \text{ before down to } S_0 e^{-\sigma\beta})\\
&= P(S_t \text{ goes up to } S_u \text{ before down to } S_d).
\end{aligned}
$$

Using (5.5.22) yields

$$P(S_t \text{ goes up to } S_u \text{ before down to } S_d) = \frac{e^{2m\beta} - 1}{e^{2m\beta} - e^{-2m\alpha}}. \qquad (5.5.23)$$

Since a computation shows that

$$
\begin{aligned}
e^{2m\beta} &= e^{-(\frac{2\mu}{\sigma^2} - 1)\ln d} = d^{1 - \frac{2\mu}{\sigma^2}} = d^\gamma\\
e^{-2m\alpha} &= e^{(-\frac{2\mu}{\sigma^2} + 1)\ln u} = u^{1 - \frac{2\mu}{\sigma^2}} = u^\gamma,
\end{aligned}
$$

formula (5.5.23) becomes

$$P(S_t \text{ goes up to } S_u \text{ before down to } S_d) = \frac{d^\gamma - 1}{d^\gamma - u^\gamma},$$

which ends the proof. ∎

This result can be used to recover the first part of Corollary 5.4.4:

Corollary 5.5.2 *Let $S_u > S_0 > 0$ be fixed. Then*

$$P(S_t \text{ hits } S_u \text{ for some } t > 0) = \left(\frac{S_0}{S_u}\right)^{1-\frac{2\mu}{\sigma^2}}.$$

Proof: Taking $d = 0$ implies $S_d = 0$. Since S_t never reaches zero,

$$
\begin{aligned}
P(S_t \text{ hits } S_u) &= P(S_t \text{ goes up to } S_u \text{ before down to } S_d = 0) \\
&= \left.\frac{d^\gamma - 1}{d^\gamma - u^\gamma}\right|_{d=0} = \frac{1}{u^\gamma} = \left(\frac{S_0}{S_u}\right)^{1-\frac{2\mu}{\sigma^2}}.
\end{aligned}
$$

■

Exercise 5.5.3 *A stock has $S_0 = \$10$, $\sigma = 0.15$, and $\mu = 0.20$. What is the probability that the stock goes up to $\$15$ before it goes down to $\$5$?*

Exercise 5.5.4 *Let $X_t = mt + W_t$, with $m > 0$, and let $\alpha, \beta > 0$. Consider the stopping time*

$$T = \inf\{t > 0; X_t \geq \alpha \text{ or } X_t \leq -\beta\},$$

when X_t exits first time the interval $(-\beta, \alpha)$.

(a) Use the Optional Stopping Theorem to show

$$\mathbb{E}[e^{cX_T - (c\mu + \frac{1}{2}c^2)T}] = 1, \qquad \forall c > 0;$$

(b) Choose a convenient value of the constant c to obtain

$$\mathbb{E}[e^{-2mX_T}] = 1;$$

(c) Denote $p_\alpha = P(X_T = \alpha)$, $p_\beta = P(X_T = -\beta)$. Show that

$$p_\alpha = \frac{e^{2m\beta} - 1}{e^{2m\beta} - e^{-2m\alpha}}, \qquad p_\beta = \frac{1 - e^{2m\alpha}}{e^{2m\beta} - e^{-2m\alpha}}.$$

In the following we denote by T the first time the stock S_t reaches either the upper bound S_u or the lower bound S_u, where

$$S_u = S_0 u, \qquad S_d = S_0 d, \qquad u > 1, \qquad 0 < d < 1.$$

In the rest of this section we are interested in finding the distribution of the random variable T. Since an explicit formula for the probability density of T is missing, we shall find a closed form formula for its Laplace transform.

We recall that the stopping time T is the first exit time of X_t from $(-\beta, \alpha)$

$$T = \inf\{t > 0; \ S_t \geq S_u \text{ or } S_t \leq S_d\} = \inf\{t > 0; \ X_t \geq \alpha \text{ or } X_t \leq -\beta\},$$

where $X_t = mt + W_t$, and

$$m = \frac{1}{\sigma}\left(\mu - \frac{\sigma^2}{2}\right), \quad \alpha = \frac{1}{\sigma}\ln u, \quad \beta = -\frac{1}{\sigma}\ln d.$$

By Exercise 5.5.4 part (a)

$$\mathbb{E}[e^{cX_T - (c\mu + \frac{1}{2}c^2)T}] = 1, \qquad \forall c > 0.$$

Since the random variables T and X_T are independent, the previous identity becomes

$$\mathbb{E}[e^{-(c\mu + \frac{1}{2}c^2)T}] = \frac{1}{E[e^{cX_T}]}.$$

Since X_T has only two states, α and $-\beta$, then

$$E[e^{cX_T}] = e^{c\alpha}p_\alpha + e^{-c\beta}(1 - p_\alpha),$$

with p_α defined in Exercise 5.5.4 part (c). Then the previous identity becomes

$$\mathbb{E}[e^{-(c\mu + \frac{1}{2}c^2)T}] = \frac{1}{e^{c\alpha}p_\alpha + e^{-c\beta}(1 - p_\alpha)}.$$

Finally, substituting $s = c\mu + c^2/2 > 0$ we obtain the moment generating function for the random variable T

$$\mathbb{E}[e^{-sT}] = \frac{1}{e^{(-\mu + \sqrt{2s + \mu^2})\alpha}p_\alpha + e^{-(-\mu + \sqrt{2s + \mu^2})\beta}(1 - p_\alpha)}. \tag{5.5.24}$$

This expression will be useful for computing the price for double barrier derivatives in Chapter 6.

5.6 Estimation of Parameters

Assume the stock price S_t, which satisfies the equation

$$dS_t = \mu S_t dt + \sigma S_t dW_t,$$

with $\mu > 0$ and $\sigma > 0$ constants, is observed n times. Denote by s_j the observed values of S_{t_j}, $j = 1, \cdots, n$. The problem is to estimate the model parameters μ and σ from the observations s_j at times t_j. This will be done in the following using the method of maximum likelihood. The process

$$X_t = \ln \frac{S_t}{S_0} = \left(\mu - \frac{1}{2}\sigma^2\right)t + \sigma W_t$$

is a Brownian motion with drift. The observations of X_t at time t_j are given by $y_j = \ln \frac{S_j}{S_0}$. Example 2.4.1 (in section 2.4.1) provides the following system of equations satisfied by the estimation of parameters from the observed values y_j

$$\mu - \frac{1}{2}\sigma^2 = \frac{\sum_{j=1}^n y_j}{\sum_{j=1}^n t_j}$$

$$\sigma^2 = \frac{1}{n}\sum_{j=1}^n \frac{(y_j - (\mu - \frac{1}{2}\sigma^2)t_j)^2}{t_j}.$$

Eliminating $\mu - \frac{1}{2}\sigma^2$ from the previous two equations, we obtain the estimation for the square of volatility

$$\sigma^{*2} = \frac{1}{n}\sum_{j=1}^n \frac{\left(y_j - \left(\frac{\sum_{k=1}^n y_k}{\sum_{k=1}^n t_k}\right)t_j\right)^2}{t_j}. \tag{5.6.25}$$

Substituting in the first equation yields the estimation for the rate of return

$$\mu^* = \frac{\sum_{j=1}^n y_j}{\sum_{j=1}^n t_j} + \frac{1}{2n}\sum_{j=1}^n \frac{\left(y_j - \left(\frac{\sum_{k=1}^n y_k}{\sum_{k=1}^n t_k}\right)t_j\right)^2}{t_j}. \tag{5.6.26}$$

5.7 Time-dependent Drift and Volatility

The model presented in this section considers the drift $\mu = \mu(t)$ and volatility $\sigma = \sigma(t)$ to be deterministic functions of time. In this case the equation (5.1.1) becomes

$$dS_t = \mu(t)S_t dt + \sigma(t)S_t dW_t. \tag{5.7.27}$$

This equation can be solved using the method of integrating factor. Multiplying by the integrating factor

$$\rho_t = e^{-\int_0^t \sigma(s)\,dW_s + \frac{1}{2}\int_0^t \sigma^2(s)\,ds}$$

the equation (5.7.27) takes the closed form $d(\rho_t S_t) = \rho_t \mu(t)S_t\,dt$. Substituting $Y_t = \rho_t S_t$ yields the deterministic equation $dY_t = \mu(t)Y_t\,dt$ with the solution

$$Y_t = Y_0 e^{\int_0^t \mu(s)\,ds}.$$

Substituting back $S_t = \rho_t^{-1} Y_t$, we obtain the closed-form solution for the equation (5.7.27)

$$S_t = S_0 e^{\int_0^t (\mu(s) - \frac{1}{2}\sigma^2(s))\,ds + \int_0^t \sigma(s)\,dW_s}. \tag{5.7.28}$$

Proposition 5.7.1 *The solution S_t given by (5.7.28) is $\mathcal{F}_t$-adapted and log-normally distributed, with mean and variance given by*

$$\mathbb{E}[S_t] = S_0 e^{\int_0^t \mu(s)\,ds}$$
$$Var[S_t] = S_0^2 e^{2\int_0^t \mu(s)\,ds}\left(e^{\int_0^t \sigma^2(s)\,ds} - 1\right).$$

Proof: Let $X_t = \int_0^t (\mu(s) - \frac{1}{2}\sigma^2(s))\,ds + \int_0^t \sigma(s)\,dW_s$. Since X_t is a sum of a deterministic integral function and a Wiener integral, then it is normally distributed, with

$$\mathbb{E}[X_t] = \int_0^t \left(\mu(s) - \frac{1}{2}\sigma^2(s)\right) ds$$
$$Var[X_t] = Var\left[\int_0^t \sigma(s)\,dW_s\right] = \int_0^t \sigma^2(s)\,ds.$$

Then the mean and variance of the log-normal random variable $S_t = S_0 e^{X_t}$ are given by

$$\mathbb{E}[S_t] = S_0 e^{\int_0^t (\mu - \frac{\sigma^2}{2})\,ds + \frac{1}{2}\int_0^t \sigma^2\,ds} = S_0 e^{\int_0^t \mu(s)\,ds}$$
$$Var[S_t] = S_0^2 e^{2\int_0^t (\mu - \frac{1}{2}\sigma^2)\,ds + \int_0^t \sigma^2\,ds}\left(e^{\int_0^t \sigma^2} - 1\right)$$
$$= S_0^2 e^{2\int_0^t \mu(s)\,ds}\left(e^{\int_0^t \sigma^2(s)\,ds} - 1\right).$$

∎

If the average drift and average squared volatility are defined as

$$\bar{\mu} = \frac{1}{t}\int_0^t \mu(s)\,ds$$
$$\bar{\sigma}^2 = \frac{1}{t}\int_0^t \sigma^2(s)\,ds,$$

the aforementioned formulas can be written in a more simple form as

$$\mathbb{E}[S_t] = S_0 e^{\bar{\mu} t}$$
$$Var[S_t] = S_0^2 e^{2\bar{\mu} t}(e^{\bar{\sigma}^2 t} - 1).$$

It is worth noting the similarity with the formulas (5.1.3)–(5.1.4).

Exercise 5.7.2 *Consider two stock prices satisfying the equations*

$$dS_1 = \mu_1(t)S_1 dt + \sigma_1(t)S_1 dW_1$$
$$dS_2 = \mu_2(t)S_2 dt + \sigma_2(t)S_2 dW_2,$$

with correlated Brownian motions, with $\rho = cor(dW_1, dW_2)$ constant.

(a) *Show that*

$$\mathbb{E}[S_1(t)S_2(t)] = S_1(0)S_2(0)e^{(\bar{\mu}_1+\bar{\mu}_2)t}e^{\rho \int_0^t \sigma_1(s)\sigma_2(s)\,ds}.$$

(b) *Verify that*

$$cov\big(S_1(t), S_2(t)\big) = S_1(0)S_2(0)e^{(\bar{\mu}_1+\bar{\mu}_2)t}\Big(e^{\rho \int_0^t \sigma_1(s)\sigma_2(s)\,ds} - 1\Big).$$

(c) *Find the correlation function* $cor\big(S_1(t), S_2(t)\big)$.

5.8 Models for Stock Price Averages

This section describes the stochastic differential equations for several types of averages on stocks. These averages are used as underlying assets in the case of Asian options.

Discretely sampled averages
  Let $S_{t_1}, S_{t_2}, \cdots, S_{t_n}$ be a sample of stock prices at n instances of time $t_1 < t_2 < \cdots < t_n$. The most common types of *discrete averages* are:
  • The *arithmetic average*

$$A(t_1, t_2, \cdots, t_n) = \frac{1}{n}\sum_{k=1}^{n} S_{t_k}.$$

  • The *geometric average*

$$G(t_1, t_2, \cdots, t_n) = \Big(\prod_{k=1}^{n} S_{t_k}\Big)^{\frac{1}{n}}.$$

  • The *harmonic average*

$$H(t_1, t_2, \cdots, t_n) = \frac{n}{\displaystyle\sum_{k=1}^{n} \frac{1}{S_{t_k}}}.$$

The well-known inequality of means states the following order relation among the means

$$H(t_1, t_2, \cdots, t_n) \leq G(t_1, t_2, \cdots, t_n) \leq A(t_1, t_2, \cdots, t_n), \qquad (5.8.29)$$

with identity in the case of constant stock prices.

In the rest of the section we shall discuss the continuous counterparts of the aforementioned types of stock averages.

The continuously sampled arithmetic average
Let $t_n = t$ and assume $t_{k+1} - t_k = \frac{t}{n}$. Using the definition of the integral as a limit of Riemann sums, we have

$$\lim_{n \to \infty} \frac{1}{n} \sum_{k=1}^{n} S_{t_k} = \lim_{n \to \infty} \frac{1}{t} \sum_{k=1}^{n} S_{t_k} \frac{t}{n} = \frac{1}{t} \int_0^t S_u \, du.$$

It follows that the *continuously sampled arithmetic average* of stock prices between 0 and t is given by

$$\boxed{A_t = \frac{1}{t} \int_0^t S_u \, du.} \qquad (5.8.30)$$

Assuming that S_t has constant drift rate and volatility, we obtain

$$A_t = \frac{S_0}{t} \int_0^t e^{(\mu - \sigma^2/2)u + \sigma W_u} \, du.$$

This integral can be computed explicitly only in the case $\sigma = 0$. In the absence of a closed form expression for A_t, one can hope to describe A_t by its probability law. For a long time the probability density of A_t was unknown, a fact that made many authors to proceed using various approximations. Fortunately, a trackable formula for the law of A_t was found by Yor, see [70], in early 90's. The fact that A_t is neither normal nor log-normal makes the price of Asian options on arithmetic averages hard to evaluate explicitly.

In the following we shall provide an elementary computation of the first two moments of A_t.

Let $I_t = \int_0^t S_u \, du$. The Fundamental Theorem of Calculus implies $dI_t = S_t \, dt$. Then the quotient rule yields

$$dA_t = d\left(\frac{I_t}{t}\right) = \frac{dI_t \, t - I_t \, dt}{t^2} = \frac{S_t t \, dt - I_t dt}{t^2} = \frac{1}{t}(S_t - A_t)dt,$$

i.e. the continuous arithmetic average A_t satisfies the equation

$$\boxed{dA_t = \frac{1}{t}(S_t - A_t)dt.}$$

If $A_t < S_t$, the right side is positive and hence $dA_t > 0$, i.e. the average A_t increases. Similarly, if $A_t > S_t$, then the average A_t decreases. This shows that the average A_t tends to trace the stock values S_t.

An application of l'Hospital's rule implies

$$A_0 = \lim_{t \to 0} \frac{I_t}{t} = \lim_{t \to 0} S_t = S_0.$$

Using that the expectation commutes with integrals, we have

$$\mathbb{E}[A_t] = \frac{1}{t} \int_0^t \mathbb{E}[S_u] \, du = \frac{1}{t} \int_0^t S_0 e^{\mu u} \, du = S_0 \frac{e^{\mu t} - 1}{\mu t}.$$

Hence, the expectation of the arithmetic average of a stock given by (5.1.1) is

$$\mathbb{E}[A_t] = \begin{cases} S_0 \dfrac{e^{\mu t} - 1}{\mu t}, & \text{if } t > 0 \\ S_0, & \text{if } t = 0. \end{cases} \tag{5.8.31}$$

In the following we shall compute the variance $Var[A_t]$. Since

$$Var[A_t] = \frac{1}{t^2} \mathbb{E}[I_t^2] - \mathbb{E}[A_t]^2, \tag{5.8.32}$$

it suffices to find $\mathbb{E}[I_t^2]$. We need first the following result:

Lemma 5.8.1 *We have*

$$\mathbb{E}[I_t S_t] = \frac{S_0^2}{\mu + \sigma^2} [e^{(2\mu + \sigma^2)t} - e^{\mu t}].$$

Proof: (*i*) Using Ito's formula

$$\begin{aligned} d(I_t S_t) &= dI_t \, S_t + I_t \, dS_t + dI_t \, dS_t \\ &= S_t^2 dt + I_t(\mu S_t dt + \sigma S_t dW_t) + \underbrace{S_t dt \, dS_t}_{=0} \\ &= (S_t^2 + \mu I_t S_t) dt + \sigma I_t S_t dW_t. \end{aligned}$$

Using $I_0 S_0 = 0$, integrating between 0 and t yields

$$I_t S_t = \int_0^t (S_u^2 + \mu I_u S_u) \, du + \sigma \int_0^t I_u S_u \, dW_u.$$

Since the expectation of the Ito integral is zero, we have

$$\mathbb{E}[I_t S_t] = \int_0^t (\mathbb{E}[S_u^2] + \mu \mathbb{E}[I_u S_u]) \, du.$$

Using Exercise 5.1.5 this becomes

$$\mathbb{E}[I_t S_t] = \int_0^t \left(S_0^2 e^{(2\mu + \sigma^2)u} + \mu \mathbb{E}[I_u S_u] \right) du.$$

If we denote $g(t) = \mathbb{E}[I_t S_t]$, differentiating yields the linear differential equation

$$g'(t) = S_0^2 e^{(2\mu+\sigma^2)t} + \mu g(t),$$

with the initial condition $g(0) = 0$. This can be solved as a linear differential equation in $g(t)$ by multiplying by the integrating factor $e^{-\mu t}$. The solution is

$$g(t) = \frac{S_0^2}{\mu + \sigma^2}[e^{(2\mu+\sigma^2)t} - e^{\mu t}].$$

∎

Remark 5.8.2 A_t and I_t are not independent. This follows from the fact that

$$\mathbb{E}[I_t S_t] \neq \mathbb{E}[I_t]\mathbb{E}[S_t] = \frac{S_0^2}{\mu}(e^{\mu t} - 1)e^{\mu t}.$$

Next we shall find $\mathbb{E}[I_t^2]$. Using $dI_t = S_t dt$, then $(dI_t)^2 = 0$ and hence Ito's formula yields

$$d(I_t^2) = 2I_t\, dI_t + (dI_t)^2 = 2I_t S_t\, dt.$$

Integrating between 0 and t and using $I_0 = 0$ leads to

$$I_t^2 = 2\int_0^t I_u S_u\, du.$$

Taking the expectation and using Lemma 5.8.1 we obtain

$$\mathbb{E}[I_t^2] = 2\int_0^t \mathbb{E}[I_u S_u]\, du = \frac{2S_0^2}{\mu + \sigma^2}\left[\frac{e^{(2\mu+\sigma^2)t} - 1}{2\mu + \sigma^2} - \frac{e^{\mu t} - 1}{\mu}\right]. \qquad (5.8.33)$$

Substituting into (5.8.32) yields

$$Var[A_t] = \frac{S_0^2}{t^2}\left\{\frac{2}{\mu + \sigma^2}\left[\frac{e^{(2\mu+\sigma^2)t} - 1}{2\mu + \sigma^2} - \frac{e^{\mu t} - 1}{\mu}\right] - \frac{(e^{\mu t} - 1)^2}{\mu^2}\right\}.$$

Concluding the previous calculations, we have the following result:

Proposition 5.8.3 *The arithmetic average A_t satisfies the stochastic equation*

$$dA_t = \frac{1}{t}(S_t - A_t)dt, \quad A_0 = S_0.$$

Its mean and variance are given by

$$\mathbb{E}[A_t] = S_0\frac{e^{\mu t} - 1}{\mu t}, \quad t > 0$$

$$Var[A_t] = \frac{S_0^2}{t^2}\left\{\frac{2}{\mu + \sigma^2}\left[\frac{e^{(2\mu+\sigma^2)t} - 1}{2\mu + \sigma^2} - \frac{e^{\mu t} - 1}{\mu}\right] - \frac{(e^{\mu t} - 1)^2}{\mu^2}\right\}.$$

Exercise 5.8.4 *Find approximative formulas for* $\mathbb{E}[A_t]$ *and* $Var[A_t]$ *for* t *small, up to the order* $O(t^2)$.

(Recall that $f(t) = O(t^2)$ *if* $\lim\limits_{t\to\infty} \dfrac{f(t)}{t^2} = c < \infty$.)

The continuously sampled geometric average First, we shall find an integral formula for the continuously sampled geometric average. Dividing the interval $[0,t]$ into equal subintervals of length $t_{k+1} - t_k = \frac{t}{n}$, we have

$$
G(t_1,\dots,t_n) = \left(\prod_{k=1}^{n} S_{t_k}\right)^{1/n} = e^{\ln\left(\prod_{k=1}^{n} S_{t_k}\right)^{1/n}}
$$

$$
= e^{\frac{1}{n}\sum_{k=1}^{n} \ln S_{t_k}} = e^{\frac{1}{t}\sum_{k=1}^{n} \ln S_{t_k}\frac{t}{n}}.
$$

Using the definition of the integral as a limit of Riemann sums

$$
G_t = \lim_{n\to\infty} \left(\prod_{k=1}^{n} S_{t_k}\right)^{1/n} = \lim_{n\to\infty} e^{\frac{1}{t}\sum_{k=1}^{n} \ln S_{t_k}\frac{t}{n}} = e^{\frac{1}{t}\int_0^t \ln S_u\, du}.
$$

Therefore, the *continuously sampled geometric average* of stock prices between time instances 0 and t is given by

$$
\boxed{G_t = e^{\frac{1}{t}\int_0^t \ln S_u\, du}.}
\tag{5.8.34}
$$

Theorem 5.8.5 G_t *has a log-normal distribution, with the mean and variance given respectively by*

$$
\mathbb{E}[G_t] = S_0 e^{(\mu-\frac{\sigma^2}{6})\frac{t}{2}}
$$

$$
Var[G_t] = S_0^2 e^{(\mu-\frac{\sigma^2}{6})t}\left(e^{\frac{\sigma^2 t}{3}} - 1\right).
$$

Proof: Since

$$
\ln S_u = \ln\left(S_0 e^{(\mu-\frac{\sigma^2}{2})u + \sigma W_u}\right) = \ln S_0 + \left(\mu - \frac{\sigma^2}{2}\right)u + \sigma W_u,
$$

then taking the logarithm on G_t yields

$$
\ln G_t = \frac{1}{t}\int_0^t \left[\ln S_0 + \left(\mu - \frac{\sigma^2}{2}\right)u + \sigma W_u\right] du = \ln S_0 + \left(\mu - \frac{\sigma^2}{2}\right)\frac{t}{2} + \frac{\sigma}{t}\int_0^t W_u\, du.
\tag{5.8.35}
$$

Using that the integrated Brownian motion $Z_t = \int_0^t W_u\, du$ is Gaussian with $Z_t \sim \mathcal{N}(0, t^3/3)$, it follows that $\ln G_t$ has a normal distribution

$$
\ln G_t \sim \mathcal{N}\left(\ln S_0 + \left(\mu - \frac{\sigma^2}{2}\right)\frac{t}{2}, \frac{\sigma^2 t}{3}\right).
\tag{5.8.36}
$$

This implies that G_t has a log-normal distribution. Its mean and variance are given by

$$
\begin{aligned}
\mathbb{E}[G_t] &= e^{\mathbb{E}[\ln G_t] + \frac{1}{2} Var[\ln G_t]} = e^{\ln S_0 + (\mu - \frac{\sigma^2}{2})\frac{t}{2} + \frac{\sigma^2 t}{6}} = S_0 e^{(\mu - \frac{\sigma^2}{6})\frac{t}{2}}. \\
Var[G_t] &= e^{2\mathbb{E}[\ln G_t] + Var[\ln G_t]} \left(e^{Var[\ln G_t]} - 1 \right) \\
&= e^{2\ln S_0 + (\mu - \frac{\sigma^2}{6})t} \left(e^{\frac{\sigma^2 t}{3}} - 1 \right) = S_0^2 e^{(\mu - \frac{\sigma^2}{6})t} \left(e^{\frac{\sigma^2 t}{3}} - 1 \right).
\end{aligned}
$$

∎

Corollary 5.8.6 *The geometric average G_t is given by the closed-form formula*

$$
\boxed{G_t = S_0 e^{(\mu - \frac{\sigma^2}{2})\frac{t}{2} + \frac{\sigma}{t} \int_0^t W_u \, du}}.
$$

Proof: Take the exponential in the formula (5.8.35). ∎

An important consequence of the fact that G_t is log-normal is that Asian options on geometric averages have closed-form solutions.

Exercise 5.8.7 (a) *Show that $\ln G_t$ satisfies the stochastic differential equation*

$$
d(\ln G_t) = \frac{1}{t} \left(\ln S_t - \ln G_t \right) dt.
$$

(b) *Show that G_t satisfies*

$$
dG_t = \frac{1}{t} G_t \left(\ln S_t - \ln G_t \right) dt.
$$

The continuously sampled harmonic average Let S_{t_k} be the values of a stock evaluated at the sampling dates t_k, $i = 1, \ldots, n$. Their *harmonic average* is defined by the inverse of the arithmetic average of the inverses of S_{t_k}

$$
H(t_1, \cdots, t_n) = \frac{n}{\displaystyle\sum_{k=1}^{n} \frac{1}{S_{t_k}}}.
$$

Consider $t_k = \frac{kt}{n}$. Then the continuously sampled harmonic average is obtained by taking the limit as $n \to \infty$ in the aforementioned relation

$$
\lim_{n \to \infty} \frac{n}{\displaystyle\sum_{k=1}^{n} \frac{1}{S_{t_k}}} = \lim_{n \to \infty} \frac{t}{\displaystyle\sum_{k=1}^{n} \frac{1}{S_{t_k}} \frac{t}{n}} = \frac{t}{\displaystyle\int_0^t \frac{1}{S_u} \, du}.
$$

Hence, the *continuously sampled harmonic average* is defined by

$$H_t = \frac{t}{\int_0^t \frac{1}{S_u}\, du}.$$

We may also write $H_t = \dfrac{t}{I_t}$, where $I_t = \displaystyle\int_0^t \frac{1}{S_u}\, du$ satisfies

$$dI_t = \frac{1}{S_t}\, dt, \qquad I_0 = 0, \qquad d\left(\frac{1}{I_t}\right) = -\frac{1}{S_t I_t^2}\, dt.$$

From the l'Hospital's rule we get

$$H_0 = \lim_{t \searrow 0} H_t = \lim_{t \searrow 0} \frac{t}{I_t} = S_0.$$

Using the product rule we obtain the following:

$$
\begin{aligned}
dH_t &= t\, d\left(\frac{1}{I_t}\right) + \frac{1}{I_t}\, dt + dt\, d\left(\frac{1}{I_t}\right) \\
&= \frac{1}{I_t}\left(1 - \frac{t}{S_t I_t}\right) dt = \frac{1}{t} H_t \left(1 - \frac{H_t}{S_t}\right) dt,
\end{aligned}
$$

so

$$dH_t = \frac{1}{t} H_t \left(1 - \frac{H_t}{S_t}\right) dt. \qquad (5.8.37)$$

If at the instance t we have $H_t < S_t$, it follows from the equation that $dH_t > 0$, i.e. the harmonic average increases. Similarly, if $H_t > S_t$, then $dH_t < 0$, i.e H_t decreases. The converses are also true. The random variable H_t is not normally distributed nor log-normally distributed.

Exercise 5.8.8 *Show that* $\dfrac{H_t}{t}$ *is a decreasing function of* t. *What is its limit as* $t \to \infty$?

Exercise 5.8.9 *Show that a continuous analog of inequality (5.8.29) holds:*

$$H_t \leq G_t \leq A_t.$$

Exercise 5.8.10 *Let* S_{t_k} *be the stock price at time* t_k. *Consider the power* α *of the arithmetic average of* $S_{t_k}^\alpha$

$$A^\alpha(t_1, \cdots, t_n) = \left[\frac{\sum_{k=1}^n S_{t_k}^\alpha}{n}\right]^\alpha.$$

(a) *Show that the aforementioned expression tends to*

$$A_t^\alpha = \left[\frac{1}{t} \int_0^t S_u^\alpha \, du \right]^\alpha,$$

as $n \to \infty$.

(b) *Find the stochastic differential equation satisfied by A_t^α.*

(c) *What does A_t^α become in the particular cases $\alpha = \pm 1$?*

Exercise 5.8.11 *The stochastic average of stock prices between 0 and t is defined by*

$$X_t = \frac{1}{t} \int_0^t S_u \, dW_u,$$

where W_u is a Brownian motion process.

(a) *Find dX_t, $\mathbb{E}[X_t]$ and $Var[X_t]$.*

(b) *Show that $\sigma X_t = R_t - \mu A_t$, where $R_t = \dfrac{S_t - S_0}{t}$ is the "raw average" of the stock price and $A_t = \frac{1}{t} \int_0^t S_u \, du$ is the continuous arithmetic average.*

Remark 5.8.12 The approximative evaluation of options and forward contracts of harmonic averages have been worked out in [16]. Another evaluation as well as the role of contracts written on the harmonic average of the underlying price in the foreign exchange markets are emphasized in Vecer [67].

5.9   Stock Prices with Rare Events

The stock price models considered so far were time continuous, and did not accomodate market crashes. In order to model the stock price when rare events are taken into account, we shall combine the effect of two stochastic processes:

- the Brownian motion process W_t, which models regular market fluctuations given by infinitesimal changes in the price, and which is a continuous process;

- the Poisson process N_t, which is discontinuous and models sporadic jumps in the stock price that corresponds to shocks in the market, see Appendix C.1.

If λ is the rate of the Poisson process, then $\mathbb{E}[dN_t] = \lambda dt$, i.e. the process N_t has a positive drift, with the rate λ. The process N_t can be "compensated" by subtracting λt from N_t. The resulting process, $M_t = N_t - \lambda t$, is a martingale,

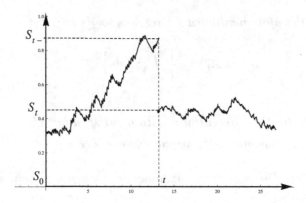

Figure 5.7: *The lateral limits S_{t-} and $S_t = S_{t+}$ of a right continuous stock price.*

called the compensated Poisson process, that models unpredictable jumps of size 1 at a constant rate λ. It is worth noting that the processes W_t and M_t involved in modeling the stock price are assumed to be independent. This can be roughly stated by saying that one cannot forecast a market crash only from the information provided by the daily regular market fluctuations.

Since in this model the stock price S_t is discontinuous at certain time instances, it makes sense to introduce and work with the lateral limits of the stock price. Let $S_{t-} = \lim_{u \nearrow t} S_u$ denote the value of the stock before a possible jump occurence at time t. The stock is assumed to be a right-continuous function, i.e., $S_t = S_{t+} = \lim_{u \searrow t} S_u$, see Fig.5.7.

To set up the model, we assume that the instantaneous return on the stock, $\dfrac{dS_t}{S_{t-}}$, is the sum of the following three components:

- the deterministic drift part, μdt;
- the noisy part due to unexpected news, σdW_t;
- the rare events part due to random jumps, ρdM_t,

where μ, σ and ρ are constants, corresponding to the drift rate of the stock, volatility and instantaneous return jump size.

Superpossing the aforementioned effects yields the equation

$$\frac{dS_t}{S_{t-}} = \mu dt + \sigma dW_t + \rho dM_t.$$

Hence, the dynamics of a stock price, subject to rare events, is modeled by the following stochastic differential equation

$$dS_t = \mu S_{t-} dt + \sigma S_{t-} dW_t + \rho S_{t-} dM_t. \tag{5.9.38}$$

It is worth noting that in the case of jumps of size zero, $\rho = 0$, the previous equation becomes the classical stochastic equation (5.1.1).

The last two terms in equation (5.9.38) model fluctuations of zero mean implied by the martingale properties

$$\mathbb{E}[\rho S_{t-} dM_t | \mathcal{F}_t] = \rho S_{t-} \mathbb{E}[dM_t | \mathcal{F}_t] = 0,$$
$$\mathbb{E}[\sigma S_{t-} dW_t | \mathcal{F}_t] = \sigma S_{t-} \mathbb{E}[dW_t | \mathcal{F}_t] = 0.$$

The term $\sigma S_{t-} dW_t$ captures regular events of insignificant size, while $\rho S_{t-} dM_t$ captures rare events of large size. The "rare events" term, $\rho S_{t-} dM_t$, incorporates jumps proportional to the stock price and is given in terms of the Poisson process N_t as

$$\rho S_{t-} dM_t = \rho S_{t-} dN_t - \lambda \rho S_{t-} dt.$$

Substituting into equation (5.9.38) yields

$$dS_t = (\mu - \lambda\rho) S_{t-} dt + \sigma S_{t-} dW_t + \rho S_{t-} dN_t. \tag{5.9.39}$$

The constant λ represents the rate at which the jumps of the Poisson process N_t occur. This is the same as the rate of rare events in the market, and can be determined from historical data. We note that the term $\lambda\rho$ (frequency times jump size) has the dimensionality of a drift.

The following result provides an explicit solution for the stock price when rare events are taken into account.

Proposition 5.9.1 *The solution of the stochastic equation (5.9.39) is given by*

$$S_t = S_0 e^{(\mu - \lambda\rho - \frac{\sigma^2}{2})t + \sigma W_t} (1 + \rho)^{N_t}, \tag{5.9.40}$$

where:
μ is the stock price drift rate;
σ is the volatility of the stock;
λ is the rate at which rare events occur;
ρ is the size of jump in the expected return when a rare event occurs.

Proof: In the following we shall verify that (5.9.40) is a solution of the stochastic differential equation (5.9.39). If t_k denotes the kth jump time, then $N_{t_k} = k$. If denote

$$U_t = S_0 e^{(\mu - \lambda\rho - \frac{\sigma^2}{2})t + \sigma W_t}, \qquad V_t = (1 + \rho)^{N_t},$$

we have $S_t = U_t V_t$ and hence Ito's lemma provides

$$dS_t = V_t dU_t + U_t dV_t + dU_t dV_t. \tag{5.9.41}$$

We already know that U_t verifies

$$
\begin{aligned}
dU_t &= (\mu - \lambda\rho)U_t dt + \sigma U_t dW_t \\
&= (\mu - \lambda\rho)U_{t-} dt + \sigma U_{t-} dW_t,
\end{aligned}
$$

since U_t is a continuous process, i.e., $U_t = U_{t-}$. Then the first term of (5.9.41) becomes

$$
V_t dU_t = (\mu - \lambda\rho)S_{t-} dt + \sigma S_{t-} dW_t.
$$

In order to compute the second term of (5.9.41) we write

$$
\begin{aligned}
dV_t = V_t - V_{t-} &= (1+\rho)^{N_t} - (1+\rho)^{N_{t-}} \\
&= \begin{cases} (1+\rho)^{1+N_{t-}} - (1+\rho)^{N_{t-}}, & \text{if } t = t_k \\ (1+\rho)^{N_t} - (1+\rho)^{N_t}, & \text{if } t \neq t_k \end{cases} \\
&= \begin{cases} \rho(1+\rho)^{N_{t-}}, & \text{if } t = t_k \\ 0, & \text{if } t \neq t_k \end{cases} \\
&= \rho(1+\rho)^{N_{t-}} \begin{cases} 1, & \text{if } t = t_k \\ 0, & \text{if } t \neq t_k \end{cases} \\
&= \rho(1+\rho)^{N_{t-}} dN_t,
\end{aligned}
$$

so $U_t dV_t = \rho U_t (1+\rho)^{N_{t-}} dN_t = \rho S_{t-} dN_t$. Since $dt dN_t = dN_t dW_t = 0$, see the multiplication table in Appendix C.2, the last term of (5.9.41) becomes $dU_t dV_t = 0$. Substituting back into (5.9.41) yields the equation

$$
dS_t = (\mu - \lambda\rho)S_{t-} dt + \sigma S_{t-} dW_t + \rho S_{t-} dN_t.
$$

∎

Formula (5.9.40) provides the stock price at time t if exactly N_t jumps have occurred and all jumps in the return of the stock are equal to ρ.

Remark 5.9.2 It is worth noting that if denote $\rho_k = \dfrac{S_{t_k} - S_{t_k-}}{S_{t_k-}}$, where t_k denotes the kth jump, a similar proof leads to the formula

$$
S_t = S_0 e^{(\mu - \lambda\rho - \frac{\sigma^2}{2})t + \sigma W_t} \prod_{k=1}^{N_t} (1 + \rho_k),
$$

where $\rho = \mathbb{E}[\rho_k]$ is the expected value of the jump in the stock return. The random variables ρ_k are assumed independent and identically distributed and also independent of W_t and N_t.

If denote the jump ratios by $Y_k = \frac{S_{t_k}}{S_{t_k-}}$, then the above formula can be written equivalently as

$$S_t = S_0 e^{(\mu - \lambda\rho - \frac{\sigma^2}{2})t + \sigma W_t} \prod_{k=1}^{N_t} Y_k, \qquad (5.9.42)$$

where $\rho = \mathbb{E}[Y_k] - 1$. It is worth noting that

$$Y_k - 1 = \frac{S_{t_k} - S_{t_k-}}{S_{t_k-}} = \frac{dS_{t_k}}{S_{t_k-}}$$

is the percentage of the stock jump. For more details the reader is referred to Merton [54].

The next result computes the forecast of the stock at time T given its value at a prior time t, assuming a rare events environment. It is interesting to remark that jumps do not affect the expectation.

Proposition 5.9.3 *The conditional expectation of the stock as of time t is given by*

$$\mathbb{E}[S_T | \mathcal{F}_t] = S_t e^{\mu(T-t)}.$$

Proof: Writing formula (5.9.40) at time instances t and T and then dividing, we obtain

$$S_T = S_t e^{(\mu - \lambda\rho - \frac{\sigma^2}{2})(T-t) + \sigma(W_T - W_t)} (1 + \rho)^{N_T - N_t}. \qquad (5.9.43)$$

Conditioning over the number of jumps in the time interval $(t, T]$ we have the following expectation

$$\begin{aligned}
\mathbb{E}[S_T | S_t = x] &= \sum_{k \geq 0} P(N_T - N_t = k) \mathbb{E}[S_T | S_t = x, N_T - N_t = k] \\
&= x e^{(\mu - \lambda\rho - \frac{\sigma^2}{2})(T-t)} \sum_{k \geq 0} \frac{e^{-\lambda(T-t)} \lambda^k (T-t)^k}{k!} (1 + \rho)^k \mathbb{E}[e^{\sigma W_{T-t}}] \\
&= x e^{(\mu - \lambda\rho - \frac{\sigma^2}{2})(T-t)} e^{-\lambda(T-t)} e^{\lambda(1+\rho)(T-t)} e^{\frac{1}{2}\sigma^2(T-t)} \\
&= x e^{\mu(T-t)},
\end{aligned}$$

where we used the stationarity property of the Brownian motion $\mathbb{E}[e^{\sigma(W_T - W_t)}] = \mathbb{E}[e^{\sigma W_{T-t}}] = e^{\frac{1}{2}\sigma^2(T-t)}$ and the power series of the exponential $e^z = \sum_{k \geq 0} \frac{z^k}{k!}$.

The desired conditional expectation is obtained replacing x by S_t

$$\mathbb{E}[S_T | \mathcal{F}_t] = S_t e^{\mu(T-t)}.$$

■

We note that this result will be useful in the sequel when pricing forward contracts in a rare events environment.

In the following we shall find the probability density for Z_T. Taking the log in formula (5.9.43) and denoting by $Z_t = \ln S_t$, we have

$$Z_T = Z_t + \left(\mu - \lambda\rho - \frac{\sigma^2}{2}\right)(T - t) + \sigma(W_T - W_t) + \ln(1 + \rho)(N_T - N_t).$$

Under the condition that there are exactly n jumps during the time interval $(t, T]$, i.e. $N_T - N_t = n$, the variable Z_T is normally distributed with

$$Z_T \sim \mathcal{N}\left(Z_t + \left(\mu - \lambda\rho - \frac{\sigma^2}{2}\right)(T - t) + \ln(1 + \rho)^n, \sigma^2(T - t)\right).$$

Multiplying by the probability of having exactly n jumps,

$$P(N_T - N_t = n) = \frac{e^{-\lambda(T-t)}\lambda^n(T - t)^n}{n!},$$

and summing over n we obtain the unconditioned distribution of Z_T:

$$Z_T \sim \sum_{n \geq 0} \frac{e^{-\lambda(T-t)}\lambda^n(T - t)^n}{n!} \mathcal{N}\left(Z_t + \left(\mu - \lambda\rho - \frac{\sigma^2}{2}\right)(T-t) + \ln(1+\rho)^n, \sigma^2(T-t)\right).$$

The exact interpretation of the previous formula is that the probability density of Z_T, conditional to Z_t, is given by the following series

$$p_{Z_T|Z_t}(z|z_0) = \frac{e^{-\lambda(T-t)}}{\sqrt{2\pi(T - t)}\,\sigma} \sum_{n \geq 0} \frac{\lambda^n(T - t)^n}{n!} e^{-\frac{[z-z_0-(\mu-\lambda\rho-\sigma^2/2)(T-t)-n\ln(1+\rho)]^2}{2\sigma^2(T-t)}}.$$

$$(5.9.44)$$

This formula will be used when pricing derivatives in a rare events environment in section 6.17.

Remark 5.9.4 The crash market of 1987 as well as other more recent market crashes proved the inability of Samuelson's geometric Brownian motion model to capture large market movements. The fact that jump-diffusion processes are a better fit for markets when shocks are present, explains the increased interest in these type of models. In addition to this, jump-diffusion processes started to appear more recently in modeling energy markets, whose behavior exhibits spikes in a natural way. For more applications regarding energy and power markets, see the book of Eydeland and Wolyniec [32].

For the following exercises assume that S_t is given by formula (5.9.40).

Exercise 5.9.5 *Find* $\mathbb{E}[S_t]$ *and* $Var[S_t]$.

Exercise 5.9.6 *Compute* $\mathbb{E}[\ln S_t]$ *and* $Var[\ln S_t]$.

Exercise 5.9.7 *Find the conditional expectation* $\mathbb{E}[S_t|\mathcal{F}_u]$ *for* $u < t$.

Exercise 5.9.8 *Compute* $\mathbb{E}[S_t]$ *in the case when* S_t *is given by formula* (5.9.42).

5.10 Dividend Paying Stocks

If the stock pays dividends at a continuous rate $\delta > 0$, then the stochastic differential equation followed by the stock is given by

$$\boxed{dS_t = (\mu - \delta)S_t dt + \sigma S_t dW_t.}$$ (5.10.45)

We notice that the dividend rate weakens the rate of return. The solution is given by the geometric brownian motion

$$S_t = S_0 e^{(\mu - \delta - \frac{\sigma^2}{2})t + \sigma W_t}.$$

The stock is log-normally distributed with the mean and variance given by

$$\mathbb{E}[S_t] = S_0 e^{(\mu - \delta)t}$$ (5.10.46)
$$Var[S_t] = S_0^2 e^{2(\mu - \delta)t}(e^{\sigma^2 t} - 1).$$ (5.10.47)

For the sake of simplicity we shall assume most of the time that the stock does not pay any dividends. Most pricing formulas that involve risk-neutral valuation are obtained in the case of dividends paying stocks by just replacing the risk-free rate r with the difference $r - \delta$. However, in the case of American call options the difference between these two types of stock makes a substantial difference and in that case they will be treated separately.

5.11 Currencies

Consider two currencies, the Japanese yen, Y_t, and the US dollar, U_t, which satisfy the following geometric Brownian motions

$$dY_t = \mu_1 Y_t dt + \sigma_1 Y_t dW_t^1$$
$$dU_t = \mu_2 U_t dt + \sigma_2 U_t dW_t^2,$$

with the correlation $cor(dW_t^1, dW_t^2) = \rho$. We would like to estimate the drift rates, μ_i and volatilities σ_i, but historical estimation does not work since Y_t and U_t cannot be observed directly. What can we actually observe in the market is the exchange rate of one currency with respect to the other. For instance, $Q_t = \dfrac{Y_t}{U_t}$ represents the conversion rate of the yen into US dollar.

We shall find the process satisfied by Q_t applying Ito's lemma to the process $Q_t = F(Y_t, U_t) = \dfrac{Y_t}{U_t}$. Using that

$$\frac{\partial F}{\partial Y} = \frac{1}{U}, \quad \frac{\partial^2 F}{\partial Y^2} = 0, \quad \frac{\partial F}{\partial U} = -\frac{Y}{U^2}, \quad \frac{\partial^2 F}{\partial U^2} = \frac{2Y}{U^3}, \quad \frac{\partial^2 F}{\partial U \partial Y} = -\frac{1}{U^2},$$

we have after simplifications

$$
\begin{aligned}
dQ_t &= \frac{\partial F}{\partial Y} dY + \frac{\partial F}{\partial U} dU + \frac{1}{2}\frac{\partial^2 F}{\partial Y^2}(dY)^2 + \frac{1}{2}\frac{\partial^2 F}{\partial U^2}(dU)^2 + \frac{\partial^2 F}{\partial U \partial Y} dU\, dY \\
&= Q_t(\mu_1 - \mu_2 + \sigma_2^2 - \rho\sigma_1\sigma_2)dt + Q_t(\sigma_1 dW_t^1 - \sigma_2 dW_t^2).
\end{aligned}
$$

If denote the drift rate of the exchange by

$$\mu_{12} = \mu_1 - \mu_2 + \sigma_2^2 - \rho\sigma_1\sigma_2,$$

then Q_t satisfies

$$dQ_t = \mu_{12} Q_t dt + Q_t(\sigma_1 dW_t^1 - \sigma_2 dW_t^2). \tag{5.11.48}$$

We shall estimate in the following the drift rate μ_{12}. Integrating in the previous equation we have

$$Q_t = Q_0 + \mu_{12} \int_0^t Q_s\, ds + \sigma_1 \int_0^t Q_s\, dW_s^1 - \sigma_2 \int_0^t Q_s\, dW_s^2.$$

Taking the expectation and using that Ito integrals have zero mean we get

$$\mathbb{E}[Q_t] = Q_0 + \mu_{12} \int_0^t \mathbb{E}[Q_s]\, ds.$$

Let $f(t) = \mathbb{E}[Q_t]$. Then $f(t)$ satisfies the integral equation

$$f(t) = Q_0 + \mu_{12} \int_0^t f(s)\, ds,$$

which by differentiation becomes a linear differential equation

$$f'(t) = \mu_{12} f(t), \qquad f(0) = Q_0,$$

with the solution $f(t) = Q_0 e^{\mu_{12} t}$. Therefore, $\mathbb{E}[Q_t] = Q_0 e^{\mu_{12} t}$, which implies the following estimation for μ_{12} in terms of the expected exchange rate

$$\mu_{12} = \frac{1}{t} \ln \frac{\mathbb{E}[Q_t]}{Q_0}.$$

The expectation $\mathbb{E}[Q_t]$ is estimated by statistical methods applied to time series, for instance, using the *trend* feature.

However, the volatilities σ_i cannot be estimated from the process (5.11.48). In order to do this, we shall make the extra assumption (not very supportive for a global economy, though) that the noise in the Japanese and US markets are uncorrelated, i.e. $\rho = 0$. This implies that there is a Brownian motion W_t such that the exchange rate Q_t satisfies the geometric Brownian motion process

$$dQ_t = \mu_{12}Q_t dt + Q_t\sqrt{\sigma_1^2 + \sigma_2^2}dW_t.$$

In this case we can use historical data to estimate the volatility

$$\widehat{\sigma_{12}}^2 = \sigma_1^2 + \sigma_2^2.$$

However, this relation is not enough for estimating separately each volatility σ_i, $i = 1, 2$. However, this can be achieved in the case of more than two currencies, as we shall see next.

Consider three currencies, the Japanese yen, Y_t, the US dollar, U_t, and the Euro, $\mathcal{E}_t$, satisfying the geometric Brownian motion processes

$$
\begin{aligned}
dY_t &= \mu_1 Y_t dt + \sigma_1 Y_t dW_t^1 \\
dU_t &= \mu_2 U_t dt + \sigma_2 U_t dW_t^2, \\
d\mathcal{E}_t &= \mu_3 \mathcal{E}_t dt + \sigma_3 \mathcal{E}_t dW_t^3,
\end{aligned}
$$

with the zero correlation coefficients, $cor(dW_t^i, dW_t^j) = 0$. Consider the exchange rates

$$Q_t = \frac{Y_t}{U_t}, \qquad R_t = \frac{U_t}{\mathcal{E}_t}, \qquad T_t = \frac{\mathcal{E}_t}{Y_t}.$$

From the previous analysis we have

$$
\begin{aligned}
dQ_t &= \mu_{12}Q_t dt + \sigma_{12}Q_t dW_t \\
dR_t &= \mu_{23}R_t dt + \sigma_{23}R_t dW_t \\
dT_t &= \mu_{31}T_t dt + \sigma_{31}T_t dW_t.
\end{aligned}
$$

The exchange rates volatilities σ_{ij} can be estimated from historical data, and denote by $\widehat{\sigma_{ij}}$ their estimators. According to the previous analysis we have the estimations

$$
\begin{aligned}
\widehat{\sigma_{12}}^2 &= \sigma_1^2 + \sigma_2^2 \\
\widehat{\sigma_{23}}^2 &= \sigma_2^2 + \sigma_3^2 \\
\widehat{\sigma_{31}}^2 &= \sigma_3^2 + \sigma_1^2.
\end{aligned}
$$

This is a linear system with 3 equations and 3 unknowns, which can be solved for σ_j^2 in terms of $\widehat{\sigma_{ij}}^2$.

The next step is to estimate the drift rates μ_i. The drift rates μ_{ij} for the exchange rates can be estimated from the data, with estimators $\widehat{\mu_{ij}}$. These can be used to determine the difference between the drifts of the currencies as follows

$$
\begin{aligned}
\mu_1 - \mu_2 &= \widehat{\mu_{12}} - \sigma_2^2 \\
\mu_2 - \mu_3 &= \widehat{\mu_{23}} - \sigma_3^2 \\
\mu_3 - \mu_1 &= \widehat{\mu_{31}} - \sigma_1^2 .
\end{aligned}
$$

Unfortunately, this linear system does not have a unique solution; its solution depends on an additive benchmark. These relative differences are useful for comparing the drift rates between any two currencies without providing an absolute value for the μ_i. One of the currencies (the one with the lowest μ) can be chosen as the benchmark.

Chapter 6

Risk-Neutral Valuation

In this chapter we shall present and use the method of risk-neutral valuation for a large number of European options from simple to complex. The main assumption is that the interest rate and volatility are constant and the stock is log-normally distributed.

6.1 The Method of Risk-Neutral Valuation

This valuation method of derivatives is based on the *risk-neutral valuation principle*, which states that the price of a derivative on an asset S_t is not affected by the risk preference of the market participants. Consequently, when pricing derivatives, we may assume that the market participants share the same risk-aversion. In this case, the valuation of the price of an European type derivative[1], f_t, at time t will be done following a few steps. A complete explanation of this procedure will be provided later in Chapter 7. The risk-neutral valuation principle involves the following steps:

1. Assume the expected return of the asset S_t is the risk-free rate, $\mu = r$.

2. Calculate the conditional expectation of the payoff of the derivative as of time t, under condition 1.

3. Discount at the risk-free rate, r, from time T to time t.

The first two steps require considering the expectation as of time t in a risk-neutral world. This expectation will be denoted by $\widehat{\mathbb{E}}_t[\cdot]$ and has the meaning of a conditional expectation given simbolically by $E[\,\cdot\,|\mathcal{F}_t, \mu = r]$. The method states that if a derivative has the payoff f_T, then its price at any

---

[1]A derivative is of European type if can be exercised only at the expiration time.

time t prior to maturity T is given by

$$f_t = e^{-r(T-t)}\widehat{\mathbb{E}}_t[f_T].$$ (6.1.1)

How can one explain this formula? If the payoff f_T were deterministic, the price would be obtained by discounting to the time t as $e^{-r(T-t)}f_T$. Since this value is stochastic, we take the best approximation of it given the available information $\mathcal{F}_t$ in the market at time t in a risk-neutral world, which is $f_t = \mathbb{E}[e^{-r(T-t)}f_T \mid \mathcal{F}_t, \mu = r] = e^{-r(T-t)}\widehat{\mathbb{E}}_t[f_T]$.

We had considered the rate r constant, but the method can be easily adapted for time dependent rates as

$$f_t = \widehat{\mathbb{E}}_t[e^{-\int_t^T r(s)\, ds} f_T].$$

In the following we shall present explicit computations for the most common European type derivative prices using the risk-neutral valuation method, under the assumption that r is constant and the stock S_t follows a geometric Brownian motion with μ and σ constants.

Zero-Coupon Bond A contract that pays at maturity T the cash amount $f_T = V$ is a *zero-coupon bond* with face value V. Its value at any time t, prior to T, is given by

$$f_t = e^{-r(T-t)}\widehat{\mathbb{E}}_t[f_T] = e^{-r(T-t)}\widehat{\mathbb{E}}_t[V] = e^{-r(T-t)}V.$$ (6.1.2)

Log-contract A financial security that pays at maturity the amount $f_T = \ln S_T$ is called a *log-contract*. Here S_T is the stock value at time T. Using the log-normality of the stock

$$\ln S_T \sim \mathcal{N}\left(\ln S_t + (\mu - \frac{\sigma^2}{2})(T-t), \sigma^2(T-t)\right),$$

the risk-neutral expectation at time t is

$$\widehat{\mathbb{E}}_t[f_T] = \mathbb{E}[\ln S_T | \mathcal{F}_t, \mu = r] = \ln S_t + (r - \frac{\sigma^2}{2})(T-t).$$

Discounting we obtain the price of the log-contract

$$f_t = e^{-r(T-t)}\widehat{\mathbb{E}}_t[f_T] = e^{-r(T-t)}\left(\ln S_t + (r - \frac{\sigma^2}{2})(T-t)\right).$$ (6.1.3)

We note that log-contracts are not traded on a daily basis and we have included them here just for the sake of completeness.

Exercise 6.1.1 *Find the price at time t of a square log-contract, whose payoff is given by* $f_T = (\ln S_T)^2$.

Power-contract The financial derivative which pays at maturity the nth power of the stock price, S_T^n, is called a *power contract*. Since S_T has a lognormal distribution with

$$S_T = S_t e^{(\mu - \frac{\sigma^2}{2})(T-t) + \sigma(W_T - W_t)},$$

the nth power of the stock, S_T^n, is also log-normally distributed, with

$$g_T = S_T^n = S_t^n e^{n(\mu - \frac{\sigma^2}{2})(T-t) + n\sigma(W_T - W_t)}.$$

Then the expectation at time t in the risk-neutral world is

$$
\begin{aligned}
\widehat{\mathbb{E}}_t[g_T] &= \mathbb{E}[g_T|\mathcal{F}_t, \mu = r] = S_t^n e^{n(r - \frac{\sigma^2}{2})(T-t)} \mathbb{E}[e^{n\sigma(W_T - W_t)}|\mathcal{F}_t] \\
&= S_t^n e^{n(r - \frac{\sigma^2}{2})(T-t)} \mathbb{E}[e^{n\sigma(W_T - W_t)}] = S_t^n e^{n(r - \frac{\sigma^2}{2})(T-t)} \mathbb{E}[e^{n\sigma W_{T-t}}] \\
&= S_t^n e^{n(r - \frac{\sigma^2}{2})(T-t)} e^{\frac{1}{2}n^2\sigma^2(T-t)},
\end{aligned}
$$

where we took out the measurable part and used the stationarity property of W_t and that the jump of the Brownian motion $W_T - W_t$ is independent of the information $\mathcal{F}_t$.

The price of the power-contract is obtained by discounting to time t

$$
\begin{aligned}
g_t &= e^{-r(T-t)} \mathbb{E}[g_T|\mathcal{F}_t, \mu = r] = S_t^n e^{-r(T-t)} e^{n(r - \frac{\sigma^2}{2})(T-t)} e^{\frac{1}{2}n^2\sigma^2(T-t)} \\
&= S_t^n e^{(n-1)(r + \frac{n\sigma^2}{2})(T-t)}.
\end{aligned}
$$

Hence the value of a power contract at time t is given by the expression

$$\boxed{g_t = S_t^n e^{(n-1)(r + \frac{n\sigma^2}{2})(T-t)}.} \tag{6.1.4}$$

It is worth noting that if $n = 1$, i.e. if the payoff is $g_T = S_T$, then the price of the contract at any time $t \leq T$ is $g_t = S_t$, i.e. the stock price itself. This will be used shortly when valuing forward contracts.

In the case $n = 2$, i.e. if the contract pays S_T^2 at maturity, then the price is $g_t = S_t^2 e^{(r+\sigma^2)(T-t)}$.

Forward Contract on the Stock A *forward contract* pays at maturity the difference between the stock price, S_T, and the delivery price of the asset, K, i.e. the payoff is $f_T = S_T - K$. First, we compute the risk-neutral expectation of the stock

$$
\begin{aligned}
\widehat{\mathbb{E}}_t[S_T] &= \mathbb{E}[S_T | \mathcal{F}_t, \mu = r] \\
&= \mathbb{E}[S_t e^{(\mu - \frac{1}{2}\sigma^2)(T-t) + \sigma(W_T - W_t)} | \mathcal{F}_t, \mu = r] \\
&= S_t e^{(r - \frac{1}{2}\sigma^2)(T-t)} \mathbb{E}[e^{\sigma(W_T - W_t)} | \mathcal{F}_t] \\
&= S_t e^{(r - \frac{1}{2}\sigma^2)(T-t)} \mathbb{E}[e^{\sigma(W_T - W_t)}] = S_t e^{(r - \frac{1}{2}\sigma^2)(T-t)} \mathbb{E}[e^{\sigma W_{T-t}}] \\
&= S_t e^{(r - \frac{1}{2}\sigma^2)(T-t)} e^{\frac{1}{2}\sigma^2(T-t)} = S_t e^{r(T-t)}.
\end{aligned}
$$

Then the price of the forward contract at time t is

$$
\begin{aligned}
f_t &= e^{-r(T-t)} \widehat{\mathbb{E}}_t[S_T - K] = e^{-r(T-t)} \widehat{\mathbb{E}}_t[S_T] - e^{-r(T-t)} K \\
&= S_t - e^{-r(T-t)} K, \hspace{4cm} (6.1.5)
\end{aligned}
$$

where we used that K is a constant.

Remark 6.1.2 If a derivatives pays the stock S_T at time T, then the price of the derivative at time t is S_t.

Exercise 6.1.3 *Find the price of the power forward contract that pays at maturity $f_T = (S_T - K)^n$ for $n \in \{2, 3\}$.*

Call Option A *call option* is a contract which gives the buyer the right to buy the stock at time T for the price K. The time T is called *maturity time* or *expiration date* and K is called the *strike price*. It is worth noting that a call option is a *right* (not an obligation!) of the buyer, which means that if the price of the stock at maturity, S_T, is less than the strike price, K, then the buyer may choose not to exercise the option. If the price S_T exceeds the strike price K, then the buyer exercises the right to buy the stock, since he pays K dollars for something which worth S_T in the market. Buying at price K and selling at S_T yields a profit of $S_T - K$. Hence, the payoff is $f_T = \max(S_T - K, 0)$, see Fig.6.1 a. The price of the call at any prior time t, $0 \le t \le T$, is given by the expectation in the risk-neutral world as

$$
c(t) = \widehat{\mathbb{E}}_t[e^{-r(T-t)} f_T] = e^{-r(T-t)} \widehat{\mathbb{E}}_t[f_T]. \hspace{2cm} (6.1.6)
$$

If we let $x = \ln(S_T / S_t)$, using the log-normality of the stock price

$$
S_T = S_t e^{(\mu - \frac{\sigma^2}{2})(T-t) + \sigma(W_T - W_t)},
$$

it follows that x has the normal distribution

$$
x \sim \mathcal{N}\left((\mu - \frac{\sigma^2}{2})(T - t), \ \sigma^2(T - t) \right).
$$

Then the density function of x is given by

$$p(x) = \frac{1}{\sigma\sqrt{2\pi(T-t)}} e^{-\frac{\left(x - (\mu - \frac{\sigma^2}{2})(T-t)\right)^2}{2\sigma^2(T-t)}}.$$

The desity function corresponding to the risk neutral world is obtained replacing μ by r

$$\hat{p}(x) = \frac{1}{\sigma\sqrt{2\pi(T-t)}} e^{-\frac{\left(x - (r - \frac{\sigma^2}{2})(T-t)\right)^2}{2\sigma^2(T-t)}}.$$

We can write the expectation as

$$
\begin{aligned}
\widehat{\mathbb{E}}_t[f_T] &= \widehat{\mathbb{E}}_t[\max(S_T - K, 0)] = \widehat{\mathbb{E}}_t[\max(S_t e^x - K, 0)] \\
&= \int_{-\infty}^{\infty} \max(S_t e^x - K, 0)\hat{p}(x)\, dx = \int_{\ln(K/S_t)}^{\infty} (S_t e^x - K)\hat{p}(x)\, dx \\
&= I_2 - I_1,
\end{aligned}
\tag{6.1.7}
$$

with notations

$$I_1 = \int_{\ln(K/S_t)}^{\infty} K\hat{p}(x)\, dx, \qquad I_2 = \int_{\ln(K/S_t)}^{\infty} S_t e^x \hat{p}(x)\, dx.$$

With the substitution $y = \frac{x - (r - \frac{\sigma^2}{2})(T-t)}{\sigma\sqrt{T-t}}$, the first integral becomes

$$
\begin{aligned}
I_1 &= K \int_{\ln(K/S_t)}^{\infty} \hat{p}(x)\, dx = K \int_{-d_2}^{\infty} \frac{1}{\sqrt{2\pi}} e^{-\frac{y^2}{2}}\, dy \\
&= K \int_{-\infty}^{d_2} \frac{1}{\sqrt{2\pi}} e^{-\frac{y^2}{2}}\, dy = KN(d_2),
\end{aligned}
$$

where

$$d_2 = \frac{\ln(S_t/K) + (r - \frac{\sigma^2}{2})(T-t)}{\sigma\sqrt{T-t}},$$

and

$$N(u) = \frac{1}{\sqrt{2\pi}} \int_{-\infty}^{u} e^{-z^2/2}\, dz$$

denotes the standard normal distribution function.

Using the aforementioned substitution the second integral can be computed

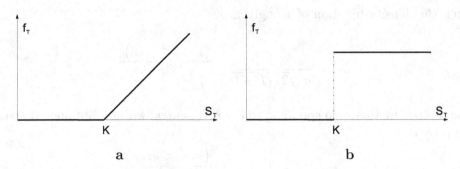

Figure 6.1: **a** *The payoff of a call option;* **b** *The payoff of a cash-or-nothing contract.*

by completing the square

$$
\begin{aligned}
I_2 &= S_t \int_{\ln(K/S_t)}^{\infty} e^x \hat{p}(x)\, dx = S_t \int_{-d_2}^{\infty} \frac{1}{\sqrt{2\pi}} e^{-\frac{1}{2}y^2 + y\sigma\sqrt{T-t} + (r-\frac{\sigma^2}{2})(T-t)}\, dx \\
&= S_t \int_{-d_2}^{\infty} \frac{1}{\sqrt{2\pi}} e^{-\frac{1}{2}(y-\sigma\sqrt{T-t})^2} e^{r(T-t)}\, dy \qquad (\text{let } z = y - \sigma\sqrt{T-t}) \\
&= S_t e^{r(T-t)} \int_{-d_2 - \sigma\sqrt{T-t}}^{\infty} \frac{1}{\sqrt{2\pi}} e^{-\frac{z^2}{2}}\, dz = S_t e^{r(T-t)} \int_{-\infty}^{d_1} \frac{1}{\sqrt{2\pi}} e^{-\frac{z^2}{2}}\, dz \\
&= S_t e^{r(T-t)} N(d_1),
\end{aligned}
$$

where

$$
d_1 = d_2 + \sigma\sqrt{T-t} = \frac{\ln(S_t/K) + (r + \frac{\sigma^2}{2})(T-t)}{\sigma\sqrt{T-t}}.
$$

Substituting back into (6.1.7) and then into (6.1.6) yields

$$
\begin{aligned}
c(t) &= e^{-r(T-t)}(I_2 - I_1) = e^{-r(T-t)}[S_t e^{r(T-t)} N(d_1) - K N(d_2)] \\
&= S_t N(d_1) - K e^{-r(T-t)} N(d_2).
\end{aligned}
$$

We have obtained the well-known formula of Black and Scholes:

Proposition 6.1.4 *The price of a European call option at time t is given by*

$$
\boxed{c(t) = S_t N(d_1) - K e^{-r(T-t)} N(d_2).}
$$

Exercise 6.1.5 *Show that*

(a) $\displaystyle \lim_{S_t \to \infty} \frac{c(t)}{S_t} = 1$;

(b) $\displaystyle \lim_{S_t \to 0} c(t) = 0$;

(c) $c(t) \sim S_t - K e^{-r(T-t)}$ *for S_t large.*

Exercise 6.1.6 *Show that* $\dfrac{dc(t)}{dS_t} = N(d_1)$. *This expression is called the delta of a call option.*

Exercise 6.1.7 *Let* $n \geq 1$ *be an integer. Find the price of a power call option whose payoff is given by* $f_T = \max(S_T^n - K, 0)$.

Cash-or-nothing contract A financial security that pays 1 dollar if the stock price $S_T \geq K$ and 0 otherwise, is called a *bet contract*, or a *cash-or-nothing contract*, see Fig.6.1 b. The payoff can be written as

$$f_T = \begin{cases} 1, & \text{if } S_T \geq K \\ 0, & \text{if } S_T < K. \end{cases}$$

Substituting $S_t = e^{X_t}$, the payoff becomes

$$f_T = \begin{cases} 1, & \text{if } X_T \geq \ln K \\ 0, & \text{if } X_T < \ln K, \end{cases}$$

where X_T has the normal distribution

$$X_T \sim N\left(\ln S_t + (\mu - \frac{\sigma^2}{2})(T - t), \sigma^2(T - t) \right).$$

The expectation in the risk-neutral world as of time t is

$$
\begin{aligned}
\widehat{\mathbb{E}}_t[f_T] &= \mathbb{E}[f_T | \mathcal{F}_t, \mu = r] = \int_{-\infty}^{\infty} f_T(x) p(x) \, dx \\
&= \int_{\ln K}^{+\infty} \frac{1}{\sqrt{2\pi}\sigma\sqrt{T-t}} e^{-\frac{[x - \ln S_t - (r - \frac{\sigma^2}{2})(T-t)]^2}{2\sigma^2(T-t)}} \, dx \\
&= \int_{-d_2}^{\infty} \frac{1}{\sqrt{2\pi}} e^{-\frac{y^2}{2}} \, dy = \int_{-\infty}^{d_2} \frac{1}{\sqrt{2\pi}} e^{-\frac{y^2}{2}} \, dy = N(d_2),
\end{aligned}
$$

where we used the substitution $y = \dfrac{x - \ln S_t - (r - \frac{\sigma^2}{2})(T-t)}{\sigma\sqrt{T-t}}$ and the notation

$$d_2 = \frac{\ln S_t - \ln K + (r - \frac{\sigma^2}{2})(T - t)}{\sigma\sqrt{T - t}}.$$

The price at time t of a bet contract is

$$\boxed{f_t = e^{-r(T-t)}\widehat{\mathbb{E}}_t[f_T] = e^{-r(T-t)} N(d_2).} \tag{6.1.8}$$

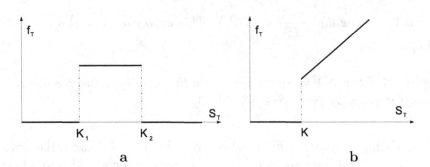

Figure 6.2: **a** *The payoff of a box-bet;* **b** *The payoff of an asset-or-nothing contract.*

Exercise 6.1.8 *Let* $0 < K_1 < K_2$. *Find the price of a financial derivative which pays at maturity* \$1 *if* $K_1 \leq S_t \leq K_2$ *and zero otherwise, see Fig.6.2 a. This is a "box-bet" and its payoff is given by*

$$f_T = \begin{cases} 1, & \text{if } K_1 \leq S_T \leq K_2 \\ 0, & \text{otherwise.} \end{cases}$$

Exercise 6.1.9 *An asset-or-nothing contract pays* S_T *if* $S_T > K$ *at maturity time* T, *and pays* 0 *otherwise, see Fig.6.2 b. Show that the price of the contract at time* t *is* $f_t = S_t N(d_1)$.

Exercise 6.1.10 *(a) Find the price at time* t *of a derivative which pays at maturity*

$$f_T = \begin{cases} S_T^n, & \text{if } S_T \geq K \\ 0, & \text{otherwise.} \end{cases}$$

(b) Show that the value of the contract can be written as

$$f_t = g_t N(d_2 + n\sigma\sqrt{T-t}),$$

where g_t *is the value at time* t *of a power contract given by (6.1.4).*
(c) Recover the result of Exercise 6.1.9 in the case $n = 1$.

6.2   The Superposition Principle and Applications

If the payoff of a derivative, f_T, can be written as a linear combination of n payoffs $h_{i,T}$ as

$$f_T = \sum_{i=1}^{n} c_i h_{i,T}$$

with c_i constants, then the price of the derivative at time t is given by

$$f_t = \sum_{i=1}^{n} c_i h_{i,t}$$

where $h_{i,t}$ is the price at time t of a derivative that pays at maturity $h_{i,T}$. We shall successfully use this method in the situation when the payoff f_T can be decomposed into simpler payoffs, for which we can evaluate directly the price of the associate derivative. In this case the price of the initial derivative, f_t, is obtained as a combination of the prices of the easier to valuate derivatives.

The reason underlying the aforementioned superposition principle is the linearity of the expectation operator $\widehat{\mathbb{E}}_t$:

$$
\begin{aligned}
f_t &= e^{-r(T-t)}\widehat{\mathbb{E}}_t[f_T] = e^{-r(T-t)}\widehat{\mathbb{E}}_t\Big[\sum_{i=1}^{n} c_i h_{i,T}\Big] \\
&= e^{-r(T-t)} \sum_{i=1}^{n} c_i \widehat{\mathbb{E}}_t[h_{i,T}] = \sum_{i=1}^{n} c_i h_{i,t}.
\end{aligned}
$$

This principle is also connected with the absence of arbitrage opportunities[2] in the market. Consider two portfolios of derivatives with equal values at the maturity time T

$$\sum_{i=1}^{n} c_i h_{i,T} = \sum_{j=1}^{m} a_j g_{j,T}.$$

If we take this common value to be the payoff of a derivative, f_T, then by the aforementioned principle, the portfolios have the same value at any prior time $t \leq T$

$$\sum_{i=1}^{n} c_i h_{i,t} = \sum_{j=1}^{m} a_j g_{j,t}.$$

The last identity also results from the absence of arbitrage opportunities. If there is a time t at which the identity fails, then buying the cheaper portfolio and selling the more expensive one will lead to an arbitrage profit.

The superposition principle can be used to price *package derivatives* such as spreads, straddles, strips, straps and strangles. We shall deal with these type of derivatives in the proposed exercises.

---

[2]An arbitrage opportunity deals with the practice of making profits by taking simultaneous long and short positions in the market.

General Contract on the Stock By a *general contract* on the stock we mean a derivative that pays at maturity the amount $g_T = G(S_T)$, where G is a given analytic function.

For instance, if $G(S_T) = S_T^n$, we have a power contract. If $G(S_T) = \ln S_T$, we have a log-contract. If choose $G(S_T) = S_T - K$, we obtain a forward contract. Since G is analytic we shall write $G(x) = \sum_{n \geq 0} c_n x^n$, with the coefficient $c_n = G^{(n)}(0)/n!$. Decomposing into power contracts and using the superposition principle we have

$$
\begin{aligned}
\widehat{\mathbb{E}}_t[g_T] &= \widehat{\mathbb{E}}_t[G(S_T)] = \widehat{\mathbb{E}}_t\Big[\sum_{n \geq 0} c_n S_T^n\Big] \\
&= \sum_{n \geq 0} c_n \widehat{\mathbb{E}}_t[S_T^n] \\
&= \sum_{n \geq 0} c_n S_t^n e^{n(r - \frac{\sigma^2}{2})(T-t)} e^{\frac{1}{2}n^2\sigma^2(T-t)} \\
&= \sum_{n \geq 0} c_n \Big[S_t e^{(r - \frac{\sigma^2}{2})(T-t)}\Big]^n \Big[e^{\frac{1}{2}\sigma^2(T-t)}\Big]^{n^2}.
\end{aligned}
$$

Hence, the value at time t of a general contract is given by the series

$$
g_t = e^{-r(T-t)} \sum_{n \geq 0} c_n \Big[S_t e^{(r - \frac{\sigma^2}{2})(T-t)}\Big]^n \Big[e^{\frac{1}{2}\sigma^2(T-t)}\Big]^{n^2}.
$$

Exercise 6.2.1 *Find the value at time t of an exponential contract that pays at maturity the amount:*

(a) $f_T = e^{S_T}$;

(b) $f_T = \cosh S_T$.

Call Option In the following we price a European call using the superposition principle. The payoff of a call option can be decomposed as

$$
c_T = \max(S_T - K, 0) = h_{1,T} - K h_{2,T},
$$

with

$$
h_{1,T} = \begin{cases} S_T, & \text{if } S_T \geq K \\ 0, & \text{if } S_T < K, \end{cases} \qquad h_{2,T} = \begin{cases} 1, & \text{if } S_T \geq K \\ 0, & \text{if } S_T < K. \end{cases}
$$

These are the payoffs of *asset-or-nothing* and of *cash-or-nothing* derivatives. From section 6.1 and Exercise 6.1.9 we have $h_{1,t} = S_t N(d_1)$, $h_{2,t} = e^{-r(T-t)} N(d_2)$. By superposition we get the price of a call at time t

$$
c_t = h_{1,t} - K h_{2,t} = S_t N(d_1) - K e^{-r(T-t)} N(d_2).
$$

Exercise 6.2.2 (Put option) (a) Consider the payoff $h_{1,T} = \begin{cases} 1, & \text{if } S_T \leq K \\ 0, & \text{if } S_T > K. \end{cases}$

Show that

$$h_{1,t} = e^{-r(T-t)} N(-d_2), \qquad t \leq T.$$

(b) Consider the payoff $h_{2,T} = \begin{cases} S_T, & \text{if } S_T \leq K \\ 0, & \text{if } S_T > K. \end{cases}$ Show that

$$h_{2,t} = S_t N(-d_1), \qquad t \leq T.$$

(c) The payoff of a put is $p_T = \max(K - S_T, 0)$. Verify that

$$p_T = K h_{1,T} - h_{2,T}$$

and use the superposition principle to find the price p_t of a put as

$$p_t = K e^{-r(T-t)} N(-d_2) - S N(-d_1).$$

Put-Call Parity This is a relation between the value of a call, a put, and the underlying asset. Consider the payoffs of a call and of a put

$$c_T = \max\{S_T - K, 0\}, \qquad p_T = \max\{K - S_T, 0\},$$

and consider a derivative whose payoff is

$$f_T = c_T - p_T.$$

By the superposition principle, at any time $t \leq T$, we have the relation

$$f_t = c_t - p_t. \tag{6.2.9}$$

A computation shows that

$$f_T = \max\{S_T - K, 0\} - \max\{K - S_T, 0\} = S_T - K,$$

which is the payoff of a forward contract. Since the value of a forward contract at time t is given by (6.1.5) as

$$f_t = S_t - K e^{-r(T-t)},$$

it follows that relation (6.2.9) becomes

$$S_t - K e^{-r(T-t)} = c_t - p_t.$$

This can be written equivalently as

$$\boxed{c_t + K e^{-r(T-t)} = p_t + S_t,} \tag{6.2.10}$$

and is called the *put-call parity*. It states that the value of a put plus the price of the underlying stock is the same as the value of a call plus the discounted value of a cash payment K payable at time T. This relation can be used to find the price of a put from the price of the corresponding call and vice-versa.

Exercise 6.2.3 *Find the price of a put using the price of a call and the put-call parity.*

General Options on the Stock Let G be an increasing analytic function with the inverse G^{-1} and K be a positive constant. A *general call option* is a contract with the payoff

$$f_T = \begin{cases} G(S_T) - K, & \text{if } S_T \geq G^{-1}(K) \\ 0, & \text{otherwise.} \end{cases} \tag{6.2.11}$$

We note the payoff function f_T is continuous. We shall work out the value of the contract at time t, f_t, using the superposition method. Since the payoff can be written as the linear combination

$$f_T = h_{1,T} - K h_{2,T},$$

with

$$h_{1,T} = \begin{cases} G(S_T), & \text{if } S_T \geq G^{-1}(K) \\ 0, & \text{otherwise,} \end{cases} \qquad h_{2,T} = \begin{cases} 1, & \text{if } S_T \geq G^{-1}(K) \\ 0, & \text{otherwise,} \end{cases}$$

then

$$f_t = h_{1,t} - K h_{2,t}. \tag{6.2.12}$$

We had already computed the value $h_{2,t}$. In this case we have $h_{2,t} = e^{-r(T-t)} N(d_2^G)$, where d_2^G is obtained by replacing K with $G^{-1}(K)$ in the formula of d_2

$$d_2^G = \frac{\ln S_t - \ln\left(G^{-1}(K)\right) + (r - \frac{\sigma^2}{2})(T - t)}{\sigma\sqrt{T - t}}. \tag{6.2.13}$$

We shall compute in the following $h_{1,t}$. Let $G(S_T) = \sum_{n \geq 0} c_n S_T^n$ with $c_n = G^{(n)}(0)/n!$. Then $h_{1,T} = \sum_{n \geq 0} c_n f_T^{(n)}$, where

$$f_T^{(n)} = \begin{cases} S_T^n, & \text{if } S_T \geq G^{-1}(K) \\ 0, & \text{otherwise.} \end{cases}$$

By Exercise 6.1.10 the price at time t for a contract with the payoff $f_T^{(n)}$ is

$$f_t^{(n)} = S_t^n e^{(n-1)(r + \frac{n\sigma^2}{2})(T-t)} N(d_2^G + n\sigma\sqrt{T - t}).$$

The value at time t of $h_{1,T}$ is given by the series

$$\begin{aligned} h_{1,t} &= \sum_{n \geq 0} c_n f_t^{(n)} \\ &= \sum_{n \geq 0} c_n S_t^n e^{(n-1)(r + \frac{n\sigma^2}{2})(T-t)} N(d_2^G + n\sigma\sqrt{T - t}). \end{aligned}$$

Substituting in (6.2.12) we obtain the following value at time t of a general call option with the payoff (6.2.11)

$$f_t = \sum_{n \geq 0} c_n S_t^n e^{(n-1)(r+\frac{n\sigma^2}{2})(T-t)} N(d_2^G + n\sigma\sqrt{T-t}) - Ke^{-r(T-t)} N(d_2^G),$$

$$(6.2.14)$$

with $c_n = \frac{G^{(n)}}{n!}$ and d_2^G given by (6.2.13).

It is worth noting that in the case $G(S_T) = S_T$ we have $n = 1$, $d_2 = d_2^G$, $d_1 = d_2 + \sigma(T-t)$ and formula (6.2.14) becomes the value of a plain vanilla call option

$$f_t = S_t N(d_1) - Ke^{-r(T-t)}.$$

6.3 Packages

Packages are derivatives whose payoffs are linear combinations of payoffs of options, cash and underlying asset. They can be priced using the superposition principle. Some of these packages are used in hedging or speculation.

The Bull Spread Let $0 < K_1 < K_2$. A derivative with the payoff

$$f_T = \begin{cases} 0, & \text{if } S_T \leq K_1 \\ S_T - K_1, & \text{if } K_1 < S_T \leq K_2 \\ K_2 - K_1, & \text{if } K_2 < S_T \end{cases}$$

is called a bull spread, see Fig.6.3 a. A market participant should enter a *bull spread* position when the stock price is expected to increase. The payoff f_T can be written as the difference of the payoffs of two calls with strike prices K_1 and K_2:

$$f_T = c_1(T) - c_2(T),$$

where

$$c_1(T) = \begin{cases} 0, & \text{if } S_T \leq K_1 \\ S_T - K_1, & \text{if } K_1 < S_T \end{cases} \qquad c_2(T) = \begin{cases} 0, & \text{if } S_T \leq K_2 \\ S_T - K_2, & \text{if } K_2 < S_T. \end{cases}$$

Using the superposition principle, the price of a bull spread at time t is

$$\begin{aligned} f_t &= c_1(t) - c_2(t) \\ &= S_t N(d_1(K_1)) - K_1 e^{-r(T-t)} N(d_2(K_1)) \\ &\quad - \left(S_t N(d_1(K_2)) - K_2 e^{-r(T-t)} N(d_2(K_2)) \right) \\ &= S_t [N(d_1(K_1)) - N(d_1(K_2))] - e^{-r(T-t)} [K_1 N(d_2(K_1)) - K_2 N(d_2(K_2))], \end{aligned}$$

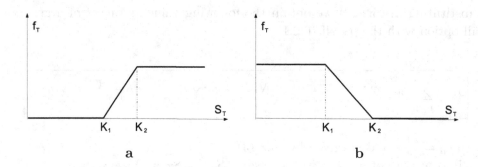

Figure 6.3: **a** *The payoff of a bull spread;* **b** *The payoff of a bear spread.*

with

$$d_2(K_i) = \frac{\ln S_t - \ln K_i + (r - \frac{\sigma^2}{2})(T - t)}{\sigma\sqrt{T - t}}, \quad d_1(K_i) = d_2(K_i) + \sigma\sqrt{T - t}, \ i = 1, 2.$$

The Bear Spread Let $0 < K_1 < K_2$. A derivative with the payoff

$$f_T = \begin{cases} K_2 - K_1, & \text{if } S_T \leq K_1 \\ K_2 - S_T, & \text{if } K_1 < S_T \leq K_2 \\ 0, & \text{if } K_2 < S_T \end{cases}$$

is called a *bear spread*, see Fig.6.3 b. A long position in this derivative leads to profits when the stock price is expected to decrease.

Exercise 6.3.1 *Find the price of a bear spread at time t, with $t < T$.*

The Butterfly Spread Let $0 < K_1 < K_2 < K_3$, with $K_2 = (K_1 + K_3)/2$. A *butterfly spread* is a derivative with the payoff given in Fig.6.4 a

$$f_T = \begin{cases} 0, & \text{if } S_T \leq K_1 \\ S_T - K_1, & \text{if } K_1 < S_T \leq K_2 \\ K_3 - S_T, & \text{if } K_2 \leq S_T < K_3 \\ 0, & \text{if } K_3 \leq S_T. \end{cases}$$

A short position in a butterfly spread leads to profits when a small move in the stock price occurs.

Exercise 6.3.2 *Find the price of a butterfly spread at time t, with $t < T$.*

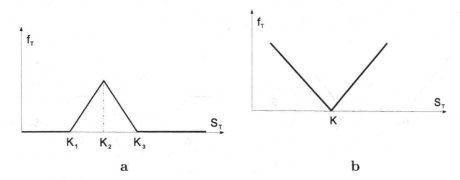

Figure 6.4: **a** *The payoff of a butterfly spread (long position);* **b** *The payoff of a straddle.*

Straddles A derivative with the payoff $f_T = |S_T - K|$ is called a *straddle*, see Fig.6.4 b. A long position in a straddle leads to profits when a move in any direction of the stock price occurs.

Exercise 6.3.3 *(a) Show that the payoff of a straddle can be written as*

$$f_T = \begin{cases} K - S_T, & \text{if } S_T \leq K \\ S_T - K, & \text{if } K < S_T. \end{cases}$$

(b) Find the price of a straddle at time t, with $t < T$.

Strangles Let $0 < K_1 < K_2$ and $K = (K_2 + K_1)/2$, $K' = (K_2 - K_1)/2$. A derivative with the payoff $f_T = \max(|S_T - K| - K', 0)$ is called a *strangle*, see Fig.6.5 a. A long position in a strangle leads to profits when a large move in any direction of the stock price occurs.

Exercise 6.3.4 *(a) Show that the payoff of a strangle can be written as*

$$f_T = \begin{cases} K_1 - S_T, & \text{if } S_T \leq K_1 \\ 0, & \text{if } K_1 < S_T \leq K_2 \\ S_T - K_2, & \text{if } K_2 \leq S_T. \end{cases}$$

(b) Find the price of a strangle at time t, with $t < T$;
(c) Which is cheaper: a straddle or a strangle?

Exercise 6.3.5 *Let $0 < K_1 < K_2 < K_3 < K_4$, with $K_4 - K_3 = K_2 - K_1$. Find the value at time t of a derivative with the payoff f_T given in Fig.6.5 b.*

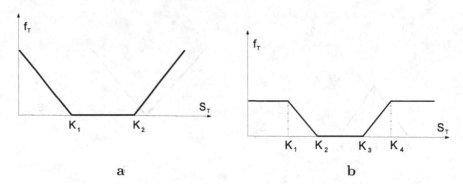

Figure 6.5: **a** *The payoff for a strangle;* **b** *The payoff for a condor.*

6.4   Strike Sensitivity

In this section we consider a few European options whose payoffs depend on the strike price K, and whose derivatives with respect to K are easily computable. Then using the risk-neutral valuation we obtain relations between the sensitivities of the option price with respect to K and the probability density of the stock.

Cash-or-nothing option In this case the derivative is a bet on the stock, with the payoff given by

$$f_T(S) = \mathcal{H}(S - K) = \begin{cases} 1, & \text{if } S \geq K \\ 0, & \text{otherwise,} \end{cases}$$

where $\mathcal{H}(x) = \begin{cases} 0, & \text{if } x < 0 \\ 1, & \text{if } x \geq 0 \end{cases}$ is the unit jump at 0, called the *Heaviside function*. Its main relation with the Dirac delta function is given by $\mathcal{H}'(x) = \delta(x)$. This identity does not hold in the classical sense, but in a generalized sense, which we shall not get into details here. Heuristically, one may notice that $\mathcal{H}'(x)$ is zero for $x \neq 0$ and infinity at $x = 0$, which suggests the "spike" shape of $\mathcal{H}'(x)$.

Differentiating with respect to K we obtain

$$\frac{\partial f_T}{\partial K} = -\mathcal{H}'(S - K) = -\delta(S - K).$$

The price of the option from the risk-neutral valuation is

$$f_t = e^{-r(T-t)}\widehat{\mathbb{E}}_t[f_T] = e^{-r(T-t)} \int_{\mathbb{R}} f_T(x)p(x)\,dx,$$

where $p(x) = p(x|S_t = S)$ is the probability density of the stock conditioned by $S_t = S$. Interchanging the derivative with the integral, we compute the sensitivity of the price f_t with respect to strike as

$$\frac{\partial f_t}{\partial K} = e^{-r(T-t)} \int_{\mathbb{R}} \frac{\partial f_T(x)}{\partial K} p(x)\, dx$$

$$= -e^{-r(T-t)} \int_{\mathbb{R}} \delta(x - K) p(x)\, dx = -e^{-r(T-t)} p(K),$$

where we used the well-known formula

$$\int \delta(x - a) p(x)\, dx = p(a), \qquad \forall a \in \mathbb{R}.$$

Solving for the density, we have

$$\boxed{p(K) = -e^{r(T-t)} \frac{\partial f_t}{\partial K}.} \qquad (6.4.15)$$

This formula states that the underlying stock probability density evaluated at the strike, K, is proportional with the rate of change of the derivative price with respect to K.

Remark 6.4.1 It is worth noting that substituting the formula for f_t given by (6.1.8)

$$f_t = e^{-r(T-t)} N(d_2)$$

into equation (6.4.15) yields

$$p(K) = -\frac{\partial}{\partial K} N(d_2). \qquad (6.4.16)$$

This formula can be also verified directly computing the derivative of the right term in two ways: directly by brute force, or using the interpretation of $N(d_2)$ as a probability. For pedagogical reasons we shall show both ways.

(*i*) We note first that

$$\frac{\partial d_2}{\partial K} = \frac{\partial}{\partial K} \left(\frac{\ln S - \ln K + (r - \sigma^2/2)(T - t)}{\sigma\sqrt{T - t}} \right) = -\frac{1}{K\sigma\sqrt{T - t}},$$

so the chain rule provides

$$\frac{\partial}{\partial K} N(d_2) = N'(d_2) \frac{\partial d_2}{\partial K} = -\frac{1}{\sqrt{2\pi}} e^{-\frac{d_2^2}{2}} \frac{1}{K\sigma\sqrt{T - t}}$$

$$= -\frac{1}{\sigma K\sqrt{2\pi(T - t)}} e^{-\frac{(\ln S - \ln K + (r - \sigma^2/2)(T - t))^2}{2\sigma^2(T-t)}}.$$

Substituting into (6.4.16) yields

$$p(K) = \frac{1}{\sigma K \sqrt{2\pi(T-t)}} e^{-\frac{(\ln S - \ln K + (r - \sigma^2/2)(T-t))^2}{2\sigma^2(T-t)}}. \tag{6.4.17}$$

This formula recovers the log-normal distribution of a stock evaluated at K, conditioned by $S_t = S$.

(ii) Another way of direct verification for formula (6.4.16) is to use the probabilistic interpretation of $N(d_2)$, see Proposition 6.9.1 part (b). Then

$$N(d_2) = P(S_T \geq K) = 1 - P(S_T < K) = 1 - F(K),$$

where $F(x) = F(x|S_t = S)$ is the cumulative distribution function of the stock, given the initial condition $S_t = S$. Differentiating and using the relation between cumulative functions and probability densities, we get

$$\frac{\partial}{\partial K} N(d_2) = -F'(K) = -p(K),$$

which recovers (6.4.16).

Call option In this case the payoff is

$$f_T(S) = \begin{cases} S - K, & \text{if } S \geq K \\ 0, & \text{if } S < K \end{cases}$$

with the first two derivatives with respect to K given by

$$\frac{\partial f_T}{\partial K} = \begin{cases} -1, & \text{if } S \geq K \\ 0, & \text{if } S < K \end{cases} \quad \text{and} \quad \frac{\partial^2 f_T}{\partial K^2} = \delta(S - K).$$

The value of a call at time t is given by the risk-neutral valuation as

$$c(t, s) = e^{-r(T-t)} \int_{\mathbb{R}} f_T(x) p(x) \, dx.$$

Differentiating twice in K and interchanging the derivatives with the integral provides

$$\begin{aligned}
\frac{\partial^2 c}{\partial K^2} &= e^{-r(T-t)} \int_{\mathbb{R}} \frac{\partial^2 f_T(x)}{\partial K^2} p(x) \, dx \\
&= e^{-r(T-t)} \int_{\mathbb{R}} \delta(x - K) p(x) \, dx = e^{-r(T-t)} p(K).
\end{aligned}$$

Hence

$$\boxed{p(K) = e^{r(T-t)} \frac{\partial^2 c}{\partial K^2}.} \tag{6.4.18}$$

We note that if $c = c(t, S, K)$ is given by the Black-Scholes formula, then $p(K)$ is the log-normal density (6.4.17). In addition to its theoretical value, formula (6.4.18) has also a practical value, since the curvature of the observed call data prices provides the empirical density of the underlying stock.

Exercise 6.4.2 *Prove relation* (6.4.18) *directly by differentiating in the Black-Scholes formula*

$$c(t, S, K) = SN(d_1) - Ke^{-r(T-t)}N(d_2).$$

Exercise 6.4.3 *Find a similar relation to* (6.4.18) *involving the price of a put in the following ways:*

(a) starting from the put call parity;

(b) by a similar procedure involving differentiation in the risk-neutral formula.

Exercise 6.4.4 *Prove a relation analog to* (6.4.18) *for a European option with the payoff*

$$f_T(S) = \begin{cases} (S - K)^2, & \text{if } S \geq K \\ 0, & \text{otherwise.} \end{cases}$$

6.5 Volatility Sensitivity

This section deals with the sensitivity of a call or put with respect to volatility and presents some consequences of this computation. Differentiating in the put-call parity relation

$$c + Ke^{-r(T-t)} = p + S$$

with respect to the volatility parameter σ, we obtain

$$\frac{\partial c}{\partial \sigma} = \frac{\partial p}{\partial \sigma}.$$

Hence, both calls and puts have the same sensitivities with respect to σ, while all other parameters, S, K, r, and $T - t$ are held fixed. This rate of change is called the *vega sensitivity* and will be denoted by $\mathcal{V}$.

Proposition 6.5.1 *The vega sensitivity of a call (or put) is given by*

$$\mathcal{V} = S\sqrt{T - t}N'(d_1),$$

where $N'(x) = \dfrac{1}{\sqrt{2\pi}}e^{-x^2/2}$.

Proof: Differentiating in the Black-Scholes formula of a call, and using that $d_1 = d_2 + \sigma\sqrt{T-t}$, we have

$$
\begin{aligned}
\mathcal{V} &= \frac{\partial c}{\partial \sigma} = SN'(d_1)d_1' - Ke^{-r(T-t)}N'(d_2)d_2' \\
&= SN'(d_1)(d_2' + \sqrt{T-t}) - Ke^{-r(T-t)}N'(d_2)d_2' \\
&= S\sqrt{T-t}N'(d_1) + d_2'\left[SN'(d_1) - Ke^{-r(T-t)}N'(d_2)\right].
\end{aligned}
$$

It suffices to show that the expression in the brackets vanishes. This is provided by the following algebraic computation:

$$
\begin{aligned}
SN'(d_1) - Ke^{-r(T-t)}N'(d_2) &= \frac{1}{\sqrt{2\pi}}\left\{ Se^{-\frac{(d_2+\sigma\sqrt{T-t})^2}{2}} - Ke^{-r(T-t)}e^{-\frac{d_2^2}{2}} \right\} \\
&= \frac{1}{\sqrt{2\pi}}\left\{ Se^{-\frac{d_2^2}{2}}e^{-\frac{\sigma^2(T-t)}{2}}e^{-\sigma\sqrt{T-t}\,d_2} - Ke^{-r(T-t)}e^{-\frac{d_2^2}{2}} \right\} \\
&= \frac{1}{\sqrt{2\pi}}e^{-\frac{d_2^2}{2}}\left\{ Se^{-\frac{\sigma^2(T-t)}{2}}e^{-\ln\frac{S}{K}}e^{-\left(r-\frac{\sigma^2}{2}\right)(T-t)} - Ke^{-r(T-t)} \right\} \\
&= \frac{K}{\sqrt{2\pi}}e^{-\frac{d_2^2}{2}}\left\{ e^{-\frac{\sigma^2(T-t)}{2}}e^{-r(T-t)}e^{\frac{\sigma^2(T-t)}{2}} - e^{-r(T-t)} \right\} \\
&= 0.
\end{aligned}
$$

∎

6.6 Implied Volatility

One important consequence of the previous result is the existence of the *implied volatility*. Considering the variables S, K, r, and $\tau = T - t$ fixed, the call price given by the Black-Scholes formula becomes a function of σ only, $c = c(\sigma)$. Since

$$
c'(\sigma) = \frac{S\sqrt{T-t}}{\sqrt{2\pi}}e^{-d_1^2/2} > 0,
$$

the call price is an increasing function of σ, and hence it is one-to-one. Then it makes sense to consider the inverse function

$$
\sigma_{imp}^{call} = \sigma(c),
$$

called the *implied volatility from the call prices.* It is worth noting that the function σ_{imp}^{call} is also increasing with respect to the call price, c, having the rate of change

$$
\sigma_{imp}^{call\,\prime}(c) = \frac{\sqrt{2\pi}}{S\sqrt{T-t}}e^{d_1^2/2} > 0,
$$

where d_1 is considered as a function of c.

A similar analysis can be applied for the put. In this case, the function $p = p(\sigma)$ is increasing, with the inverse

$$\sigma_{imp}^{put} = \sigma(c)$$

called the *implied volatility from the put prices*.

It makes sense to ask whether the two kind of implied volatilities are actually equal. The next result is a consequence of the put-call parity.

Proposition 6.6.1 *The implied volatility from the call price is equal to the implied volatility from the put price, i.e.*

$$\sigma_{imp}^{call} = \sigma_{imp}^{put},$$

as long as both options are written on the same stock, have the same strike and time to maturity, and the market conditions required by the Black-Scholes analysis are satisfied.

Proof: Denote by d_j^{call} the value of d_j, where σ is replaced by σ_{imp}^{call}. A similar notation also applies to d_j^{put}, $j = 1, 2$. Substituting the Black-Scholes prices for the call and put

$$
\begin{aligned}
c &= SN(d_1^{call}) - Ke^{-r(T-t)}N(d_2^{call}) \\
p &= Ke^{-r(T-t)}N(-d_2^{put}) - SN(d_1^{put})
\end{aligned}
$$

into the put call-parity relation

$$c - p = S - Ke^{-r(T-t)},$$

and using that $N(-d_j^{put}) = 1 - N(d_j^{put})$, after simplifying terms we obtain

$$S[N(d_1^{call}) - N(d_1^{put})] + Ke^{-r(T-t)}[N(d_2^{call}) - N(d_2^{put})] = 0.$$

Since this relation holds for any value of the stock, S, the bracket coefficients must vanish

$$
\begin{aligned}
N(d_1^{call}) &= N(d_1^{put}) \\
N(d_2^{call}) &= N(d_2^{put}).
\end{aligned}
$$

Since the normal cumulative function is one-to-one, it follows that

$$
\begin{aligned}
d_1^{call} &= d_1^{put} \\
d_2^{call} &= d_2^{put}.
\end{aligned}
$$

Subtracting, and using the relation $d_1 - d_2 = \sigma\sqrt{T-t}$, yields $\sigma_{imp}^{call} = \sigma_{imp}^{put}$.

∎

Hence, it makes sense to talk about the *implied volatility*,

$$\sigma_{imp} = \sigma_{imp}^{call} = \sigma_{imp}^{put},$$

without specifying whether it was implied from the calls or from the puts.

Volatility smile The implied volatility is a result of a market consensus regarding the prices of traded options. This means that, if at a given time to maturity, one measures the price of two calls (puts) with two distinct strikes, then their implied volatilities should be the same. However, empirical evidence shows that this is not the case. It can be seen that when plotting implied volatilities, σ_{imp}, versus strike price, K, the plots usually lie on a convex curve, referred to as the *volatility smile*. This is strong evidence that the stock price does not have a perfect log-normal distribution. This contradiction screams for the need of better models, which can account for volatility smiles. A great success in this direction has been accomplished by stochastic volatility models, which will be covered in the last part of the book, in Chapters 11, 12, and 13.

Exercise 6.6.2 *Starting from the Black-Scholes formula for a put, compute the vega sensitivity* $\dfrac{\partial p}{\partial \sigma}$.

Remark 6.6.3 A similar computation shows that the formula for the vega for a divident-paying stock is $\mathcal{V} = S\sqrt{T-t}N'(d_1)e^{-\delta(T-t)}$, where δ denotes the continuous dividend rate.

6.7   Asian Forward Contracts

In the following we shall compute the value of a few Asian derivatives, i.e. contracts which depend on the average value of the stock price during the life span of the contract.

Forward Contracts on the Arithmetic Average Let $A_T = \dfrac{1}{T}\displaystyle\int_0^T S_u\,du$ denote the continuous arithmetic average of the asset price between 0 and T. It makes sense sometimes for two parties to make a contract in which one party pays the other at the delivery time, T, the difference between the average price of the asset, A_T, and a fixed delivery price K. The payoff for a forward contract on the arithmetic average is

$$\boxed{f_T = A_T - K.} \tag{6.7.19}$$

For instance, if the asset is natural gas and the delivery time is in the winter, then it makes sense to make a deal on the average price of the asset,

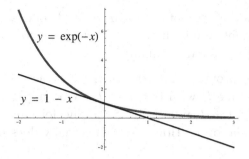

Figure 6.6: *The inequality $e^{-x} \geq 1 - x$.*

rather than on the asset itself, since the price is volatile and can become expensive during the cold season.

Since the risk-neutral expectation at time $t = 0$, $\widehat{\mathbb{E}}_0[A_T]$, is obtained by replacing μ by r in the formula of $\mathbb{E}[A_T]$, the price of the forward contract at $t = 0$ is obtained as

$$
\begin{aligned}
f_0 &= e^{-rT}\widehat{\mathbb{E}}_0[f_T] = e^{-rT}(\widehat{\mathbb{E}}_0[A_T] - K) \\
&= e^{-rT}\left(S_0\frac{e^{rT} - 1}{rT} - K\right) = S_0\frac{1 - e^{-rT}}{rT} - e^{-rT}K \\
&= \frac{S_0}{rT} - e^{-rT}\left(\frac{S_0}{rT} + K\right).
\end{aligned}
$$

Hence, the value of a forward contract on the arithmetic average at time $t = 0$ is given by

$$
\boxed{f_0 = \frac{S_0}{rT} - e^{-rT}\left(\frac{S_0}{rT} + K\right).} \tag{6.7.20}
$$

It is worth noting that the price of a forward contract on the arithmetic average is cheaper than the price of a usual forward contract on the asset. This makes sense heuristically, since the average A_t is less volatile than the stock itself; therefore the contract written on A_t will involve less risk than the one written on S_t, and hence it will be cheaper, and hence, more appealing to the buyer. To see this mathematically, we substitute $x = rT$ in the inequality $e^{-x} > 1 - x$, for $x > 0$, see Fig.6.6, to get

$$
\frac{1 - e^{-rT}}{rT} < 1.
$$

This implies the inequality

$$
S_0\frac{1 - e^{-rT}}{rT} - e^{-rT}K < S_0 - e^{-rT}K.
$$

Since the left side is the price of an Asian forward contract on the arithmetic average, while the right side is the price of a forward contract on a stock, see (6.1.5), we obtain the desired inequality.

Formula (6.7.20) provides the price of the contract at time $t = 0$. What is the price at any time t, with $0 < t < T$? One might be tempted to say that replacing T by $T - t$ and S_0 by S_t in the formula of f_0 leads to the corresponding formula for f_t. However, this actually does not happen, as the next result shows:

Proposition 6.7.1 *The value at time t of a contract that pays at maturity $f_T = A_T - K$ is given by*

$$f_t = e^{-r(T-t)}\left(\frac{t}{T}A_t - K\right) + \frac{1}{rT}S_t\left(1 - e^{-r(T-t)}\right). \tag{6.7.21}$$

Proof: We start by computing the risk-neutral expectation $\widehat{\mathbb{E}}_t[A_T]$. Splitting the integral into two parts, we have

$$
\begin{aligned}
\widehat{\mathbb{E}}_t[A_T] &= \widehat{\mathbb{E}}\left[\frac{1}{T}\int_0^T S_u\,du\Big|\mathcal{F}_t\right] \\
&= \widehat{\mathbb{E}}\left[\frac{1}{T}\int_0^t S_u\,du + \frac{1}{T}\int_t^T S_u\,du\Big|\mathcal{F}_t\right] \\
&= \frac{1}{T}\int_0^t S_u\,du + \widehat{\mathbb{E}}\left[\frac{1}{T}\int_t^T S_u\,du\Big|\mathcal{F}_t\right] \\
&= \frac{1}{T}\int_0^t S_u\,du + \frac{1}{T}\int_t^T \widehat{\mathbb{E}}[S_u|\mathcal{F}_t]\,du \\
&= \frac{1}{T}\int_0^t S_u\,du + \frac{1}{T}\int_t^T S_t e^{r(u-t)}\,du \\
&= \frac{t}{T}A_t + \frac{1}{T}S_t e^{-rt}\frac{e^{rT} - e^{rt}}{r} \\
&= \frac{t}{T}A_t + \frac{1}{rT}S_t\left(e^{r(T-t)} - 1\right). \tag{6.7.22}
\end{aligned}
$$

Using (6.7.22), the risk-neutral valuation provides

$$
\begin{aligned}
f_t &= e^{-r(T-t)}\widehat{\mathbb{E}}_t[A_T - K] = e^{-r(T-t)}\widehat{\mathbb{E}}_t[A_T] - e^{-r(T-t)}K \\
&= e^{-r(T-t)}\left(\frac{t}{T}A_t - K\right) + \frac{1}{rT}S_t\left(1 - e^{-r(T-t)}\right).
\end{aligned}
$$

∎

Exercise 6.7.2 *Show that* $\lim_{t\to 0} f_t = f_0$, *where f_t is given by (6.7.21) and f_0 by (6.7.20).*

Exercise 6.7.3 *Find the value at time t of a contract that pays at maturity the difference between the asset price and its arithmetic average, $f_T = S_T - A_T$.*

Forward Contracts on the Geometric Average We shall consider in the following Asian forward contracts on the geometric average. This is a derivative that pays at maturity the difference $f_T = G_T - K$, where G_T is the continuous geometric average of the asset price between 0 and T and K is a fixed delivery price.

We shall work out first the value of the contract at $t = 0$. Substituting $\mu = r$ in the first relation provided by Theorem 5.8.5, the risk-neutral expectation of G_T as of time $t = 0$ is

$$\widehat{\mathbb{E}}_0[G_T] = S_0 e^{\frac{1}{2}(r - \frac{\sigma^2}{6})T}.$$

Then

$$
\begin{aligned}
f_0 &= e^{-rT}\widehat{\mathbb{E}}_0[G_T - K] = e^{-rT}\widehat{\mathbb{E}}_0[G_T] - e^{-rT}K \\
&= e^{-rT}S_0 e^{\frac{1}{2}(r - \frac{\sigma^2}{6})T} - e^{-rT}K \\
&= S_0 e^{-\frac{1}{2}(r + \frac{\sigma^2}{6})T} - e^{-rT}K.
\end{aligned}
$$

Thus, the price of a forward contract on the geometric average at $t = 0$ is given by

$$\boxed{f_0 = S_0 e^{-\frac{1}{2}(r + \frac{\sigma^2}{6})T} - e^{-rT}K.} \tag{6.7.23}$$

As in the case of forward contracts on arithmetic average, the value at any time $0 < t < T$ cannot be obtained from (6.7.23) by replacing blindly T and S_0 by $T - t$ and S_t, respectively. The correct relation is given by the following result:

Proposition 6.7.4 *The value at time t of a contract which pays at maturity $G_T - K$ is*

$$\boxed{f_t = G_t^{\frac{t}{T}} S_t^{1 - \frac{t}{T}} e^{-r(T-t) + (r - \frac{\sigma^2}{2})\frac{(T-t)^2}{2T} + \frac{\sigma^2}{T^2}\frac{(T-t)^3}{6}} - e^{-r(T-t)}K.} \tag{6.7.24}$$

Proof: Since for $t < u$

$$\ln S_u = \ln S_t + (\mu - \frac{\sigma^2}{2})(u - t) + \sigma(W_u - W_t),$$

we have

$$\int_0^T \ln S_u \, du = \int_0^t \ln S_u \, du + \int_t^T \ln S_u \, du$$

$$= \int_0^t \ln S_u \, du + \int_t^T \left(\ln S_t + (\mu - \frac{\sigma^2}{2})(u - t) + \sigma(W_u - W_t) \right) du$$

$$= \int_0^t \ln S_u \, du + (T - t) \ln S_t + (\mu - \frac{\sigma^2}{2}) \left(\frac{T^2 - t^2}{2} - t(T - t) \right)$$

$$+ \sigma \int_t^T (W_u - W_t) \, du.$$

The geometric average becomes

$$G_T = e^{\frac{1}{T} \int_0^T \ln S_u \, du}$$

$$= e^{\frac{1}{T} \int_0^t \ln S_u \, du} S_t^{1 - \frac{t}{T}} e^{\frac{1}{T}(\mu - \frac{\sigma^2}{2})(\frac{T+t}{2} - t)(T-t)} e^{\frac{\sigma}{T} \int_t^T (W_u - W_t) \, du}$$

$$= G_t^{\frac{t}{T}} S_t^{1 - \frac{t}{T}} e^{(\mu - \frac{\sigma^2}{2})\frac{(T-t)^2}{2T}} e^{\frac{\sigma}{T} \int_t^T (W_u - W_t) \, du}, \qquad (6.7.25)$$

where we used that

$$e^{\frac{1}{T} \int_0^T \ln S_u \, du} = e^{\frac{t}{T} \ln G_t} = G_t^{\frac{t}{T}}.$$

Relation (6.7.25) provides G_T in terms of G_t and S_t. Taking the measurable part out and replacing μ by r we have

$$\widehat{\mathbb{E}}_t[G_T] = G_t^{\frac{t}{T}} S_t^{1 - \frac{t}{T}} e^{(r - \frac{\sigma^2}{2})\frac{(T-t)^2}{2T}} \mathbb{E}\left[e^{\frac{\sigma}{T} \int_t^T (W_u - W_t) \, du} \middle| \mathcal{F}_t \right]. \qquad (6.7.26)$$

Since the jump $W_u - W_t$ is independent of the information set $\mathcal{F}_t$, the condition can be dropped

$$\mathbb{E}\left[e^{\frac{\sigma}{T} \int_t^T (W_u - W_t) \, du} \middle| \mathcal{F}_t \right] = \mathbb{E}\left[e^{\frac{\sigma}{T} \int_t^T (W_u - W_t) \, du} \right].$$

Integrating by parts yields

$$\int_t^T (W_u - W_t) \, du = \int_t^T W_u \, du - (T - t) W_t$$

$$= TW_T - tW_t - \int_t^T u \, dW_u - TW_t + tW_t$$

$$= T(W_T - W_t) - \int_t^T u \, dW_u = \int_t^T T \, dW_u - \int_t^T u \, dW_u$$

$$= \int_t^T (T - u) \, dW_u,$$

which is a Wiener integral. This is normal distributed with mean 0 and variance

$$\int_t^T (T-u)^2 \, du = \frac{(T-t)^3}{3}.$$

Then $\int_t^T (W_u - W_t) \, du \sim \mathcal{N}\left(0, \frac{(T-t)^3}{3}\right)$ and hence

$$\mathbb{E}\left[e^{\frac{\sigma}{T} \int_t^T (W_u - W_t) \, du}\right] = \mathbb{E}\left[e^{\frac{\sigma}{T} \int_t^T (T-u) \, dW_u}\right] = e^{\frac{1}{2}\frac{\sigma^2}{T^2}\frac{(T-t)^3}{3}} = e^{\frac{\sigma^2}{T^2}\frac{(T-t)^3}{6}}.$$

Substituting into (6.7.26) yields

$$\widehat{\mathbb{E}}_t[G_T] = G_t^{\frac{t}{T}} S_t^{1-\frac{t}{T}} e^{(r-\frac{\sigma^2}{2})\frac{(T-t)^2}{2T}} e^{\frac{\sigma^2}{T^2}\frac{(T-t)^3}{6}}. \qquad (6.7.27)$$

Hence, the value of the contract at time t is given by

$$f_t = e^{-r(T-t)}\widehat{\mathbb{E}}_t[G_T - K] = G_t^{\frac{t}{T}} S_t^{1-\frac{t}{T}} e^{-r(T-t)+(r-\frac{\sigma^2}{2})\frac{(T-t)^2}{2T}+\frac{\sigma^2}{T^2}\frac{(T-t)^3}{6}} - e^{-r(T-t)} K.$$

■

Exercise 6.7.5 *Show that* $\lim_{t \to 0} f_t = f_0$, *where* f_t *is given by (6.7.24) and* f_0 *by (6.7.23).*

Exercise 6.7.6 *Which is cheaper: an Asian forward contract on* A_t *or an Asian forward contract on* G_t?

Exercise 6.7.7 *Using Corollary 5.8.6 find a formula for* G_T *in terms of* G_t, *and then compute the risk-neutral world expectation* $\widehat{\mathbb{E}}_t[G_T]$.

6.8 Asian Options

There are several types of Asian options depending on how the payoff is related to the average stock price:

- *Average Price Options:*

    - Call: $f_T = \max(S_{ave} - K, 0)$
    - Put: $f_T = \max(K - S_{ave}, 0)$.

- *Average Strike Options:*

    - Call: $f_T = \max(S_T - S_{ave}, 0)$
    - Put: $f_T = \max(S_{ave} - S_T, 0)$.

The average asset price S_{ave} can be either the arithmetic or the geometric average of the asset price between 0 and T. The average strike options are more complicated to price than the average price options and their valuation will be deferred until section 9.3.

Geometric Average Price Options When the asset is the geometric average, G_T, we shall obtain closed form formulas for average price options. Since G_T is log-normally distributed, the pricing procedure is similar with the one used for the usual options on a stock. We shall do this by using the superposition principle and the following two results. The first one is a cash-or-nothing type contract where the underlying asset is the geometric mean of the stock between 0 and T.

Lemma 6.8.1 *The value at time $t = 0$ of a derivative, which pays at maturity $\$1$ if the geometric average $G_T \geq K$ and 0 otherwise, is given by*

$$h_0 = e^{-rT} N(\widetilde{d_2}),$$

where

$$\widetilde{d_2} = \frac{\ln S_0 - \ln K + (\mu - \frac{\sigma^2}{2})\frac{T}{2}}{\sigma\sqrt{T/3}}.$$

Proof: The payoff can be written as

$$h_T = \begin{cases} 1, & \text{if } G_T \geq K \\ 0, & \text{if } G_T < K \end{cases} = \begin{cases} 1, & \text{if } X_T \geq \ln K \\ 0, & \text{if } X_T < \ln K, \end{cases}$$

where $X_T = \ln G_T$ has the normal distribution

$$X_T \sim N\left[\ln S_0 + \left(\mu - \frac{\sigma^2}{2}\right)\frac{T}{2}, \frac{\sigma^2 T}{3}\right],$$

see formula (5.8.36). Let $\hat{p}(x)$ be the probability density of the random variable X_T in the risk neutral world (obtained replacing μ by r)

$$\hat{p}(x) = \frac{1}{\sqrt{2\pi}\sigma\sqrt{T/3}} e^{-[x - \ln S_0 - (r - \frac{\sigma^2}{2})\frac{T}{2}]^2 / (\frac{2\sigma^2 T}{3})}. \tag{6.8.28}$$

The risk neutral expectation of the payoff at time $t = 0$ is

$$\widehat{\mathbb{E}}_0[h_T] = \int_{-\infty}^{\infty} h_T(x)\hat{p}(x)\, dx = \int_{\ln K}^{\infty} \hat{p}(x)\, dx$$

$$= \frac{1}{\sqrt{2\pi}\sigma\sqrt{T/3}} \int_{\ln K}^{\infty} e^{-[x - \ln S_0 - (r - \frac{\sigma^2}{2})\frac{T}{2}]^2 / (\frac{2\sigma^2 T}{3})}\, dx.$$

Substituting

$$y = \frac{x - \ln S_0 - (r - \frac{\sigma^2}{2})\frac{T}{2}}{\sigma\sqrt{T/3}}, \tag{6.8.29}$$

yields

$$\begin{aligned}
\widehat{\mathbb{E}}_0[h_T] &= \frac{1}{\sqrt{2\pi}} \int_{-\tilde{d}_2}^{\infty} e^{-y^2/2}\,dy = \frac{1}{\sqrt{2\pi}} \int_{-\infty}^{\tilde{d}_2} e^{-y^2/2}\,dy \\
&= N(\tilde{d}_2),
\end{aligned}$$

where

$$\tilde{d}_2 = \frac{\ln S_0 - \ln K + (r - \frac{\sigma^2}{2})\frac{T}{2}}{\sigma\sqrt{T/3}}.$$

Discounting to the free interest rate yields the price at time $t = 0$

$$h_0 = e^{-rT}\widehat{\mathbb{E}}_0[h_T] = e^{-rT}N(\tilde{d}_2).$$

∎

The following result deals with the price of an average-or-nothing derivative on the geometric average.

Lemma 6.8.2 *The value at time $t = 0$ of a derivative, which pays at maturity G_T if $G_T \geq K$ and 0 otherwise, is given by the formula*

$$g_0 = e^{-\frac{1}{2}(r+\frac{\sigma^2}{6})T}S_0 N(\tilde{d}_1),$$

where

$$\tilde{d}_1 = \frac{\ln S_0 - \ln K + (r + \frac{\sigma^2}{6})\frac{T}{2}}{\sigma\sqrt{T/3}}.$$

Proof: Since the payoff can be written as

$$g_T = \begin{cases} G_T, & \text{if } G_T \geq K \\ 0, & \text{if } G_T < K \end{cases} = \begin{cases} e^{X_T}, & \text{if } X_T \geq \ln K \\ 0, & \text{if } X_T < \ln K, \end{cases}$$

with $X_T = \ln G_T$, the risk neutral expectation of the payoff at the time $t = 0$ is given by

$$\widehat{\mathbb{E}}_0[g_T] = \int_{-\infty}^{\infty} g_T(x)\hat{p}(x)\,dx = \int_{\ln K}^{\infty} e^x \hat{p}(x)\,dx,$$

where $\hat{p}(x)$ is given by (6.8.28), with μ replaced by r. Using the substitution (6.8.29) and completing the square yields

$$\widehat{\mathbb{E}}_0[g_T] = \frac{1}{\sqrt{2\pi}\sigma\sqrt{T/3}} \int_{\ln K}^{\infty} e^x e^{-[x - \ln S_0 - (r - \frac{\sigma^2}{2})\frac{T}{2}]^2/(\frac{2\sigma^2 T}{3})} \, dx$$

$$= \frac{1}{\sqrt{2\pi}} S_0 e^{\frac{1}{2}(r - \frac{\sigma^2}{6})T} \int_{-\tilde{d}_2}^{\infty} e^{-\frac{1}{2}[y - \sigma\sqrt{T/3}]^2} \, dy$$

If we let

$$\tilde{d}_1 = \tilde{d}_2 + \sigma\sqrt{T/3}$$

$$= \frac{\ln S_0 - \ln K + (r - \frac{\sigma^2}{2})\frac{T}{2}}{\sigma\sqrt{T/3}} + \sigma\sqrt{T/3}$$

$$= \frac{\ln S_0 - \ln K + (r + \frac{\sigma^2}{6})\frac{T}{2}}{\sigma\sqrt{T/3}},$$

the previous integral becomes, after substituting $z = y - \sigma\sqrt{T/3}$,

$$\frac{1}{\sqrt{2\pi}} S_0 e^{\frac{1}{2}(r - \frac{\sigma^2}{6})T} \int_{-\tilde{d}_1}^{\infty} e^{-\frac{1}{2}z^2} \, dz = S_0 e^{\frac{1}{2}(r - \frac{\sigma^2}{6})T} N(\tilde{d}_1).$$

Then the risk-neutral expectation of the payoff is

$$\widehat{\mathbb{E}}_0[g_T] = S_0 e^{\frac{1}{2}(r - \frac{\sigma^2}{6})T} N(\tilde{d}_1).$$

The value of the derivative at time t is obtained by discounting at the interest rate r

$$g_0 = e^{-rT}\widehat{\mathbb{E}}_0[g_T] = e^{-rT} S_0 e^{\frac{1}{2}(r - \frac{\sigma^2}{6})T} N(\tilde{d}_1)$$

$$= e^{-\frac{1}{2}(r + \frac{\sigma^2}{6})T} S_0 N(\tilde{d}_1).$$

∎

Proposition 6.8.3 *The value at time $t = 0$ of a geometric average price call option is*

$$\boxed{f_0 = e^{-\frac{1}{2}(r + \frac{\sigma^2}{6})T} S_0 N(\tilde{d}_1) - K e^{-rT} N(\tilde{d}_2).}$$

Proof: Since the payoff $f_T = \max(G_T - K, 0)$ can be decomposed as

$$f_T = g_T - K h_T,$$

with

$$g_T = \begin{cases} G_T, & \text{if } G_T \geq K \\ 0, & \text{if } G_T < K, \end{cases} \qquad h_T = \begin{cases} 1, & \text{if } G_T \geq K \\ 0, & \text{if } G_T < K, \end{cases}$$

applying the superposition principle and Lemmas 6.8.1 and 6.8.2 yields

$$
\begin{aligned}
f_0 &= g_0 - K h_0 \\
&= e^{-\frac{1}{2}(r+\frac{\sigma^2}{6})T} S_0 N(\tilde{d}_1) - K e^{-rT} N(\tilde{d}_2).
\end{aligned}
$$

■

Exercise 6.8.4 *Find the value at time $t = 0$ of a price put option on a geometric average, i.e. a derivative with the payoff $f_T = \max(K - G_T, 0)$.*

Arithmetic Average Price Options There is no simple closed-form solution for a call or a put on the arithmetic average A_t. This is caused by the complexity of the underlying density function of A_t. However, there is an approximate solution based on computing exactly the first two moments of the distribution of A_t, and applying the risk-neutral valuation assuming that the distribution is log-normal with the same two moments. This idea was developed by Turnbull and Wakeman [64], and works well for volatilities up to about 20%.

The following result provides the mean and variance of a normal distribution in terms of the first two moments of the associated log-normal distribution.

Proposition 6.8.5 *Let Y be a log-normally distributed random variable, having the first two moments given by*

$$
m_1 = \mathbb{E}[Y], \qquad m_2 = \mathbb{E}[Y^2].
$$

Then $\ln Y$ has the normal distribution $\ln Y \sim \mathcal{N}(\mu, \sigma^2)$, with the mean and variance given respectively by

$$
\mu = \ln \frac{m_1^2}{\sqrt{m_2}}, \qquad \sigma^2 = \ln \frac{m_2}{m_1^2}. \tag{6.8.30}
$$

Proof: Consider the normal distribution $X \sim \mathcal{N}(\mu, \sigma^2)$, and let $Y = e^X$. Using the moment generating function formula for X, we have

$$
m_1 = \mathbb{E}[Y] = e^{\mu + \frac{\sigma^2}{2}}, \qquad m_2 = \mathbb{E}[Y^2] = e^{2\mu + 2\sigma^2}.
$$

Taking the logarithm we obtain the system of equations

$$
\mu + \frac{\sigma^2}{2} = \ln m_1, \qquad 2\mu + 2\sigma^2 = \ln m_2.
$$

Solving for μ and σ^2 leads to formulas (6.8.30). ■

Assume the arithmetic average $A_t = \frac{I_t}{t}$ has a log-normal distribution, where $I_t = \int_0^t S_u\, du$. Then $\ln A_t = \ln I_t - \ln t$ is normal, so $\ln I_t$ is also normal, and hence I_t is log-normally distributed. Since $I_T = \int_0^T S_u\, du$, using (5.8.33) yields

$$
\begin{aligned}
m_1 &= \mathbb{E}[I_T] = \int_0^T \mathbb{E}[S_u]\, du = S_0 \frac{e^{\mu T} - 1}{\mu} \\
m_2 &= \mathbb{E}[I_T^2] = \frac{2S_0^2}{\mu + \sigma^2}\left[\frac{e^{(2\mu+\sigma^2)T} - 1}{2\mu + \sigma^2} - \frac{e^{\mu T} - 1}{\mu}\right].
\end{aligned}
$$

Using Proposition 6.8.5 it follows that $\ln A_T$ is normally distributed, with

$$
\boxed{\ln A_T \sim \mathcal{N}\left(\ln \frac{m_1^2}{\sqrt{m_2}} - \ln T,\ \ln \frac{m_2}{m_1^2}\right).} \tag{6.8.31}
$$

Relation (6.8.31) represents the normal approximation of $\ln A_T$. We shall price the arithmetic average price call under this condition.

In the next two exercises we shall assume that the distribution of A_T is given by the log-normal distribution (6.8.31).

Exercise 6.8.6 *Using a method similar to the one used in Lemma 6.8.1, show that an approximate value at time 0 of a derivative, which pays at maturity $1 if the arithmetic average $A_T \geq K$ and 0 otherwise, is given by*

$$
h_0 = e^{-rT} N(\check{d}_2),
$$

with

$$
\check{d}_2 = \frac{\ln(m_1^2/\sqrt{m_2}) - \ln K - \ln T}{\sqrt{\ln(m_2/m_1^2)}}, \tag{6.8.32}
$$

where in the expressions of m_1 and m_2 we replaced μ by r.

Exercise 6.8.7 *Using a method similar to the one used in Lemma 6.8.2, show that the approximate value at time 0 of a derivative, which pays at maturity A_T if $A_T \geq K$ and 0 otherwise, is given by the formula*

$$
a_0 = S_0 \frac{1 - e^{-rT}}{rT} N(\check{d}_1),
$$

where

$$
\check{d}_1 = \sqrt{\ln\left(\frac{m_2}{m_1^2}\right)} + \check{d}_2, \tag{6.8.33}
$$

where in the expressions of m_1 and m_2 we replaced μ by r.

Proposition 6.8.8 *The approximate value at $t = 0$ of an arithmetic average price call is given by*

$$f_0 = \frac{S_0(1 - e^{-rT})}{rT} N(\check{d}_1) - Ke^{-rT}N(\check{d}_2),$$

with $\check{d}_1$ and $\check{d}_2$ given by formulas (6.8.32) and (6.8.33).

Exercise 6.8.9 (*a*) *Prove Proposition 6.8.8.*
(*b*) *How does the formula change if the value is taken at time t instead of time 0?*

6.9 The d_k Notations

Pricing formulas for European-type options can often be written in terms of $N(d_k)$, where $N(x) = \int_{-\infty}^{x} \frac{1}{\sqrt{2\pi}} e^{-\frac{x^2}{2}} dx$ is the standard normal cumulative probability function and d_k are some expressions involving maturity T, interest rate r, initial stock price S_0, barrier b, and strike price K. The expressions of d_k are provided in the following:

$$d_1 = \frac{\ln \frac{S_0}{K} + (r + \frac{\sigma^2}{2})T}{\sigma\sqrt{T}} \tag{6.9.34}$$

$$d_2 = \frac{\ln \frac{S_0}{K} + (r - \frac{\sigma^2}{2})T}{\sigma\sqrt{T}} \tag{6.9.35}$$

$$d_3 = \frac{\ln \frac{b^2}{S_0 K} + (r + \frac{\sigma^2}{2})T}{\sigma\sqrt{T}} \tag{6.9.36}$$

$$d_4 = \frac{\ln \frac{b^2}{S_0 K} + (r - \frac{\sigma^2}{2})T}{\sigma\sqrt{T}} \tag{6.9.37}$$

$$d_5 = \frac{\ln \frac{S_0}{b} + (r + \frac{\sigma^2}{2})T}{\sigma\sqrt{T}} \tag{6.9.38}$$

$$d_6 = \frac{\ln \frac{S_0}{b} + (r - \frac{\sigma^2}{2})T}{\sigma\sqrt{T}} \tag{6.9.39}$$

$$d_7 = \frac{\ln \frac{b}{S_0} + (r + \frac{\sigma^2}{2})T}{\sigma\sqrt{T}} \tag{6.9.40}$$

$$d_8 = \frac{\ln \frac{b}{S_0} + (r - \frac{\sigma^2}{2})T}{\sigma\sqrt{T}}. \tag{6.9.41}$$

The next result deals with the probabilistic interpretations of d_k, $1 \leq k \leq 8$.

Proposition 6.9.1 *Let S_T denote the stock price at time T. The following identities hold in the risk-neutral world*

(a) $N(d_1) = P(S_T \geq Ke^{-\sigma^2 T})$;

(b) $N(d_2) = P(S_T \geq K)$;

(c) $N(d_3) = P\left(S_T < \dfrac{b^2}{K}e^{2rT}\right)$;

(d) $N(d_4) = P\left(S_T < \dfrac{b^2}{K}e^{(2r-\sigma^2)T}\right)$;

(e) $N(d_5) = P(S_T \geq be^{-\sigma^2 T})$;

(f) $N(d_6) = P(S_T \geq b)$;

(g) $N(d_7) = P(S_T < be^{2rT})$;

(h) $N(d_8) = P\left(S_T < be^{(2r-\sigma^2)T}\right)$.

Proof: (a) Since $W_T/\sqrt{T}$ is a standard normal variable, the following sequence of equalities hold

$$
\begin{aligned}
N(d_1) &= N\left(\frac{\ln\frac{S_0}{K} + (r + \frac{\sigma^2}{2})T}{\sigma\sqrt{T}}\right) = P\left(\frac{W_T}{\sqrt{T}} < \frac{\ln\frac{S_0}{K} + (r + \frac{\sigma^2}{2})T}{\sigma\sqrt{T}}\right) \\
&= P\left(\sigma W_T < \ln\frac{S_0}{K} + (r + \frac{\sigma^2}{2})T\right) = P\left(\sigma W_T > \ln\frac{K}{S_0} - (r + \frac{\sigma^2}{2})T\right) \\
&= P\left(\sigma W_T + (r - \frac{\sigma^2}{2})T > \ln\frac{K}{S_0} - \sigma^2 T\right) \\
&= P\left(e^{\sigma W_T + (r - \frac{\sigma^2}{2})T} > \frac{K}{S_0}e^{-\sigma^2 T}\right) = P(S_T > Ke^{-\sigma^2 T}).
\end{aligned}
$$

(b) We write the event $\{S_T \geq K\}$ in terms of the standard normal variable $W_T/\sqrt{T}$ as follows

$$
\begin{aligned}
\{S_T \geq K\} &= \{S_0 e^{(r - \frac{\sigma^2}{2})T + \sigma W_T} \geq K\} = \left\{(r - \frac{\sigma^2}{2})T + \sigma W_T \geq \ln\frac{K}{S_0}\right\} \\
&= \left\{W_T \geq -\frac{\ln\frac{S_0}{K} + (r - \frac{\sigma^2}{2})T}{\sigma}\right\} = \left\{\frac{W_T}{\sqrt{T}} \geq -d_2\right\}.
\end{aligned}
$$

Since $\dfrac{W_T}{\sqrt{T}} \sim \mathcal{N}(0,1)$, then

$$
P(S_T \geq K) = P\left(\frac{W_T}{\sqrt{T}} \geq -d_2\right) = P\left(\frac{W_T}{\sqrt{T}} < d_2\right) = N(d_2).
$$

(c) Following a similar computation to (a) we have

$$
N(d_3) = N\left(\frac{\ln \frac{b^2}{S_0 K} + (r + \frac{\sigma^2}{2})T}{\sigma\sqrt{T}}\right) = P\left(\frac{W_T}{\sqrt{T}} < \frac{\ln \frac{b^2}{S_0 K} + (r + \frac{\sigma^2}{2})T}{\sigma\sqrt{T}}\right)
$$

$$
= P\left(\sigma W_T + (r - \frac{\sigma^2}{2})T < \ln \frac{b^2}{S_0 K} + 2rT\right) = P\left(S_T < \frac{b^2}{K}e^{2rT}\right).
$$

(d) The computation is similar with the one done in part (c).
(e) Similar to part (a); just replace K to b.
(f) Similar to part (b); just replace K to b.
(g) Similar to part (c).
(h) Similar to part (d).   ∎

Remark 6.9.2 Notations d_1 and d_2 appear in the Black-Scholes pricing formulas of a plain vanilla European option, while the other d_k appear in pricing different types of barrier or lookback options.

6.10   All-or-nothing Look-back Options

An *all-or-nothing look-back option* is a contract that pays \$1 at maturity T if the stock S_t did ever reach or exceed level $b > 0$ until time T, and the amount 0 otherwise. The payoff is given by

$$
V_T = \begin{cases} 1, & \text{if } \bar{S}_T \geq b \\ 0, & \text{otherwise,} \end{cases} \tag{6.10.42}
$$

where $\bar{S}_T$ stands for the running maximum on the stock, see Exercise 5.3.8. The value at time $t = 0$ is obtained using the risk-neutral valuation and Proposition 5.4.3

$$
V_0 = e^{-rT}\widehat{\mathbb{E}}[V_T] = e^{-rT}\widehat{\mathbb{E}}[V_T | \bar{S}_T \geq b]\, P(\bar{S}_T \geq b) = e^{-rT}\, P(\bar{S}_T \geq b)
$$

$$
= e^{-rT} N(d_6) + e^{-rT}\left(\frac{b}{S_0}\right)^{\frac{2r}{\sigma^2}-1} N(-d_8),
$$

with d_6 and d_8 given by (6.9.39) and (6.9.41), respectively. Hence the price of an all-or-nothing look-back option at time $t = 0$ is

$$
\boxed{V_0 = e^{-rT} N(d_6) + e^{-rT}\left(\frac{b}{S_0}\right)^{\frac{2r}{\sigma^2}-1} N(-d_8).} \tag{6.10.43}
$$

Remark 6.10.1 Now we shall look more closely at the previous formula with the help of Proposition 6.9.1, parts (f) and (h). The first term, $e^{-rT} N(d_6)$, is the value at time $t = 0$ of an all-or-nothing contract, whose payoff is

$$f_T = \begin{cases} 1, & \text{if } S_T \geq b \\ 0, & \text{otherwise.} \end{cases}$$

Since $N(-d_8) = 1 - N(d_8) = P(S_T \geq be^{(2r-\sigma^2)T})$, then the second term becomes

$$e^{-rT} \left(\frac{b}{S_0}\right)^{\frac{2r}{\sigma^2}-1} P(S_T \geq be^{(2r-\sigma^2)T}),$$

which is the value at $t = 0$ of an all-or-nothing contract with payoff

$$g_T = \begin{cases} \left(\dfrac{b}{S_0}\right)^{\frac{2r}{\sigma^2}-1}, & \text{if } S_T \geq be^{(2r-\sigma^2)T} \\ \\ 0, & \text{otherwise.} \end{cases}$$

Hence, the value of an all-or-nothing look-back option can be written as a sum of two all-or-nothing contracts on the stock.

Remark 6.10.2 The payoff of an all-or-nothing look-back option pays \$1 as long as a certain barrier has been hit. If T_b is the first time the stock reaches the barrier b, its payoff can be also formalized as

$$V_T = \begin{cases} 1, & \text{if } T_b < T \\ 0, & \text{if } T_b \geq T. \end{cases}$$

Since the payment is made at the expiration time T, this contract is also called a *deferred rebate option*.

6.11 Asset-or-nothing Look-back Options

This is a contract that pays at expiration the maximum value of the stock during the life of the contract if the stock have exceeded the barrier b, and zero otherwise. Its payoff can be written as a piece-wise function

$$f_T = \begin{cases} \bar{S}_T, & \text{if } \bar{S}_T \geq b \\ 0, & \text{if } \bar{S}_T < b. \end{cases} \tag{6.11.44}$$

Consider the process $X_t = \ln \frac{S_t}{S_0} = mt + \sigma W_t$, where $m = \mu - \frac{\sigma^2}{2}$, and define the running maximum $\bar{X}_T = \max_{0 \leq t \leq T} X_t$. Since $\bar{S}_T = S_0 e^{\bar{X}_T}$, the payoff (6.11.44) can be written in terms of $\bar{X}_T$ as

$$
f_T = \begin{cases} S_0 e^{\bar{X}_T}, & \text{if } \bar{X}_T \geq \ln \frac{b}{S_0} \\[2mm] 0, & \text{if } \bar{X}_T < \ln \frac{b}{S_0}. \end{cases}
$$

If $\hat{g}(x)$ is the probability density of $\bar{X}_T$ in the risk-neutral world, the value of the option at time $t = 0$ is given by

$$
f_0 = e^{-rT} \widehat{\mathbb{E}}_0[f_T] = e^{-rT} S_0 \int_{\ln \frac{b}{S_0}}^{\infty} e^x \hat{g}(x) \, dx. \tag{6.11.45}
$$

The next result provides the desired probability density.

Lemma 6.11.1 *Let* $\bar{X}_T = \max_{0 \leq t \leq T} (mt + \sigma W_t).$

(a) *The probability function of* $\bar{X}_T$ *is given by*

$$
G(x) = P(\bar{X}_T \leq x) = N\left(\frac{x - mT}{\sigma\sqrt{T}}\right) - e^{\frac{2mx}{\sigma^2}} N\left(\frac{-x - mT}{\sigma\sqrt{T}}\right).
$$

(b) *The probability density of* $\bar{X}_T$ *is*

$$
g(x) = \frac{2}{\sigma\sqrt{2\pi T}} e^{-\frac{(x-mT)^2}{2\sigma^2 T}} - \frac{2m}{\sigma^2} e^{\frac{2mx}{\sigma^2}} N\left(\frac{-x - mT}{\sigma\sqrt{T}}\right).
$$

Proof: (a) Substituting $\gamma = -m/\sigma$ and applying Lemma 5.4.1 we have

$$
\begin{aligned}
G(x) &= P(\bar{X}_T \leq x) = P\left(\max_{0 \leq t \leq T} \left(\frac{m}{\sigma}t + W_t\right) \leq \frac{x}{\sigma}\right) \\[2mm]
&= 1 - P\left(\max_{0 \leq t \leq T} (W_t - \gamma t) \geq \frac{x}{\sigma}\right) \\[2mm]
&= N\left(\gamma\sqrt{T} + \frac{x}{\sigma\sqrt{T}}\right) - e^{-2x\gamma/\sigma} N\left(\gamma\sqrt{T} - \frac{x}{\sigma\sqrt{T}}\right) \\[2mm]
&= N\left(\frac{x - mT}{\sigma\sqrt{T}}\right) - e^{\frac{2mx}{\sigma^2}} N\left(\frac{-x - mT}{\sigma\sqrt{T}}\right).
\end{aligned}
$$

(b) Differentiating yields the density

$$
\begin{aligned}
g(x) = G'(x) &= \frac{1}{\sigma\sqrt{2\pi T}} e^{-\frac{(x-mT)^2}{2\sigma^2 T}} - \frac{2m}{\sigma^2} e^{\frac{2mx}{\sigma^2}} N\left(\frac{-x - mT}{\sigma\sqrt{T}}\right) \\[2mm]
&\quad + \frac{1}{\sigma\sqrt{2\pi T}} e^{\frac{2mx}{\sigma^2}} e^{-\frac{(x+mT)^2}{2T\sigma^2}} \\[2mm]
&= \frac{2}{\sigma\sqrt{2\pi T}} e^{-\frac{(x-mT)^2}{2\sigma^2 T}} - \frac{2m}{\sigma^2} e^{\frac{2mx}{\sigma^2}} N\left(\frac{-x - mT}{\sigma\sqrt{T}}\right).
\end{aligned}
$$

The risk-neutral density $\hat{g}(x)$ is obtained from $g(x)$ replacing μ by r. Then using Lemma 6.11.1 we can continue the computation of the integral (6.11.45) writing it as

$$f_0 = e^{-rT} S_0 (I_1 - I_2), \tag{6.11.46}$$

with

$$I_1 = \frac{2}{\sigma \sqrt{2\pi T}} \int_{\ln \frac{b}{S_0}}^{\infty} e^x e^{-\frac{(x-mT)^2}{2\sigma^2 T}} \, dx,$$

$$I_2 = \frac{2m}{\sigma^2} \int_{\ln \frac{b}{S_0}}^{\infty} e^x e^{\frac{2mx}{\sigma^2}} N\left(\frac{-x-mT}{\sigma \sqrt{T}}\right) \, dx,$$

where $m = r - \frac{\sigma^2}{2}$. Next we shall compute these integrals in a convenient way. Using the substitutions $y = \frac{x-mT}{\sigma \sqrt{T}}$, $z = y - \sigma \sqrt{T}$ and completing the square, the first integral transforms as

$$
\begin{aligned}
I_1 &= \frac{2}{\sqrt{2\pi}} \int_{\frac{\ln(b/S_0)-mT}{\sigma\sqrt{T}}}^{\infty} e^{-\frac{1}{2}(y-\sigma\sqrt{T})^2} e^{mT+\sigma^2 T/2} \, dy \\
&= 2e^{rT} \int_{\frac{\ln(b/S_0)-mT}{\sigma\sqrt{T}}-\sigma\sqrt{T}}^{\infty} \frac{1}{\sqrt{2\pi}} e^{-\frac{1}{2}z^2} \, dz \\
&= 2e^{rT} \int_{-d_1}^{\infty} \frac{1}{\sqrt{2\pi}} e^{-\frac{1}{2}z^2} \, dz = 2e^{rT} N(d_5), \tag{6.11.47}
\end{aligned}
$$

where

$$d_5 = \frac{\ln \frac{S_0}{b} + (r + \frac{\sigma^2}{2})T}{\sigma \sqrt{T}}.$$

The integral I_2 is computed using integration by parts. Using the risk-neutral relation $\frac{2m}{\sigma^2} = \frac{2r}{\sigma^2} - 1$ and Exercise 6.11.2, we have

$$
\begin{aligned}
I_2 &= \frac{2m}{\sigma^2} \int_{\ln \frac{b}{S_0}}^{\infty} e^x e^{\frac{2mx}{\sigma^2}} N\left(\frac{-x-mT}{\sigma\sqrt{T}}\right) \, dx \\
&= \left(\frac{2r}{\sigma^2}-1\right) \int_{\ln \frac{b}{S_0}}^{\infty} e^{\frac{2rx}{\sigma^2}} \left\{1 - N\left(\frac{x+mT}{\sigma\sqrt{T}}\right)\right\} \\
&= -\left(1 - \frac{\sigma^2}{2r}\right) \left(\frac{b}{S_0}\right)^{\frac{2r}{\sigma^2}} N\left(\frac{\ln(S_0/b)-mT}{\sigma\sqrt{T}}\right) \\
&\quad + \underbrace{\left(\frac{2r}{\sigma^2}-1\right) \frac{\sigma^2}{2r} \int_{\ln \frac{b}{S_0}}^{\infty} e^{\frac{2rx}{\sigma^2}} \frac{1}{\sqrt{2\pi}} e^{-\frac{(x+mT)^2}{2\sigma^2 T}} \frac{1}{\sigma\sqrt{T}} \, dx}_{=U}. \tag{6.11.48}
\end{aligned}
$$

Next we compute the integral U. Completing to a square and using the substitution $y = x - (r + \frac{\sigma^2}{2})T$, we have

$$
\begin{aligned}
U &= \int_{\ln \frac{b}{S_0}}^{\infty} e^{\frac{2rx}{\sigma^2}} \frac{1}{\sqrt{2\pi}} e^{-\frac{(x+mT)^2}{2\sigma^2 T}} \frac{1}{\sigma\sqrt{T}} dx \\
&= \frac{1}{\sigma\sqrt{T}} \int_{\ln \frac{b}{S_0}}^{\infty} \frac{1}{\sqrt{2\pi}} e^{rT} e^{-\frac{[x-(r+\frac{\sigma^2}{2})T]^2}{2\sigma^2 T}} dx \\
&= e^{rT} \int_{\frac{\ln \frac{b}{S_0} - (r+\frac{\sigma^2}{2})T}{\sigma\sqrt{T}}}^{\infty} \frac{1}{\sqrt{2\pi}} e^{-\frac{y^2}{2}} dy \\
&= e^{rT} \int_{-\infty}^{\frac{\ln \frac{S_0}{b} + (r+\frac{\sigma^2}{2})T}{\sigma\sqrt{T}}} \frac{1}{\sqrt{2\pi}} e^{-\frac{y^2}{2}} dy \\
&= e^{rT} N(d_5), \tag{6.11.49}
\end{aligned}
$$

where $d_5 = \dfrac{\ln \frac{S_0}{b} + (r + \frac{\sigma^2}{2})T}{\sigma\sqrt{T}}$. Since

$$
\frac{\ln(S_0/b) - mT}{\sigma\sqrt{T}} = -\frac{\ln(b/S_0) + (r - \frac{\sigma^2}{2})T}{\sigma\sqrt{T}} = -d_8, \tag{6.11.50}
$$

substituting (6.11.50) and (6.11.49) into (6.11.48) yields

$$
I_2 = -\left(1 - \frac{\sigma^2}{2r}\right)\left(\frac{b}{S_0}\right)^{\frac{2r}{\sigma^2}} N(-d_8) + \left(1 - \frac{\sigma^2}{2r}\right) e^{rT} N(d_5). \tag{6.11.51}
$$

Substitute the expressions for I_1 and I_2 given by (6.11.47) and (6.11.51) into the value of the option (6.11.46) and get

$$
\begin{aligned}
f_0 &= S_0 e^{-rT} I_1 - S_0 e^{-rT} I_2 \\
&= 2S_0 N(d_5) + S_0 e^{-rT}\left(1 - \frac{\sigma^2}{2r}\right)\left(\frac{b}{S_0}\right)^{\frac{2r}{\sigma^2}} N(-d_8) - S_0\left(1 - \frac{\sigma^2}{2r}\right)N(d_5) \\
&= S_0\left(1 + \frac{\sigma^2}{2r}\right)N(d_5) + S_0 e^{-rT}\left(1 - \frac{\sigma^2}{2r}\right)\left(\frac{b}{S_0}\right)^{\frac{2r}{\sigma^2}} N(-d_8).
\end{aligned}
$$

Hence, the asset-or-nothing look-back option defined by the payoff (6.11.44) has the value at time $t = 0$ given by

$$
\boxed{f_0 = S_0\left(1 + \frac{\sigma^2}{2r}\right)N(d_5) + S_0 e^{-rT}\left(1 - \frac{\sigma^2}{2r}\right)\left(\frac{b}{S_0}\right)^{\frac{2r}{\sigma^2}} N(-d_8).} \tag{6.11.52}
$$

Exercise 6.11.2 *Show that* $\lim\limits_{x\to\infty} e^x\big(1 - N(x)\big) = 0$.

6.12 Look-back Call Options

A *look-back call option* is a contract which offers the right to buy at expiration the maximum value of the stock for the price b; equivalently, this derivative pays off the maximum amount by which the stock ever exceeded the given barrier b in the time interval $[0, T]$, and zero otherwise

$$g_T = \max\{\bar{S}_T - b, 0\} = \begin{cases} \bar{S}_T - b, & \text{if } \bar{S}_T \geq b \\ 0, & \text{if } \bar{S}_T < b. \end{cases} \quad (6.12.53)$$

This contract can be priced by the well-known superposition principle. Note that $g_T = f_T - bV_T$, where f_T and V_T are the payoffs of an asset-or-nothing and all-or-nothing look-back options, respectively, see formulas (6.11.44) and (6.10.42). Then, at time $t = 0$, the price of the call is $g_0 = f_0 - bV_0$. Using expressions (6.11.52) and (6.10.43) an algebraic computation provides

$$
\begin{aligned}
g_0 &= S_0\left(1 + \frac{\sigma^2}{2r}\right)N(d_5) + S_0 e^{-rT}\left(1 - \frac{\sigma^2}{2r}\right)\left(\frac{b}{S_0}\right)^{\frac{2r}{\sigma^2}} N(-d_8) \\
&\quad - b e^{-rT} N(d_6) - S_0 e^{-rT}\left(\frac{b}{S_0}\right)^{\frac{2r}{\sigma^2}} N(-d_8) \\
&= S_0 N(d_5) - b e^{-rT} N(d_6) + S_0 \frac{\sigma^2}{2r} N(d_5) \\
&\quad + S_0 e^{-rT}\left(\frac{b}{S_0}\right)^{\frac{2r}{\sigma^2}} N(-d_8)\left(1 - \frac{\sigma^2}{2r} - 1\right) \\
&= S_0 N(d_5) - b e^{-rT} N(d_6) + S_0 \frac{\sigma^2}{2r}\left[N(d_5) - e^{-rT}\left(\frac{b}{S_0}\right)^{\frac{2r}{\sigma^2}} N(-d_8)\right].
\end{aligned}
$$

Hence, the look-back call option defined by the payoff (6.12.53) has the value at $t = 0$ given by

$$\boxed{g_0 = S_0 N(d_5) - b e^{-rT} N(d_6) + S_0 \frac{\sigma^2}{2r}\left[N(d_5) - e^{-rT}\left(\frac{b}{S_0}\right)^{\frac{2r}{\sigma^2}} N(-d_8)\right].}$$

$$(6.12.54)$$

It is worth noting that the first two terms in the pricing formula, $S_0 N(d_5) - b e^{-rT} N(d_6)$, represent the Black-Scholes price for a call option with the strike price b.

6.13 Forward Look-back Contracts

We ask first what is the value at time $t = 0$ of a contract that pays at time T the amount $\bar{S}_T$? This value is given by the risk-neutral valuation as

$$e^{-rT}\widehat{\mathbb{E}}_0[\bar{S}_T] = e^{-rT} S_0 \int_0^\infty e^x \hat{g}(x)\, dx,$$

where $\hat{g}(x)$ is the risk-neutral density function of $\bar{S}_T$, see Lemma 6.11.1 (b). Instead of computing the integral directly, we rather make use of a previous computation. Taking the limit in (6.11.45) and using the result (6.11.52), we have

$$
e^{-rT}S_0 \int_0^\infty e^x \hat{g}(x)\, dx = \lim_{b \to S_0} e^{-rT} S_0 \int_{\ln \frac{b}{S_0}}^\infty e^x \hat{g}(x)\, dx
$$

$$
= S_0\left(1 + \frac{\sigma^2}{2r}\right) N(\lim_{b \to S_0} d_5) + S_0 e^{-rT}\left(1 - \frac{\sigma^2}{2r}\right) N(-\lim_{b \to S_0} d_8)
$$

$$
= S_0\left(1 + \frac{\sigma^2}{2r}\right) N\left(\left(1 + \frac{\sigma^2}{2r}\right)\frac{r\sqrt{T}}{\sigma}\right) + S_0 e^{-rT}\left(1 - \frac{\sigma^2}{2r}\right) N\left(-\left(1 - \frac{\sigma^2}{2r}\right)\frac{r\sqrt{T}}{\sigma}\right).
$$

$$(6.13.55)$$

Now, we are able to compute the value at $t = 0$ of a *forward look-back contract*, which has the payoff $F_T = \bar{S}_T - K$. This is a contract that delivers for price K the maximum of the stock until time T, i.e. a contract that enforces you to buy for the price K the best performance of the stock during $[0, T]$. Using (6.13.55), the risk-neutral valuation provides

$$
\boxed{F_0 = S_0\left(1 + \frac{\sigma^2}{2r}\right) N(\alpha_+) + S_0 e^{-rT}\left(1 - \frac{\sigma^2}{2r}\right) N(-\alpha_-) - e^{-rT} K,} \quad (6.13.56)
$$

where

$$
\alpha_\pm = \left(1 \pm \frac{\sigma^2}{2r}\right)\frac{r\sqrt{T}}{\sigma}.
$$

Exercise 6.13.1 (put-call parity) *Let $L_T^c = \max\{\bar{S}_T - K, 0\}$, $L_T^p = \max\{K - \bar{S}_T, 0\}$ and $F_T = \bar{S}_T - K$ be the payoffs of a look-back call, look-back put and forward contract, respectively.*

(a) *Show that $L_T^c - L_T^p = F_T$;*

(b) *Assuming there are no arbitrage opportunities, deduce that $L_0^c - L_0^p = F_0$;*

(c) *Find a close-form expression for L_0^p.*

Exercise 6.13.2 *Evaluate the following strike look-back options*

(a) $V_T = S_T - \underline{S}_T$ *(call)*

(b) $V_T = \bar{S}_T - S_T$ *(put)*

(c) *Why is the payoff not given as a piece-wise function?*

For more closed form formulas for derivatives the reader is refered to the book of McDonald [53].

6.14   Immediate Rebate Options

Assume $S_0 < b$. A contract that pays \$1 at the time T_b, when the stock hits for the first time the barrier b, is called an *immediate rebate option*. The discounted value at time $t = 0$ is given by

$$f_0 = \begin{cases} e^{-rT_b}, & \text{if } T_b \leq T \\ 0, & \text{otherwise}, \end{cases}$$

where T_b is a random variable with the inverse Gaussian distribution $p_b(\tau)$ given by (5.3.12), and T is the expiration time of the option. The price of the contract at $t = 0$, denoted by R_0, is obtained taking the risk-neutral expectation of f_0

$$R_0 = \widehat{\mathbb{E}}[f_0] = \int_0^T e^{-r\tau} \widehat{p_b}(\tau) \, d\tau,$$

where $\widehat{p_b}(\tau)$ is obtained from $p_b(\tau)$ replacing μ by r.

Proposition 6.14.1 *The price of an immediate rebate option at time $t = 0$ is given by*

$$R_0 = \left(\frac{b}{S_0}\right)^{n_1} N(-d_7) + \left(\frac{b}{S_0}\right)^{n_2} N(d_5), \tag{6.14.57}$$

where $n_1 = \dfrac{2r}{\sigma^2}$ and $n_2 = -1$.

Proof: For the sake of simplicity, we employ the following notations

$$\alpha = \frac{1}{\sigma} \ln \frac{b}{S_0}, \quad \gamma = \frac{1}{\sigma}\left(r + \frac{\sigma^2}{2}\right) = \frac{r}{\sigma} + \frac{\sigma}{2}.$$

Then

$$d_5 = -\frac{\alpha}{\sqrt{T}} + \gamma\sqrt{T}$$

$$d_7 = \frac{\alpha}{\sqrt{T}} + \gamma\sqrt{T}.$$

If denote

$$f(T) = \int_0^T e^{-r\tau} \widehat{p_b}(\tau) \, d\tau$$

$$g(T) = \left(\frac{b}{S_0}\right)^{n_1} N(-d_7) + \left(\frac{b}{S_0}\right)^{n_2} N(d_5)$$

then to show (6.14.57) it suffices to verify that $f(T) = g(T)$, for $T > 0$. This is implied by the conditions

$$f(0) = g(0)$$
$$f'(T) = g'(T).$$

The first one follows from

$$f(0) = 0$$
$$g(0) = \lim_{T \searrow 0} g(T) = 0.$$

The second condition will be proved by direct computation. By an algebraic computation the left side can be written as

$$
\begin{aligned}
f'(T) &= e^{-rT} \widehat{p}_b(T) = \frac{\alpha}{\sqrt{2\pi}T^{3/2}} e^{-rT} e^{-\frac{1}{2T}\left(\alpha - (\gamma - \sigma)T\right)^2} \\
&= \frac{\alpha}{\sqrt{2\pi}T^{3/2}} e^{-\frac{\alpha^2}{2T}} e^{-\frac{\gamma^2 T}{2}} e^{\alpha(\gamma - \sigma)} \\
&= \frac{\alpha}{\sqrt{2\pi}T^{3/2}} e^{\Phi},
\end{aligned}
\tag{6.14.58}
$$

with $\Phi = -\frac{\alpha^2}{2T} - \frac{\gamma^2 T}{2} + \alpha(\gamma - \sigma)$. Noting that

$$\left(\frac{b}{S_0}\right)^{n_j} = e^{n_j \frac{b}{S_0}} = e^{n_j \sigma \alpha},$$

then differentiating in

$$g(T) = e^{n_1 \sigma \alpha} N(-d_7) + e^{n_2 \sigma \alpha} N(d_5)$$

we obtain

$$
\begin{aligned}
g'(T) &= -e^{n_1 \sigma \alpha} \frac{1}{\sqrt{2\pi}} e^{-d_7^2/2} d_7'(T) + e^{n_2 \sigma \alpha} \frac{1}{\sqrt{2\pi}} e^{-d_5^2/2} d_5'(T) \\
&= \frac{1}{\sqrt{2\pi}} \left\{ e^{n_1 \sigma \alpha - d_7^2/2} \left(\frac{\alpha}{2T^{3/2}} - \frac{\gamma}{2T^{1/2}} \right) + e^{n_2 \sigma \alpha - d_5^2/2} \left(\frac{\alpha}{2T^{3/2}} + \frac{\gamma}{2T^{1/2}} \right) \right\} \\
&= \frac{1}{\sqrt{2\pi}} \left\{ e^{\Phi_1} \left(\frac{\alpha}{2T^{3/2}} - \frac{\gamma}{2T^{1/2}} \right) + e^{\Phi_2} \left(\frac{\alpha}{2T^{3/2}} + \frac{\gamma}{2T^{1/2}} \right) \right\},
\end{aligned}
$$

where

$$\Phi_1 = n_1 \sigma \alpha - d_7^2/2$$
$$\Phi_2 = n_2 \sigma \alpha - d_5^2/2.$$

Assuming now that n_1 and n_2 are chosen such that

$$\Phi_1 = \Phi = \Phi_2, \tag{6.14.59}$$

then

$$g'(T) = \frac{\alpha}{\sqrt{2\pi}T^{3/2}} e^{\Phi},$$

which matches the expression of $f'(T)$ given by (6.14.58).

The last step is to find n_1 and n_2 that satisfy the assumption (6.14.59). A computation provides

$$
\begin{aligned}
\Phi_1 &= n_1 \sigma \alpha - \frac{1}{2}\left(\frac{\alpha}{\sqrt{T}} + \gamma\sqrt{T}\right)^2 \\
&= -\frac{\alpha^2}{2T} - \frac{\gamma^2 T}{2} + \alpha(n_1\sigma - \gamma).
\end{aligned}
$$

The identity $\Phi_1 = \Phi$ provides an equation for n_1

$$
\alpha(\gamma - \sigma) = \alpha(n_1\sigma - \gamma)
$$

with solution

$$
n_1 = \frac{2\gamma}{\sigma} - 1 = \frac{2r}{\sigma^2}.
$$

Similarly,

$$
\begin{aligned}
\Phi_2 &= n_2 \sigma \alpha - \frac{1}{2}\left(-\frac{\alpha}{\sqrt{T}} + \gamma\sqrt{T}\right)^2 \\
&= -\frac{\alpha^2}{2T} - \frac{\gamma^2 T}{2} + \alpha(\gamma + n_2\sigma),
\end{aligned}
$$

and the identity $\Phi_2 = \Phi$ provides the equation

$$
\alpha(\gamma - \sigma) = \alpha(\gamma + n_2\sigma)
$$

with solution $n_2 = -1$.   ■

Remark 6.14.2 The option discussed previously is called an *up-rebate option*, since the stock reaches the barrier b from below. If $S_0 > b$, then we have a *down-rebate option*, whose pricing formula is similar to the one developed in this section, namely

$$
D_0 = \left(\frac{b}{S_0}\right)^{n_1} N(d_7) + \left(\frac{b}{S_0}\right)^{n_2} N(-d_5),
$$

with the same values for n_1 and n_2.

Exercise 6.14.3 *Show that an immediate up-rebate option can be written as the sum of two all-or-nothing contracts with strikes $K_1 = be^{-\sigma^2 T}$, $K_2 = be^{2rT}$ and that pay $\dfrac{S_0}{b}$ and $\left(\dfrac{b}{S_0}\right)^{n_1}$, respectively.*

6.15 Perpetual Look-back Options

A look-back option that has no expiration date (i.e. has an infinite time horizon) is called a *perpetual look-back option*. This contract pays \$1 at the time the stock price S_t reaches barrier b for the first time. The payment occurs at time T_b, where $T_b = \inf\{t > 0; S_t \geq b\}$. Assume $S_0 < b$. The value at time $t = 0$ is obtained by discounting at the risk-free rate r and taking the risk-neutral expectation

$$V_0 = \widehat{\mathbb{E}}\big[e^{-rT_b}\big] = \left(\frac{S_0}{b}\right)^{h(r)} = \frac{S_0}{b}, \tag{6.15.60}$$

where we used formula (5.3.11) and Exercise 5.3.1 (*b*) with $\mu = r$.

Exercise 6.15.1 *A perpetual look-back option can be considered as an immediate rebate option with no expiration date. Find the pricing formula for a perpetual look-back option taking the limit $T \to \infty$ in formula (6.14.57).*

Exercise 6.15.2 *Find the price of a contract that pays \$1 when the initial stock price doubles its value.*

6.16 Double Barrier Immediate Option

Consider the barriers S_u and S_d, with $0 < S_d < S_0 < S_u$. A double barrier immediate option pays \$1 at the time when the stock S_t reaches for the first time either the barrier S_u or the barrier S_d. Let T be the time when the stock reaches for the first time either one of the previous barriers

$$T = \inf\{t > 0; \ S_t \geq S_u \text{ or } S_t \leq S_d\}.$$

Then the price of the contract at time $t = 0$ is obtained as

$$f_0 = \widehat{\mathbb{E}}[e^{-rT}].$$

The formula on the right side can be obtained from expression (5.5.24) by making $s = \mu = r$:

$$\widehat{\mathbb{E}}[e^{-rT}] = \frac{1}{e^{(-r+\sqrt{2r+r^2})\alpha}p_\alpha + e^{-(-r+\sqrt{2r+r^2})\beta}(1 - p_\alpha)}.$$

Using that

$$\alpha = \frac{1}{\sigma}\ln u, \quad \beta = -\frac{1}{\sigma}\ln d,$$

we obtain

$$f_0 = \frac{1}{u^{(-r+\sqrt{2r+r^2})/\sigma}p_\alpha + d^{(-r+\sqrt{2r+r^2})/\sigma}(1 - p_\alpha)},$$

with

$$p_\alpha = \frac{e^{2r\beta} - 1}{e^{2r\beta} - e^{-2r\alpha}} = \frac{d^{-2r/\sigma} - 1}{d^{-2r/\sigma} - u^{-2r/\sigma}}.$$

6.17 Pricing in a Rare Events Environment

Pricing derivatives in a rare events environment faces the delicate problem of hedging the risk provided by an infinite number of shocks in the market. This will involve the creation of a hedging portfolio formed from an infinite number of independently traded products, which neutralize both the random shocks dN_t and the diffusive noise, dW_t. This type of approach was used by Jones [44] and Bates [7].

   However, we shall follow the assumptions of Merton [54] under which the risk-neutral valuation can be applied. This critical assumption is that the jumps represent nonsystematic risk, i.e. they are not related to the movements of the whole market. In this case, we are allowed to replace μ by the risk-free rate, r, in formula (5.9.39) to obtain the process followed by the stock, S_t, in the risk-neutral world

$$dS_t = (r - \lambda\rho)S_{t-}dt + \sigma S_{t-}dW_t + \rho S_{t-}dN_t, \qquad (6.17.61)$$

where we employ all notations introduced in section 5.9, *Stock Prices with Rare Events*. The solution is given by formula (5.9.40)

$$S_t = S_0 e^{(r - \lambda\rho - \frac{\sigma^2}{2})t + \sigma W_t}(1 + \rho)^{N_t}, \qquad (6.17.62)$$

where: σ is the volatility of the stock; λ is the rate at which rare events occur; ρ is the size of the jump in the expected return when a rare event occurs; and N_t is a Poisson process with rate λ, which models the market shocks.

   Denote $Z_T = \ln S_T$. Then the probability of Z_T, conditioned by Z_t, in the risk-neutral world follows from (5.9.44)

$$p_{Z_T|Z_t}(z|z_0) = \frac{e^{-\lambda(T-t)}}{\sqrt{2\pi(T-t)}\,\sigma} \sum_{n \geq 0} \frac{\lambda^n(T-t)^n}{n!} e^{-\frac{[z - z_0 - (r - \lambda\rho - \sigma^2/2)(T-t) - n\ln(1+\rho)]^2}{2\sigma^2(T-t)}}.$$

$$(6.17.63)$$

Pricing with respect to the density $p_{Z_T|Z_t}(z|z_0)$ reduces to a linear combinations of Black-Scholes prices, as can be seen from the linearity property of the integral. This can be formalized as in the following. If V_T is the payoff of a European derivative, its price at time t is computed by the risk-neutral valuation as

$$
\begin{aligned}
V_t &= e^{-r(T-t)}\mathbb{E}[V_T(S_T)|\mathcal{F}_t] = e^{-r(T-t)}\int V_T(z)p(z|S_t)\,dz \\
&= e^{-\lambda(T-t)}\sum_{n\geq 0}\frac{\lambda^n(T-t)^n}{n!} \\
&\quad\times \frac{e^{-r(T-t)}}{\sqrt{2\pi(T-t)}\,\sigma}\int V_T(z)e^{-\frac{[z-S_t-(r-\lambda\rho-\sigma^2/2)(T-t)-n\ln(1+\rho)]^2}{2\sigma^2(T-t)}}\,dz \\
&= e^{-\lambda(T-t)}\sum_{n\geq 0}\frac{\lambda^n(T-t)^n}{n!}V_{t,n}^{BS},
\end{aligned}
\tag{6.17.64}
$$

where $V_{t,n}^{BS}$ is the price of a derivative with payoff V_T, obtained *a la Black-Scholes*, with respect to the normal probability density

$$
\mathcal{N}\left(S_t + \left(r - \lambda\rho + \frac{n}{T-t}\ln(1+\rho) - \frac{\sigma^2}{2}\right)(T-t),\ \sigma^2(T-t)\right).
$$

Comparing with the regular Black-Scholes case, when the normal density is just

$$
\mathcal{N}\left(S_t + \left(r - \frac{\sigma^2}{2}\right)(T-t),\ \sigma^2(T-t)\right),
$$

we notice that for computing $V_{t,n}^{BS}$ one has to replace r by $r - \lambda\rho + \frac{n}{T-t}\ln(1+\rho)$ in the Black-Scholes formula.

6.17.1 Pricing a Call

As an application, we provide the formula of a European call in a rare events environment. Recall first the Black-Scholes formula for the European call with strike K and maturity T

$$
C_{BS}(t,S,T,K,r,\sigma) = SN(d_1) - Ke^{-r(T-t)}N(d_2).
$$

Then (6.17.64) provides the formula for the call in the case of market jumps as

$$
\boxed{C_J(t) = e^{-\lambda(T-t)}\sum_{n\geq 0}\frac{\lambda^n(T-t)^n}{n!}C_{BS}\left(t,S,T,K,r-\lambda\rho+\frac{n\ln(1+\rho)}{T-t},\sigma\right).}
$$

Remark 6.17.1 Merton proved a more general result allowing for jumps of stochastic magnitude. In this case, equation (5.9.39) is replaced by

$$
dS_t = (\mu - \lambda\rho)S_t dt + \sigma S_t dW_t + (Y_t - 1)dN_t,
$$

where $Y_t - 1$ represents the magnitude of jumps price returns, with the expectation $\mathbb{E}[Y_t] = \rho + 1$. If Assuming that Y_t is log-normally distributed, with $\ln Y_t \sim \mathcal{N}(\gamma, \delta^2)$, then the price of a call in the presence of market jumps is given by a formula similar with the previous one

$$C_J(t) = e^{-\lambda(T-t)} \sum_{n \geq 0} \frac{\lambda^n (T-t)^n}{n!} C_{BS}\left(t, S, T, K, r - \lambda\rho + \frac{n \ln(1+\rho)}{T-t}, \sqrt{\sigma^2 + \frac{n\delta^2}{T-t}}\right).$$

Exercise 6.17.2 *Write the expression of a put option starting from the formula* (6.17.64).

Exercise 6.17.3 *Starting from the formula* (6.17.64) *find the price of an all-or-nothing option.*

6.17.2   Pricing a Forward Contract

Another application is to price a forward contract, whose payoff is given by $V_T = S_T - K$. Since the value of a forward contract evaluated without considering the effect of jumps is independent of n

$$V_{t,n}^{BS} = S_t - e^{r(T-t)}K,$$

then formula (6.17.64) provides its value in a market with shocks as

$$\begin{aligned}
V_t &= e^{-\lambda(T-t)} \sum_{n \geq 0} \frac{\lambda^n (T-t)^n}{n!} V_{t,n}^{BS} \\
&= (S_t - e^{r(T-t)}K)e^{-\lambda(T-t)} \sum_{n \geq 0} \frac{\lambda^n (T-t)^n}{n!} \\
&= S_t - e^{r(T-t)}K,
\end{aligned}$$

since the probabilities of the Poisson distribution sum up to 1. We notice that this formula is independent of the return jump magnitude, ρ, and shocks frequency, λ.

In the following we shall prove, for pedagogical reasons, the same formula using a direct computation. It will be clear from what it follows that the use of formula (6.17.64) considerably shortens the computations.

Recall that the stock follows a stochastic process with rare events, where the number of events $n = N_T$ until time T is assumed to be Poisson distributed. As usual, T denotes the maturity of the forward contract. The stock price at maturity is given by formula (5.9.40)

$$S_T = S_0 e^{(\mu - \lambda\rho - \frac{1}{2}\sigma^2)T + \sigma W_T}(1 + \rho)^{N_T}.$$

Writing a similar formula for S_t after a division yields

$$S_T = S_t e^{(\mu - \lambda\rho - \frac{1}{2}\sigma^2)(T-t) + \sigma(W_T - W_t)} (1 + \rho)^{(N_T - N_t)}.$$

The risk neutral expectation of the stock as of time t is

$$
\begin{aligned}
\widehat{\mathbb{E}}_t[S_T] &= \mathbb{E}[S_T | \mathcal{F}_t, \mu = r] \\
&= S_t e^{(r - \lambda\rho - \frac{1}{2}\sigma^2)(T-t)} \mathbb{E}[e^{\sigma(W_T - W_t)} | \mathcal{F}_t] \, \mathbb{E}[(1 + \rho)^{(N_T - N_t)} | \mathcal{F}_t] \\
&= S_t e^{(r - \lambda\rho - \frac{1}{2}\sigma^2)(T-t)} \mathbb{E}[e^{\sigma(W_T - W_t)}] \, \mathbb{E}[(1 + \rho)^{(N_T - N_t)}], \quad (6.17.65)
\end{aligned}
$$

where we used that the information $\mathcal{F}_t$ is independent of the jumps $W_T - W_t$ and $N_T - N_t$. Using the stationarity property

$$\mathbb{E}[e^{\sigma(W_T - W_t)}] = \mathbb{E}[e^{\sigma W_{T-t}}] = e^{\frac{1}{2}\sigma^2(T-t)}.$$

$$\mathbb{E}[(1 + \rho)^{(N_T - N_t)}] = \mathbb{E}[(1 + \rho)^{N_{T-t}}] = e^{\lambda\rho(T-t)}.$$

The last expression follows by conditioning over $N_{T-t} = n$ as in the following

$$
\begin{aligned}
\mathbb{E}[(1 + \rho)^{N_{T-t}}] &= \sum_{n \geq 0} \mathbb{E}[(1 + \rho)^{N_{T-t}} | N_{T-t} = n] P(N_{T-t} = n) \\
&= \sum_{n \geq 0} (1 + \rho)^n P(N_{T-t} = n) \\
&= \sum_{n \geq 0} (1 + \rho)^n \frac{\lambda^n (T - t)^n}{n!} e^{-\lambda(T-t)} \\
&= e^{\lambda(1+\rho)(T-t)} e^{-\lambda(T-t)} = e^{\lambda\rho(T-t)}.
\end{aligned}
$$

Substituting in (6.17.65) we obtain

$$
\begin{aligned}
\widehat{\mathbb{E}}_t[S_T] &= S_t e^{(r - \lambda\rho - \frac{1}{2}\sigma^2)(T-t)} e^{\frac{1}{2}\sigma^2(T-t)} e^{\lambda\rho(T-t)} \qquad (6.17.66) \\
&= S_t e^{r(T-t)}.
\end{aligned}
$$

The payoff of a forward contract is $V_T = S_T - K$. Using (6.17.67), the value of the forward contract at time t is given by

$$V_t = e^{-r(T-t)} \widehat{\mathbb{E}}_t[f_T] = e^{-r(T-t)} (\widehat{\mathbb{E}}_t[S_T] - K) = S_t - K e^{-r(T-t)}.$$

Exercise 6.17.4 *Starting from the formula (6.17.64), find the price of a square of a forward contract, with the payoff $V_T = (S_T - K)^2$.*

6.18   Pricing with Pareto Distribution

As it has been shown, stocks are seldom log-normally distributed. In most cases, due to occurrences of extreme events, the upper tail of the stock distribution is heavier than the tail of a log-normal distribution. Several heavy tail distributions have been used to model this behavior, such as the one-sided Pareto, Fisher and Gumbel distributions. This section shows that exact pricing formulas can be obtained for derivatives whose underlying asset is distributed according to a Pareto distribution.

Pareto distribution A *standard one-sided Pareto distribution* is defined by

$$f_\xi(x) = \begin{cases} (1 + \xi x)^{-(1+\frac{1}{\xi})}, & \text{if } x > 0 \\ 0, & \text{otherwise.} \end{cases}$$

where $\xi > 0$ is a parameter that describes the heaviness of the upper tail. One way to describe this fact is to estimate the probability of the upper tail of the stock S_T using Markov's inequality

$$P(S_T \geq b) \leq \frac{\mathbb{E}[S_T^p]}{b^p}, \qquad p \geq 1. \tag{6.18.67}$$

For $b > 1$ large enough, the event $\{S_T \geq b\}$ is considered an extreme event. It can be shown that if X is a random variable which is Pareto distributed, then the size of parameter ξ controls the number of finite moments:

$$\mathbb{E}[X^p] < \infty \leftrightarrow p < \frac{1}{\xi}.$$

More precisely, when ξ gets small, the number of finite moments increases, and hence the right side of (6.18.67) holds for a larger p, case in which the probability on the left side gets smaller since $\lim_{p \to \infty} b^p = \infty$.

A standard one-sided Pareto distribution can be shifted and scaled. This way, we obtain a *general Pareto distribution*, which depends on three parameters, m, λ and ξ, describing respectively location, scale and shape:

$$f_{m,\lambda,\xi}(x) = \begin{cases} \frac{1}{\lambda}(1 + \frac{\xi}{\lambda}(x - m))^{-(1+\frac{1}{\xi})}, & \text{if } x > m \\ 0, & \text{otherwise.} \end{cases}$$

An integration yields the following cumulative distribution function

$$F_{m,\lambda,\xi}(x) = \int_{-\infty}^{x} f_{m,\lambda,\xi}(y)\, dy = \begin{cases} 1 - \left(1 + \frac{\xi}{\lambda}(x - m)\right)^{-\frac{1}{\xi}}, & \text{if } x > m \\ 0, & \text{otherwise.} \end{cases}$$

The parameter m is estimated to be the minimum of stock prices, $\xi \in (0,1)$ describes the heaviness of the tail (depending on the frequency of extreme events), while λ depends on the relative size of stock prices. These parameters can be estimated using statistical methods such as the L-moments method or the maximum likelihood method. The reader is referred to the book of Carmona [20] and the R library therein for these estimations. Next we shall price some derivatives under the assumption that in the risk-neutral world S_T has a Pareto distribution.

Forward contract Assume the asset price, S_T, at time T has a Pareto distribution with parameters $m > 0$, $\lambda > 0$ and $\xi \in (0,1)$. The expected value of S_T with respect to this distribution is given by

$$
\begin{aligned}
\int_m^\infty x f_{m,\lambda,\xi}(x)\, dx &= \int_m^\infty x \frac{1}{\lambda}\left(1 + \frac{\xi}{\lambda}(x-m)\right)^{-(1+\frac{1}{\xi})} dx \\
&= \int_0^\infty (m+v)\frac{1}{\lambda}\left(1 + \frac{\xi}{\lambda}v\right)^{-(1+\frac{1}{\xi})} dv \\
&= m \int_0^\infty f_{0,\lambda,\xi}(v)\, dv + \frac{\lambda}{\xi^2}\int_1^\infty \left(u^{-\frac{1}{\xi}} - u^{-1-\frac{1}{\xi}}\right) du \\
&= m + \frac{\lambda}{\xi^2}\left(\frac{-\xi}{\xi-1} - \xi\right) = m + \frac{\lambda}{1-\xi},
\end{aligned}
$$

where we used the substitutions $v = x - m$, $u = 1 + \frac{\xi}{\lambda}v$ and that $\xi \in (0,1)$. Discounted at the free-interest rate we obtain the value of the forward contract at time $t = 0$ given by

$$
V_0 = e^{-rT}\left(m + \frac{\lambda}{1-\xi}\right). \tag{6.18.68}
$$

We notice that the value of the forward contract, V_0, increases unbounded as the distribution tail gets heavier, satisfying $\xi \nearrow 1$.

All-or-nothing contract We need to find the expectation of the payoff

$$
f_T = \begin{cases} 1, & \text{if } S_T > K \\ 0, & \text{if } S_T \le K, \end{cases}
$$

given that the underlying asset, S_T, is Pareto distributed. This can be computed using the cumulative function $F_{m,\lambda,\xi}$ as in the following:

$$
\mathbb{E}[f_T] = \int_\mathbb{R} f_T(x) f_{m,\lambda,\xi}(x)\, dx = \int_K^\infty f_{m,\lambda,\xi}(x)\, dx.
$$

The value of the previous integral depends on the relative position of m and K. We investigate two cases:

(i) Case $K < m$. This models the case when the smallest value of the stock S_T, which estimated by m, is larger than the strike K, i.e. the option is "in the money" all the time. Hence, the payoff value is $f_T = 1$ and $\mathbb{E}[f_T] = 1$.

(ii) Case $K \geq m$. The computation of the expectation continues in this case as

$$
\begin{aligned}
\mathbb{E}[f_T] &= \int_K^\infty f_{m,\lambda,\xi}(x)\, dx = 1 - \int_m^K f_{m,\lambda,\xi}(x)\, dx \\
&= 1 - F_{m,\lambda,\xi}(K) = \left(1 + \frac{\xi}{\lambda}(K - m)\right)^{-\frac{1}{\xi}}.
\end{aligned}
$$

The value of the all-or-nothing option at time $t = 0$ is obtained by discounting at the free-interest rate

$$
V_0 = e^{-rT}\mathbb{E}[f_T] = \begin{cases} e^{-rT}\left(1 + \frac{\xi}{\lambda}(K - m)\right)^{-\frac{1}{\xi}}, & \text{if } K \geq m \\ e^{-rT}, & \text{if } K < m. \end{cases} \tag{6.18.69}
$$

Asset-or-nothing contract The payoff

$$
h_T = \begin{cases} S_T, & \text{if } S_T > K \\ 0, & \text{if } S_T \leq K, \end{cases}
$$

has the expectation

$$
\begin{aligned}
\mathbb{E}[h_T] &= \int_{\mathbb{R}} x f_{m,\lambda,\xi}(x)\, dx = \int_{\max\{m,K\}}^\infty x f_{m,\lambda,\xi}(x)\, dx \\
&= \begin{cases} \displaystyle\int_K^\infty x f_{m,\lambda,\xi}(x)\, dx, & \text{if } K \geq m \\ \displaystyle\int_m^\infty x f_{m,\lambda,\xi}(x)\, dx, & \text{if } K < m. \end{cases}
\end{aligned}
$$

(i) Case $K < m$. Using the same computation done when we had valued the forward contract, we have

$$
\mathbb{E}[f_T] = \int_m^\infty x f_{m,\lambda,\xi}(x)\, dx = m + \frac{\lambda}{1 - \xi}. \tag{6.18.70}
$$

(ii) Case $K \geq m$. We compute using substitution and the cumulative distribution

$$
\begin{aligned}
\mathbb{E}[f_T] &= \int_K^\infty x f_{m,\lambda,\xi}(x)\, dx \\
&= \int_K^\infty (x-m) f_{m,\lambda,\xi}(x)\, dx + m \int_K^\infty f_{m,\lambda,\xi}(x)\, dx \\
&= \frac{\lambda}{\xi^2} \int_{1+\frac{\xi}{m}(K-m)}^\infty (u-1) u^{-1-\frac{1}{\xi}}\, du + m\left(1 - F_{m,\lambda,\xi}(K)\right) \\
&= \frac{\lambda}{\xi}\left\{ \frac{1}{1-\xi}\left(1+\frac{\xi}{\lambda}(K-m)\right)^{1-\frac{1}{\xi}} - \left(1+\frac{\xi}{\lambda}(K-m)\right)^{-\frac{1}{\xi}} \right\} \\
&\quad + m\left(1+\frac{\xi}{\lambda}(K-m)\right)^{-\frac{1}{\xi}}.
\end{aligned}
$$

Factoring out $\left(1+\frac{\xi}{\lambda}(K-m)\right)^{-\frac{1}{\xi}}$ and then working out the tedious algebra it follows that

$$
\mathbb{E}[f_T] = \frac{(K+\lambda-m\xi)}{1-\xi}\left(1+\frac{\xi}{\lambda}(K-m)\right)^{-\frac{1}{\xi}}. \tag{6.18.71}
$$

Using (6.18.70) and (6.18.71) we conclude with the following value for the asset-or-nothing contract at $t=0$

$$
W_0 = e^{-rT}\mathbb{E}[f_T] = \begin{cases} e^{-rT}\frac{(K+\lambda-m\xi)}{1-\xi}\left(1+\frac{\xi}{\lambda}(K-m)\right)^{-\frac{1}{\xi}}, & \text{if } K \geq m \\[2mm] e^{-rT}\left(m+\frac{\lambda}{1-\xi}\right), & \text{if } K < m. \end{cases} \tag{6.18.72}
$$

Pricing a European call We shall use the fact that the value of a European call with strike K can be written in terms of the values of the asset-or-nothing and all-or-nothing contracts valued in the previous paragraphs as

$$
C_0 = W_0 - K V_0.
$$

We shall perform the calculations in two cases:

(i) Case $K < m$. Using (6.18.72) and (6.18.69) yields

$$
C_0 = W_0 - K V_0 = e^{-rT}\left(m+\frac{\lambda}{1-\xi}\right) - K e^{-rT} = e^{-rT}\left(m+\frac{\lambda}{1-\xi}-K\right).
$$

(ii) Case $K \geq m$. Again, relations (6.18.72) and (6.18.69) provide

$$
\begin{aligned}
C_0 = W_0 - KV_0 &= e^{-rT}\frac{(K+\lambda-m\xi)}{1-\xi}\left(1+\frac{\xi}{\lambda}(K-m)\right)^{-\frac{1}{\xi}} \\
&\quad -Ke^{-rT}\left(1+\frac{\xi}{\lambda}(K-m)\right)^{-\frac{1}{\xi}} \\
&= e^{-rT}\left(1+\frac{\xi}{\lambda}(K-m)\right)^{-\frac{1}{\xi}}\left\{\frac{K+\lambda-m\xi}{1-\xi}-K\right\} \\
&= e^{-rT}\left(1+\frac{\xi}{\lambda}(K-m)\right)^{-\frac{1}{\xi}}\frac{\lambda}{1-\xi}\left(1+\frac{\xi}{\lambda}(K-m)\right) \\
&= \frac{\lambda}{1-\xi}e^{-rT}\left(1+\frac{\xi}{\lambda}(K-m)\right)^{1-\frac{1}{\xi}}.
\end{aligned}
$$

We conclude with the following piece-wise formula for the price of a call option under the assumption that the asset price at expiration, S_T, is Pareto distributed with parameters $m > 0$, $\lambda > 0$ and $\xi \in (0,1)$:

$$
C_0 = \begin{cases}
e^{-rT}\left(m+\frac{\lambda}{1-\xi}-K\right), & \text{if } K < m \\[2mm]
\frac{\lambda}{1-\xi}e^{-rT}\left(1+\frac{\xi}{\lambda}(K-m)\right)^{1-\frac{1}{\xi}}, & \text{if } K \geq m.
\end{cases} \tag{6.18.73}
$$

It is worth noting that in both cases $\lim\limits_{\xi \to 1} C_0 = \infty$. This can be stated by saying that when the tails get heavier, or when the extreme events occur more often, the value of the call increases unboundedly.

Exercise 6.18.1 *What type of distribution is obtained if we let $\xi \to 0$ in the Pareto distribution?*

Exercise 6.18.2 *Find the price of a put option, given that S_T is distributed as a Pareto with parameters $m > 0$, $\lambda > 0$ and $\xi \in (0,1)$. What is the price of the put for $\xi \to 1$?*

Chapter 7

Martingale Measures

7.1 Martingale Measures

An $\mathcal{F}_t$-adapted stochastic process X_t on the probability space $(\Omega, \mathcal{F}, P)$ is not always a P-martingale. However, it might become a Q-martingale with respect to another probability measure Q on $\mathcal{F}$. This measure Q is called a *martingale measure*. The main result of this section is to find a martingale measure with respect to which the discounted stock price becomes a martingale. This measure plays an important role in the mathematical explanation of the risk-neutral valuation and its mathematical construction is based on Girsanov theorem.

7.1.1 Is the stock price S_t a martingale?

Since the stock price S_t is an $\mathcal{F}_t$-adapted and integrable process on $[0, T]$, the only condition which needs to be satisfied to be an $\mathcal{F}_t$-martingale is

$$\mathbb{E}[S_t | \mathcal{F}_u] = S_u, \qquad \forall u < t. \tag{7.1.1}$$

Heuristically speaking, this means that given all the market information at time u, the expected future stock prices are the same as the price of the stock at time u. This does not make sense, since in this case the investor would prefer investing the money in a bank at a risk-free interest rate, rather than buying a stock with a zero return. Then (7.1.1) does not hold in real life (i.e. with respect to the real world measure[1] P). The next result shows how to fix this problem.

---

[1] Also known as the physical measure or the historical measure.

Proposition 7.1.1 *Let μ be the rate of return of the stock S_t. Then*

$$\mathbb{E}[e^{-\mu t}S_t|\mathcal{F}_u] = e^{-\mu u}S_u, \qquad \forall u < t, \qquad (7.1.2)$$

i.e. $M_t = e^{-\mu t}S_t$ is an $\mathcal{F}_t$-martingale.

Proof: The process $e^{-\mu t}S_t$ is integrable since

$$\mathbb{E}[|e^{-\mu t}S_t|] = e^{-\mu t}\mathbb{E}[S_t] = e^{-\mu t}S_0 e^{\mu t} = S_0 < \infty.$$

Since S_t is $\mathcal{F}_t$-adapted (i.e. the stock price S_t can be measured given the information at time t), so will be $e^{-\mu t}S_t$. Using formula (5.1.2) and taking out the measurable part yields

$$\begin{aligned}
\mathbb{E}[S_t|\mathcal{F}_u] &= \mathbb{E}[S_0 e^{(\mu - \frac{1}{2}\sigma^2)t + \sigma W_t}|\mathcal{F}_u] \\
&= \mathbb{E}[S_0 e^{(\mu - \frac{1}{2}\sigma^2)u + \sigma W_u} e^{(\mu - \frac{1}{2}\sigma^2)(t-u) + \sigma(W_t - W_u)}|\mathcal{F}_u] \\
&= \mathbb{E}[S_u e^{(\mu - \frac{1}{2}\sigma^2)(t-u) + \sigma(W_t - W_u)}|\mathcal{F}_u] \\
&= S_u e^{(\mu - \frac{1}{2}\sigma^2)(t-u)}\mathbb{E}[e^{\sigma(W_t - W_u)}|\mathcal{F}_u]. \qquad (7.1.3)
\end{aligned}$$

Since the increment $W_t - W_u$ is independent of all values W_s, $s \leq u$, then it will also be independent of $\mathcal{F}_u$. In this case the condition can be dropped and the conditional expectation becomes the usual expectation

$$\mathbb{E}[e^{\sigma(W_t - W_u)}|\mathcal{F}_u] = \mathbb{E}[e^{\sigma(W_t - W_u)}].$$

Since $\sigma(W_t - W_u) \sim \mathcal{N}\left(0, \sigma^2(t-u)\right)$, from the properties of log-normal distributions we have

$$\mathbb{E}[e^{\sigma(W_t - W_u)}] = e^{\frac{1}{2}\sigma^2(t-u)}.$$

Substituting back into (7.1.3) yields

$$\mathbb{E}[S_t|\mathcal{F}_u] = S_u e^{(\mu - \frac{1}{2}\sigma^2)(t-u)} e^{\frac{1}{2}\sigma^2(t-u)} = S_u e^{\mu(t-u)},$$

which is equivalent to

$$\mathbb{E}[e^{-\mu t}S_t|\mathcal{F}_u] = e^{-\mu u}S_u,$$

and hence $e^{-\mu t}S_t$ is a martingale. ∎

The conditional expectation $\mathbb{E}[S_t|\mathcal{F}_u]$ can be expressed in terms of the conditional density function as

$$\mathbb{E}[S_t|\mathcal{F}_u] = \int S_t\, p(x|\mathcal{F}_u)\, dx. \qquad (7.1.4)$$

Exercise 7.1.2 (a) *Find the formula for conditional density function, $p(x|\mathcal{F}_u)$, defined by (7.1.4).*

(b) *Verify the formula*
$$\mathbb{E}[S_t|\mathcal{F}_0] = \mathbb{E}[S_t]$$
in two different ways, either by using part (a), or by using the independence of S_t with respect to $\mathcal{F}_0$.

The martingale relation (7.1.2) can be written equivalently as
$$\int e^{-\mu t} S_t\, p(S_t|\mathcal{F}_u)\, dS_t = e^{-\mu u} S_u, \qquad u < t.$$

This way, $dP(x) = p(x|\mathcal{F}_u)\, dx$ becomes a *martingale measure* for the process $e^{-\mu t} S_t$.

The next section deals with a similar problem where μ is replaced by r.

7.1.2 Risk-neutral World and Martingale Measure

Since the stock's rate of return μ might not be known from the beginning, and it depends on each individual stock, a meaningful question would be:

Under what martingale measure does the discounted stock price, $M_t = e^{-rt} S_t$, become a martingale?

The constant r denotes, as usual, the risk-free interest rate. Assume such a martingale measure exists. Then we must have
$$\widehat{\mathbb{E}}_u[e^{-rt} S_t] = \widehat{\mathbb{E}}[e^{-rt} S_t|\mathcal{F}_u] = e^{-ru} S_u, \qquad u < t.$$

where $\widehat{\mathbb{E}}$ denotes the expectation with respect to the requested martingale measure. Using the linearity of the expectation operator, the previous relation can also be written as
$$e^{-r(t-u)} \widehat{\mathbb{E}}[S_t|\mathcal{F}_u] = S_u, \qquad u < t.$$

This states that the discounted expectation at the risk-free interest rate for the time interval $t - u$ is the price of the stock, S_u. Since this does not involve any of the riskiness of the stock, we might think of it as an expectation in the risk-neutral world. The aforementioned formula can be written in the compound mode as
$$\widehat{\mathbb{E}}[S_t|\mathcal{F}_u] = S_u e^{r(t-u)}, \qquad u < t. \tag{7.1.5}$$

It is worthy to note that this formula can be obtained from the conditional expectation $\mathbb{E}[S_t|\mathcal{F}_u] = S_u e^{\mu(t-u)}$ by substituting $\mu = r$ and replacing $\mathbb{E}$ by $\widehat{\mathbb{E}}$, which corresponds to the definition of the expectation in a risk-neutral world. Therefore, the evaluation of derivatives in Chapter 6 is done using the aforementioned martingale measure under which $e^{-rt}S_t$ becomes a martingale. In the next section we shall determine this measure explicitly.

Exercise 7.1.3 *Consider the following two games that consist in flipping a fair coin and taking the following decisions:*

A. *If the coin lands Heads, you win \$2; otherwise you loose \$1.*

B. *If the coin lands Heads, you win \$20,000; otherwise you loose \$10,000.*

(a) *Which game involves more risk? Explain your answer.*

(b) *Which game would you choose to play, and why?*

(c) *Are you risk-neutral in your decision?*

The *risk-neutral measure* is the measure with respect to which investors have the same degree of risk-aversion. In the case of the previous exercise, it is the measure with respect to which all players are indiferent whether they choose option A or option B.

Each market participant has its own system of specific beliefs regarding the evolution of the market. This individual beliefs can be mathematically modeled by the concept of probability measure. The price of a contract is not biased to any of the particular market participant expectations; on the contrary, the price depends on a measure with respect to which all market participants are treated equally from the risk point of view. This is nothing but the risk-neutral measure.

7.1.3 Finding the Risk-Neutral Measure

We have seen that the solution of the stochastic differential equation of the stock price

$$dS_t = \mu S_t dt + \sigma S_t dW_t$$

is a geometric Brownian motion given by

$$S_t = S_0 e^{\mu t} e^{\sigma W_t - \frac{1}{2}\sigma^2 t}.$$

Then $e^{-\mu t}S_t = S_0 e^{\sigma W_t - \frac{1}{2}\sigma^2 t}$ is an *exponential process*, see Appendix C.3. It is known that this process is an $\mathcal{F}_t$-martingale, where $\mathcal{F}_t = \sigma\{W_u; u \leq t\}$ is the information available in the market until time t (the σ-field generated by the Brownian motion until time t). Hence $e^{-\mu t}S_t$ is a martingale, which is a result proved also by Proposition 7.1.1. The probability space where this martingale exists is $(\Omega, \mathcal{F}, P)$.

In the following we shall change the rate of return μ into the risk-free rate r and change the probability measure such that the discounted stock price becomes a martingale. The discounted stock price can be expressed in terms of the Brownian motion with drift (a hat was added in order to distinguish it from the former Brownian motion W_t)

$$\widehat{W}_t = \frac{\mu - r}{\sigma} t + W_t \tag{7.1.6}$$

as in the following

$$e^{-rt} S_t = e^{-rt} S_0 e^{\mu t} e^{\sigma W_t - \frac{1}{2} \sigma^2 t} = S_0 e^{\sigma \widehat{W}_t - \frac{1}{2} \sigma^2 t}.$$

If we let $\lambda = \dfrac{\mu - r}{\sigma}$ in Girsanov's theorem (see Appendix C.5), it follows that $\widehat{W}_t$ becomes a Brownian motion on the probability space $(\Omega, \mathcal{F}, Q)$, where

$$dQ = e^{-\frac{1}{2}(\frac{\mu - r}{\sigma})^2 T - \lambda W_T} \, dP.$$

As an exponential process, $e^{\sigma \widehat{W}_t - \frac{1}{2} \sigma^2 t}$ becomes a martingale on this space. Consequently $e^{-rt} S_t$ is a martingale process with respect to the probability measure Q. This means

$$\mathbb{E}^Q[e^{-rt} S_t | \mathcal{F}_u] = e^{-ru} S_u, \qquad u < t.$$

where $\mathbb{E}^Q[\,\cdot\,|\mathcal{F}_u]$ denotes the conditional expectation in the measure Q, and it is given by

$$\mathbb{E}^Q[X_t | \mathcal{F}_u] = \mathbb{E}^P[X_t e^{-\frac{1}{2}(\frac{\mu - r}{\sigma})^2 T - \lambda W_T} | \mathcal{F}_u].$$

The measure Q is called the *equivalent martingale measure*, or the *risk-neutral measure*.[2] The expectation taken with respect to this measure is called the *expectation in the risk-neutral world*. We customarily use the following notations

$$\widehat{\mathbb{E}}[e^{-rt} S_t] = \mathbb{E}^Q[e^{-rt} S_t]$$
$$\widehat{\mathbb{E}}_u[e^{-rt} S_t] = \mathbb{E}^Q[e^{-rt} S_t | \mathcal{F}_u].$$

It is worth noting that $\widehat{\mathbb{E}}[e^{-rt} S_t] = \widehat{\mathbb{E}}_0[e^{-rt} S_t]$, since $e^{-rt} S_t$ is independent of the initial information set $\mathcal{F}_0$.

The importance of the process $\widehat{W}_t$ is contained in the following useful result.

---

[2] A trader in Wall Street who would invest people's pension plan money would think from the point of view of measure Q, while an individual who invests his own money has a position described by the measure P.

Proposition 7.1.4 *The probability measure Q that makes the discounted stock price, $e^{-rt}S_t$, a martingale changes the rate of return μ into the risk-free interest rate r, i.e*

$$dS_t = rS_t dt + \sigma S_t d\widehat{W}_t.$$

Proof: The proof is a straightforward verification using (7.1.6)

$$
\begin{aligned}
dS_t &= \mu S_t dt + \sigma S_t dW_t = rS_t dt + (\mu - r)S_t dt + \sigma S_t dW_t \\
&= rS_t dt + \sigma S_t \Big(\frac{\mu - r}{\sigma} dt + dW_t \Big) \\
&= rS_t dt + \sigma S_t d\widehat{W}_t.
\end{aligned}
$$

We note that the solution of the previous stochastic equation is

$$S_t = S_0 e^{rt} e^{\sigma \widehat{W}_t - \frac{1}{2}\sigma^2 t}.$$

■

Exercise 7.1.5 *Assume $\mu \neq r$ and let $u < t$.*

(a) *Find $\mathbb{E}^P[e^{-rt}S_t | \mathcal{F}_u]$ and show that $e^{-rt}S_t$ is not a martingale w.r.t. the probability measure P.*

(b) *Find $\mathbb{E}^Q[e^{-\mu t}S_t | \mathcal{F}_u]$ and show that $e^{-\mu t}S_t$ is not a martingale w.r.t. the probability measure Q.*

Exercise 7.1.6 *Use the reduction formulas given in Appendix C.5 to prove:*

(a) $\mathbb{E}[g(S_t)] = e^{-\lambda^2 t/2} \mathbb{E}[g(S_0 e^{\sigma W_t}) e^{\lambda W_t}]$

(b) $\mathbb{E}[g(S_t) e^{-\lambda W_t}] = e^{\lambda^2 t/2} \mathbb{E}[g(S_0 e^{\sigma W_t})]$,

where $\lambda = \mu/\sigma - \sigma/2$ and g is a measurable function.

Exercise 7.1.7 (a) *Find a formula for $\mathbb{E}[W_t e^{W_t}]$ choosing $g(x) = \ln x$ in Exercise 7.1.6 (a).*

(b) *Show that $\mathbb{E}[S_t] = S_0^{\mu t}$ by choosing $g(x) = x$ in Exercise 7.1.6 (a).*

7.2   Risk-neutral World Density Functions

The purpose of this section is to establish formulas for the densities of Brownian motions W_t and $\widehat{W}_t$ with respect to both probability measures P and Q, and discuss their relationship. This will clear some confusions that appear in practical applications when we need to choose the right probability density.

The densities of W_t and $\widehat{W}_t$ with respect to P and Q will be denoted respectively by p_P, p_Q and $\widehat{p}_P$, $\widehat{p}_Q$. Since W_t and $\widehat{W}_t$ are Brownian motions on the spaces $(\Omega, \mathcal{F}, P)$ and $(\Omega, \mathcal{F}, Q)$, respectively, they have the following normal probability densities

$$p_P(x) = \frac{1}{\sqrt{2\pi t}} e^{-\frac{x^2}{2t}} = p(x);$$

$$\widehat{p}_Q(x) = \frac{1}{\sqrt{2\pi t}} e^{-\frac{x^2}{2t}} = p(x).$$

The associated distribution functions are

$$F^P_{W_t}(x) = P(W_t \le x) = \int_{\{W_t \le x\}} dP(\omega) = \int_{-\infty}^{x} p(u)\, du;$$

$$F^Q_{\widehat{W}_t}(x) = Q(\widehat{W}_t \le x) = \int_{\{\widehat{W}_t \le x\}} dQ(\omega) = \int_{-\infty}^{x} p(u)\, du.$$

Expressing W_t in terms of $\widehat{W}_t$ and using that $\widehat{W}_t$ is normally distributed with respect to Q, we get the distribution function of W_t with respect to Q as

$$
\begin{aligned}
F^Q_{W_t}(x) &= Q(W_t \le x) = Q(\widehat{W}_t - \eta t \le x) \\
&= Q(\widehat{W}_t \le x + \eta t) = \int_{\{\widehat{W}_t \le x + \eta t\}} dQ(\omega) \\
&= \int_{-\infty}^{x+\eta t} p(y)\, dy = \int_{-\infty}^{x+\eta t} \frac{1}{\sqrt{2\pi t}} e^{-\frac{y^2}{2t}}\, dy.
\end{aligned}
$$

Differentiating yields the density function

$$p_Q(x) = \frac{d}{dx} F^Q_{W_t}(x) = \frac{1}{\sqrt{2\pi t}} e^{-\frac{1}{2t}(y+\eta t)^2}. \qquad (7.2.7)$$

It is worth noting that $p_Q(x)$ can be decomposed as

$$p_Q(x) = e^{-\eta x - \frac{1}{2}\eta^2 t} p(x),$$

which makes the connection with Girsanov theorem.

The distribution function of $\widehat{W}_t$ with respect to P can be worked out in a similar way

$$
\begin{aligned}
F^P_{\widehat{W}_t}(x) &= P(\widehat{W}_t \le x) = P(W_t + \eta t \le x) \\
&= P(W_t \le x - \eta t) = \int_{\{W_t \le x - \eta t\}} dP(\omega) \\
&= \int_{-\infty}^{x-\eta t} p(y)\, dy = \int_{-\infty}^{x-\eta t} \frac{1}{\sqrt{2\pi t}} e^{-\frac{y^2}{2t}}\, dy,
\end{aligned}
$$

so the density function is

$$p_P(x) = \frac{d}{dx}F^P_{\widehat{W}_t}(x) = \frac{1}{\sqrt{2\pi t}}e^{-\frac{1}{2t}(y-\eta t)^2}. \tag{7.2.8}$$

7.3 Self-financing Portfolios

The value of a *portfolio* that contains $\theta_j(t)$ units of security $S_j(t)$ at time t is given by

$$V(t) = \sum_{j=1}^n \theta_j(t)S_j(t). \tag{7.3.9}$$

The value process $V(t)$ is $\mathcal{F}_t$-adapted, since $\theta_j(t)$ and $S_j(t)$ are known given the information at time t. Furthermore, the process $\theta_j(t)$ is left continuous in time t. The portfolio is called *self-financing* if

$$dV(t) = \sum_{j=1}^n \theta_j(t)dS_j(t). \tag{7.3.10}$$

This can be stated by saying that an infinitesimal change in the value of the portfolio is due to infinitesimal changes in security values. This implies that there is no inflow of capital (no additional units of security are bought or sold).

Since an application of the product rule for stochastic processes provides

$$dV(t) = \sum_{j=1}^n \theta_j(t)dS_j(t) + \sum_{j=1}^n [d\theta_j(t)S_j(t) + d\theta_j(t)\,dS_j(t)],$$

then an equivalent condition for self-financing is

$$\sum_{j=1}^n d\theta_j(t)\big(S_j(t) + dS_j(t)\big) = 0. \tag{7.3.11}$$

This equation can be used to solve for the processes $\theta_i(t)$.

Example 7.3.1 Consider two securities, $S_1 = e^t$ and $S_2 = t$. The portfolio value is given by $V(t) = \theta_1(t)e^t + \theta_2(t)t$. The portfolio is self-financing if $\theta_1(t)$ and $\theta_2(t)$ satisfy

$$e^t d\theta_1(t)(1 + dt) + t\,d\theta_2(t) = 0.$$

The solution is not unique. One of the solutions is $\theta_1(t) = t^2/2$ and $\theta_2(t) = -e^t$. Another one is $\theta_1(t) = (1 + t)e^{-t}$ and $\theta_2(t) = t$. Hence, there are several self-financing portfolios that can be set up on the same securities.

Exercise 7.3.2 Let $S_1 = 1$ and $S_2 = W_t$. Show that $\theta_1(t) = -(t + W_t^2)$ and $\theta_2(t) = 2W_t$ form a self-financing portfolio.

7.4 The Sharpe Ratio

If μ is the expected rate of return on the stock S_t, the *risk premium* is defined as the difference $\mu - r$, where r is the risk-free interest rate. The *Sharpe ratio*, η, is defined as the quotient between the risk premium and stock price volatility

$$\boxed{\eta = \frac{\mu - r}{\sigma}.} \tag{7.4.12}$$

The following result shows that the Sharpe ratio is an important invariant for the family of stocks driven by the same uncertainty source. It is also known under the name of the *market price of risk* for assets. η measures in σ units how much the return μ exceeds the interest rate r.

Proposition 7.4.1 *Let S_1 and S_2 be two stocks satisfying equations* (5.2.8) − (5.2.9). *Then their Sharpe ratio are equal*

$$\frac{\mu_1 - r}{\sigma_1} = \frac{\mu_2 - r}{\sigma_2}. \tag{7.4.13}$$

Proof: Eliminating the term dW_t from equations

$$
\begin{aligned}
dS_1 &= \mu_1 S_1 dt + \sigma_1 S_1 dW_t \\
dS_2 &= \mu_2 S_2 dt + \sigma_2 S_2 dW_t
\end{aligned}
$$

yields

$$\frac{\sigma_2}{S_1} dS_1 - \frac{\sigma_1}{S_2} dS_2 = (\mu_1 \sigma_2 - \mu_2 \sigma_1) dt. \tag{7.4.14}$$

Consider the portfolio $P(t) = \theta_1(t) S_1(t) - \theta_2(t) S_2(t)$, with $\theta_1(t) = \dfrac{\sigma_2(t)}{S_1(t)}$ and $\theta_2(t) = \dfrac{\sigma_1(t)}{S_2(t)}$. Using the properties of self-financing portfolios, we have

$$
\begin{aligned}
dP(t) &= \theta_1(t) dS_1(t) - \theta_2(t) dS_2(t) \\
&= \frac{\sigma_2}{S_1} dS_1 - \frac{\sigma_1}{S_2} dS_2.
\end{aligned}
$$

Substituting in (7.4.14) yields $dP = (\mu_1 \sigma_2 - \mu_2 \sigma_1) dt$, i.e. P is a risk-less portfolio (it does not have an uncertainty term). Since the portfolio earns interest at the risk-free interest rate, we have $dP = rPdt$. Then equating the coefficients of dt yields

$$\mu_1 \sigma_2 - \mu_2 \sigma_1 = rP(t).$$

Using the definition of $P(t)$, the previous relation becomes

$$\mu_1 \sigma_2 - \mu_2 \sigma_1 = r\theta_1 S_1 - r\theta_2 S_2,$$

that can be transformed into

$$\mu_1\sigma_2 - \mu_2\sigma_1 = r\sigma_2 - r\sigma_1,$$

which is equivalent with relation (7.4.13). ∎

Using Proposition 7.1.4, relations (5.2.8) − (5.2.9) can be written as

$$\begin{aligned}
dS_1 &= rS_1 dt + \sigma_1 S_1 d\widehat{W}_t \\
dS_2 &= rS_2 dt + \sigma_2 S_2 d\widehat{W}_t,
\end{aligned}$$

where the risk-neutral process $d\widehat{W}_t$ is the same in both equations

$$d\widehat{W}_t = \frac{\mu_1 - r}{\sigma_1} dt + dW_t = \frac{\mu_2 - r}{\sigma_2} dt + dW_t = \eta dt + dW_t.$$

This shows that the process $\widehat{W}_t$ is a Brownian motion with drift, where the drift is the Sharpe ratio. Under the risk-neutral measure the process $\widehat{W}_t$ becomes a Brownian motion B_t and the price of the stock in the risk-neutral world becomes

$$S_t = S_0 e^{(r - \sigma^2/2)t + \sigma B_t}.$$

Remark 7.4.2 Proposition 7.4.1 implies the following statement: Two stocks that are driven by the same risk source have equal rates of return if they have equal volatilities, and vice-versa. We may also restate this by saying that higher return profits are the result of undertaking higher risk.

Exercise 7.4.3 *The rate of return of two stocks are $\mu_1 = 10\%$ and $\mu_2 = 12\%$ and their volatilities are given by $\sigma_1 = 15\%$ and $\sigma_2 = 25\%$, respectively. What is the risk-free interest rate r?*

Which stock should you buy? Assume we have the choice to buy for the same price one of the two stocks, S_1 and S_2, with rates of return μ_1, μ_2 and volatilities σ_1, σ_2, respectively. A legitimate question is: *which one is a better buy?*

First, we should be sure that buying the stock is more beneficial than keeping the same amount of money in a bank, so we assume the following inequalities $\mu_1 > r$ and $\mu_2 > r$ hold, where r is the risk-free rate.

Now, we make our choice based on the simultaneous optimization of the following two objectives: (i) maximize return profit; (ii) minimize risk.

If we go just by (i), then the best choice is the stock that has the difference $\mu_i - r$ largest. On the other side, optimizing (ii) yields to the choice of the stock with the smallest volatility σ_i. Considering both criteria, one would like to choose the stock with the largest quotient $\eta_i = \frac{\mu_i - r}{\sigma_i}$, i.e. the largest

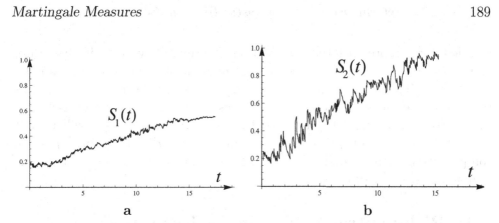

Figure 7.1: *Two stocks with the same Sharpe ratios:* **a.** *Buying stock $S_1(t)$ undertakes a small risk and provides a small return profit;* **b.** *Buying stock $S_2(t)$ provides a larger return profit, but also assumes a larger risk.*

return profit measured in units of risk. This is nothing but the Sharpe ratio. According to Proposition 7.4.1, if the stocks are driven by the same risk source, dW_t, then their Sharpe ratios are equal. Therefore, in this case, we cannot distinguish between choices, see Fig.7.1. However, if the stocks are driven by different risk factors, then the Sharpe rations might be different, and choosing the stock with the largest η_i provides the answer to the previously stated question.

Why the name of market price of risk? When one buys a stock S, he or she expects the stock value to increase during the time interval dt by more than $rSdt$, which is the profit in a risk-free world (one can do this by selling the stock and reinvesting the money into a bank at the risk-free rate r for the time interval dt). By having such expectation, one assumes a risk. What is the compensation for assuming this risk?

We can compute the *excess return above the risk-free rate* in the time dt as the difference between the earnings in the stock and the amount made in a bank at the risk-free rate

$$dS - rSdt = \mu Sdt + \sigma SdW_t - rSdt = (\mu - r)Sdt + \sigma SdW_t.$$

Dividing by S, we compute the percentage change as

$$\frac{dS - rSdt}{S} = (\mu - r)dt + \sigma dW_t = \sigma\left[\frac{\mu - r}{\mu}dt + dW_t\right],$$

which can be written in terms of the Sharpe ration as

$$\frac{dS - rSdt}{S} = \sigma(\eta dt + dW_t).$$

The left term represents the rate of excess return above the risk-free rate, and is computed as a sum between a deterministic amount, $\sigma \eta dt$, and a risky quantity, σdW_t, which is the white noise induced by the market. This has the interpretation of a compensation for assuming risk. More precisely, for each extra unit of risk dW_t assumed by the investor, he or she will benefit of an extra earning of ηdt during the time interval dt. Therefore, η is the compensation for assuming a unit of risk, per unit of time. Hence, the name of market price of risk for η.

7.5   Risk-neutral Valuation for Derivatives

The risk-neutral process $\widehat{dW_t}$ plays an important role in the risk neutral valuation of derivatives. In this section we shall prove that if f_T is the price of a derivative at the maturity time, then $f_t = \widehat{\mathbb{E}}[e^{-r(T-t)} f_T | \mathcal{F}_t]$ is the price of the derivative at the time t, for any $t < T$.

In other words, the discounted price of a derivative in the risk-neutral world is the price of the derivative at the new instance of time. This is based on the fact that $e^{-rt} f_t$ is an $\mathcal{F}_t$-martingale with respect to the risk-neutral measure Q introduced previously.

In particular, the idea of the proof can be applied for the stock S_t. Applying the product rule

$$d(e^{-rt} S_t) = d(e^{-rt}) S_t + e^{-rt} dS_t + \underbrace{d(e^{-rt}) dS_t}_{=0}$$

$$= -r e^{-rt} S_t dt + e^{-rt} (r S_t dt + \sigma S_t \widehat{dW_t})$$

$$= e^{-rt} (r S_t dt + \sigma S_t \widehat{dW_t}).$$

If $u < t$, integrating between u and t we have

$$e^{-rt} S_t = e^{-ru} S_u + \int_u^t \sigma e^{-rs} S_s \widehat{dW_s},$$

and taking the risk-neutral expectation with respect to the information set $\mathcal{F}_u$ yields

$$\widehat{\mathbb{E}}[e^{-rt} S_t | \mathcal{F}_u] = \widehat{\mathbb{E}}[e^{-ru} S_u + \int_u^t \sigma e^{-rs} S_s \widehat{dW_s} \mid \mathcal{F}_u]$$

$$= e^{-ru} S_u + \widehat{\mathbb{E}}\Big[\int_u^t \sigma e^{-rs} S_s \widehat{dW_s} | \mathcal{F}_u \Big]$$

$$= e^{-ru} S_u + \widehat{\mathbb{E}}\Big[\int_u^t \sigma e^{-rs} S_s \widehat{dW_s} \Big]$$

$$= e^{-ru} S_u,$$

since $\int_u^t \sigma e^{-rs} S_s d\widehat{W}_s$ is independent of $\mathcal{F}_u$ and the expectation of a Ito integral is zero. It follows that $e^{-rt} S_t$ is an $\mathcal{F}_t$-martingale in the risk-neutral world. The following fundamental result can be shown using a similar proof as the one encountered previously:

Theorem 7.5.1 *If $f_t = f(t, S_t)$ is the price of a derivative at time t, then $e^{-rt} f_t$ is an $\mathcal{F}_t$-martingale in the risk-neutral world, i.e.*

$$\widehat{\mathbb{E}}[e^{-rt} f_t | \mathcal{F}_u] = e^{-ru} f_u, \qquad 0 < u < t.$$

Proof: Using Ito's formula, see Appendix C.6, and the risk neutral process $dS = rSdt + \sigma Sd\widehat{W}_t$, the process followed by f_t is

$$
\begin{aligned}
df_t &= \frac{\partial f}{\partial t} dt + \frac{\partial f}{\partial S} dS + \frac{1}{2} \frac{\partial^2 f}{\partial S^2} (dS)^2 \\
&= \frac{\partial f}{\partial t} dt + \frac{\partial f}{\partial S} (rSdt + \sigma Sd\widehat{W}_t) + \frac{1}{2} \sigma^2 S^2 \frac{\partial^2 f}{\partial S^2} dt \\
&= \left(\frac{\partial f}{\partial t} + rS \frac{\partial f}{\partial S} + \frac{1}{2} \sigma^2 S^2 \frac{\partial^2 f}{\partial S^2} \right) dt + \sigma S \frac{\partial f}{\partial S} d\widehat{W}_t \\
&= rf dt + \sigma S \frac{\partial f}{\partial S} d\widehat{W}_t,
\end{aligned}
$$

where in the last identity we used that f satisfies the Black-Scholes equation (see equation (8.8.18)). Applying the product rule we obtain

$$
\begin{aligned}
d(e^{-rt} f_t) &= d(e^{-rt}) f_t + e^{-rt} df_t + \underbrace{d(e^{-rt}) df_t}_{=0} \\
&= -re^{-rt} f_t dt + e^{-rt} \left(rf_t dt + \sigma S \frac{\partial f}{\partial S} d\widehat{W}_t \right) \\
&= e^{-rt} \sigma S \frac{\partial f}{\partial S} d\widehat{W}_t.
\end{aligned}
$$

Integrating between u and t we get

$$e^{-rt} f_t = e^{-ru} f_u + \int_u^t e^{-rs} \sigma S \frac{\partial f_s}{\partial S} d\widehat{W}_s,$$

which assures that $e^{-rt} f_t$ is a martingale, since $\widehat{W}_s$ is a Brownian motion process. Using that $e^{-ru} f_u$ is $\mathcal{F}_u$-adapted, and $\int_u^t e^{-rs} \sigma S \frac{\partial f_s}{\partial S} d\widehat{W}_s$ is independent of the information set $\mathcal{F}_u$ and has zero mean, we have

$$\widehat{\mathbb{E}}[e^{-rt} f_t | \mathcal{F}_u] = e^{-ru} f_u + \widehat{\mathbb{E}}\left[\int_u^t e^{-rs} \sigma S \frac{\partial f_s}{\partial S} d\widehat{W}_s \right] = e^{-ru} f_u.$$

■

Exercise 7.5.2 *Show the following:*

(a) $\widehat{\mathbb{E}}[e^{\sigma(W_t - W_u)}|\mathcal{F}_u] = e^{(r-\mu+\frac{1}{2}\sigma^2)(t-u)}, \quad u < t;$

(b) $\widehat{\mathbb{E}}\left[\frac{S_t}{S_u}|\mathcal{F}_u\right] = e^{(\mu-\frac{1}{2}\sigma^2)(t-u)}\widehat{\mathbb{E}}\left[e^{\sigma(W_t - W_u)}|\mathcal{F}_u\right], \quad u < t;$

(c) $\widehat{\mathbb{E}}\left[\frac{S_t}{S_u}|\mathcal{F}_u\right] = e^{r(t-u)}, \quad u < t.$

Exercise 7.5.3 *Find the following risk-neutral world conditional expectations:*

(a) $\widehat{\mathbb{E}}[\int_0^t S_u\, du|\mathcal{F}_s], \quad s < t;$

(b) $\widehat{\mathbb{E}}[S_t \int_0^t S_u\, du|\mathcal{F}_s], \quad s < t;$

(c) $\widehat{\mathbb{E}}[\int_0^t S_u\, dW_u|\mathcal{F}_s], \quad s < t;$

(d) $\widehat{\mathbb{E}}[S_t \int_0^t S_u\, dW_u|\mathcal{F}_s], \quad s < t;$

(e) $\widehat{\mathbb{E}}[\left(\int_0^t S_u\, du\right)^2|\mathcal{F}_s], \quad s < t;$

Exercise 7.5.4 *Use risk-neutral valuation to find the price of a derivative that pays at maturity the following payoffs:*

(a) $f_T = TS_T;$

(b) $f_T = \int_0^T S_u\, du;$

(c) $f_T = \left(\int_0^T S_u\, du\right)^2.$

Part IV

PDE Approach

Chapter 8

Black-Scholes Analysis

This chapter provides an alternative way to price derivative securities using tools of partial differential equations. The idea is to eliminate the stochastic news factor and to obtain a deterministic partial differential equation, which was considered for the first time by Black, Scholes and Merton in the early 1970s.

8.1 Heat Equation

This section is devoted to a basic discussion about the heat equation. Its importance resides in the remarkable fact that the Black-Scholes equation, which is the main differential equation satisfied by the value of a derivative, can be reduced to this type of equation.

Let $u(\tau, x)$ denote the temperature at point x and time τ in an infinite rod. In the absence of exterior heat sources the heat diffuses according to the following parabolic differential equation

$$\frac{\partial u}{\partial \tau} - \frac{\partial^2 u}{\partial x^2} = 0, \tag{8.1.1}$$

called *the heat equation*. If the initial heat distribution is known and is given by $u(0, x) = f(x)$, then we have an initial value problem for the heat equation.

Solving this equation involves a convolution between the initial temperature $f(x)$ and the fundamental solution of the heat equation $G(\tau, x)$, which will be defined next.

Definition 8.1.1 *The function*

$$G(\tau, x) = \frac{1}{\sqrt{4\pi\tau}} e^{-\frac{x^2}{4\tau}}, \qquad \tau > 0,$$

is called the fundamental solution of the heat equation (8.1.1).

We recall the most important properties of the function $G(\tau, x)$.

- $G(\tau, x)$ has the properties of a probability density[1], i.e.

  1. $G(\tau, x) > 0, \qquad \forall x \in \mathbb{R}, \tau > 0$;

  2. $\displaystyle\int_{\mathbb{R}} G(\tau, x) \, dx = 1, \qquad \forall \tau > 0.$

- it satisfies the heat equation

$$\frac{\partial G}{\partial \tau} - \frac{\partial^2 G}{\partial x^2} = 0, \quad \tau > 0.$$

- G tends to the Dirac delta function $\delta(x)$ as τ gets closer to the initial time

$$\lim_{\tau \searrow 0} G(\tau, x) = \delta(x).$$

We recall that the Dirac delta function is defined as a measure using integration as

$$\int_{\mathbb{R}} \varphi(x)\delta(x)dx = \varphi(0),$$

for any smooth function with compact support φ. Consequently, using a change of variables, we also have

$$\int_{\mathbb{R}} \varphi(x)\delta(x - y)dx = \varphi(y).$$

One can think of $\delta(x)$ as a measure with infinite value at $x = 0$ and zero for the rest of the values, and with the integral equal to 1, see Fig.8.1.

The physical significance of the fundamental solution $G(\tau, x)$ is that it describes the heat evolution in the infinite rod after an initial heat impulse of infinite size is applied at $x = 0$.

Proposition 8.1.2 *The solution of the initial value heat equation*

$$\frac{\partial u}{\partial \tau} - \frac{\partial^2 u}{\partial x^2} = 0$$
$$u(0, x) = f(x)$$

is given by the convolution between the fundamental solution and the initial temperature

$$u(\tau, x) = \int_{\mathbb{R}} G(\tau, y - x)f(y) \, dy, \qquad \tau > 0.$$

---

[1]In fact it is a Gaussian probability density.

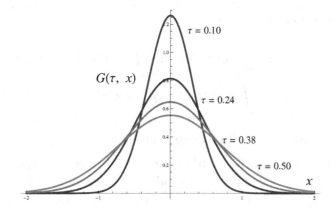

$G(\tau,\ x)$

$\tau = 0.10$

$\tau = 0.24$

$\tau = 0.38$

$\tau = 0.50$

x

Figure 8.1: *The function $G(\tau, x)$ tends to the Dirac measure $\delta(x)$ as $\tau \searrow 0$, and flattens out as $\tau \to \infty$.*

Proof: Substituting $z = y - x$, the solution can be written as

$$u(\tau, x) = \int_{\mathbb{R}} G(\tau, z) f(x + z)\, dz. \qquad (8.1.2)$$

Differentiating under the integral yields

$$\frac{\partial u}{\partial \tau} = \int_{\mathbb{R}} \frac{\partial G(\tau, z)}{\partial \tau} f(x + z)\, dz,$$

$$\frac{\partial^2 u}{\partial x^2} = \int_{\mathbb{R}} G(\tau, z) \frac{\partial^2 f(x + z)}{\partial x^2}\, dz = \int_{\mathbb{R}} G(\tau, z) \frac{\partial^2 f(x + z)}{\partial z^2}\, dz$$

$$= \int_{\mathbb{R}} \frac{\partial^2 G(\tau, z)}{\partial z^2} f(x + z)\, dz,$$

where we applied integration by parts twice and the fact that

$$\lim_{z \to \infty} G(\tau, z) = \lim_{z \to \infty} \frac{\partial G(\tau, z)}{\partial z} = 0.$$

Since G satisfies the heat equation, we have

$$\frac{\partial u}{\partial \tau} - \frac{\partial^2 u}{\partial x^2} = \int_{\mathbb{R}} \left[\frac{\partial G(\tau, z)}{\partial \tau} - \frac{\partial^2 G(\tau, z)}{\partial z^2} \right] f(x + z)\, dz = 0.$$

Since the limit and the integral commute[2], using the properties of the Dirac

---

[2]This is allowed by the dominated convergence theorem.

delta function, we have

$$
\begin{aligned}
u(0, x) &= \lim_{\tau \searrow 0} u(\tau, x) = \lim_{\tau \searrow 0} \int_{\mathbb{R}} G(\tau, z) f(x + z) \, dz \\
&= \int_{\mathbb{R}} \delta(z) f(x + z) \, dz = f(x).
\end{aligned}
$$

Hence (8.1.2) satisfies the initial value heat equation. ∎

It is worth noting that the solution $u(\tau, x) = \int_{\mathbb{R}} G(\tau, y - x) f(y) \, dy$ provides the temperature at each point in the rod x at any time $\tau > 0$, but it cannot provide the temperature for $\tau < 0$. This is because of the singularity the fundamental solution exhibits at $\tau = 0$. We can reformulate this by saying that the heat equation is *semi-deterministic*, in the sense that given the present, we can know the future but not the past.

The semi-deterministic character of diffusion phenomena can be exemplified with a drop of ink which starts diffusing in a bucket of water at time $t = 0$. We can determine the density of the ink at any time $t > 0$ at any point x in the bucket. However, given the density of ink at a given time $t > 0$, it is not possible to trace back in time the ink distribution density and to find the initial point where the drop started its diffusion.

The semi-deterministic behavior occurs in the study of derivatives too. In the case of the Black-Scholes equation, which is a backwards heat equation[3], given the present value of the derivative, we can find the past values but not the future ones. This is the capital difficulty in foreseeing the prices of stock market instruments from the present prices. This difficulty will be overcame by working backward the price from the given final condition, which is the payoff at maturity.

8.2 What is a Portfolio?

A *portfolio* is a position in the market that consists in long and short positions in one or more stocks and other securities. The value of a portfolio can be represented algebraically as a linear combination of stock prices and other securities' values:

$$
P = \sum_{j=1}^{n} a_j S_j + \sum_{k=1}^{m} b_k F_k. \tag{8.2.3}
$$

---

[3]This comes from the fact that at some point τ becomes $T - t$ due to a substitution.

The market participant holds a_j units of stock S_j and b_k units of derivative F_k. The coefficients are positive for long positions and negative for short positions. For instance, a portfolio given by $P = 2F - 3S$ means that we buy 2 securities and sell 3 units of stock (this is a position with 2 securities long and 3 stocks short).

We recall from section 7.3 that the portfolio is self-financing if

$$dP = \sum_{j=1}^{n} a_j dS_j + \sum_{k=1}^{m} b_k dF_k,$$

i.e. the change in the portfolio's value is caused only by the change in the values of the asset prices and securities values and is not due to any capital inflows or outflows.

8.3 Risk-less Portfolios

A portfolio P is called *risk-less* if the increments dP are completely predictable. In this case the increments' value dP should equal the interest earned in the time interval dt on the portfolio P. This can be written as

$$dP = rPdt, \tag{8.3.4}$$

where r denotes the risk-free interest rate. For the sake of simplicity the rate r will be assumed constant, unless specified otherwise.

We assume now that the portfolio P depends on only one stock, S, and one derivative, F, whose underlying asset is S. The portfolio depends also on time t, since the coefficients a_j, b_k are time-dependent, so

$$P = P(t, S, F).$$

We are interested in deriving the stochastic differential equation satisfied by the portfolio value P. We note that at this moment the portfolio is not assumed risk-less. By Ito's formula, see Appendix C.6, we get

$$dP = \frac{\partial P}{\partial t}dt + \frac{\partial P}{\partial S}dS + \frac{\partial P}{\partial F}dF + \frac{1}{2}\frac{\partial^2 P}{\partial S^2}dS^2 + \frac{1}{2}\frac{\partial^2 P}{\partial F^2}(dF)^2. \tag{8.3.5}$$

The stock S is assumed to follow the geometric Brownian motion

$$dS = \mu S dt + \sigma S dW_t, \tag{8.3.6}$$

where the expected return rate on the stock μ and the stock's volatility σ are constants. Since the derivative F depends on time and underlying stock, we can write $F = F(t, S)$. Applying Ito's formula, see Appendix C.6, yields

$$
\begin{aligned}
dF &= \frac{\partial F}{\partial t}dt + \frac{\partial F}{\partial S}dS + \frac{1}{2}\frac{\partial^2 F}{\partial S^2}(dS)^2 \\
&= \left(\frac{\partial F}{\partial t} + \mu S\frac{\partial F}{\partial S} + \frac{1}{2}\sigma^2 S^2\frac{\partial^2 F}{\partial S^2}\right)dt + \sigma S\frac{\partial F}{\partial S}dW_t, \quad (8.3.7)
\end{aligned}
$$

where we have used (8.3.6). Taking the square in relations (8.3.6) and (8.3.7), and using the stochastic relations $(dW_t)^2 = dt$ and $dt^2 = dW_t dt = 0$, we get

$$
\begin{aligned}
(dS)^2 &= \sigma^2 S^2 dt \\
(dF)^2 &= \sigma^2 S^2\left(\frac{\partial F}{\partial S}\right)^2 dt.
\end{aligned}
$$

Substituting back in (8.3.5), and collecting the deterministic and stochastic parts, yields

$$
\begin{aligned}
dP &= \left[\frac{\partial P}{\partial t} + \mu S\left(\frac{\partial P}{\partial S} + \frac{\partial P}{\partial F}\frac{\partial F}{\partial S}\right) + \frac{\partial P}{\partial F}\left(\frac{\partial F}{\partial t} + \frac{1}{2}\sigma^2 S^2\frac{\partial^2 F}{\partial S^2}\right)\right. \\
&\quad \left. + \frac{1}{2}\sigma^2 S^2\left(\frac{\partial^2 P}{\partial S^2} + \frac{\partial^2 P}{\partial F^2}\left(\frac{\partial F}{\partial S}\right)^2\right)\right]dt \\
&\quad + \sigma S\left(\frac{\partial P}{\partial S} + \frac{\partial P}{\partial F}\frac{\partial F}{\partial S}\right)dW_t. \quad (8.3.8)
\end{aligned}
$$

We have the following characterization result:

Proposition 8.3.1 *The portfolio P is risk-less if and only if $\dfrac{dP}{dS} = 0$.*

Proof: A portfolio P is risk-less if and only if its unpredictable component, which is the coefficient of dW_t, is identically zero, i.e.

$$
\frac{\partial P}{\partial S} + \frac{\partial P}{\partial F}\frac{\partial F}{\partial S} = 0.
$$

Since the total derivative of P is given by the chain rule

$$
\frac{dP}{dS} = \frac{\partial P}{\partial S} + \frac{\partial P}{\partial F}\frac{\partial F}{\partial S},
$$

the previous relation becomes $\dfrac{dP}{dS} = 0$. ∎

Definition 8.3.2 *The amount $\Delta_P = \dfrac{dP}{dS}$ is called the delta of the portfolio P.*

The previous result can be reformulated by stating that a portfolio is risk-less if and only if its delta vanishes. In practice this can hold only for a short amount of time, so the portfolio needs to be re-balanced periodically. The process of making a portfolio risk-less involves a procedure called delta hedging, through which the portfolio's delta becomes zero or very close to this value.

Assume that P is a risk-less portfolio, so

$$\frac{dP}{dS} = \frac{\partial P}{\partial S} + \frac{\partial P}{\partial F}\frac{\partial F}{\partial S} = 0. \tag{8.3.9}$$

Then equation (8.3.8) simplifies to

$$dP = \left[\frac{\partial P}{\partial t} + \frac{\partial P}{\partial F}\left(\frac{\partial F}{\partial t} + \frac{1}{2}\sigma^2 S^2 \frac{\partial^2 F}{\partial S^2}\right)\right.$$
$$\left. + \frac{1}{2}\sigma^2 S^2\left(\frac{\partial^2 P}{\partial S^2} + \frac{\partial^2 P}{\partial F^2}\left(\frac{\partial F}{\partial S}\right)^2\right)\right] dt. \tag{8.3.10}$$

Comparing with (8.3.4) yields

$$\boxed{\frac{\partial P}{\partial t} + \frac{\partial P}{\partial F}\left(\frac{\partial F}{\partial t} + \frac{1}{2}\sigma^2 S^2 \frac{\partial^2 F}{\partial S^2}\right) + \frac{1}{2}\sigma^2 S^2\left(\frac{\partial^2 P}{\partial S^2} + \frac{\partial^2 P}{\partial F^2}\left(\frac{\partial F}{\partial S}\right)^2\right) = rP.}$$
$$\tag{8.3.11}$$

This is the equation satisfied by a risk-free financial instrument, $P = P(t, S, F)$, that depends on time t, stock S and derivative price F.

8.4 Black-Scholes Equation

This section deals with a parabolic partial differential equation satisfied by all European-type securities, called the *Black-Scholes* equation. This was initially used by Black and Scholes to find the value of options. This is a deterministic equation obtained by eliminating the unpredictable component of the derivative by making a risk-less portfolio. The main reason for this being possible is the fact that both the derivative F and the stock S are driven by the same source of uncertainty.

The next result holds in a market with the following restrictive conditions:

- the risk-free rate r and stock volatility σ are constant.

- there are no arbitrage opportunities.

- no transaction costs.

Proposition 8.4.1 *If $F(t, S)$ is a derivative defined for $t \in [0, T]$, then*

$$\boxed{\frac{\partial F}{\partial t} + rS\frac{\partial F}{\partial S} + \frac{1}{2}\sigma^2 S^2 \frac{\partial^2 F}{\partial S^2} = rF.} \qquad (8.4.12)$$

Proof: The equation (8.3.11) works under the general hypothesis that $P = P(t, S, F)$ is a risk-free financial instrument that depends on time t, stock S and derivative F. We shall consider P to be the following particular portfolio

$$P = F - \lambda S.$$

This means to take a long position in derivative and a short position in λ units of stock (if assuming λ positive). The partial derivatives in this case are

$$\frac{\partial P}{\partial t} = 0, \qquad \frac{\partial P}{\partial F} = 1, \qquad \frac{\partial P}{\partial S} = -\lambda,$$

$$\frac{\partial^2 P}{\partial F^2} = 0, \qquad \frac{\partial^2 P}{\partial S^2} = 0.$$

From the risk-less property (8.3.9) we get $\lambda = \dfrac{\partial F}{\partial S}$. Substituting into equation (8.3.11) yields

$$\frac{\partial F}{\partial t} + \frac{1}{2}\sigma^2 S^2 \frac{\partial^2 F}{\partial S^2} = rF - rS\frac{\partial F}{\partial S},$$

which is equivalent to the desired equation. ∎

Remark 8.4.2 It is striking that the value of a derivative $F(t, S)$ satisfies a deterministic equation, which is of parabolic type, and which does not depend on the stock's rate of return μ.

However, the Black-Scholes equation is derived most often in a more heuristic way. This is based on the assumption that the number $\lambda = \frac{\partial F}{\partial S}$, which appears in the formula of the risk-less portfolio $P = F - \lambda S$, is considered constant for the time interval Δt. If we consider the increments over the time interval Δt given by

$$\begin{aligned}
\Delta W_t &= W_{t+\Delta t} - W_t \\
\Delta S &= S_{t+\Delta t} - S_t \\
\Delta F &= F(t + \Delta t, S_t + \Delta S) - F(t, S),
\end{aligned}$$

then Ito's formula yields

$$\begin{aligned}
\Delta F = {}& \left(\frac{\partial F}{\partial t}(t, S) + \mu S\frac{\partial F}{\partial S}(t, S) + \frac{1}{2}\sigma^2 S^2 \frac{\partial^2 F}{\partial S^2}(t, S) \right)\Delta t \\
&+ \sigma S\frac{\partial F}{\partial S}(t, S)\Delta W_t.
\end{aligned}$$

On the other side, the increments in the stock are given by

$$\Delta S = \mu S \Delta t + \sigma S \Delta W_t.$$

Since both increments ΔF and ΔS are driven by the same uncertainly source, ΔW_t, we can eliminate it by multiplying the latter equation by $\frac{\partial F}{\partial S}$ and subtracting it from the former

$$\Delta F - \frac{\partial F}{\partial S}(t, S)\Delta S = \left(\frac{\partial F}{\partial t}(t, S) + \frac{1}{2}\sigma^2 S^2 \frac{\partial^2 F}{\partial S^2}(t, S)\right)\Delta t.$$

The left side can be regarded as the increment ΔP in the portfolio

$$P = F - \frac{\partial F}{\partial S}S$$

during the time interval Δt. This portfolio is risk-less because its increments are totally deterministic, so it must also satisfy $\Delta P = rP\Delta t$. The number $\frac{\partial F}{\partial S}$ is assumed constant for small intervals of time Δt. Even if this assumption is not rigorous enough, the procedure still leads to the right equation. This is obtained by equating the coefficients of Δt in the last two equations

$$\frac{\partial F}{\partial t}(t, S) + \frac{1}{2}\sigma^2 S^2 \frac{\partial^2 F}{\partial S^2}(t, S) = r\left(F - \frac{\partial F}{\partial S}S\right),$$

which is equivalent to the Black-Scholes equation.

8.5 Delta Hedging

The proof for the Black-Scholes' equation is based on the fact that the portfolio $P = F - \frac{\partial F}{\partial S}S$ is risk-less. Since the delta of the derivative F is

$$\Delta_F = \frac{dF}{dS} = \frac{\partial F}{\partial S},$$

then the portfolio $P = F - \Delta_F S$ is risk-less. This leads to the *delta-hedging* procedure, by which selling Δ_F units of the underlying stock S yields a risk-less investment.

Example 8.5.1 The value of the risk-free portfolio $P = F - \Delta_F S$ in the case when F is a call option is

$$P = F - \Delta_F S = c - N(d_1)S = -Ke^{-r(T-t)},$$

which is the value of a short bond, and hence risk-less.

8.6 Tradable Securities

A derivative $F(t, S)$ that is a solution of the Black-Scholes equation is called a *tradable security*. Its name comes from the fact that it can be traded (either on an exchange or over-the-counter). The Black-Scholes equation constitutes the equilibrium relation that provides the traded price of the derivative. We shall deal next with a few examples of tradable securities.

Example 8.6.1 (i) It is easy to show that $F = S$ is a solution of the Black-Scholes equation. Hence the stock is a tradable security.

(ii) If K is a constant, then $F = e^{rt}K$ is a tradable security. Also the price of a bond, $e^{-r(T-t)}K$, is a tradable security.

(iii) If S is the stock price, then $F = e^S$ is not a tradable security, since F does not satisfy equation (8.4.12).

Exercise 8.6.1 *Show that $F = \ln S$ is not a tradable security.*

Example 8.6.2 We shall find all constants α such that S^α is tradable. Substituting $F = S^\alpha$ into equation (8.4.12) we obtain

$$rS\alpha S^{\alpha-1} + \frac{1}{2}\sigma^2 S^2 \alpha(\alpha - 1)S^{\alpha-2} = rS^\alpha.$$

Dividing by S^α yields $r\alpha + \frac{1}{2}\sigma^2\alpha(\alpha - 1) = r$. This can be factorized as

$$\frac{1}{2}\sigma^2(\alpha - 1)(\alpha + \frac{2r}{\sigma^2}) = 0,$$

with two distinct solutions $\alpha_1 = 1$ and $\alpha_2 = -\dfrac{2r}{\sigma^2}$. Hence there are only two tradable securities that are powers of the stock: the stock itself, S, and S^{-2r/σ^2}. In particular, S^2 is not tradable, since $-2r/\sigma^2 \neq 2$ (the left side is negative). The role of these two cases will be clarified by the next result.

Proposition 8.6.2 *The general form of a traded derivative, which does not depend explicitly on time, is given by*

$$F(S) = C_1 S + C_2 S^{-2r/\sigma^2}, \tag{8.6.13}$$

with C_1, C_2 constants.

Proof: If the derivative depends solely on the stock, $F = F(S)$, then the Black-Scholes equation becomes the ordinary differential equation

$$rS\frac{dF}{dS} + \frac{1}{2}\sigma^2 S^2 \frac{d^2F}{dS^2} = rF. \tag{8.6.14}$$

This is an Euler-type equation, which can be solved by using the substitution $S = e^x$. The derivatives $\dfrac{d}{dS}$ and $\dfrac{d}{dx}$ are related by the chain rule

$$\frac{d}{dx} = \frac{dS}{dx}\frac{d}{dS}.$$

Since $\dfrac{dS}{dx} = \dfrac{de^x}{dx} = e^x = S$, it follows that $\dfrac{d}{dx} = S\dfrac{d}{dS}$. Using the product rule,

$$\frac{d^2}{dx^2} = S\frac{d}{dS}\left(S\frac{d}{dS}\right) = S\frac{d}{dS} + S^2\frac{d^2}{dS^2},$$

and hence

$$S^2\frac{d^2}{dS^2} = \frac{d^2}{dx^2} - \frac{d}{dx}.$$

Substituting into (8.6.14) yields

$$\frac{1}{2}\sigma^2\frac{d^2G(x)}{dx^2} + \left(r - \frac{1}{2}\sigma^2\right)\frac{dG(x)}{dx} = rG(x)$$

where $G(x) = G(e^x) = F(S)$. The associated indicial equation

$$\frac{1}{2}\sigma^2\alpha^2 - \left(r - \frac{1}{2}\sigma^2\right)\alpha = r$$

has solutions $\alpha_1 = 1$, $\alpha_2 = -r/\sigma^2$, so the general solution has the form

$$G(x) = C_1 e^x + C_2 e^{-\frac{r}{\sigma^2}x},$$

which is equivalent to (8.6.13) after subtituting back $S = e^x$. ∎

Exercise 8.6.3 *Show that the price of a forward contract, which is given by $F(t, S) = S - Ke^{-r(T-t)}$, satisfies the Black-Scholes equation, i.e. a forward contract is a tradable derivative.*

Exercise 8.6.4 *Show that the bond $F(t) = e^{-r(T-t)}K$ is a tradable security.*

Exercise 8.6.5 *Let d_1 and d_2 be given by*

$$d_1 = d_2 + \sigma\sqrt{T-t}$$
$$d_2 = \frac{\ln(S_t/K) + (r - \frac{\sigma^2}{2})(T-t)}{\sigma\sqrt{T-t}}.$$

Show that the following functions satisfy the Black-Scholes equation:
(a) $F_1(t, S) = SN(d_1)$
(b) $F_2(t, S) = e^{-r(T-t)}N(d_2)$
(c) $F_2(t, S) = SN(d_1) - Ke^{-r(T-t)}N(d_2)$.
(d) To which well-known derivatives do these formulas correspond to?

8.7 Risk-less Investment Revised

A risk-less investment is given by a portfolio $P(t, S, F)$, which depends on time t, stock price S, and derivative F with the underlying asset S, which satisfies equation (8.3.11). Using that the derivative F verifies the Black-Scholes equation

$$\frac{\partial F}{\partial t} + \frac{1}{2}\sigma^2 S^2 \frac{\partial^2 F}{\partial S^2} = rF - rS\frac{\partial F}{\partial S},$$

equation (8.3.11) becomes

$$\frac{\partial P}{\partial t} + \frac{\partial P}{\partial F}\left(rF - rS\frac{\partial F}{\partial S}\right) + \frac{1}{2}\sigma^2 S^2\left(\frac{\partial^2 P}{\partial S^2} + \frac{\partial^2 P}{\partial F^2}\left(\frac{\partial F}{\partial S}\right)^2\right) = rP.$$

Using the risk-less condition (8.3.9)

$$\frac{\partial P}{\partial S} = -\frac{\partial P}{\partial F}\frac{\partial F}{\partial S}, \tag{8.7.15}$$

the previous equation becomes

$$\boxed{\frac{\partial P}{\partial t} + rS\frac{\partial P}{\partial S} + rF\frac{\partial P}{\partial F} + \frac{1}{2}\sigma^2 S^2\left[\frac{\partial^2 P}{\partial S^2} + \frac{\partial^2 P}{\partial F^2}\left(\frac{\partial F}{\partial S}\right)^2\right] = rP.} \tag{8.7.16}$$

In the following we shall find an equivalent expression for the term in the brackets on the left side. Differentiating in (8.7.15) with respect to F yields

$$\frac{\partial^2 P}{\partial F \partial S} = -\frac{\partial^2 P}{\partial F^2}\frac{\partial F}{\partial S} - \frac{\partial P}{\partial F}\frac{\partial^2 F}{\partial F \partial S}$$
$$= -\frac{\partial^2 P}{\partial F^2}\frac{\partial F}{\partial S},$$

where we used

$$\frac{\partial^2 F}{\partial F \partial S} = \frac{\partial}{\partial S}\frac{\partial F}{\partial F} = \frac{\partial}{\partial S}(1) = 0.$$

Multiplying by $\dfrac{\partial F}{\partial S}$ implies

$$\frac{\partial^2 P}{\partial F^2}\left(\frac{\partial F}{\partial S}\right)^2 = -\frac{\partial^2 P}{\partial F \partial S}\frac{\partial F}{\partial S}.$$

Substituting in the aforementioned equation yields the Black-Scholes equation for portfolios $P = P(t, S, F)$

$$\boxed{\frac{\partial P}{\partial t} + rS\frac{\partial P}{\partial S} + rF\frac{\partial P}{\partial F} + \frac{1}{2}\sigma^2 S^2\left[\frac{\partial^2 P}{\partial S^2} - \frac{\partial^2 P}{\partial F \partial S}\frac{\partial F}{\partial S}\right] = rP.} \tag{8.7.17}$$

We note that if F does not appear as a variable, i.e. if $P = P(t, S)$, then the equation (8.7.17) becomes the regular Black-Scholes equation (8.4.12).

We have seen in section 8.4 that $P = F - \frac{\partial F}{\partial S} S$ is a risk-less investment, in fact a risk-less portfolio. We shall discuss in the following another risk-less investment.

Example 8.7.1 If a risk-less investment P has the variables S and F separable, with $P(S, F) = f(F) + g(S)$, and f and g smooth functions, then

$$P(S, F) = F + c_1 S + c_2 S^{-2r/\sigma^2},$$

with c_1, c_2 constants. Furthermore, the derivative F is given by the formula

$$F(t, S) = -c_1 S - c_2 S^{-2r/\sigma^2} + c_3 e^{rt}, \qquad c_3 \in \mathbb{R}.$$

Proof: Since P has separable variables, the mixed derivative term vanishes, and the equation (8.7.17) becomes

$$Sg'(S) + \frac{\sigma^2}{2r} S^2 g''(S) - g(S) = f(F) - Ff'(F).$$

There is a separation constant C such that

$$f(F) - Ff'(F) = C$$
$$Sg'(S) + \frac{\sigma^2}{2r} S^2 g''(S) - g(S) = C.$$

Dividing the first equation by F^2 yields the exact equation

$$\left(\frac{1}{F} f(F)\right)' = -\frac{C}{F^2},$$

with the solution $f(F) = c_0 F + C$. To solve the second equation, let $\kappa = \frac{\sigma^2}{2r}$. Then the substitution $S = e^x$ leads to the ordinary differential equation with constant coefficients

$$\kappa h''(x) + (1 - \kappa)h'(x) - h(x) = C,$$

where $h(x) = g(e^x) = g(S)$. The associated characteristic equation

$$\kappa \lambda^2 + (1 - \kappa)\lambda - 1 = 0$$

has solutions $\lambda_1 = 1$, $\lambda_2 = -\frac{1}{\kappa}$. The general solution is the sum between the particular solution $h_p(x) = -C$ and the solution of the associated homogeneous equation, which is $h_0(x) = c_1 e^x + c_2 e^{-\frac{1}{\kappa}x}$. Then

$$h(x) = c_1 e^x + c_2 e^{-\frac{1}{\kappa}x} - C.$$

Going back to the variable S, we get the general form of $g(S)$

$$g(S) = c_1 S + c_2 S^{-2r/\sigma^2} - C,$$

with c_1, c_2 constants. Since the constant C cancels by addition, we have the following formula for the risk-less investment with separable variables F and S:

$$P(S, F) = f(F) + g(S) = c_0 F + c_1 S + c_2 S^{-2r/\sigma^2}.$$

Dividing by c_0, we may assume $c_0 = 1$. We shall find the derivative $F(t, S)$ which enters the previous formula. Substituting in (8.7.15) yields

$$-\frac{\partial F}{\partial S} = c_1 - \frac{2r}{\sigma^2} c_2 S^{-1 - 2r/\sigma^2},$$

which after partial integration in S gives

$$F(t, S) = -c_1 S - c_2 S^{-2r/\sigma^2} + \phi(t),$$

where the integration constant $\phi(t)$ is a function of t. The sum of the first two terms is the derivative given by formula (8.6.13). The remaining function $\phi(t)$ has also to satisfy the Black-Scholes equation, and hence it is of the form $\phi(t) = c_3 e^{rt}$, with c_3 constant. Then the derivative F is given by

$$F(t, S) = -c_1 S - c_2 S^{-2r/\sigma^2} + c_3 e^{rt}.$$

It is worth noting that substituting in the formula of P yields $P = c_3 e^{rt}$, which agrees with the formula of a risk-less investment. ∎

Example 8.7.1 Find the function $g(S)$ such that the product $P = Fg(S)$ is a risk-less investment, with $F = F(t, S)$ derivative. Find the expression of the derivative F in terms of S and t.

Substituting $P = Fg(S)$ into equation (8.7.16) and simplifying by rF yields

$$S \frac{dg(S)}{dS} + \frac{\sigma^2}{2r} S^2 \frac{d^2 g(S)}{dS^2} = 0.$$

Substituting $S = e^x$, and $h(x) = g(e^x) = g(S)$ yields

$$h''(x) + \left(\frac{2r}{\sigma^2} - 1 \right) h'(x) = 0.$$

Integrating leads to the solution

$$h(x) = C_1 + C_2 e^{(1 - \frac{2r}{\sigma^2})x}.$$

Going back to variable S we have

$$g(S) = h(\ln S) = C_1 + C_2 e^{(1 - \frac{2r}{\sigma^2}) \ln S} = C_1 + C_2 S^{1 - \frac{2r}{\sigma^2}}.$$

Next we shall find F. Substituting the partial derivatives

$$\frac{\partial P}{\partial S} = C_2 \left(1 - \frac{2r}{\sigma^2}\right) FS^{-\frac{2r}{\sigma^2}}$$

$$\frac{\partial P}{\partial F} = C_1 + S^{1 - \frac{2r}{\sigma^2}}$$

into the risk-less condition (8.3.9) yields

$$\frac{\partial F}{\partial S} = -\frac{\partial P/\partial S}{\partial P/\partial F} = -C_2 \left(1 - \frac{2r}{\sigma^2}\right) \frac{FS^{-\frac{2r}{\sigma^2}}}{C_1 + C_2 S^{1 - \frac{2r}{\sigma^2}}}.$$

Let $Y = \ln F$. Then

$$\frac{\partial Y}{\partial S} = \frac{1}{F} \frac{\partial F}{\partial S} = \frac{C_2 \left(\frac{2r}{\sigma^2} - 1\right)}{C_1 S^{\frac{2r}{\sigma^2}} + C_2 S}.$$

Integrating we have

$$Y = \int \frac{C_2 \left(\frac{2r}{\sigma^2} - 1\right)}{C_1 S^{\frac{2r}{\sigma^2}} + C_2 S} \, dS + \varphi(t) = \ln \frac{S^{\frac{2r}{\sigma^2}}}{C_2 S + C_1 S^{\frac{2r}{\sigma^2}}} + \varphi(t)$$

and hence

$$F = e^Y = \frac{S^{\frac{2r}{\sigma^2}}}{C_2 S + C_1 S^{\frac{2r}{\sigma^2}}} e^{\varphi(t)} = \frac{e^{\varphi(t)}}{g(S)}.$$

Since $P = Fg(S) = e^{\varphi(t)}$ satisfies the Black-Scholes equation, it follows that $\varphi(t) = rt + C$.

8.8 Solving the Black-Scholes

In this section we shall solve the Black-Scholes equation and show that its solution coincides with the one provided by the risk-neutral valuation in Chapter 6. This way, the Black-Scholes equation provides a variant approach for valuing European-type derivatives by using partial differential equations instead of expectations.

Consider a European-type derivative F, with the payoff at maturity T given by f_T, which is a function of the stock price at maturity, S_T. Then $F(t, S)$ satisfies the following final condition partial differential equation

$$\frac{\partial F}{\partial t} + rS\frac{\partial F}{\partial S} + \frac{1}{2}\sigma^2 S^2 \frac{\partial^2 F}{\partial S^2} = rF \tag{8.8.18}$$

$$F(T, S_T) = f_T(S_T). \tag{8.8.19}$$

This means the solution is known at the final time T and we need to find its expression at any time t prior to T, i.e.

$$f_t = F(t, S_t), \qquad 0 \le t < T.$$

First we shall transform the equation into an equation with constant coefficients. Substituting $S = e^x$, and using the identities

$$S\frac{\partial}{\partial S} = \frac{\partial}{\partial x}, \qquad S^2\frac{\partial^2}{\partial S^2} = \frac{\partial^2}{\partial x^2} - \frac{\partial}{\partial x}$$

the equation becomes

$$\frac{\partial V}{\partial t} + \frac{1}{2}\sigma^2 \frac{\partial^2 V}{\partial x^2} + \left(r - \frac{1}{2}\sigma^2\right)\frac{\partial V}{\partial x} = rV, \tag{8.8.20}$$

where $V(t, x) = F(t, e^x)$. Using the time scaling $\tau = \frac{1}{2}\sigma^2(T - t)$, the chain rule provides

$$\frac{\partial}{\partial t} = \frac{\partial\tau}{\partial t}\frac{\partial}{\partial\tau} = -\frac{1}{2}\sigma^2\frac{\partial}{\partial\tau}.$$

Denote $k = \frac{2r}{\sigma^2}$. Substituting in the aforementioned equation yields

$$\frac{\partial W}{\partial\tau} = \frac{\partial^2 W}{\partial x^2} + (k - 1)\frac{\partial W}{\partial x} - kW, \tag{8.8.21}$$

where $W(\tau, x) = V(t, x)$. Next we shall get rid of the last two terms on the right side of the equation by using a crafted substitution.

Consider $W(\tau, x) = e^\varphi u(\tau, x)$, where $\varphi = \alpha x + \beta\tau$, with α, β constants that will be determined such that the equation satisfied by $u(\tau, x)$ has on the right side only the second derivative in x. Since

$$\frac{\partial W}{\partial x} = e^\varphi\left(\alpha u + \frac{\partial u}{\partial x}\right)$$

$$\frac{\partial^2 W}{\partial x^2} = e^\varphi\left(\alpha^2 u + 2\alpha\frac{\partial u}{\partial x} + \frac{\partial u}{\partial x^2}\right)$$

$$\frac{\partial W}{\partial\tau} = e^\varphi\left(\beta u + \frac{\partial u}{\partial\tau}\right),$$

substituting in (8.8.21), dividing by e^{φ} and collecting the derivatives yields

$$\frac{\partial u}{\partial \tau} = \frac{\partial^2 u}{\partial x^2} + (2\alpha + k - 1)\frac{\partial u}{\partial x} + (\alpha^2 + \alpha(k-1) - k - \beta)u = 0.$$

The constants α and β are chosen such that the coefficients of $\dfrac{\partial u}{\partial x}$ and u vanish

$$2\alpha + k - 1 = 0$$
$$\alpha^2 + \alpha(k-1) - k - \beta = 0.$$

Solving yields

$$\alpha = -\frac{k-1}{2}$$

$$\beta = \alpha^2 + \alpha(k-1) - k = -\frac{(k+1)^2}{4}.$$

The function $u(\tau, x)$ satisfies the heat equation

$$\frac{\partial u}{\partial \tau} = \frac{\partial^2 u}{\partial x^2}$$

with the initial condition expressible in terms of f_T

$$u(0,x) = e^{-\varphi(0,x)}W(0,x) = e^{-\alpha x}V(T,x)$$
$$= e^{-\alpha x}F(T,e^x) = e^{-\alpha x}f_T(e^x).$$

From the general theory of heat equation (see Proposition 8.1.2), the solution can be expressed as the convolution between the fundamental solution and the initial condition

$$u(\tau, x) = \int_{-\infty}^{\infty} \frac{1}{\sqrt{4\pi\tau}} e^{-\frac{(y-x)^2}{4\tau}} u(0,y)\, dy.$$

The previous substitutions yield the following relation between F and u

$$F(t,S) = F(t,e^x) = V(t,x) = W(\tau,x) = e^{\varphi(\tau,x)}u(\tau,x),$$

so $F(T,e^x) = e^{\alpha x}u(0,x)$. This implies

$$\begin{aligned}
F(t,e^x) &= e^{\varphi(\tau,x)}u(\tau,x) = e^{\varphi(\tau,x)}\int_{-\infty}^{\infty} \frac{1}{\sqrt{4\pi\tau}} e^{-\frac{(y-x)^2}{4\tau}} u(0,y)\, dy \\
&= e^{\varphi(\tau,x)}\int_{-\infty}^{\infty} \frac{1}{\sqrt{4\pi\tau}} e^{-\frac{(y-x)^2}{4\tau}} e^{-\alpha y}F(T,e^y)\, dy.
\end{aligned}$$

With the substitution $y = x = s\sqrt{2\tau}$ this becomes

$$F(t, e^x) = e^{\varphi(\tau, x)} \int_{-\infty}^{\infty} \frac{1}{\sqrt{2\pi}} e^{-\frac{s^2}{2} - \alpha(x + s\sqrt{2\tau})} F(T, e^{x + s\sqrt{2\tau}}) \, dy.$$

Completing the square as

$$-\frac{s^2}{2} - \alpha(x + s\sqrt{2\tau}) = \frac{1}{2}\left(s - \frac{k-1}{2}\sqrt{2\tau}\right)^2 + \frac{(k-1)^2\tau}{4} + \frac{k-1}{2}x,$$

after cancelations, the previous integral becomes

$$F(t, e^x) = e^{-\frac{(k+1)^2}{4}\tau} \frac{1}{\sqrt{2\pi}} \int_{-\infty}^{\infty} e^{-\frac{1}{2}\left(s - \frac{k-1}{2}\sqrt{2\tau}\right)^2} e^{\frac{(k-1)^2}{4}\tau} F(T, e^{x + s\sqrt{2\tau}}) \, ds.$$

Using

$$e^{-\frac{(k+1)^2}{4}\tau} e^{\frac{(k-1)^2}{4}\tau} = e^{-k\tau} = e^{-r(T-t)},$$

$$(k-1)\tau = (r - \frac{1}{2}\sigma^2)(T - t),$$

after the substitution $z = x + s\sqrt{2\tau}$ we get

$$F(t, e^x) = e^{-r(T-t)} \frac{1}{\sqrt{2\pi}} \int_{-\infty}^{\infty} e^{-\frac{1}{2}\frac{(z - x - (k-1)\tau)^2}{2\tau}} F(T, e^z) \frac{1}{\sqrt{2\pi}} \, dz$$

$$= e^{-r(T-t)} \frac{1}{\sqrt{2\pi\sigma^2(T-t)}} \int_{-\infty}^{\infty} e^{-\frac{[z - x - (r - \frac{1}{2}\sigma^2)(T-t)]^2}{2\sigma^2(T-t)}} F(T, e^z) \, dz.$$

Since $e^x = S_t$, considering the probability density

$$p(z) = \frac{1}{\sqrt{2\pi\sigma^2(T-t)}} e^{-\frac{[z - \ln S_t - (r - \frac{1}{2}\sigma^2)(T-t)]^2}{2\sigma^2(T-t)}},$$

the previous expression becomes

$$F(t, S_t) = e^{-r(T-t)} \int_{-\infty}^{\infty} p(z) f_T(e^z) \, dz = e^{-r(T-t)} \widehat{\mathbb{E}}_t[f_T],$$

with $f_T(S_T) = F(T, S_T)$ and $\widehat{\mathbb{E}}_t$ the risk-neutral expectation operator as of time t, which was introduced and used in Chapter 6.

8.9   Black-Scholes and Risk-neutral Valuation

The conclusion of the computation of the last section is of capital importance for derivatives calculus. It shows the equivalence between the Black-Scholes equation and the risk-neutral valuation. It turns out that instead of computing the risk-neutral expectation of the payoff, as in the case of the risk-neutral valuation, we may have the choice to solve the Black-Scholes equation directly, and impose the final condition to be the payoff.

In many cases solving a partial differential equation is simpler than evaluating the expectation integral. This is due to the fact that we may look for a solution dictated by the particular form of the payoff f_T. We shall apply that later in finding put-call parities for different types of derivatives.

Consequently, all derivatives evaluated by the risk-neutral valuation are solutions of the Black-Scholes equation. The only distinction is in the expression of their payoff. The payoff is the one which defines the type of the European option. A few cases are given in the next example.

Example 8.9.1 (*a*) The price of a European call option is the solution $F(t, S)$ of the Black-Scholes equation satisfying

$$f_T(S_T) = \max(S_T - K, 0).$$

(*b*) The price of a European put option is the solution $F(t, S)$ of the Black-Scholes equation with the final condition

$$f_T(S_T) = \max(K - S_T, 0).$$

(*c*) The value of a forward contract is the solution the Black-Scholes equation with the final condition

$$f_T(S_T) = S_T - K.$$

It is worth noting that the superposition principle discussed in Chapter 6 can be explained now by the fact that the solution space of the Black-Scholes equation is a linear space. This means that a linear combination of solutions is also a solution.

Another interesting feature of the Black-Scholes equation is its independence of the stock drift rate μ. Then its solutions must have the same property. This explains why, in the risk-neutral valuation, the value of μ does not appear explicitly in the solution.

We note that Asian options (path dependent options) satisfy similar Black-Scholes equations, with small differences, as we shall see in Chapter 9.

8.10 Boundary Conditions

We have solved the Black-Scholes equation for a call option, under the assumption that there is a unique solution. The Black-Scholes equation is an equation of first order in the time variable t and of second order in the stock

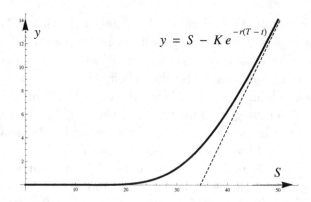

Figure 8.2: *The graph of the option price before maturity in the case* $K = 40$, $\sigma = 30\%$, $r = 8\%$, *and* $T - t = 1$.

variable S, so it needs one final condition at $t = T$ and two boundary conditions for $S = 0$ and $S \to \infty$. These boundary solutions restrict the set of possible solutions to only one solution, which corresponds to the price of the derivative.

In the case of a call option, the final condition is given by the following payoff:

$$F(T, S_T) = \max\{S_T - K, 0\}.$$

When $S \to 0$, the option does not get exercised, so the initial boundary condition writes as

$$F(t, 0) = 0.$$

When $S \to \infty$ the price becomes linear in S (see Exercise 6.1.5)

$$F(t, S) \sim S - Ke^{-r(T-t)},$$

and the graph of $F(t, \cdot)$ has a slant asymptote, see Fig.8.2.

8.11   The Black-Scholes Operator

Consider the partial differential operator

$$\mathcal{L}_{BS} = \partial_t + rx\partial_x + \frac{1}{2}\sigma^2 x^2 \partial_x^2 - r, \tag{8.11.22}$$

with r and σ positive constants. $\mathcal{L}_{BS}$ is defined on the space

$$\mathcal{C}^{2,1} = \{f : (0, \infty) \times (0, \infty) \to [0, \infty), \partial_x^2 f \text{ and } \partial_t f \text{ exist and are continuous}\}.$$

Then the Black-Scholes equation can be written as $\mathcal{L}_{BS}V(t,x) = 0$. The operator $\mathcal{L}_{BS}$ is linear, i.e.

$$\mathcal{L}_{BS}(f+g) = \mathcal{L}_{BS}f + \mathcal{L}_{BS}g,$$
$$\mathcal{L}_{BS}(cf) = c\mathcal{L}_{BS}f,$$

for any functions $f, g \in C^{2,1}$ and any constant c. However, the operator $\mathcal{L}_{BS}$ is not multiplicative, i.e. $\mathcal{L}_{BS}(fg) \neq \mathcal{L}_{BS}f \cdot \mathcal{L}_{BS}g$.

Proposition 8.11.1 *Let $f = f(t,x)$ and $g = g(t,x)$ be two functions in $C^{2,1}$. Then*

$$\mathcal{L}_{BS}(fg) = f\,\mathcal{L}_{BS}(g) + g\,\mathcal{L}_{BS}(f) + (f,g),$$

where

$$(f,g) = rfg + \sigma^2 x^2 \partial_x f\, \partial_x g.$$

Proof: The proof is a direct application of the chain rule

$$
\begin{aligned}
\mathcal{L}_{BS}(fg) &= f\,\partial_t g + g\,\partial_t g + rx\partial_x g\, f + rx\partial_x f\, g \\
&\quad + \frac{1}{2}\sigma^2 x^2 \left(\partial_x^2 f\, g + 2\partial_x f\, \partial_x g + f\, \partial_x^2 g \right) \\
&\quad - rfg - rfg + rfg \\
&= f\mathcal{L}_{BS}(g) + g\mathcal{L}_{BS}(f) + rfg + \sigma^2 x^2 \partial_x f\, \partial_x g.
\end{aligned}
$$

■

Remark 8.11.2 The operator $(\,,\,)$ has the properties of a scalar product, see Exercise 8.11.3. The space $C^{2,1}$ endowed with $(\,,\,)$ becomes a Hilbert space.

Exercise 8.11.3 *Let $f, g \in C^{2,1}$. Show the following properties*
(a) $(f,g) = (g,f)$ *(symmetric);*
(b) $(f,f) \geq 0$ *(non-negative);*
(c) *If $(f,f) = 0$, then $f = 0$ (non-degenerate);*
(d) $(f_1 + f_2, g) = (f_1, g) + (f_2, g)$ *(additive);*
(e) $(cf, g) = c(f,g)$, $\forall c \in \mathbb{R}$ *(homogeneous).*

Given a function $f \in C^{2,1}$, we denote the associate normal subspace by

$$f^{\perp} = \{g \in C^{2,1};\, (f,g) = 0\}.$$

It can be easily verified that $f^{\perp}$ is a vectorial subspace of $C^{2,1}$, see Exercise 8.11.4.

Exercise 8.11.4 *For any $g_1, g_2 \in f^{\perp}$ show that*

$$c_1 g_1 + c_2 g_2 \in f^{\perp}, \quad \forall c_1, c_2 \in \mathbb{R}.$$

Exercise 8.11.5 *Find the following normal subspaces*

(a) $0^{\perp}$;

(b) $c^{\perp}$, where $c \neq 0$ is a constant.

Exercise 8.11.6 *Show that*

$$x^{\perp} = \{\psi(t) x^{-r/\sigma^2} ; \psi(t) \text{ differentiable}\}.$$

Exercise 8.11.7 *If f is a nonzero function that satisfies the Black-Scholes equation, show that $\mathcal{L}_{BS}(f^2) > 0$.*

Exercise 8.11.8 *Let $g(x)$ be a nonzero solution of the Black-Scholes equation. Then $F(t, x) = f(t)g(x)$ is also a solution if and only if $f(t)$ is a constant function.*

Exercise 8.11.9 *(a) Show that if $\mathcal{L}_{BS}(f(t)) = 0$, then $f(t) = Ce^{rt}$, where C is a constant.*

(b) Find $g(x)$ such that $\mathcal{L}_{BS}(e^{rt}g(x)) = 0$.

It is worth noting that in the case of a dividend-paying stock the Black-Scholes operator changes into

$$\mathcal{L}_{BS} = \partial_t + (r - \delta)x\partial_x + \frac{1}{2}\sigma^2 x^2 \partial_x^2 - r, \tag{8.11.23}$$

where δ is the continuous divident payment rate.

8.12   Hedging and Black-Scholes

In this section we shall hedge one derivative by another and find in each case a market invariant, which measures the risk price of the market. We shall do this for stocks, derivatives, bonds, and swaps. As a consequence, we shall derive the Black-Scholes equation for bonds and we shall evaluate the bond in the case when the interest rate satisfies the Vasicek model. We notice that this deduction use the tool of partial differential equations, while similar results have been obtained in Chapter 4 using expectations. We shall also derive a Black-Scholes equation for swaps and solve it.

8.12.1   Hedging Stocks

We shall start, for the sake of simplicity, with two stocks, S_1 and S_2, which are driven by the same source of uncertainty

$$
\begin{aligned}
dS_1 &= \mu_1 S_1 dt + \sigma_1 S_1 dW_t \\
dS_2 &= \mu_2 S_2 dt + \sigma_2 S_2 dW_t,
\end{aligned}
$$

and consider the portfolio $\Pi = S_1 - \alpha S_2$, consisting in a long position in the stock S_1 and α units short in the stock S_2. Since

$$
\begin{aligned}
d\Pi &= dS_1 dt - \alpha dS_2 \\
&= (\mu_1 S_1 - \alpha \mu_2 S_2)dt + (\sigma_1 S_1 - \alpha \sigma_2 S_2)dW_t,
\end{aligned}
$$

then the choice $\alpha = (\sigma_1 S_1 / \sigma_2 S_2)$ cancels the white noise term, and hence the portfolio becomes risk-less

$$
d\Pi = (\mu_1 S_1 - \alpha \mu_2 S_2)dt.
$$

Since the risk-less condition can be also written as $d\Pi = r\Pi dt$, equating the coefficients of dt in the previous two expressions yields

$$
\mu_1 S_1 - \alpha \mu_2 S_2 = r(S_1 - \alpha S_2).
$$

Substituting the expression for α, and then separating the terms involving S_1 from the terms involving S_2 we arrive at

$$
\frac{\mu_1 S_1 - r S_1}{\sigma_1 S_1} = \frac{\mu_2 S_2 - r S_2}{\sigma_2 S_2},
$$

which after simplification becomes

$$
\frac{\mu_1 - r}{\sigma_1} = \frac{\mu_2 - r}{\sigma_2}. \tag{8.12.24}
$$

This identity states that any two stocks, which are driven by the same uncertainty source, have the same market price of risk. It is worth noting that in the risk neutral world, where we take $\mu = r$, the market price of risk is zero. More details about this measure have been presented in section 7.4.

The goal of the next sections is to apply a hedging procedure to find similar market invariants for more general cases.

8.12.2   Hedging Derivatives

Considering now the case of two derivatives, $V(t, S)$ and $U(t, S)$ written on the same underlying stock S. This implies that the derivatives are driven by the dynamics of the same underlying asset, which follows the process

$$dS = \mu S dt + \sigma S dW_t.$$

This fact will make possible to hedge the risk away between the derivatives. In order to do this, we form the portfolio $\Pi = V - \alpha U$, and select the value of α for which Π becomes risk-less. Using the self-financing condition of Π, as well as Ito's lemma, see Appendix C.6, we have

$$
\begin{aligned}
d\Pi &= dV - \alpha dU \\
&= \frac{\partial V}{\partial t} dt + \frac{\partial V}{\partial S} dS + \frac{1}{2} \frac{\partial^2 V}{\partial S^2} (dS)^2 - \alpha \left(\frac{\partial U}{\partial t} dt + \frac{\partial U}{\partial S} dS + \frac{1}{2} \frac{\partial^2 U}{\partial S^2} (dS)^2 \right) \\
&= \left(\frac{\partial V}{\partial t} + \frac{1}{2} \sigma^2 S^2 \frac{\partial^2 V}{\partial S^2} \right) dt + \frac{\partial V}{\partial S} dS - \alpha \left(\frac{\partial U}{\partial t} + \frac{1}{2} \sigma^2 S^2 \frac{\partial^2 U}{\partial S^2} \right) dt \\
&\quad - \alpha \frac{\partial U}{\partial S} dS \\
&= (L(V) - \alpha L(U)) dt + \left(\frac{\partial V}{\partial S} - \alpha \frac{\partial U}{\partial S} \right) dS,
\end{aligned}
$$

where we used the shorthand notation $L(V) = \dfrac{\partial V}{\partial t} + \dfrac{1}{2} \sigma^2 S^2 \dfrac{\partial^2 V}{\partial S^2}$. The portfolio is hedged against movements of the underlying stock by choosing

$$\alpha = \frac{\partial V / \partial S}{\partial U / \partial S} = \frac{\Delta_V}{\Delta_U},$$

which is just the Deltas ratio of derivatives V and U. With this choice of α, the change in the portfolio during the time interval dt writes as

$$d\Pi = (L(V) - \alpha L(U)) dt.$$

Comparing with the portfolio risk-less condition

$$d\Pi = r\Pi dt = r(V - \alpha U) dt$$

we obtain $L(V) - \alpha L(U) = r(V - \alpha U)$. Substituting the expression of α and then separating V from U yields a similar expression to (8.12.24)

$$\frac{L(V) - rV}{\Delta_V} = \frac{L(U) - rU}{\Delta_U}. \tag{8.12.25}$$

Since relation (8.12.25) holds for any two derivatives V and U, it follows that the ratio

$$\phi(t, S) = \frac{L(V) - rV}{\Delta_V}$$

is independent of the derivative V. We shall find the expression of this "market invariant" choosing a convenient derivative V. For $V = S$, the stock itself, it is easy to see that $L(V) = 0$ and $\Delta_V = 1$. Hence

$$\phi(t, S) = -rS.$$

Substituting in the previous expression yields

$$-rS = \frac{L(V) - rV}{\Delta_V}$$

and then cross multiplying we recover the Black-Scholes equation

$$L(V) + rS\Delta_V = rV.$$

Therefore, hedging a derivative with another derivative, both having the same underlying asset, also leads to the Black-Scholes equation.

8.12.3 Hedging Bonds

In order to find a Black-Scholes type equation satisfied by bonds, one might be tempted to hedge the bond with the underlying spot rate r_t, mimicking the procedure for derivatives with underlying assets. Unfortunately, *the rate r_t is not a tradable security*, and hence this type of approach is fated to fallacy. The working procedure in this situation is to hedge a bond with another bond, contingent to the same spot rate r_t.

We shall start by clarifying some notations. We have used $P(t, T)$ to denote the value of a bond at time t, with maturity T. If dependency with respect to the rate is needed, we shall denote the bond by $P(r, t, T)$. If no confusion arises, the bond is denoted just by P.

The bond depends on the spot rate $r = r_t$, which is assumed to satisfy the generic stochastic differential equation

$$dr = a(r)dt + \sigma(r)dW_t,$$

with functions $a(r)$ and $\sigma(r)$ to be specified later. Since the bond value is a function of the stochastic process r_t, Ito's lemma provides

$$
\begin{aligned}
dP &= \frac{\partial P}{\partial t}dt + \frac{\partial P}{\partial r}dr + \frac{1}{2}\frac{\partial^2 P}{\partial r^2}(dr)^2 \\
&= \left(\frac{\partial P}{\partial t} + a(r)\frac{\partial P}{\partial r} + \frac{1}{2}\sigma(r)^2\frac{\partial^2 P}{\partial r^2}\right)dt + \sigma(r)\frac{\partial P}{\partial r}dW_t \\
&= L(P)dt + \sigma(r)\frac{\partial P}{\partial r}dW_t, \tag{8.12.26}
\end{aligned}
$$

where we used the shorthand notation

$$L(P) = \frac{\partial P}{\partial t} + a(r)\frac{\partial P}{\partial r} + \frac{1}{2}\sigma(r)^2\frac{\partial^2 P}{\partial r^2}.$$

We shall consider next two zero-coupon bonds, $P_1 = P(r, t, T_1)$ and $P_2 = P(r, t, T_2)$, with maturities T_1 and T_2, and the same underlying rate r. Since the stochastic differential equation (8.12.26) is independent on T, applying it to the previous bonds we get

$$dP_1 = L(P_1)dt + \sigma(r)\frac{\partial P_1}{\partial r}\,dW_t$$

$$dP_2 = L(P_2)dt + \sigma(r)\frac{\partial P_2}{\partial r}\,dW_t.$$

Now we form the hedging portfolio $\Pi = P_1 - \gamma P_2$, with γ chosen such that Π is risk-less. Since Π is self-financing, we obtain

$$d\Pi = dP_1 - \gamma dP_2$$

$$= \Big(L(P_1) - \gamma L(P_2)\Big)dt + \sigma(r)\Big[\frac{\partial P_1}{\partial r} - \gamma\frac{\partial P_2}{\partial r}\Big]dW_t.$$

The risk-less condition can be quantized as

$$\gamma = \frac{\partial P_1/\partial r}{\partial P_2/\partial r},$$

which means that γ is the ratio of the durations of the two bonds. Comparing the risk-less conditions

$$d\Pi = \Big(L(P_1) - \gamma L(P_2)\Big)dt$$

$$d\Pi = r\Pi dt$$

yields $L(P_1) - \gamma L(P_2) = r(P_1 - \gamma P_2)$. Substituting for γ and then separating P_1 from P_2, the expression can be written as

$$\frac{L(P_1) - rP_1}{\partial P_1/\partial r} = \frac{L(P_2) - rP_2}{\partial P_2/\partial r}. \tag{8.12.27}$$

The left term of (8.12.27) is a function of maturity T_1, while the right term depends on T_2. The only case when this can be satisfied is when there exists a separation function $\phi(r, t)$ independent of T and P, which is equal to both terms of the equation (8.12.27). Therefore, for any bond $P(r, t, T)$ we have

$$\frac{L(P) - rP}{\partial P/\partial r} = \phi(r, t), \tag{8.12.28}$$

i.e. the ratio in the left term is an invariant for the bonds market.

We shall address in the following two consequences of the formula (8.12.28):

(*i*) Cross multiplying, we obtain the partial differential equation

$$L(P) - \phi(r,t)\frac{\partial P}{\partial r} = rP.$$

Substituting for $L(P)$ yields the following Black-Scholes equation for bonds that pay \$1 at maturity

$$\frac{\partial P}{\partial t} + [a(r) - \phi(r,t)]\frac{\partial P}{\partial r} + \frac{1}{2}\sigma(r)^2\frac{\partial^2 P}{\partial r^2} = rP \qquad (8.12.29)$$

$$P(r,T,T) = 1. \qquad (8.12.30)$$

Unfortunately, the term $\phi(r,t)$ cannot be found as in the previous case, where it sufficed to choose a simple derivative (the stock) to compute the quotient. In this case, we have only one bond expression (depending on maturity T), and this is the unknown solution of the above Black-Scholes equation. Knowing the value of P is equivalent to knowing the ratio ϕ. We shall show how this difficulty can be overcome by changing to a risk-neutral world point of view.

(*ii*) Assume that the bond satisfies the stochastic differential equation

$$dP = m(r,t)Pdt + \nu(r,t)PdW_t, \qquad (8.12.31)$$

where $m(r,t)$ is the drift rate and $\nu(r,t)$ is the bond volatility. These coefficients can be expressed in terms of P as in the following. Comparing with (8.12.26)

$$dP = L(P)dt + \sigma(r)\frac{\partial P}{\partial r}dW_t,$$

we obtain

$$m(r,t) = \frac{L(P)}{P}, \qquad \nu(r,t) = \frac{\sigma(r)\frac{\partial P}{\partial r}}{P}.$$

These relations enable us to write the ratio ϕ in terms of m and ν as follows

$$\phi(r,t) = \frac{L(P) - rP}{\partial P/\partial r} = \frac{\frac{L(P)}{P} - r}{\frac{\partial P/\partial r}{P}} = \frac{m(r,t) - r}{\nu(r,t)/\sigma(r)}.$$

Then the *Sharpe ratio* for bonds, defined as the excess of bond return over the risk-free rate measured in volatility units, can be computed as

$$\eta = \frac{m(r,t) - r}{\nu(r,t)} = \frac{\phi(r,t)}{\sigma(r)}. \qquad (8.12.32)$$

Therefore, the Black-Scholes equation (8.12.29) takes the following form dependent of Sharpe ratio

$$\frac{\partial P}{\partial t} + [a(r) - \sigma(r)\eta(r,t)]\frac{\partial P}{\partial r} + \frac{1}{2}\sigma(r)^2\frac{\partial^2 P}{\partial r^2} = rP. \qquad (8.12.33)$$

The Sharpe ratio $\eta(r,t)$ does not depend on the maturity time T. In order to make this more clear, if denote by $m(r,t,T)$ and $\nu(r,t,T)$ the drift rate and volatility of a bond with maturity T, then for any two bonds $P(r,t,T_1)$ and $P(r,t,T_2)$ we have

$$\frac{m(r,t,T_1) - r}{\nu(r,t,T_1)} = \frac{m(r,t,T_2) - r}{\nu(r,t,T_2)}.$$

The ratio η cannot be determined exactly, since it quantizes the bond market risk that depends of different market points of view or market beliefs. We shall show, however, that in the risk-neutral world, i.e. when the belief is that all market participants have the same risk aversion, the ratio η vanishes.

This can be naively achieved, just replacing the drift rate $m(r,t)$ by the risk-free rate r in (8.12.32), a fact that implies $\eta = 0$. In this case the bond process (8.12.31) becomes

$$dP = rPdt + \nu(r,t)PdW_t,$$

which retrieves formula (4.1.4).

A more elaborate explanation of this fact is based on Girsanov theorem and it will be given in the following. Adding and subtracting a term, we have

$$\begin{aligned}
dP &= m(r,t)Pdt + \nu(r,t)PdW_t \\
&= (m(r,t) - r)Pdt + \nu(r,t)PdW_t + rPdt \\
&= rPdt + P\nu(r,t)[\eta(r,t)dt + dW_t].
\end{aligned}$$

By Girsanov theorem, see Appendix C.5, there is a measure (the risk-neutral measure) with respect to which the process $\widehat{W}_t$, given by $d\widehat{W}_t = \eta(r,t)dt + dW_t$, becomes a Brownian motion. Therefore, in the risk-neutral world, we have

$$dP = rPdt + P\nu(r,t)d\widehat{W}_t.$$

This is equivalent by stating that the Sharpe ratio, η, in the risk-neutral world vanishes. Hence, the Black-Scholes equation for bonds in the risk-neutral world takes the form

$$\frac{\partial P}{\partial t} + a(r)\frac{\partial P}{\partial r} + \frac{1}{2}\sigma(r)^2\frac{\partial^2 P}{\partial r^2} = rP \qquad (8.12.34)$$

$$P(r,T,T) = 1. \qquad (8.12.35)$$

We shall attempt to solve the equation (8.12.34)-(8.12.35) for different choices of model functions $a(r)$ and $\sigma(r)$. The solution is unique as long as some boundary conditions are satisfied: when the interest rate is large, the zero-coupon bond worth nothing

$$\lim_{r\to\infty} P(r,t,T) = 0,$$

and when the interest rate is small, the bond does not worth more than its face value:

$$\lim_{r\to 0} P(r,t,T) < \infty.$$

8.12.4 Particular Cases

(1) Consider $a(r) = 0$ and $\sigma(r) = \sigma$ positive constants. In this case the spot rate is given by the diffusion $r_t = r_0 + \sigma W_t$. The Black-Scholes equation becomes

$$\frac{\partial P}{\partial t} + \frac{1}{2}\sigma^2\frac{\partial^2 P}{\partial r^2} = rP.$$

We shall look for a solution in the form

$$P(r,t,T) = A(t,T)e^{-rB(t,T)}.$$

The final condition $P(r,T,T) = 1$ implies

$$A(T,T) = 1, \qquad B(T,T) = 0.$$

The boundary conditions for r large and small are also satisfied. Substituting in the Black-Scholes equation and collecting terms we obtain

$$r(B'+1) + \frac{A'}{A} + \frac{1}{2}\sigma^2 B^2 = 0,$$

where $A' = \frac{dA}{dt}$, $B' = \frac{dB}{dt}$. Treating r as an independent variable, we equate the coefficients to zero, obtaining the following problems

$$B' + 1 = 0, \qquad B(T,T) = 0$$
$$A' + \frac{1}{2}\sigma^2 B^2 A = 0, \qquad A(T,T) = 1.$$

It is easy to see by direct integration that the first one has the solution $B(t,T) = T - t$. In order to solve the second equation, we multiply it by the integrating factor $\rho = e^{\frac{1}{2}\sigma^2 \int_0^t (T-u)^2\, du}$ and obtain $(\rho A)' = 0$. Integrating, we get $A = \rho^{-1}C$, with the constant $C = e^{\frac{1}{2}\sigma^2 \int_0^T (T-u)^2\, du}$ obtained from the condition $A(T,T) = 1$. Therefore

$$A = \rho^{-1}C = e^{\frac{1}{2}\sigma^2 \int_t^T (T-u)^2\, du} = e^{\frac{\sigma^2}{6}(T-t)^3}.$$

Putting the pieces together, we arrive at the following bond formula

$$P(r, t, T) = Ae^{-rB} = e^{-r(T-t)+\frac{\sigma^2}{6}(T-t)^3}, \tag{8.12.36}$$

which recovers the formula obtained in section 4.7 using expectations.

(2) Consider $a(r) = a(b - r)$, with $a, b > 0$ and $\sigma(r) = \sigma$, positive constant. This corresponds to the Vasicek's model

$$dr = a(b - r)dt + \sigma dW_t,$$

which was introduced in section 3.3.2. The Black-Scholes equation in this case takes the form

$$\frac{\partial P}{\partial t} + a(b - r)\frac{\partial P}{\partial r} + \frac{1}{2}\sigma^2\frac{\partial^2 P}{\partial r^2} = rP.$$

Again, looking for a solution of the type

$$P(r, t, T) = A(t, T)e^{-rB(t,T)},$$

after substituting into the Black-Scholes equation, simplifying and collecting terms, we arrive at the following vanishing linear function in r

$$r(B' - aB + 1) - \left(\frac{A'}{A} - abB + \frac{1}{2}\sigma^2 B^2\right) = 0.$$

Equating the coefficients to zero, we obtain that A and B satisfy the following problems

$$B' - aB = -1, \qquad B(T, T) = 0$$
$$\frac{A'}{A} - abB + \frac{1}{2}\sigma^2 B^2 = 0, \qquad A(T, T) = 1.$$

The first one can be solved using the integrating factor $\rho = e^{-at}$. Its solution is given by

$$B(t, T) = \frac{e^{-a(T-t)} - 1}{a}.$$

The second equation, after multiplying by A, becomes a homogeneous linear differential equation, which can be also solved by standard methods. A straightforward, but tedious computation, leads to the solution

$$A = e^{(B(t,T)-T+t)(b-\frac{\sigma^2}{2a^2})-\frac{\sigma^2}{4a}B^2(t,T)},$$

which retrieves the bond formula given in Proposition 4.9.1.

Exercise 8.12.1 *Consider three stocks, S_1, S_2, and S_3, driven by two independent sources of uncertainty*

$$\begin{aligned}
dS_1 &= \mu_1 S_1 dt + S_1(\sigma_{11}dW_1 + \sigma_{12}dW_2) \\
dS_2 &= \mu_2 S_2 dt + S_2(\sigma_{21}dW_1 + \sigma_{22}dW_2) \\
dS_3 &= \mu_3 S_3 dt + S_3(\sigma_{31}dW_1 + \sigma_{32}dW_2),
\end{aligned}$$

with μ_i and σ_{ij} constants and W_j Brownian motions. Assume the determinants

$$\Sigma_1 = \begin{vmatrix} \sigma_{21} & \sigma_{31} \\ \sigma_{22} & \sigma_{32} \end{vmatrix}, \qquad \Sigma_2 = \begin{vmatrix} \sigma_{11} & \sigma_{31} \\ \sigma_{12} & \sigma_{32} \end{vmatrix}, \qquad \Sigma_3 = \begin{vmatrix} \sigma_{21} & \sigma_{11} \\ \sigma_{22} & \sigma_{12} \end{vmatrix}$$

are nonzero.

(a) Consider the self-financing portfolio

$$\Pi = S_1 - \alpha S_2 - \beta S_3.$$

Show that the portfolio Π is risk-less for

$$\alpha = \frac{S_1}{S_2}\frac{\Sigma_2}{\Sigma_1}, \qquad \beta = \frac{S_1}{S_3}\frac{\Sigma_3}{\Sigma_1}.$$

(b) Comparing the risk-less portfolio with the condition $d\Pi = r\Pi dt$, show the following relation

$$\frac{\mu_1 - r}{\Sigma_2 \Sigma_3} = \frac{\mu_2 - r}{\Sigma_1 \Sigma_3} + \frac{\mu_3 - r}{\Sigma_1 \Sigma_2}.$$

Exercise 8.12.2 *Consider the Rendleman and Bartter model for the spot rate*

$$dr = \mu r dt + \sigma r dW_t,$$

with μ and σ constants.

(a) Write the Black-Scholes equation satisfied by the associated bond $P(r, t, T)$;

(b) Solve the equation and find the price of the bond in the form $P(r, t, T) = A(t, T)e^{-rB(t,T)}$.

Exercise 8.12.3 *Consider the spot rate given by the Brownian motion with drift, $r_t = at + \sigma W_t$, with a and σ constants.*

(a) Write the Black-Scholes equation satisfied by the associated bond $P(r, t, T)$;

(b) Solve the equation and find the price of the bond.

(c) Take the limit $a \to 0$ in the result obtained at part (a) and retrieve formula (8.12.36).

8.12.5   The Bond Formula

We have seen in Chapter 4 that the value at time t of a bond, which pays off
\$1 at maturity T, is given by the formula

$$P(t,T) = \mathbb{E}[e^{\int_t^T r_s\, ds}|\mathcal{F}_t], \qquad (8.12.37)$$

and we have used it there to price bonds for a few particular cases of interest
rates (such as Vasicek's model, etc.).

   In section 8.12.3 we have proved that if the spot rate, r_t, satisfies the
stochastic differential equation

$$dr = a(r)dt + \sigma(r)dW_t,$$

then the bond price, $P(t,T)$, verifies in the risk-neutral world (i.e. taking
$\phi(r,t)=0$) the following partial differential equation

$$\frac{\partial P}{\partial t} + a(r)\frac{\partial P}{\partial r} + \frac{1}{2}\sigma(r)^2\frac{\partial^2 P}{\partial r^2} \;=\; rP \qquad (8.12.38)$$

$$P(r,T,T) \;=\; 1. \qquad (8.12.39)$$

The goal of this section is to show that the solution of the problem (8.12.38)-
(8.12.39) is given by the formula (8.12.37).

   Let $X_t = -\int_t^T r_s\, ds$ and consider the exponential process

$$Y_t = U(t,r) = e^{-\int_t^T r_s\, ds} = e^{X_t}.$$

The plan of the proof is the following: we compute dY_t in two equivalent ways,
by the Fundamental Theorem of Calculus and by Ito's lemma, and then take
the conditional expectation $\mathbb{E}[\cdot|\mathcal{F}_t]$ and equate the answers. What we will
obtain is the differential equation (8.12.38).

Using $dX_t = r_t dt$, the Fundamental Theorem of Calculus yields

$$dY_t = r_t Y_t\, dt.$$

On the other hand, Ito's lemma applied to $Y_t = U(t,r)$ provides

$$\begin{aligned}
dY_t &= \frac{\partial U}{\partial t}dt + \frac{\partial U}{\partial r}dr + \frac{1}{2}\frac{\partial^2 U}{\partial r^2}(dr_t)^2 \\
&= \left(\frac{\partial U}{\partial t} + a(r)\frac{\partial U}{\partial r} + \frac{1}{2}\sigma(r)^2\frac{\partial^2 U}{\partial r^2}\right)dt + \sigma(r)\frac{\partial U}{\partial r}dW_t.
\end{aligned}$$

Equating the right rides of the last two equations yields

$$\left(\frac{\partial U}{\partial t} + a(r)\frac{\partial U}{\partial r} + \frac{1}{2}\sigma(r)^2\frac{\partial^2 U}{\partial r^2}\right)dt + \sigma(r)\frac{\partial U}{\partial r}dW_t = r_t U.$$

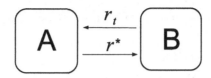

Figure 8.3: *Typical swap, by which A pays a fixed rate r^* to B and receives in return a floating rate r_t.*

Applying the conditional expectation $\mathbb{E}[\,\cdot\,|\mathcal{F}_t]$ on both sides, and using that $\mathbb{E}[dW_t|\mathcal{F}_t] = 0$, we obtain

$$\left(\frac{\partial P}{\partial t} + a(r)\frac{\partial P}{\partial r} + \frac{1}{2}\sigma(r)^2\frac{\partial^2 P}{\partial r^2}\right)dt = rP,$$

where $P = P(t,T) = \mathbb{E}[U(t,r_t)|\mathcal{F}_t] = \mathbb{E}[U(t,r_t)|r_t = r]$. Hence, $P(t,T)$, given by (8.12.37), satisfies the equation (8.12.38). We end the proof by noting that the boundary condition (8.12.39) is an obvious check:

$$P(T,T) = \mathbb{E}[e^{\int_T^T r_s\, ds}|\mathcal{F}_t] = e^0 = 1.$$

Remark 8.12.4 It is worth to note that there are two main methods for computing prices for bonds with stochastic underling interest rates. One involves formula (8.12.37), which can be computed by a Monte-Carlo simulation. The other, deals with a numerical method for the partial differential equation (8.12.38)-(8.12.39). We have seen that in certain particular cases, see section 8.12.4, these bond prices are given by closed form formulas.

8.13 Interest Rate Swaps

An *interest rate swap* is a contract by which two parties agree to exchange the interest payments on a certain amount, Z, for a given interval of time, T. One party agrees to pay the other a fixed rate of interest, r^*, and to receive in return a variable (floating) interest rate payment, r_t. The value of this contract at time t is denoted by $V(t,T,r,r^*)$, or for short, $V(t)$, if no confusions arise, see Fig.8.3.

8.13.1 Case of Deterministic Rates

Even if in real life conditions the floating rate, r_t, is stochastic, it is convenient to assume for the time being, that $r_t = r(t)$ is a deterministic time-dependent rate. We shall find the value of the swap $V(t)$ at any time t, for $0 \leq t \leq T$.

From the party A point of view, during the time step Δt, we have the following balances:

- Party A receives $r(t)Z\Delta t$
- Party A pays $r^*Z\Delta t$
- the contract appreciates by $r(t)V(t)\Delta t$ (the risk-free condition).

Then the change in the value of the contract, $\Delta V(t) = V(t + \Delta t) - V(t)$, during the time step Δt can be written as

$$\Delta V(t) = rV(t)\Delta t + Z(r(t) - r^*)\Delta t.$$

Taking $\Delta t \to 0$, the equation becomes

$$dV(t) = r(t)V(t)dt + Z(r(t) - r^*)dt,$$

which is equivalent to the following linear differential equation

$$V'(t) - rV(t) = Z(r(t) - r^*).$$

Multiplying by $\rho(t) = e^{-\int_0^t r(u)\,du}$, we obtain an exact equation, which after integration yields

$$\rho(t)V(t) = C_0 + Z\int_0^t (r(s) - r^*)\rho(s)\,ds,$$

with C_0 constant. Since the swap contract is void at maturity, $V(T) = 0$, making $t = T$ in the previous equation we obtain the value of the constant

$$C_0 = -Z\int_0^T (r(s) - r^*)\rho(s)\,ds.$$

Substituting back into the previous equation, we obtain

$$\rho(t)V(t) = -Z\int_t^T (r(s) - r^*)\rho(s)\,ds.$$

Dividing by $\rho(t)$, and using that

$$\frac{\rho(s)}{\rho(t)} = e^{-\int_t^s r(u)\,du},$$

we obtain the following value for the swap contract

$$V(t) = -Z\int_t^T e^{-\int_t^s r(u)\,du}(r(s) - r^*)\,ds.$$

Denoting the discount factor by $D(t, s) = e^{-\int_t^s r(u)\, du}$, and changing the order of the terms, we get to the following integral swap formula

$$V(t) = Z \int_t^T D(t, s)(r(s) - r^*)\, ds. \qquad (8.13.40)$$

This formula provides the value of the swap as the product between the principal Z and the cumulative discounted value at time t of the excess return of the floating rate over the fixed rate. The formula holds as long as:

- the floating rate, $r(t)$, is deterministic;
- the rates $r(t)$ and r^* are continuous. In real life applications the rate payments are made regularly at certain times, a fact that transforms the previous integral into a finite sum.

Exercise 8.13.1 *A mortgage lender, A, lends a house buyer, B, the principal $Z = \$300,000$. The buyer, B, agrees to pay back the lender, A, fixed rate payments at rate r^* for the next $T = 30$ years. Assume the lender is well equipped to take the risk of a floating rate r_t on the principal Z. Work out the fair mortgage rate r^* in the case when $r_t = r_0 e^{0.01t}$.*

Exercise 8.13.2 *A company has a bank account with the initial balance $V(0)$. The account earns interest at the continuous rate r. If spending occurs from this account at the constant continuous rate r^*, assuming there is a money inflow in the account at the constant continuous rate ρ, find the initial balance $V(0)$, given that the balance in the account at time T is zero.*

8.13.2 A Black-Scholes Type Equation

In this section we shall develop the partial differential equation satisfied by an interest rate swap under the assumption that the floating interest rate, $r = r_t$, satisfies the stochastic differential equation

$$dr = a(r)dt + \sigma(r)dW_t.$$

Since the swap value $V = V(t, r)$ is a function of the stochastic process r_t and time t, an application of Ito's lemma yields

$$
\begin{aligned}
dV &= Z(r - r^*)dt + \frac{\partial V}{\partial t}dt + \frac{\partial V}{\partial r}dr + \frac{1}{2}\frac{\partial^2 V}{\partial r^2}(dr)^2 \\
&= \left(Z(r - r^*) + \frac{\partial V}{\partial t} + a(r)\frac{\partial V}{\partial r} + \frac{1}{2}\sigma(r)^2\frac{\partial^2 V}{\partial r^2} \right)dt + \sigma(r)\frac{\partial V}{\partial r}dW_t \\
&= L(P)dt + \sigma(r)\frac{\partial V}{\partial r}dW_t, \qquad (8.13.41)
\end{aligned}
$$

where we used the convenient notation

$$L(P) = Z(r - r^*) + \frac{\partial V}{\partial t} + a(r)\frac{\partial V}{\partial r} + \frac{1}{2}\sigma(r)^2\frac{\partial^2 V}{\partial r^2}.$$

The first term, $Z(r - r^*)dt$, has been added to compensate company A for the inflow premium of the floating rate payment over the fixed rate payment during the time step dt.

We now consider two swaps, $V_1 = V_1(t, r, T_1)$ and $V_2 = V_2(t, r, T_2)$, with the same underlying rates but different maturity dates T_1 and T_2. Using these we form the self-financing portfolio $\Pi = V_1 - \alpha V_2$ and compute the change over the time step dt

$$\begin{aligned} d\Pi &= dV_1 - \alpha dV_2 \\ &= (L(V_1) - \alpha L(V_2))dt + \left(\frac{\partial V_1}{\partial r} - \alpha\frac{\partial V_2}{\partial r}\right)dr. \end{aligned}$$

Hedging against the change in the floating interest rate r requires

$$\alpha = \frac{\partial V_1/\partial r}{\partial V_2/\partial r},$$

a case in which the change in the portfolio value simplifies to

$$d\Pi = (L(V_1) - \alpha L(V_2))dt.$$

On the other side the risk-less condition provides

$$d\Pi = r\Pi dt = r(V_1 - \alpha V_2)dt.$$

Equating the right sides of the last two equations and separating V_1 from V_2 yields

$$L(V_1) - rV_1 = \alpha\Big(L(V_2) - rV_2\Big),$$

which after substitution of α becomes

$$\frac{L(V_1) - rV_1}{\partial V_1/\partial r} = \frac{L(V_2) - rV_2}{\partial V_2/\partial r}.$$

The left term is a function of T_1, while the right term depends on T_2. This fact implies the existence of a separation function, $f(t, r)$, such that for any swap $V = V(t, r)$ we have

$$\frac{L(V) - rV}{\partial V/\partial r} = f(t, r).$$

Using the previous expression for $L(V)$ the above relation writes as a partial differential equation

$$\frac{\partial V}{\partial t} + (a(r) - f(t,r))\frac{\partial V}{\partial r} + \frac{1}{2}\sigma(r)^2\frac{\partial^2 V}{\partial r^2} + Z(r - r^*) = rV. \qquad (8.13.42)$$

The equation is considered together with the final condition $V(T,r) = 0$, which states that the contract is void at maturity.

It is worth noting that for any prior time to maturity, $t < T$, the swap value $V(t,r)$ might take positive or negative values, since r_t might be larger or smaller than r^*.

The boundary conditions, which guarantee the solution uniqueness are given in the following. If the floating rate r becomes very large, the swap value tends to be large as well, in order to compensate for the large payments, so

$$\lim_{r \to \infty} V(t,r) = \infty.$$

If the floating rate r becomes very small, $r \to 0$, then the value of the swap, $V(t,0)$, at any time t has to be equal to the remaining future payments at rate r^*, i.e.

$$\lim_{r \to 0} V(t,r) = -r^* Z(T - t),$$

where the minus sign has been added to show its negative value. We notice that the value $V(t,0)$ is proportional with the time horizon $T - t$ and with the continuous payments $r^* Z$. See Exercise 8.13.4 for a deduction of this formula, starting from the differential equation.

Another boundary condition occurs for $r \to r^*$. In this case we swap a rate for the same rate, so the value of the contract is null, i.e.

$$\lim_{r \to r^*} V(t,r) = V(t,r^*) = 0.$$

Exercise 8.13.3 (a) *Solve equation (8.13.42) for the case $Z = 0$.*
(b) *Show that the swap price is given by $V(t,r) = ZW(t,r)$, where $W(t,r)$ satisfies*

$$\frac{\partial W}{\partial t} + (a(r) - f(t,r))\frac{\partial W}{\partial r} + \frac{1}{2}\sigma(r)^2\frac{\partial^2 W}{\partial r^2} + r - r^* = rW.$$

Exercise 8.13.4 (a) *Taking $r \to 0$ in equation (8.13.42), show that $V(t,0)$ satisfies*

$$V'(t,0) = r^* Z, \qquad V(T,0) = 0.$$

(b) *Show that the solution of the equation obtained in part (a) is given by $V(t,0) = -r^* Z(T - t)$.*
(c) *Take $r(s) = 0$ in the swap formula (8.13.40) and show that we obtain the same formula as in part (b).*

8.13.3   Solving the Equation

The partial differential equation (8.13.42) together with the boundary condition $V(T, r) = 0$ can be always solved numerically. However, to be in line with the idea of this book, we shall provide next a closed form solution for the swap price in the risk-neutral world, which can be computed by a Monte-Carlo simulation.

It suffices to find a formula for the case when the principal $Z = \$1$, as Exercise 8.13.3 (b) suggests.

Proposition 8.13.5 *The solution of the problem*

$$\frac{\partial V}{\partial t} + a(r)\frac{\partial V}{\partial r} + \frac{1}{2}\sigma(r)^2 \frac{\partial^2 V}{\partial r^2} + r - r^* \;=\; rV \qquad (8.13.43)$$

$$V(T, r) \;=\; 0 \qquad (8.13.44)$$

satisfying the boundary values

$$V(T, r) \;=\; 0 \qquad (8.13.45)$$
$$V(t, r^*) \;=\; 0 \qquad (8.13.46)$$
$$V(t, 0) \;=\; -r^*(T - t) \qquad (8.13.47)$$
$$V(t, +\infty) \;=\; +\infty. \qquad (8.13.48)$$

is given by

$$\boxed{V(t, r) = \mathbb{E}\left[\int_t^T e^{-\int_t^s r_u\, du}(r_s - r^*)\, ds \,\Big|\, \mathcal{F}_t\right].} \qquad (8.13.49)$$

Proof: The solution uniqueness is stated by standard properties of boundary value problems. What is left is to verify that (8.13.49) satisfies the equation (8.13.43), as well as the final condition (8.13.44) and boundary conditions (8.13.45)-(8.13.48).

The plan of the proof is the following. We consider the process

$$Y_t = U(t, r_t) = \int_t^T e^{-\int_t^s r_u\, du}(r_s - r^*)\, ds,$$

and compute dY_t in two different ways: by Ito's lemma and by Leibnitz' formula of differentiating integrals

$$\frac{d}{dx}\int_{a(x)}^{b(x)} f(x, t)\, dt = f(x, b(x))b'(x) - f(x, a(x))a'(x) + \int_{a(x)}^{b(x)} \frac{\partial}{\partial x} f(x, t)\, dt.$$

Then we apply the conditional expectation to both cases and equate the resulting expressions, a fact that leads to the differential equation (8.13.43). Ito's lemma applied to $Y_t = U(t, r)$ provides

$$
\begin{aligned}
dY_t &= \frac{\partial U}{\partial t} dt + \frac{\partial U}{\partial r} dr + \frac{1}{2} \frac{\partial^2 U}{\partial r^2} (dr_t)^2 \\
&= \left(\frac{\partial U}{\partial t} + a(r) \frac{\partial U}{\partial r} + \frac{1}{2} \sigma(r)^2 \frac{\partial^2 U}{\partial r^2} \right) dt + \sigma(r) \frac{\partial U}{\partial r} dW_t.
\end{aligned}
$$

On the other side, applying Leibnitz' formula, we have

$$
\begin{aligned}
dY_t &= \left[-e^{-\int_t^t r_u \, du} (r_t - r^*) + \int_t^T \frac{d}{dt} e^{-\int_t^s r_u \, du} (r_s - r^*) \, ds \right] dt \\
&= -(r_t - r^*) \, dt + \left[\int_t^T r_t e^{-\int_t^s r_u \, du} (r_s - r^*) \, ds \right] dt \\
&= -(r_t - r^*) \, dt + r_t Y_t \, dt \\
&= -(r_t - r^*) \, dt + r_t U(t, r_t) \, dt.
\end{aligned}
$$

Equating the right sides of the last two equations yields

$$
\left(\frac{\partial U}{\partial t} + a(r) \frac{\partial U}{\partial r} + \frac{1}{2} \sigma(r)^2 \frac{\partial^2 U}{\partial r^2} \right) dt + \sigma(r) \frac{\partial U}{\partial r} dW_t = -(r_t - r^*) \, dt + r_t U(t, r_t) \, dt.
$$

Applying the conditional expectation $\mathbb{E}[\cdot | \mathcal{F}_t]$ to both sides, using that we have $\mathbb{E}[dW_t | \mathcal{F}_t] = 0$, and then dividing by dt, we obtain

$$
\frac{\partial V}{\partial t} + a(r) \frac{\partial V}{\partial r} + \frac{1}{2} \sigma(r)^2 \frac{\partial^2 V}{\partial r^2} = -(r_t - r^*) + rV,
$$

where we used the notation $V(t, r) = \mathbb{E}[U(t, r_t) | \mathcal{F}_t]$. Thus, we have checked that formula (8.13.49) satisfies the equation (8.13.43).
The final condition is obviously verified:

$$
V(T, r) = \mathbb{E}\left[\int_T^T e^{-\int_T^s r_u \, du} (r_s - r^*) \, ds \Big| \mathcal{F}_t \right] = 0.
$$

The verifications of the other boundary conditions are left to the reader, see Exercise 8.13.6.

■

It is worth noting that (8.13.49) is similar to formula (8.13.40), which is obtained in the case when r_t is deterministic.

Exercise 8.13.6 *Show that $V(t, r)$ given by the formula (8.13.49) satisfies the boundary conditions (8.13.45)-(8.13.48).*

8.13.4 Some Particular Cases

The swap formula (8.13.49) can be computed explicitly in very few cases. We shall encounter next a couple of simple cases.

1. Case $a(r) = a$, constant, and $\sigma(r) = 0$.
This choice of the coefficients corresponds to the process $dr_t = a\,dt$, which implies that $r_t = r_0 + at$ is a deterministic spot rate, with linear increase. The associated Black-Scholes equation is

$$\frac{\partial V}{\partial t} + a\frac{\partial V}{\partial r} + r - r^* = rV.$$

Even if this equation can be solved explicitly by the method of characteristics, we shall compute in the following the swap price by the formula (8.13.49). Since everything is deterministic, this formula is the same as formula (8.13.40). An elementary manipulation of integrals provide

$$
\begin{aligned}
V(t,r) &= \int_t^T e^{-\int_t^T (r_0 + au)\,du}(as + r_0 - r^*)\,ds \\
&= \int_t^T e^{-r_0(s-t) - \frac{a}{2}(s^2 - t^2)}(as + r_0 - r^*)\,ds \\
&= e^{r_0 t}e^{\frac{a}{2}t^2}\int_t^T e^{-r_0 s - \frac{a}{2}s^2}(as + r_0 - r^*)\,ds \\
&= r^*\sqrt{\frac{\pi}{2a}}\,e^{\frac{1}{2a}r_0^2 + r_0 t + \frac{a}{2}t^2}\left[\mathrm{Erf}\!\left(\frac{r_t}{\sqrt{2u}}\right) - \mathrm{Erf}\!\left(\frac{r_T}{\sqrt{2a}}\right)\right] \\
&\quad + 1 - e^{-r_0(T-t) - \frac{a}{2}(T^2 - t^2)},
\end{aligned}
$$

where we used the well-known special function

$$\mathrm{Erf}(z) = \frac{2}{\sqrt{\pi}}\int_0^z e^{-t^2}\,dt.$$

2. Case $a(r) = 0$ and $\sigma(r) = \sigma$, constant.
This choice provides the equation $dr_t = \sigma\,dW_t$, with solution $r_t = r_0 + \sigma W_t$, which is a Brownian motion around the level r_0. The associated partial differential equation is

$$\frac{\partial V}{\partial t} + \frac{1}{2}\sigma^2\frac{\partial^2 V}{\partial r^2} + r - r^* = rV.$$

Again, we shall attempt to find the swap price starting from the formula (8.13.49), with the desire of a closed form expression. Substituting for r_t, we have

$$
\begin{aligned}
V(t,r) &= \mathbb{E}\left[\int_t^T e^{-\int_t^s r_u\,du}(r_s - r^*)\,ds\Big|\mathcal{F}_t\right] \\
&= \mathbb{E}\left[\int_t^T e^{-\int_t^s (r_0+\sigma W_u)\,du}(r_0 - r^* + \sigma W_s)\,ds\Big|\mathcal{F}_t\right] \\
&= \mathbb{E}\left[\int_t^T e^{-r_0(s-t)}e^{-\sigma\int_t^s W_u\,du}(r_0 - r^* + \sigma W_s)\,ds\Big|\mathcal{F}_t\right] \\
&= (r_0 - r^*)e^{r_0 t}\mathbb{E}\left[\int_t^T e^{-r_0 s}e^{-\sigma\int_t^s W_u\,du}\,ds\Big|\mathcal{F}_t\right] \\
&\quad + \sigma e^{r_0 t}\mathbb{E}\left[\int_t^T e^{-r_0 s}W_s e^{-\sigma\int_t^s W_u\,du}\,ds\Big|\mathcal{F}_t\right] \\
&= (r_0 - r^*)e^{r_0 t}\mathcal{E}_1 + \sigma e^{r_0 t}\mathcal{E}_2,
\end{aligned}
$$

where $\mathcal{E}_1$ and $\mathcal{E}_2$ denote the previous expectations. We shall compute the first expectation. Interchanging the integral with the conditional expectation (allowed by the Fubini's theorem of swapping integrals), we obtain

$$
\begin{aligned}
\mathcal{E}_1 &= \mathbb{E}\left[\int_t^T e^{-r_0 s}e^{-\sigma\int_t^s W_u\,du}\,ds\Big|\mathcal{F}_t\right] \\
&= \int_t^T e^{-r_0 s}\mathbb{E}\left[e^{-\sigma\int_t^s W_u\,du}\Big|\mathcal{F}_t\right]ds \\
&= \int_t^T e^{-r_0 s}\mathbb{E}\left[e^{-\sigma\int_0^{s-t} W_u\,du}\Big|\mathcal{F}_0\right]ds \quad \text{(by stationarity)} \\
&= \int_t^T e^{-r_0 s}\mathbb{E}\left[e^{-\sigma\int_0^{s-t} W_u\,du}\right]ds \quad \text{(by indepencence of } \mathcal{F}_0) \\
&= \int_t^T e^{-r_0 s}e^{\frac{1}{6}\sigma^2(t-s)^3}\,ds \quad \text{(since } \int_0^t W_u\,du \sim \mathcal{N}(0, t^3/3)) \\
&= e^{-r_0 t}\int_0^{T-t} e^{-r_0 v + \sigma^2 v^3/6}\,dv, \quad \text{(by substitution } v = s - t)
\end{aligned}
$$

which is an integral that cannot be computed in terms of elementary functions (for the properties of conditional expectation the reader is refered to Appendix B.4). One way to deal with this integral is to expand the integrand into a Taylor series and integrate terms by term, but also other numerical approximations exist.

The second expectation, $\mathcal{E}_2$, is even more complicated we shall omit its computation here.

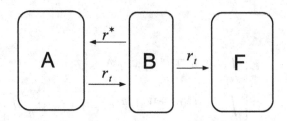

Figure 8.4: *Hedging with swaps.*

8.13.5 Hedging with Swaps

In this section we shall consider a real life example involving swaps. Assume the party B borrows a principal Z from the financial institution F, see Fig.8.4, and needs to pay back F an interest payment on Z at the market floating rate r_t. For instance, B is a home buyer and F is a mortgage lender, and the rate r_t is the ARM mortgage rate, which varies according to market fluctuations. The home buyer B tries to offset the risk, entering a swap contract with the party A, which is more equipped to manage this type of risks. By this contract B will pay a fixed rate r^* to A for the next T years in return for a floating rate r_t, which is passed to party F. This hedging procedure using swaps is applied whenever B wishes to refinance its mortgage with party A. Now the question is the following: *how much should A charge B for the refinance contract?*

The value of the contract depends on the dynamics of the floating rate r_t. We shall assume that r_t follows a CIR process

$$dr_t = a(b - r_t)dt + \sigma\sqrt{r_t}\,dW_t.$$

Then the contract price, $V(t, r)$, is given by the value of a swap with the above underlying rate r_t. This satisfies the following differential equation

$$\frac{\partial V}{\partial t} + a(b - r)\frac{\partial V}{\partial r} + \frac{1}{2}\sigma^2 r\frac{\partial^2 V}{\partial r^2} + Z(r - r^*) = rV$$
$$V(T, r) = 0.$$

The most time efficient way to solve the previous equation is by numerical methods.

Chapter 9

Black-Scholes for Asian Derivatives

In this chapter we shall develop the Black-Scholes equation in the case of Asian derivatives and we shall discuss the particular cases of options and forward contracts on weighted averages. In the case of the latter contracts we obtain closed form solutions, while for the former we apply the reduction variable method to decrease the number of variables and discuss the solution.

9.1 Weighted Averages

In many practical problems the asset price needs to be considered with a certain weight. For instance, when computing car insurance, more weight is assumed for recent accidents than for accidents that occurred several years ago.

In the following we shall define the weight function and provide several examples.

Let $\rho : [0, T] \to \mathbb{R}$ be a *weight function*, i.e. a function satisfying the following two conditions:

(*i*) $\rho > 0$;

(*ii*) $\displaystyle \int_0^T \rho(t) \, dt = 1$.

The *stock weighted average* with respect to the weight ρ is defined as

$$\boxed{S_{ave} = \int_0^T \rho(t) S_t \, dt.}$$
(9.1.1)

Example 9.1.1 (*a*) The uniform weight is obtained for $\rho(t) = \dfrac{1}{T}$. In this case

$$S_{ave} = \frac{1}{T} \int_0^T S_t \, dt \tag{9.1.2}$$

is the *continuous arithmetic average* of the stock on the time interval $[0, T]$.

(*b*) The *linear weight* is obtained if $\rho(t) = \dfrac{2t}{T^2}$. In this case the weight is the time

$$S_{ave} = \frac{2}{T^2} \int_0^T t S_t \, dt. \tag{9.1.3}$$

(*c*) The *exponential weight* is obtained for $\rho(t) = \dfrac{k e^{kt}}{e^{kT} - 1}$. If $k > 0$, the weight is increasing, so recent data are weighted more than old data; if $k < 0$, the weight is decreasing. The exponential weighted average is given by

$$S_{ave} = \frac{k}{e^{kT} - 1} \int_0^T e^{kt} S_t \, dt. \tag{9.1.4}$$

Exercise 9.1.2 *Consider the polynomial weighted average*

$$S_{ave}^{(n)} = \frac{n+1}{T^{n+1}} \int_0^T t^n S_t \, dt.$$

Find the limit $\lim_{n \to \infty} S_{ave}^{(n)}$.

In all previous examples we had $\rho(t) = \rho(t, T) = \dfrac{f(t)}{g(T)}$, with $g(T) = \int_0^T f(t) \, dt$, so $g'(T) = f(T)$ and $g(0) = 0$. The average becomes

$$S_{ave}(T) = \frac{1}{g(T)} \int_0^T f(u) S_u \, du = \frac{I_T}{g(T)},$$

with $I_t = \int_0^t f(u) S_u \, du$ satisfying $dI_t = f(t) S_t dt$. From the quotient rule we get

$$
\begin{aligned}
dS_{ave}(t) &= \frac{dI_t \, g(t) - I_t dg(t)}{g(t)^2} = \left(\frac{f(t)}{g(t)} S_t - \frac{g'(t)}{g(t)} \frac{I_t}{g(t)} \right) dt \\
&= \frac{f(t)}{g(t)} \left(S_t - \frac{g'(t)}{f(t)} S_{ave}(t) \right) dt \\
&= \frac{f(t)}{g(t)} \left(S_t - S_{ave}(t) \right) dt,
\end{aligned}
$$

since $g'(t) = f(t)$. The initial condition is

$$S_{ave}(0) = \lim_{t \searrow 0} S_{ave}(t) = \lim_{t \searrow 0} \frac{I_t}{g(t)} = \lim_{t \searrow 0} \frac{f(t)S_t}{g'(t)} = S_0 \lim_{t \searrow 0} \frac{f(t)}{g'(t)} = S_0.$$

We arrived at the following result:

Proposition 9.1.3 *The weighted average $S_{ave}(t)$ satisfies the stochastic differential equation*

$$
\begin{aligned}
dX_t &= \frac{f(t)}{g(t)}(S_t - X_t)dt \\
X_0 &= S_0.
\end{aligned}
$$

Exercise 9.1.4 *Let $x(t) = \mathbb{E}[S_{ave}(t)]$.*

(a) Show that $x(t)$ satisfies the ordinary differential equation

$$
\begin{aligned}
x'(t) &= \frac{f(t)}{g(t)}\left(S_0 e^{\mu t} - x(t)\right) \\
x(0) &= S_0.
\end{aligned}
$$

(b) Find $x(t)$.

Exercise 9.1.5 *Let $y(t) = \mathbb{E}[S_{ave}^2(t)]$.*

(a) Find the stochastic differential equation satisfied by $S_{ave}^2(t)$;

(b) Find the ordinary differential equation satisfied by $y(t)$;

(c) Solve the previous equation to get $y(t)$ and compute $Var[S_{ave}]$.

9.2 Setting up the Black-Scholes Equation

Consider an Asian derivative whose value at time t, $F(t, S_t, S_{ave}(t))$, depends on variables t, S_t, and $S_{ave}(t)$. Using the stochastic process of S_t

$$dS_t = \mu S_t dt + \sigma S_t dW_t$$

and Proposition 9.1.3, an application of Ito's formula together with the stochastic formulas

$$dt^2 = 0, \quad (dW_t)^2 = 0, \quad (dS_t)^2 = \sigma^2 S^2 dt, \quad (dS_{ave})^2 = 0$$

yield

$$
\begin{aligned}
dF \;=\;& \frac{\partial F}{\partial t}dt + \frac{\partial F}{\partial S_t}dS_t + \frac{1}{2}\frac{\partial^2 F}{\partial S_t^2}(dS_t)^2 + \frac{\partial F}{\partial S_{ave}}dS_{ave} \\
\;=\;& \left(\frac{\partial F}{\partial t} + \mu S_t \frac{\partial F}{\partial S_t} + \frac{1}{2}\sigma^2 S_t^2 \frac{\partial^2 F}{\partial S_t^2} + \frac{f(t)}{g(t)}(S_t - S_{ave})\frac{\partial F}{\partial S_{ave}}\right)dt \\
& + \sigma S_t \frac{\partial F}{\partial S_t}dW_t.
\end{aligned}
$$

Let $\Delta_F = \frac{\partial F}{\partial S_t}$. Consider the following self-financing portfolio at time t

$$
P(t) = F - \Delta_F S_t,
$$

obtained by buying one derivative F and selling Δ_F units of stock. The change in the portfolio value during the time dt does not depend on W_t

$$
\begin{aligned}
dP \;=\;& dF - \Delta_F dS_t \\
\;=\;& \left(\frac{\partial F}{\partial t} + \frac{1}{2}\sigma^2 S_t^2 \frac{\partial^2 F}{\partial S_t^2} + \frac{f(t)}{g(t)}(S_t - S_{ave})\frac{\partial F}{\partial S_{ave}}\right)dt \qquad (9.2.5)
\end{aligned}
$$

so the portfolio P is risk-less. Since no arbitrage opportunities are allowed, investing a value P at time t in a bank at the risk-free rate r for the time interval dt implies

$$
dP = rPdt = \left(rF - rS_t\frac{\partial F}{\partial S_t}\right)dt. \qquad (9.2.6)
$$

Equating (9.2.5) and (9.2.6) provide the following form of the Black-Scholes equation for Asian derivatives on weighted averages:

$$
\boxed{\frac{\partial F}{\partial t} + rS_t\frac{\partial F}{\partial S_t} + \frac{1}{2}\sigma^2 S_t^2 \frac{\partial^2 F}{\partial S_t^2} + \frac{f(t)}{g(t)}(S_t - S_{ave})\frac{\partial F}{\partial S_{ave}} = rF.} \qquad (9.2.7)
$$

9.3   Weighted Average Strike Call Option

In this section we shall use the reduction variable method to decrease the number of variables from three to two. Since $S_{ave}(t) = \dfrac{I_t}{g(t)}$, it is convenient to consider the derivative as a function of t, S_t and I_t

$$
V(t, S_t, I_t) = F(t, S_t, S_{ave}).
$$

A computation similar to the previous one yields the simpler equation

$$\frac{\partial V}{\partial t} + rS_t\frac{\partial V}{\partial S_t} + \frac{1}{2}\sigma^2 S_t^2\frac{\partial^2 V}{\partial S_t^2} + f(t)S_t\frac{\partial V}{\partial I_t} = rV. \qquad (9.3.8)$$

The payoff at maturity of an *average strike call option* can be written in the following form

$$
\begin{aligned}
V_T &= V(T, S_T, I_T) = \max\{S_T - S_{ave}(T), 0\} \\
&= \max\{S_T - \frac{I_T}{g(T)}, 0\} = S_T\max\{1 - \frac{1}{g(T)}\frac{I_T}{S_T}, 0\} \\
&= S_T L(T, R_T),
\end{aligned}
$$

where

$$R_t = \frac{I_t}{S_t}, \qquad L(t, R) = \max\{1 - \frac{1}{g(t)}R, 0\}.$$

Since at maturity the variable S_T is separated from T and R_T, we shall look for a solution of equation (9.3.8) of the same type, for any $t \le T$, i.e. $V(t, S, I) = SG(t, R)$. Since

$$
\begin{aligned}
\frac{\partial V}{\partial t} &= S\frac{\partial G}{\partial t}, \qquad \frac{\partial V}{\partial I} = S\frac{\partial G}{\partial R}\frac{1}{S} = \frac{\partial G}{\partial R}; \\
\frac{\partial V}{\partial S} &= G + S\frac{\partial G}{\partial R}\frac{\partial R}{\partial S} = G - R\frac{\partial G}{\partial R}; \\
\frac{\partial^2 V}{\partial S^2} &= \frac{\partial}{\partial S}(G - R\frac{\partial G}{\partial R}) = \frac{\partial G}{\partial R}\frac{\partial R}{\partial S} - \frac{\partial R}{\partial S}\frac{\partial G}{\partial R} - R\frac{\partial^2 G}{\partial R^2}\frac{\partial R}{\partial S}; \\
&= R\frac{\partial^2 G}{\partial R^2}\frac{I}{S}; \\
S^2\frac{\partial^2 V}{\partial S^2} &= RI\frac{\partial^2 G}{\partial R^2},
\end{aligned}
$$

substituting in (9.3.8) and using that $\dfrac{RI}{S} = R^2$, after cancelations we obtain

$$\frac{\partial G}{\partial t} + \frac{1}{2}\sigma^2 R^2\frac{\partial^2 G}{\partial R^2} + (f(t) - rR)\frac{\partial G}{\partial R} = 0. \qquad (9.3.9)$$

This is a partial differential equation in only two variables, t and R. It can be solved explicitly sometimes, depending on the form of the final condition $G(T, R_T)$ and expression of the function $f(t)$.

Note that in the case of a weighted average strike call option the final condition is

$$G(T, R_T) = \max\{1 - \frac{R_T}{g(T)}, 0\}. \qquad (9.3.10)$$

Example 9.3.1 In the case of the arithmetic average (9.1.2) the function $G(t, R)$ satisfies the partial differential equation

$$\frac{\partial G}{\partial t} + \frac{1}{2}\sigma^2 R^2 \frac{\partial^2 G}{\partial R^2} + (1 - rR)\frac{\partial G}{\partial R} = 0 \qquad (9.3.11)$$

with the final condition $G(T, R_T) = \max\{1 - \frac{R_T}{T}, 0\}$.

Example 9.3.2 In the case of the exponential average (9.1.4) the function $G(t, R)$ satisfies the equation

$$\frac{\partial G}{\partial t} + \frac{1}{2}\sigma^2 R^2 \frac{\partial^2 G}{\partial R^2} + (ke^{kt} - rR)\frac{\partial G}{\partial R} = 0 \qquad (9.3.12)$$

with the final condition $G(T, R_T) = \max\left\{1 - \frac{R_T}{e^{kT} - 1}, 0\right\}$.

Neither of the previous two final condition problems can be solved explicitly in an easy way. Theoretically, the solution $G(t, R_t)$ is obtained as a convolution between the heat kernel of the differential operator $\frac{1}{2}\sigma^2 x^2 \partial_x^2 + (f(t) - rx)\partial_x$ and the boundary condition $G(T, R_T)$. The difficulty is due to the fact that an explicit expression for the previously specified heat kernel is missing at the moment.

9.4 Boundary Conditions

The partial differential equation (9.3.9) is of first order in t and second order in R. We need to specify one condition at $t = T$ (the payoff at maturity), which is given by (9.3.10), and two conditions for $R = 0$ and $R \to \infty$, which specify the behavior of solution $G(t, R)$ at two limiting positions of the variable R.

Taking $R \to 0$ in equation (9.3.9) and using Exercise 9.4.1 yields the first boundary condition for $G(t, R)$

$$\boxed{\left(\frac{\partial G}{\partial t} + f\frac{\partial G}{\partial R}\right)\bigg|_{R=0} = 0.} \qquad (9.4.13)$$

The term $\frac{\partial G}{\partial R}\big|_{R=0}$ represents the slope of $G(t, R)$ with respect to R at $R = 0$, while $\frac{\partial G}{\partial t}\big|_{R=0}$ is the variation of the price G with respect to time t when $R = 0$.

Another boundary condition is obtained by specifying the behavior of $G(t, R)$ for large values of R. If $R_t \to \infty$, we must have $S_t \to 0$, because

$$R_t = \frac{1}{S_t}\int_0^t f(u)S_u\, du$$

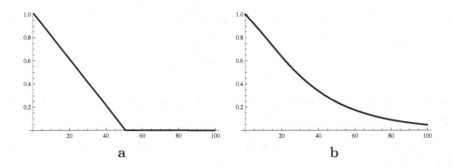

Figure 9.1: *The profile of the solution $H(t, R)$:* **a** *at expiration;* **b** *when there is $T - t$ time left before expiration.*

and $\int_0^t f(u) S_u\, du > 0$ for $t > 0$. In this case we are better off not exercising the option (since otherwise we get a negative payoff), so the boundary condition is

$$\boxed{\lim_{R \to \infty} G(R, t) = 0.}\qquad (9.4.14)$$

It can be shown in the theory of partial differential equations that equation (9.3.9) together with the final condition (9.3.10), see Fig.9.1 a, and boundary conditions (9.4.13) and (9.4.14) has a unique solution $G(t, R)$, see Fig.9.1 b.

Exercise 9.4.1 *Let f be a bounded differentiable function. Show that*

    (a) $\lim_{x \to 0} x f'(x) = 0$;

    (b) $\lim_{x \to 0} x^2 f''(x) = 0$, *if limits exist.*

Monte-Carlo simulation There is no close form solution for the weighted average strike call option. Even in the simplest case, when the average is arithmetic, the solution is just approximative, see section 6.8. In real life the price is worked out using the *Monte-Carlo simulation*. This is based on averaging a large number, n, of simulations of the process R_t in the risk-neutral world, i.e. assuming $\mu = r$. For each realization, the associated payoff $G_{T,j} = \max\left\{1 - \dfrac{R_{T,j}}{g(T)}\right\}$ is computed, with $j \leq n$. Here $R_{T,j}$ represents the value of R at time T in the jth realization. The average

$$\frac{1}{n} \sum_{j=1}^{n} G_{T,j}$$

is an unbiased approximation of the payoff expectation $\mathbb{E}[G_T]$. Discounting under the risk-free rate we get the price at time t

$$G(t, R) = e^{-r(T-t)}\left(\frac{1}{n}\sum_{j=1}^{n}G_{T,j}\right).$$

It is worth noting that the term on the right is an approximation of the risk neutral conditional expectation $\widehat{\mathbb{E}}[G_T|\mathcal{F}_t]$.

When simulating the process R_t, it is convenient to know its stochastic differential equation. Using

$$dI_t = f(t)S_t dt, \qquad d\left(\frac{1}{S_t}\right) = \frac{1}{S_t}\left((\sigma^2 - \mu)dt - \sigma dW_t\right)dt,$$

the product rule yields

$$
\begin{aligned}
dR_t &= d\left(\frac{I_t}{S_t}\right) = d\left(I_t\frac{1}{S_t}\right) \\
&= dI_t\frac{1}{S_t} + I_t d\left(\frac{1}{S_t}\right) + dI_t d\left(\frac{1}{S_t}\right) \\
&= f(t)dt + R_t\left((\sigma^2 - \mu)dt - \sigma dW_t\right).
\end{aligned}
$$

Collecting terms yields the following stochastic differential equation for R_t:

$$\boxed{dR_t = -\sigma R_t dW_t + \left(f(t) + (\sigma^2 - \mu)R_t\right)dt.} \qquad (9.4.15)$$

The initial condition is $R_0 = \frac{I_0}{S_0} = 0$ (since $I_0 = 0$) if the simulation starts at $t = 0$.

Before solving explicitly this equation, we shall find the mean and variance of R_t. The equation can be written more conveniently as

$$dR_t - (\sigma^2 - \mu)R_t dt = f(t)dt - \sigma R_t dW_t.$$

Multiplying by $e^{-(\sigma^2 - \mu)t}$ yields the exact equation

$$d\left(e^{-(\sigma^2 - \mu)t}R_t\right) = e^{-(\sigma^2 - \mu)t}f(t)dt - \sigma e^{-(\sigma^2 - \mu)t}R_t dW_t.$$

Integrating yields

$$e^{-(\sigma^2 - \mu)t}R_t = \int_0^t e^{-(\sigma^2 - \mu)u}f(u)\,du - \int_0^t \sigma e^{-(\sigma^2 - \mu)u}R_u\,dW_u.$$

The first integral is deterministic while the second is an Ito integral. Using that the expectations of Ito integrals vanish, we get

$$\mathbb{E}[e^{-(\sigma^2 - \mu)t}R_t] = \int_0^t e^{-(\sigma^2 - \mu)u}f(u)\,du$$

and hence

$$\mathbb{E}[R_t] = e^{(\sigma^2 - \mu)t} \int_0^t e^{-(\sigma^2 - \mu)u} f(u)\, du.$$

Exercise 9.4.2 *Find* $\mathbb{E}[R_t^2]$ *and* $Var[R_t]$.

Equation (9.4.15) is a linear stochastic differential equation. Multiplying by the integrating factor

$$\rho_t = e^{\sigma W_t + \frac{1}{2}\sigma^2 t}$$

the equation is transformed into an exact equation

$$d(\rho_t R_t) = \big(\rho_t f(t) + (\sigma^2 - \mu)\rho_t R_t\big) dt.$$

Substituting $Y_t = \rho_t R_t$ yields

$$dY_t = \big(\rho_t f(t) + (\sigma^2 - \mu)Y_t\big) dt,$$

which can be written as

$$dY_t - (\sigma^2 - \mu)Y_t dt = \rho_t f(t) dt.$$

Multiplying by $e^{-(\sigma^2 - \mu)t}$ yields the exact equation

$$d(e^{-(\sigma^2 - \mu)t} Y_t) = e^{-(\sigma^2 - \mu)t} \rho_t f(t) dt,$$

which can be solved by integration

$$e^{-(\sigma^2 - \mu)t} Y_t = \int_0^t e^{-(\sigma^2 - \mu)u} \rho_u f(u)\, du.$$

Going back to the variable $R_t = Y_t/\rho_t$, we obtain the following closed form expression

$$R_t = e^{-(\mu - \frac{\sigma^2}{2})t - \sigma W_t} \int_0^t e^{(\mu - \frac{\sigma^2}{2})u + \sigma W_u} f(u)\, du. \tag{9.4.16}$$

Exercise 9.4.3 *Find* $\mathbb{E}[R_t]$ *by taking the expectation in formula* (9.4.16).

It is worth noting that we can arrive at formula (9.4.16) directly, without going through solving a stochastic differential equation. We shall show this procedure in the following.

Using the well-known formulas for the stock price

$$S_u = S_0 e^{(\mu - \frac{1}{2}\sigma^2)u + \sigma W_u}, \qquad S_t = S_0 e^{(\mu - \frac{1}{2}\sigma^2)t + \sigma W_t},$$

and dividing, yields

$$\frac{S_u}{S_t} = e^{(\mu - \frac{1}{2}\sigma^2)(u-t) + \sigma(W_u - W_t)}.$$

Then we get

$$
\begin{aligned}
R_t &= \frac{I_t}{S_t} = \frac{1}{S_t}\int_0^t S_u f(u)\, du \\
&= \int_0^t \frac{S_u}{S_t} f(u)\, du = \int_0^t e^{(\mu - \frac{1}{2}\sigma^2)(u-t) + \sigma(W_u - W_t)} f(u)\, du,
\end{aligned}
$$

which is formula (9.4.16).

Exercise 9.4.4 *Find an explicit formula for R_t in terms of the integrated Brownian motion $Z_t^{(\sigma)} = \int_0^t e^{\sigma W_u}\, du$, in the case of an exponential weight with $k = \frac{1}{2}\sigma^2 - \mu$, see Example 9.1.1(c).*

Exercise 9.4.5 *(a) Find the price of a derivative G which satisfies*

$$\frac{\partial G}{\partial t} + \frac{1}{2}\sigma^2 R^2 \frac{\partial^2 G}{\partial R^2} + (1 - rR)\frac{\partial G}{\partial R} = 0$$

with the payoff $G(T, R_T) = R_T^2$.
(b) Find the value of an Asian derivative V_t on the arithmetic average, that has the payoff

$$V_T = V(T, S_T, I_T) = \frac{I_T^2}{S_T},$$

where $I_T = \int_0^T S_t\, dt$.

Exercise 9.4.6 *Use a computer simulation to find the value of an Asian arithmetic average strike option with $r = 4\%$, $\sigma = 50\%$, $S_0 = \$40$, and $T = 0.5$ years.*

9.5   Asian Forward Contracts

An *Asian forward contract* is a derivative that pays at delivery the amount $S_{ave}(T)$ for the price S_T. We still work under the hypothesis that the weighted stock average S_{ave} is given by (9.1.1). Since the payoff of this derivative is given by

$$V_T = S_T - S_{ave}(T) = S_T\left(1 - \frac{R_T}{g(T)}\right),$$

the reduction variable method suggests considering a solution of the type $V(t, S_t, I_t) = S_t G(t, R_t)$, where $G(t, T)$ satisfies equation (9.3.9) with the final condition $G(T, R_T) = 1 - \dfrac{R_T}{g(T)}$. Since this is linear in R_T, it makes sense to look for a solution $G(t, R_t)$ in the following form

$$G(t, R_t) = a(t)R_t + b(t), \qquad (9.5.17)$$

with functions $a(t)$ and $b(t)$ subject to be determined. Substituting into (9.3.9) and collecting R_t yields

$$(a'(t) - ra(t))R_t + b'(t) + f(t)a(t) = 0.$$

Since this polynomial in R_t vanishes for all values of R_t, then its coefficients are identically zero, so

$$a'(t) - ra(t) = 0, \qquad b'(t) + f(t)a(t) = 0.$$

When $t = T$ we have

$$G(T, R_T) = a(T)R_T + b(T) = 1 - \frac{R_T}{g(T)}.$$

Equating the coefficients of R_T yields the final conditions

$$a(T) = -\frac{1}{g(T)}, \qquad b(T) = 1.$$

The coefficient $a(t)$ satisfies the ordinary differential equation

$$
\begin{aligned}
a'(t) &= ra(t) \\
a(T) &= -\frac{1}{g(T)},
\end{aligned}
$$

which has the solution

$$a(t) = -\frac{1}{g(T)}e^{-r(T-t)}.$$

Similarly, the coefficient $b(t)$ satisfies the equation

$$
\begin{aligned}
b'(t) &= -f(t)a(t) \\
b(T) &= 1
\end{aligned}
$$

with the solution

$$b(t) = 1 + \int_t^T f(u)a(u)\, du.$$

Substituting in (9.5.17) yields

$$
\begin{aligned}
G(t, R) &= -\frac{1}{g(T)}e^{-r(T-t)}R_t + 1 + \int_t^T f(u)a(u)\, du \\
&= 1 - \frac{1}{g(T)}\left[R_t e^{-r(T-t)} + \int_t^T f(u)e^{-r(T-u)}\, du \right].
\end{aligned}
$$

Then going back into the variable $I_t = S_t R_t$ yields

$$
\begin{aligned}
V(t, S_t, I_t) &= S_t G(t, R_t) \\
&= S_t - \frac{1}{g(T)} \left[I_t e^{-r(T-t)} + S_t \int_t^T f(u) e^{-r(T-u)} \, du \right].
\end{aligned}
$$

Using that $\rho(u) = \dfrac{f(u)}{g(T)}$ and going back to the initial variable $S_{ave}(t) = I_t/g(t)$ yields

$$
\begin{aligned}
F(t, S_t, S_{ave}(t)) &= V(t, S_t, I_t) \\
&= S_t - \frac{g(t)}{g(T)} S_{ave}(t) e^{-r(T-t)} - S_t \int_t^T \rho(u) e^{-r(T-u)} \, du.
\end{aligned}
$$

We have arrived at the following result:

Proposition 9.5.1 *The value at time t of an Asian forward contract on a weighted average with the weight function $\rho(t)$, i.e. an Asian derivative with the payoff $F_T = S_T - S_{ave}(T)$, is given by*

$$
\boxed{F(t, S_t, S_{ave}(t)) = S_t \left(1 - \int_t^T \rho(u) e^{-r(T-u)} \, du \right) - \frac{g(t)}{g(T)} e^{-r(T-t)} S_{ave}(t).}
$$

It is worth noting that the previous price can be written as a linear combination of S_t and $S_{ave}(t)$

$$
F(t, S_t, S_{ave}(t)) = \alpha(t) S_t + \beta(t) S_{ave}(t),
$$

where

$$
\alpha(t) = 1 - \int_t^T \rho(u) e^{-r(T-u)} \, du
$$

$$
\beta(t) = -\frac{g(t)}{g(T)} e^{-r(T-t)} = -\frac{\int_0^t f(u) \, du}{\int_0^T f(u) \, du} e^{-r(T-t)}.
$$

In the first formula $\rho(u) e^{-r(T-u)}$ is the discounted weight at time u, and $\alpha(t)$ is 1 minus the total discounted weight between t and T. One can easily check that $\alpha(T) = 1$ and $\beta(T) = -1$.

Exercise 9.5.2 *Find the value at time t of an Asian forward contract on an arithmetic average $A_t = \frac{1}{t} \int_0^t S_u \, du$.*

Exercise 9.5.3 (a) *Find the value at time t of an Asian forward contract on an exponential weighted average with the weight given by Example 9.1.1 (c).*

(b) *What happens if* $k = -r$? *Why?*

Exercise 9.5.4 *Find the value at time t of an Asian power contract on the arithmetic average with the payoff* $F_T = \left(\int_0^T S_u \, du \right)^n$, $n = 1, 2$.

9.6 Put-Call Parity

This section provides relations between the values of Asian calls and puts and the underlying stock price and weighted stock average.

Put-Call parity for Asian Strike Options We shall find a relation between the prices of a call and a put for Asian strike options written on the weighted average. Denote by C_t and P_t the values at time t of a call and a put, with payoffs

$$C_T = \max\{S_T - S_{ave}(T)\}$$
$$P_T = \max\{S_{ave}(T) - S_T\}.$$

Consider the derivative $F_t = C_t - P_t$. Since its payoff is

$$F_T = C_T - P_T = S_T - S_{ave}(T),$$

it follows that F_t is an Asian forward contract whose value is provided by Proposition 9.5.1. Hence

$$C_t - P_t = S_t \left(1 - \int_t^T \rho(u) e^{-r(T-u)} \, du \right) - \frac{g(t)}{g(T)} e^{-r(T-t)} S_{ave}(t),$$

or equivalently,

$$\boxed{C_t + \frac{g(t)}{g(T)} e^{-r(T-t)} S_{ave}(t) = P_t + S_t \left(1 - \int_t^T \rho(u) e^{-r(T-u)} \, du \right).} \quad (9.6.18)$$

Exercise 9.6.1 *Write the put-call parity for Asian strike options in the case when* S_{ave} *is:*

(a) *the continuous arithmetic average;*

(b) *the exponential weighted average, see Example 9.1.1.*

Put-Call parity for Asian Price Options Let C_t and P_t be the values of average price call and put options, respectively, with the payoffs

$$C_T = \max\{S_{ave}(T) - K\}$$
$$P_T = \max\{K - S_{ave}(T)\}.$$

Consider the derivative $f_t = C_t - P_t$ that has the payoff $f_T = C_T - P_T = S_{ave}(T) - K$, and note that

$$f_T = (S_T - K) - (S_T - S_{ave}(T)). \tag{9.6.19}$$

Then consider two more derivatives, V_t and F_t, whose payoffs are

$$V_T = S_T - K, \qquad F_T = S_T - S_{ave}(T).$$

The first one corresponds to a forward contract on the stock given by (6.1.5) and the other to an Asian forward contract given by Proposition 9.5.1. The value at time t is

$$
\begin{aligned}
f_t &= V_t - F_t \\
&= S_t - e^{-r(T-t)}K - \left\{ S_t\left(1 - \int_t^T \rho(u)e^{-r(T-u)}\,du\right) - \frac{g(t)}{g(T)}e^{-r(T-t)}S_{ave}(t) \right\} \\
&= S_t \int_t^T \rho(u)e^{-r(T-u)}\,du - e^{-r(T-t)}\left\{ K - \frac{g(t)}{g(T)}S_{ave}(t) \right\}.
\end{aligned}
$$

This is equivalent with the relation

$$\boxed{C_t + e^{-r(T-t)}\left\{ K - \frac{g(t)}{g(T)}S_{ave}(t) \right\} = P_t + S_t \int_t^T \rho(u)e^{-r(T-u)}\,du.} \tag{9.6.20}$$

Exercise 9.6.2 *Write the put-call parity for Asian price options in the case when S_{ave} is given by:*

(a) the continuous arithmetic average;

(b) the exponential weighted average, see Example 9.1.1.

9.7 Valuation of Arithmetic Asian Price Options

An arithmetic average Asian price call option has the payoff $V_T = \max\{A_T - K, 0\}$, where $A_T = \frac{1}{T}\int_0^T S_u\,du$. It provides the buyer the opportunity to purchase at time T the average of the stock for the price K. This option has the advantage that it depends on the entire history of the market, and hence, it reduces the risk of price manipulations of the underlying asset at the maturity date.

Due to the complexity of the probability density of the arithmetic stock average, A_T, some authors have initially valued this option using Monte Carlo simulations (Kemna and Vorst [48]), or various approximations (see, for example, Bouaziz et al.[13], Turnbull and Wakeman [64], Vorst [68]). In section 6.8 we also had presented an approximative solution using the first two moments.

Even if it was initially believed that an exact trackable solution for an arithmetic Asian price option does not exist, Geman and Yor [35] provided in 1993 a closed form solution involving an inverse Laplace transform (see also Eydeland and Geman [32]). This section gives a summary of this solution without providing any computational proof. First, define the following substitutions:

$$\tau = \frac{1}{4}\sigma^2(T-t)$$

$$\nu = \frac{2(r-q)}{\sigma^2} - 1$$

$$\alpha = \frac{\sigma^2(KT - tA_t)}{4S_t}.$$

The value of the average price option is then given by

$$V_t = e^{-r(T-t)} \frac{4S_t}{T\sigma^2} C(\tau, \nu, \alpha). \tag{9.7.21}$$

The function $C(\tau, \nu, \alpha)$ is not given explicitly, but its Laplace transform (see Appendix A.2) is given by

$$
\begin{aligned}
\mathcal{L}(C)(p) &= \int_0^\infty C(\tau, \nu, \alpha) e^{-p\tau} \, du \\
&= \frac{1}{p(p - 2\nu - 2)\Gamma\left(\frac{\mu-\nu}{2} - 1\right)} \int_0^{\frac{1}{2\alpha}} x^{\frac{\mu-\nu}{2}-2}(1 - 2\alpha x)^{\frac{\mu+\nu}{2}+1} e^{-x} \, dx,
\end{aligned}
$$

where μ is defined by the transformation $\mu(p) = \sqrt{\nu^2 + 2p}$. As pointed out in Donati-Martin et al.[26], using the representation of the previous integral as a confluent hypergeometric function, the Laplace transform can be written as

$$\mathcal{L}(C)(p) = \frac{(2\alpha)^{\frac{-\mu+\nu}{2}+1}\Gamma\left(\frac{\mu+\nu+4}{2}\right)}{p(p-2\nu-2)\Gamma(\mu+1)} {}_1F_1\left(\frac{\mu-\nu}{2} - 1; \mu + 1; -\frac{1}{2\alpha}\right),$$

where the confluent hypergeometric function is defined by

$$ {}_1F_1(a, b; z) = \sum_{n=0}^\infty \frac{(a)^n}{(b)^n} \frac{z^n}{n!},$$

with $a^{(0)} = 1$ and $a^{(n)} = a(a+1)(a+2)\cdots(a+n-1)$. Similar definition formulas apply to $b^{(n)}$. For more details, see Appendix D.1.

Chapter 10

American Options

American options are options that are allowed to be exercised at any time before maturity. Because of this extra feature, they tend to be more expensive than their European counterparts. Exact pricing formulas exist just for perpetuities, while they are missing for finitely lived American options. However, some approximative formulas can be worked out.

10.1 Perpetual American Options

A *perpetual American option* is an American option that never expires. These contracts can be exercised at any time t, $0 \leq t \leq \infty$. Even if finding the optimal exercise time for finite maturity American options is a delicate matter, in the case of perpetual American calls and puts there is always possible to find the optimal exercise time and to derive a closed form pricing formula (see Merton, [55]).

10.1.1 Present Value of Barriers

The goal of this section is to compute the present value of a contract that pays a fixed cash amount at a stochastic time defined by the first passage of time of a stock. These types of contracts were presented in section 6.15 under the name of perpetual look-back options. We shall re-evaluate them here including both the above and below barrier cases. We obtain this way formulas that will be the main ingredient in pricing perpetual American options over the next few sections.

Reaching the barrier from below Let S_t denote the stock price with initial value S_0 and consider a positive number b such that $b > S_0$. We recall the first passage of time τ_b when the stock S_t hits for the first time the barrier

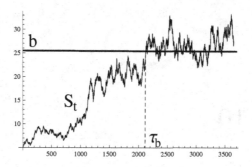

Figure 10.1: *The first passage of time when S_t hits the level b from below.*

b, see Fig. 10.1

$$\tau_b = \inf\{l > 0; S_t \geq b\}.$$

Consider a contract that pays \$1 at the time when the stock reaches the barrier b for the first time. Under the constant interest rate assumption, the value of the contract at time $t = 0$ is obtained by discounting the value of \$1 at the rate r for the period τ_b and taking the expectation in the risk neutral world

$$f_0 = \widehat{\mathbb{E}}[e^{-r\tau_b}].$$

In the following we shall compute the right side of the previous expression. Using that the stock price in the risk-neutral world is given by the expression $S_t = S_0 e^{(r-\frac{\sigma^2}{2})t+\sigma W_t}$, with $r > \frac{\sigma^2}{2}$, then

$$M_t = e^{-rt}S_t = S_0 e^{\sigma W_t - \frac{\sigma^2}{2}t}, \qquad t \geq 0$$

is a martingale (see Proposition 7.1.4). Applying the Optional Stopping Theorem (see section 1.8) yields $\widehat{\mathbb{E}}[M_{\tau_b}] = \widehat{\mathbb{E}}[M_0]$, which is equivalent to

$$\widehat{\mathbb{E}}[e^{-r\tau_b}S_{\tau_b}] = S_0.$$

Since $S_{\tau_b} = b$, the previous relation implies

$$\widehat{\mathbb{E}}[e^{-r\tau_b}] = \frac{S_0}{b},$$

where the expectation is taken in the risk-neutral world. Hence, we arrived at the following result applicable to non-dividend paying stocks:

Proposition 10.1.1 *The value at time $t = 0$ of \$1 received at the time when the stock reaches level b from below is*

$$\boxed{f_0 = \frac{S_0}{b}.}$$

Remark 10.1.2 In the previous proof we had applied the Optional Stopping Theorem. We leave as an exercise for the reader to verify the hypothesis of the theorem.

Exercise 10.1.3 *Let $0 < S_0 < b$ and assume $r > \frac{\sigma^2}{2}$.*
(a) Show that $P(S_t$ reaches $b) = 1$ in the risk-neutral world.
(b) Prove the identity $P(\tau_b < \infty) = 1$.

Remark 10.1.4 The result of Proposition 10.1.1 can also be obtained as a consequence of Exercise 5.4.7. Using the expression of the stock price in the risk-neutral world, $S_t = S_0 e^{(r-\frac{\sigma^2}{2})t + \sigma W_t}$, we have

$$
\begin{aligned}
\tau_b &= \inf\{t > 0; S_t \geq b\} = \inf\{t > 0; (r - \frac{\sigma^2}{2})t + \sigma W_t \geq \ln \frac{b}{S_0}\} \\
&= \inf\{t > 0; \mu t + \sigma W_t \geq x\},
\end{aligned}
$$

where $x = \ln \frac{b}{S_0} > 0$ and $\mu = r - \frac{\sigma^2}{2} > 0$. Then Exercise 5.4.7 yields

$$
\begin{aligned}
\mathbb{E}[e^{-r\tau_b}] &= e^{\frac{1}{\sigma^2}(\mu - \sqrt{2s\sigma^2 + \mu^2})x} = e^{\frac{1}{\sigma^2}\left(r - \frac{\sigma^2}{2} - \sqrt{2r\sigma^2 + (r - \frac{\sigma^2}{2})^2}\right) \ln \frac{b}{S_0}} \\
&= e^{-\ln \frac{b}{S_0}} = e^{\ln \frac{S_0}{b}} = \frac{S_0}{b},
\end{aligned}
$$

where we used that

$$
\begin{aligned}
r - \frac{\sigma^2}{2} - \sqrt{2r\sigma^2 + \left(r - \frac{\sigma^2}{2}\right)^2} &= r - \frac{\sigma^2}{2} - \sqrt{\left(r + \frac{\sigma^2}{2}\right)^2} \\
&= r - \frac{\sigma^2}{2} - r - \frac{\sigma^2}{2} = -\sigma^2.
\end{aligned}
$$

Exercise 10.1.5 *Let $S_0 < b$. Find the probability density function for the hitting time $\tau_b = \inf\{t > 0; S_t > b\}$.*

Exercise 10.1.6 *Assume the stock pays continuous dividends at the constant rate $\delta > 0$ and let $b > 0$ such that $S_0 < b$.*
(a) Prove that the value at time $t = 0$ of $\$1$ received at the time when the stock reaches level b from below is

$$
f_0 = \left(\frac{S_0}{b}\right)^{h_1},
$$

where

$$
h_1 = \frac{1}{2} - \frac{r - \delta}{\sigma^2} + \sqrt{\left(\frac{r - \delta}{\sigma^2} - \frac{1}{2}\right)^2 + \frac{2r}{\sigma^2}}.
$$

(b) Show that $h_1(0) = 1$ and $h_1(\delta)$ is an increasing function for $\delta > 0$.

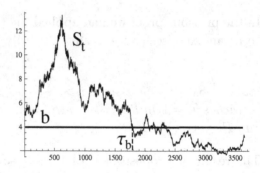

Figure 10.2: *The first passage of time when S_t hits the level b from above.*

(c) *Find the limit of the rate of change $\lim_{\delta \to \infty} h_1'(\delta)$.*

(d) *Work out a formula for the sensitivity of the value f_0 with respect to the dividend rate δ and compute the long run value of this rate.*

Reaching the barrier from above Sometimes a stock can reach a barrier b from above. Let S_0 be the initial value of the stock S_t and assume the inequality $b < S_0$. Consider again the first passage of time, see Fig.10.2

$$\tau_b = \inf\{t > 0; S_t \leq b\}.$$

In this paragraph we compute the value of a contract that pays \$1 at the time when the stock reaches the barrier b for the first time, which is given by $f_0 = \widehat{\mathbb{E}}[e^{-r\tau_b}]$. We shall work under the assumption that the interest rate r is constant and $r > \frac{\sigma^2}{2}$.

Proposition 10.1.7 *The value at time $t = 0$ of \$1 received at the time when the stock reaches level b from above is*

$$\boxed{f_0 = \left(\frac{S_0}{b}\right)^{\frac{-2r}{\sigma^2}}.}$$

Proof: We should reduce the problem from S_t to a Brownian motion with drift and then apply Proposition 4.6.6. of [15] (see also Appendix C.7, point 2). We write

$$
\begin{aligned}
\tau_b &= \inf\{t > 0; S_t \leq b\} = \inf\{t > 0; \left(r - \frac{\sigma^2}{2}\right)t + \sigma W_t \leq \ln\frac{b}{S_0}\} \\
&= \inf\{t > 0; \mu t + \sigma W_t \leq -x\},
\end{aligned}
$$

where $x = \ln\frac{S_0}{b} > 0$, $\mu = r - \frac{\sigma^2}{2}$. Choosing $s = r$ in the aforementioned Proposition yields

$$
\begin{aligned}
\mathbb{E}[e^{-r\tau_b}] &= e^{\frac{-1}{\sigma^2}(\mu+\sqrt{2r\sigma^2+\mu^2})x} = e^{\frac{-1}{\sigma^2}\left(r-\frac{\sigma^2}{2}+\sqrt{2r\sigma^2+(r-\frac{\sigma^2}{2})^2}\right)x} \\
&= e^{-\frac{2r}{\sigma^2}x} = e^{-\frac{2r}{\sigma^2}\ln\frac{S_0}{b}} = \left(\frac{S_0}{b}\right)^{-\frac{2r}{\sigma^2}}.
\end{aligned}
$$

In the previous computation we used that

$$
r - \frac{\sigma^2}{2} + \sqrt{2r\sigma^2 + \left(r - \frac{\sigma^2}{2}\right)^2} = r - \frac{\sigma^2}{2} + \sqrt{\left(r + \frac{\sigma^2}{2}\right)^2}
$$

$$
= r - \frac{\sigma^2}{2} + r + \frac{\sigma^2}{2} = 2r.
$$

■

Exercise 10.1.8 *Assume the stock pays continuous dividends at the constant rate $\delta > 0$ and let $b > 0$ such that $b < S_0$.*

(a) Use a similar method as in the proof of Proposition 10.1.7 to prove that the value at time $t = 0$ of $\$1$ received at the time when the stock reaches level b from above is

$$
f_0 = \left(\frac{S_0}{b}\right)^{h_2},
$$

where

$$
h_2 = \frac{1}{2} - \frac{r - \delta}{\sigma^2} - \sqrt{\left(\frac{r-\delta}{\sigma^2} - \frac{1}{2}\right)^2 + \frac{2r}{\sigma^2}}.
$$

(b) What is the value of the contract at any time t, with $0 \le t < \tau_b$?

Perpetual American options have simple exact pricing formulas. This is because of the "time invariance" property of their values. Since the time to expiration for these type of options is the same (i.e. infinity), the option exercise problem looks the same at every instance of time. Consequently, their value do not depend on the time to expiration.

10.1.2 Perpetual American Calls

A *perpetual American call* is a call option that never expires, i.e. is a contract that gives the holder the right to buy the stock for the price K at any instance of time $0 \le t \le +\infty$. The infinity is included here to cover the case when the option is never exercised.

When the call is exercised the holder receives $S_\tau - K$, where τ denotes the exercise time.

Pricing strategy Assume the holder has the strategy to exercise the call whenever the stock S_t reaches the barrier b, with $b > K$ subject to be determined later, such that a certain optimization property holds. Then at the exercise time τ_b the payoff is $b - K$, where

$$\tau_b = \inf\{t > 0; S_t \geq b\}.$$

We note that it makes sense to choose the barrier such that $S_0 < b$. The value of the payoff at time $t = 0$ is obtained discounting at the interest rate r and using Proposition 10.1.1

$$f(b) = \widehat{\mathbb{E}}[(b - K)e^{-r\tau_b}] = (b - K)\frac{S_0}{b} = \left(1 - \frac{K}{b}\right)S_0.$$

We need to choose the value of the barrier $b > 0$ for which $f(b)$ reaches its maximum. Since $1 - \frac{K}{b}$ is an increasing function of b, the optimum value can be evaluated as

$$\max_{b>0} f(b) = \max_{b>0}\left(1 - \frac{K}{b}\right)S_0 = \lim_{b \to \infty}\left(1 - \frac{K}{b}\right)S_0 = S_0.$$

This is reached for the optimal barrier $b^* = \infty$, which corresponds to the infinite exercise time $\tau_{b^*} = \infty$. *Hence, it is never optimal to exercise a perpetual call option on a nondividend paying stock.*

The next exercise covers the case of the dividend paying stock. The method is similar with the one described previously.

Exercise 10.1.9 *Consider a stock that pays dividends at a continuous rate* $\delta > 0$.

(a) Assume a perpetual call is exercised whenever the stock reaches the barrier b from below. Show that the discounted value at time $t = 0$ is

$$f(b) = (b - K)\left(\frac{S_0}{b}\right)^{h_1},$$

where

$$h_1 = \frac{1}{2} - \frac{r - \delta}{\sigma^2} + \sqrt{\left(\frac{r - \delta}{\sigma^2} - \frac{1}{2}\right)^2 + \frac{2r}{\sigma^2}}.$$

(b) Use differentiation to show that the maximum value of $f(b)$ is realized for

$$b^* = K\frac{h_1}{h_1 - 1}.$$

(c) Prove the price of perpetual call

$$f(b^*) = \max_{b>0} f(b) = \frac{K}{h_1 - 1}\left(\frac{h_1 - 1}{h_1}\frac{S_0}{K}\right)^{h_1}.$$

(d) Let τ_{b^*} be the exercise time of the perpetual call. When do you expect to exercise the call? (Find $\mathbb{E}[\tau_{b^*}]$).

10.1.3 Perpetual American Puts

A *perpetual American put* is a put option that never expires, i.e. is a contract that gives the holder the right to sell the stock for the price K at any instance of time $0 \le t \le \infty$.

Assume the put is exercised when S_t reaches the barrier b. Then its payoff, $K - S_t = K - b$, has a couple of noteworthy features. First, if we choose b too large, we loose option value, which eventually vanishes for $b \ge K$. Second, if we pick b too small, the chances that the stock S_t will hit b are too small (see Exercise 10.1.10), a fact that diminishes the put value. It follows that the optimum exercise barrier, b^*, is somewhere in between these two extreme values.

Exercise 10.1.10 *Let $0 < b < S_0$ and let $t > 0$ fixed.*

(a) *Show that the following inequality holds in the risk neutral world*

$$P(S_t < b) \le e^{-\frac{1}{2\sigma^2 t}[\ln(S_0/b)+(r-\sigma^2/2)t]^2}.$$

(b) *Use the Squeeze Theorem to show that $\lim_{b \to 0+} P(S_t < b) = 0$.*

Since a put is an insurance that gets exercised when the stock declines, it makes sense to assume that at the exercise time, τ_b, the stock reaches the barrier b from above, i.e. $0 < b < S_0$. Using Proposition 10.1.7 we obtain the value of the contract that pays $K - b$ at time τ_b

$$f(b) \;=\; \widehat{\mathbb{E}}[(K-b)e^{-r\tau_b}] = (K-b)\mathbb{E}[e^{-r\tau_b}] = (K-b)\left(\frac{S_0}{b}\right)^{-\frac{2r}{\sigma^2}}.$$

We need to pick the optimal value b^* for which $f(b)$ is maximum

$$f(b^*) = \max_{0<b<K} f(b).$$

It is useful to notice that the functions $f(b)$ and $g(b) = (K-b)b^{\frac{2r}{\sigma^2}}$ reach the maximum for the same value of b. For the sake of simplicity, denote $\alpha = \frac{2r}{\sigma^2}$. Then

$$\begin{aligned} g(b) &= Kb^\alpha - b^{\alpha+1} \\ g'(b) &= \alpha Kb^{\alpha-1} - (\alpha+1)b^\alpha = b^{\alpha-1}[\alpha K - (\alpha+1)b], \end{aligned}$$

and the equation $g'(b) = 0$ has the solution $b^* = \frac{\alpha}{\alpha+1}K$. Since $g'(b) > 0$ for $b < b^*$ and $g'(b) < 0$ for $b > b^*$, it follows that b^* is a maximum point for the function $g(b)$, and hence for the function $f(b)$. Substituting for the value of α, the optimal value of the barrier becomes

$$b^* = \frac{2r/\sigma^2}{2r/\sigma^2 + 1}K = \frac{K}{1 + \frac{\sigma^2}{2r}}. \tag{10.1.1}$$

The condition $b^* < K$ is obviously satisfied, while the condition $b^* < S_0$ is equivalent with

$$K < \left(1 + \frac{\sigma^2}{2r}\right)S_0.$$

The value of the perpetual put is obtained computing the value at b^*

$$
\begin{aligned}
f(b^*) \;&=\; \max f(b) = (K - b^*)\left(\frac{S_0}{b^*}\right)^{-\frac{2r}{\sigma^2}} = \left(K - \frac{K}{1 + \frac{\sigma^2}{2r}}\right)\left[\frac{S_0}{K}\left(1 + \frac{\sigma^2}{2r}\right)\right]^{-\frac{2r}{\sigma^2}} \\
&=\; \frac{K}{1 + \frac{2r}{\sigma^2}}\left[\frac{S_0}{K}\left(1 + \frac{\sigma^2}{2r}\right)\right]^{-\frac{2r}{\sigma^2}}.
\end{aligned}
$$

Hence the value of a perpetual put is

$$\frac{K}{1 + \frac{2r}{\sigma^2}}\left[\frac{S_0}{K}\left(1 + \frac{\sigma^2}{2r}\right)\right]^{-\frac{2r}{\sigma^2}}, \qquad S > b^* \tag{10.1.2}$$

and $K - S$ for $S \le b^*$.

But what is the expected exercise time of a perpetual American put? To answer this question we need to compute $\mathbb{E}[\tau_{b^*}]$. The optimal exercise time of the put, τ_{b^*}, is when the stock hits the optimal barrier

$$\tau_{b^*} = \inf\{t > 0; S_t \le b^*\} = \inf\{t > 0; \mu t + \sigma W_t \le -x\},$$

where

$$\tau = \tau_{b^*}, \qquad x = \ln\frac{S_0}{b^*}, \qquad \mu = r - \frac{\sigma^2}{2}. \tag{10.1.3}$$

Then using Appendix C.7, point 2, part (c) yields

$$
\begin{aligned}
\mathbb{E}[\tau_{b^*}] \;&=\; \frac{x}{\mu}e^{-\frac{2\mu x}{\sigma^2}} = \frac{\ln\frac{S_0}{b^*}}{r - \frac{\sigma^2}{2}}e^{-\frac{2}{\sigma^2}(r - \frac{\sigma^2}{2})\ln\frac{S_0}{b^*}} \\
&=\; \ln\left[\left(\frac{S_0}{b^*}\right)^{\frac{1}{r - \frac{\sigma^2}{2}}}\right]e^{(1 - \frac{2r}{\sigma^2})\ln\frac{S_0}{b^*}} = \ln\left[\left(\frac{S_0}{b^*}\right)^{\frac{1}{r - \frac{\sigma^2}{2}}}\right]\left(\frac{S_0}{b^*}\right)^{1 - \frac{2r}{\sigma^2}}.
\end{aligned}
$$

Hence, the expected time when the holder should exercise the put is given by the exact formula

$$\boxed{\mathbb{E}[\tau_{b^*}] = \ln\left[\left(\frac{S_0}{b^*}\right)^{\frac{1}{r - \frac{\sigma^2}{2}}}\right]\left(\frac{S_0}{b^*}\right)^{1 - \frac{2r}{\sigma^2}},}$$

with b^* given by (10.1.1). The probability density function of the optimal exercise time τ_{b^*} can be found from Appendix C.7, point 2, part (b) using substitutions (10.1.3)

$$p(\tau) = \frac{\ln \frac{S_0}{b^*}}{\sigma\sqrt{2\pi}\tau^{3/2}} e^{-\frac{\left(\ln \frac{S_0}{b^*} + \tau(r - \frac{\sigma^2}{2})\right)^2}{2\tau\sigma^2}}, \qquad \tau > 0. \tag{10.1.4}$$

Exercise 10.1.11 *Consider a stock that pays continuous dividends at rate $\delta > 0$.*

(a) Assume a perpetual put is exercised whenever the stock reaches the barrier b from above. Show that the discounted value at time $t = 0$ is

$$g(b) = (K - b)\left(\frac{S_0}{b}\right)^{h_2},$$

where

$$h_2 = \frac{1}{2} - \frac{r - \delta}{\sigma^2} - \sqrt{\left(\frac{r - \delta}{\sigma^2} - \frac{1}{2}\right)^2 + \frac{2r}{\sigma^2}}.$$

(b) Use differentiation to show that the maximum value of $g(b)$ is realized for

$$b^* = K \frac{h_2}{h_2 - 1}.$$

(c) Prove the price of perpetual put

$$g(b^*) = \max_{b>0} g(b) = \frac{K}{1 - h_2}\left(\frac{h_2 - 1}{h_2}\frac{S_0}{K}\right)^{h_2}.$$

The next couple of sections present some American perpetuities that are interesting to evaluate, even if this type of contracts are not currently traded in the market.

10.2   Perpetual American Log Contract

A *perpetual American log contract* is a contract that never expires and can be exercised at any time, providing the holder the log of the value of the stock, $\ln S_t$, at the exercise time t. It is interesting to note that these type of contracts are always optimal to be exercised, and their pricing formula is fairly uncomplicated.

We shall adopt the same valuation strategy as in the case of perpetual American options. Assume the contract is exercised when the stock S_t reaches the barrier b, where $b > S_0$. If the hitting time of the barrier b is τ_b, then its

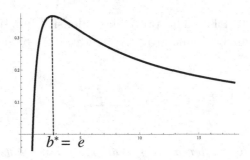

Figure 10.3: *The graph of the function* $g(b) = \frac{\ln b}{b}$, $b > 0$.

payoff is $\ln S_\tau = \ln b$. Discounting at the risk free interest rate, the value of the contract at time $t = 0$ is

$$f(b) = \widehat{\mathbb{E}}[e^{-rT_b} \ln S_\tau] = \widehat{\mathbb{E}}[e^{-rT_b} \ln b] = \frac{\ln b}{b} S_0,$$

since the barrier is assumed to be reached from below, see Proposition 10.1.1.

The function $g(b) = \dfrac{\ln b}{b}$, $b > 0$, has the derivative $g'(b) = \dfrac{1 - \ln b}{b^2}$, so $b^* = e$ is a global maximum point, see Fig.10.3. The maximum value is $g(b^*) = 1/e$. Then the optimal value of the barrier is $b^* = e$, and the price of the contract at $t = 0$ is given by

$$f_0 = \max_{b > 0} f(b) = S_0 \max_{b > 0} g(b) = \frac{S_0}{e}.$$

In order for the stock to reach the optimum barrier b^* from below we need to require the condition $S_0 < e$. Hence we arrived at the following result:

Proposition 10.2.1 *Let $S_0 < e$. Then the optimal exercise price of a perpetual American log contract is*

$$\tau = \inf\{t > 0; S_t \geq e\},$$

and its value at $t = 0$ is $\dfrac{S_0}{e}$.

Remark 10.2.2 *If $S_0 > e$, then it is optimal to exercise the perpetual log contract as soon as possible.*

Exercise 10.2.3 *Let $S_0 < e^{2n+1}$ with n natural number, and consider a perpetual American power log contract with the payoff $(\ln S_\tau)^{2n+1}$ at the exercise time τ. Show that the value of this contract at time $t = 0$ is*

$$f_0 = \frac{(2n + 1)S_0}{e^{2n+1}}.$$

Exercise 10.2.4 *Consider a stock that pays continuous dividends at a rate $\delta > 0$, and assume that $S_0 < e^{1/h_1}$, where*

$$h_1 = \frac{1}{2} - \frac{r - \delta}{\sigma^2} + \sqrt{\left(\frac{r - \delta}{\sigma^2} - \frac{1}{2}\right)^2 + \frac{2r}{\sigma^2}}.$$

(a) Assume a perpetual log contract is exercised whenever the stock reaches the barrier b from below. Show that the discounted value at time $t = 0$ is

$$f(b) = \ln b \left(\frac{S_0}{b}\right)^{h_1}.$$

(b) Use differentiation to show that the maximum value of $f(b)$ is realized for

$$b^* = e^{1/h_1}.$$

(c) Prove the price of perpetual log contract

$$f(b^*) = \max_{b>0} f(b) = \frac{S_0^{h_1}}{h_1 e}.$$

(d) Show that the higher the dividend rate δ, the lower the optimal exercise time.

10.3 Perpetual American Power Contract

A *perpetual American power contract* is a contract that never expires and can be exercised at any time, providing the holder with the α-power of the stock, $(S_t)^\alpha$, at the exercise time t, where $\alpha \neq 0$. (If $\alpha = 0$, the payoff is a constant, which is equal to 1). Again, we adopt the optimal barrier strategy as in the case of perpetual American options.

1. Case $\alpha > 0$. Since we expect the value of the payoff to increase over time, we assume the contract is exercised when the stock S_t reaches the barrier b, from below. If the hitting time of the barrier b is τ_b, then its payoff is $(S_{\tau_b})^\alpha$. Discounting at the risk-free rate, the value of the contract at time $t = 0$ is

$$f(b) = \widehat{\mathbb{E}}[e^{-r\tau_b}(S_{\tau_b})^\alpha] = \widehat{\mathbb{E}}[e^{-r\tau_b}b^\alpha] = b^\alpha \frac{S_0}{b} = b^{\alpha-1}S_0,$$

where we used Proposition 10.1.1. We shall discuss the following cases:

(*i*) If $\alpha > 1$, then the optimal barrier is $b^* = \infty$, and hence, it is never optimal to exercise the contract in this case.

(*ii*) If $0 < \alpha < 1$, the function $f(b)$ is decreasing, so its maximum is reached for $b^* = S_0$, which corresponds to $\tau_{b^*} = 0$. Hence, it is optimal to exercise the contract as soon as possible.

(*iii*) In the case $\alpha = 1$, the value of $f(b)$ is constant, and the contract can be exercised at any time (see Exercise 10.3.1).

2. Case $\alpha < 0$. The payoff value, $(S_t)^\alpha$, is expected to decrease, so we assume the exercise occurs when the stock reaches the barrier b from above. Discounting to the initial time $t = 0$ and using Proposition 10.1.7 yields

$$f(b) = \widehat{\mathbb{E}}[e^{-r\tau_b}(S_{\tau_b})^\alpha] = \widehat{\mathbb{E}}[e^{-r\tau_b}b^\alpha] = b^\alpha \left(\frac{S_0}{b}\right)^{-\frac{2r}{\sigma^2}} = b^{\alpha+\frac{2r}{\sigma^2}} S_0^{-\frac{2r}{\sigma^2}}.$$

Here we have two subcases:

(*i*) If $\alpha < -\frac{2r}{\sigma^2}$, then $f(b)$ is decreasing, so its maximum is reached for $b^* = 0$, which is a boundary that is never reached.

(*ii*) If $-\frac{2r}{\sigma^2} < \alpha < 0$, then the maximum of $f(b)$ occurs for $b^* = S_0$, so it is optimal to exercise the contract as soon as possible.

Exercise 10.3.1 *Prove using expectations that the value at time $t = 0$ of an American perpetuity that pays at the exercise time τ the price of the stock, S_τ, is S_0, and this is independent of the exercise time.*

Exercise 10.3.2 *Consider a stock that pays continuous dividends at a rate $\delta > 0$, and assume $\alpha > h_1$, with $h_1 = \frac{1}{2} - \frac{r-\delta}{\sigma^2} + \sqrt{\left(\frac{r-\delta}{\sigma^2} - \frac{1}{2}\right)^2 + \frac{2r}{\sigma^2}}$. Show that the perpetual power contract with payoff S_t^α is never optimal to be exercised.*

Exercise 10.3.3 *Consider a perpetual American-type contract with the payoff $(K - S_t)^2$, where $K > 0$ is the strike price. Find the optimal exercise time and the contract value at $t = 0$.*

10.4 Finitely Lived American Options

Exact pricing formulas are great when they exist and can be easily implemented. Even if we cherish all closed form pricing formulas we can get, there is also a time when exact formulas are not possible and approximations are most desired. If we run into a problem whose solution cannot be found explicitly, it would still be very valuable to know something about its approximate quantitative behavior. This will be the case of finitely lived American options.

10.4.1 American Call

In this section we shall distinguish between the cases of dividend-paying stock and non-dividend-paying stock. The optimal exercise time depends on this detail.

The Case of Non-dividend-Paying Stock

The holder has the right to buy a stock for the price K at any time before or at the expiration T. The strike price K and expiration T are specified at the beginning of the contract. The payoff at time t is $(S_t - K)^+ = \max\{S_t - K, 0\}$. The price of the American call at time $t = 0$ is given by

$$f_0 = \max_{0 \leq \tau \leq T} \widehat{\mathbb{E}}[e^{-r\tau}(S_\tau - K)^+],$$

where the maximum is taken over all stopping times τ less than or equal to T.

Theorem 10.4.1 *It is not optimal to exercise an American call on a non-dividend paying stock early. It is optimal to exercise the call at maturity, T, if at all. Consequently, the price of an American call is equal to the price of the corresponding European call.*

Proof: The heuristic idea of the proof is based on the observation that the difference $S_t - K$ tends to get larger as time goes on. As a result, there is always hope for a larger payoff and the later we exercise the better. In the following we shall formalize this idea mathematically using the submartingale property of the stock (see Exercise 5.1.3) together with the Optional Stopping Theorem (Theorem 1.8.1).

Let $X_t = e^{-rt}S_t$ and $f(t) = -Ke^{-rt}$. Since X_t is a martingale in the risk-neutral world, see Proposition 7.1.1, and $f(t)$ is an increasing, integrable function (see Exercise 10.4.1), then the process

$$Y_t = X_t + f(t) = e^{-rt}(S_t - K)$$

is an $\mathcal{F}_t$-submartingale, where $\mathcal{F}_t$ is the information set provided by the underlying Brownian motion W_t.

Since the hokey-stick function $\phi(x) = x^+ = \max\{x, 0\}$ is convex, then by Proposition B.6.1 (b) of Appendix B.6, the process $Z_t = \phi(Y_t) = e^{-rt}(S_t - K)^+$ is a submartingale. Applying Doob's stopping theorem (see Theorem 1.8.2) for stopping times τ and T, with $\tau \leq T$, we obtain $\widehat{E}[Z_\tau] \leq \widehat{E}[Z_T]$. This means

$$\widehat{\mathbb{E}}[e^{-r\tau}(S_\tau - K)^+] \leq \widehat{\mathbb{E}}[e^{-rT}(S_T - K)^+],$$

i.e. the maximum of the American call price is realized for the optimum exercise time $\tau^* = T$. The maximum value is given by the right side, which denotes the price of an European call option.

■

Example 10.4.1 *Show that the sum between a martingale and an increasing function of time is a submartingale.*

With a slight modification in the proof, we can treat the problem of American power contract.

Proposition 10.4.2 (American power contract) *Consider a contract with maturity date T, which pays, when exercised, S_t^n, where $n > 1$, and $t \leq T$. Then it is not optimal to exercise this contract early.*

Proof: Using that $M_t = e^{-rt} S_t$ is a martingale (in the risk neutral world), then $X_t = M_t^n$ is a submartingale. Since $Y_t = e^{-rt} S_t^n = \left(e^{-rt} S_t \right)^n e^{(n-1)rt} = X_t e^{(n-1)rt}$, then for $s < t$

$$
\begin{aligned}
\mathbb{E}[Y_t | \mathcal{F}_s] &= \mathbb{E}[X_t e^{(n-1)rt} | \mathcal{F}_s] > \mathbb{E}[X_t e^{(n-1)rs} | \mathcal{F}_s] \\
&= e^{(n-1)rs} \mathbb{E}[X_t | \mathcal{F}_s] \geq e^{(n-1)rs} X_s = Y_s,
\end{aligned}
$$

so Y_t is a submartingale. Applying Doob's stopping theorem (see Theorem 1.8.2) for stopping times τ and T, with $\tau \leq T$, we obtain $\widehat{\mathbb{E}}[Y_\tau] \leq \widehat{\mathbb{E}}[Y_T]$, or equivalently, $\widehat{\mathbb{E}}[e^{-r\tau} S_\tau^n] \leq \widehat{\mathbb{E}}[e^{-rT} S_T^n]$, which implies

$$
\max_{\tau \leq T} \widehat{\mathbb{E}}[e^{-r\tau} S_\tau^n] = \widehat{\mathbb{E}}[e^{-rT} S_T^n].
$$

Then it is optimal to exercise the contract at maturity T. ∎

Exercise 10.4.3 *Consider an American future contract with maturity date T and delivery price K, i.e. a contract with payoff at maturity $S_T - K$, which can be exercised at any time $t \leq T$. Show that it is not optimal to exercise this contract early.*

Exercise 10.4.4 *Consider an American option contract with maturity date T, and time-dependent strike price $K(t)$, i.e. a contract with payoff at maturity $S_T - K(T)$, which can be exercised at any time.*

(a) Show that it is not optimal to exercise this contract early in the following two cases:

(i) If $K(t)$ is a decreasing function;

(ii) If $K(t)$ is an increasing function with $K(t) < e^{rt}$.

(b) What happens if $K(t)$ is increasing and $K(t) > e^{rt}$?

The Case of Dividend-Paying Stock

When the stock pays dividends it is optimal to exercise the American call early. An exact solution of this problem is hard to get explicitly, or might not exist. However, there are some asymptotic solutions that are valid close to expiration (see section 10.8.5 and also Wilmott [69]) and analytic approximations given by MacMillan, Barone-Adesi and Whaley (see section 10.9).

In the following we shall discuss why it is difficult to find an exact optimal exercise time for an American call on a dividend-paying stock. First, consider two contracts:

1. Consider a contract by which one can acquire a stock, S_t, at any time before or at time T. This can be seen as a contract with expiration T, that pays when exercised the stock price, S_t. Assume the stock pays dividends at a continuous rate $\delta > 0$. When should the contract be exercised in order to maximize its value?

The value of the contract at time $t = 0$ is $\max_{\tau \leq T} \{e^{-r\tau} S_\tau\}$, where the maximum is taken over all stopping times τ less than or equal to T. Since the stock price, which pays dividends at rate δ, is given in the risk neutral world by

$$S_t = S_0 e^{(r-\delta-\frac{\sigma^2}{2})t+\sigma W_t},$$

then $M_t = e^{-(r-\delta)t} S_t$ is a martingale (in the risk-neutral world). Therefore $e^{-rt} S_t = e^{-\delta t} M_t$. Let $X_t = e^{-rt} S_t$. Then for $0 < s < t$

$$
\begin{aligned}
\widehat{\mathbb{E}}[X_t | \mathcal{F}_s] &= \widehat{\mathbb{E}}[e^{-\delta t} M_t | \mathcal{F}_s] < \widehat{\mathbb{E}}[e^{-\delta s} M_t | \mathcal{F}_s] \\
&= e^{-\delta s} \widehat{\mathbb{E}}[M_t | \mathcal{F}_s] = e^{-\delta s} M_s = e^{-rs} S_s = X_s,
\end{aligned}
$$

so X_t is a supermartingale (i.e. $-X_t$ is a submartingale). Applying the Optional Stopping Theorem for the stopping time τ we obtain

$$\widehat{\mathbb{E}}[X_\tau] \leq \widehat{\mathbb{E}}[X_0] = S_0.$$

Hence it is optimal to exercise the contract at the initial time $t = 0$. This makes sense, since in this case we have a longer period of time during which dividends are collected.

2. Consider a contract by which one has to pay the amount of cash K at any time before or at time T. Given the time value of money

$$e^{-rt} K > e^{-rT} K, \qquad t < T,$$

it is always optimal to defer the payment until time T.

3. Now consider a combination of the previous two contracts. This new contract pays $S_t - K$ and can be exercised at any time t, with $t \leq T$. Since it is not clear when it is optimal to exercise the contract, we shall consider two limiting cases. Let τ^* denote its optimal exercise time.

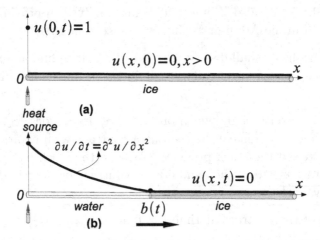

Figure 10.4: *The temperature distribution at two time instances for the one-phase Stefan problem:* **a.** *At* $t = 0$ *there is only one phase (solid ice);* **b.** *For* $t > 0$ *there are two phases with the free-boundary* $b(t)$ *moving towards right.*

(i) When $K \to 0^+$, then $\tau^* \to 0^+$, because we approach the conditions of case 1, and also assume continuity conditions on the price.

(ii) If $K \to \infty$, then the latter the pay day, the better, i.e. $\tau^* \to T^-$.

The optimal exercise time, τ^*, is *somewhere* between 0 and T, with the tendency of moving towards T as K gets large.

In order to understand how the pricing mechanism for this type of option works, we shall present in the next section a famous Physics problem whose solution is simpler but very instructive in this direction. Using this method, which is borrowed from physicists, we shall be able to determine approximative solutions for finitely lived American options.

10.5   One-phase Stefan Problem

For an easier understanding of the pricing mechanism of an American option, we shall present first a similar problem, the so-called *one-phase Stefan problem*. This deals with the description of the change-of-phase between ice and water along a boundary that modifies in time, called *free-boundary*.

Consider a semi-infinite pipe filled with ice and a heat source situated at the left end of the pipe. Consider the x-axis along the pipe, and denote by $u(x, t)$ the temperature in the pipe at the point x and time t. Assume the heat source heats up the left end of the pipe to the constant temperature $u(0, t) = 1$, for all $t > 0$, initiating a melting process in the ice. Consequently, the water-ice interface, $b(t)$, moves to the right. Therefore, at any time $t > 0$, the region $[0, b(t)]$ is liquid while the rest of the pipe $(b(t), \infty)$ is ice-solid, see Fig.10.4. We are interested in finding the following quantities:

(i) the equation for the free-boundary, $b(t)$, between water and ice as a function of time t.

(ii) the temperature at all points in the pipe, $u(x,t)$, for all $t > 0$.

We shall show that this problem can be modeled by the following partial differential equation with free-boundary:

$$\frac{\partial u}{\partial t} = \frac{\partial^2 u}{\partial x^2}, \quad 0 < x < b(t) \tag{10.5.5}$$

$$u\big(b(t), t\big) = 0, \quad t > 0 \tag{10.5.6}$$

$$\frac{\partial u}{\partial x}\big(b(t), t\big) = -b'(t), \quad t > 0 \tag{10.5.7}$$

$$u(0, t) = 1, \quad t > 0 \tag{10.5.8}$$

$$u(x, t) = 0, \quad x > b(t). \tag{10.5.9}$$

We shall explain next the reasoning behind each of the previous equations and conditions. Equation (10.5.5) states that the water temperature $u(x,t)$ in the pipe segment $[0, b(t)]$ evolves according to the heat equation (8.1.1). The boundary condition $u(0,t) = 1$ given by (10.5.8) says that the heat source produces unit temperature at the left end of the pipe all the time. The fact that the ice, which is situated in the pipe interval $(b(t), \infty)$, has zero temperature (in Celsius degrees) is described by (10.5.9). The equation (10.5.6) assures the temperature continuity across the interface $b(t)$ and that the ice melts at zero temperature. The dynamics of the free-boundary $b(t)$ is described by (10.5.7) and is a consequence of the *law of conservation of energy*, as it will be described shortly.

First, let's introduce a few notations. Let $q(x,t)$ be the heat flux across the transversal section of the pipe, k the thermal conductivity of water, $u(x,t)$ the temperature of water, and A the cross section of the pipe. The heat flux (which is roughly the number of particles that cross the section in unit time) is related to the temperature by the *Fourier's law*

$$q(x,t) = -k\frac{\partial u(x,t)}{\partial x}, \quad k > 0, \tag{10.5.10}$$

where the negative sign is consistent with the fact that the heat moves from a high temperature region towards a low temperature region, see Fig.10.5 a. The heat flux during the time step Δt transfers energy into the interface $b(t)$, this being equal to

$$q\big(b(t), t\big)\Delta t.$$

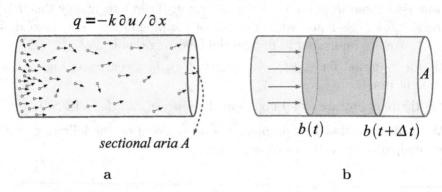

Figure 10.5: **a.** *The flux across the pipe sectional area A;* **b.** *The pipe segment from* $b(t)$ *to* $b(t + \Delta t)$.

At time t the water-ice interface is at $b(t)$ and at time $t + \Delta t$ it has moved to the right, at $b(t + \Delta t)$. This is due to the fact that some energy has been absorbed by the pipe element $[b(t), b(t + \Delta t)]$, transforming ice into water. This is equal to $\lambda \cdot$ mass, where λ is the latent heat of water. This means that the amount of energy needed to melt the ice is proportional with the mass of water resulted. This mass is the product between the water density and pipe element volume

$$\text{mass} = \rho A\big(b(t + \Delta t) - b(t)\big).$$

Balancing the absorbed energy against the transferred energy by the heat flux during the time interval Δt, we get

$$q\big(b(t), t\big)\Delta t = \lambda \rho A\big(b(t + \Delta t) - b(t)\big).$$

Dividing by Δt, then making $\Delta t \to 0$, and using Fourier's law (10.5.10) yields

$$-k\frac{\partial u(x, t)}{\partial x} = \lambda \rho A b'(t).$$

Scaling the units of measure such that $\dfrac{\lambda \rho A}{k} = 1$, leads to $\dfrac{\partial u(b(t), t)}{\partial x} = -b'(t)$, which is relation (10.5.7). It is worth noting that $u(x, t)$ is not tangent to the x-axis, since the slope at $b(t)$ is equal to the negative rate of change of $b(t)$, which is non-zero.

The system (10.5.5-10.5.9) models a one-phase problem, since there is only one function, the water temperature $u(x, t)$, to be determined. The other phase, the ice, has a constant zero temperature. We shall solve this problem explicitly finding the temperature $u(x, t)$ and the free-boundary $b(t)$.

Given that equation (10.5.5) is of the first order in t and the second order in x, it makes sense to look for a solution which depends on the quotient $x/\sqrt{t}$

as follows

$$u(x,t) = w\left(\frac{x}{2\sqrt{t}}\right),$$

with the function $w(\cdot)$ smooth enough (twice differentiable with w'' continuous). Substituting $\xi = \frac{x}{2\sqrt{t}}$, the chain rule provides

$$\frac{\partial u}{\partial t} = -\frac{x}{4}t^{-3/2}w'(\xi)$$

$$\frac{\partial u}{\partial x} = \frac{1}{\sqrt{2t}}w'(\xi)$$

$$\frac{\partial^2 u}{\partial x^2} = \frac{1}{4t}w''(\xi).$$

Using the aforementioned formulas, the heat equation $\frac{\partial u}{\partial t} = \frac{\partial^2 u}{\partial x^2}$ can be written in terms of w as follows

$$w''(\xi) + 2\xi w'(\xi) = 0.$$

Making $v = w'$ we obtain a linear differential equation in v

$$v'(\xi) + 2\xi v(\xi) = 0,$$

which after multiplication by the integrating factor e^{ξ^2} becomes the exact equation

$$\left(e^{\xi^2} v(\xi)\right)' = 0.$$

Integrating yields $v(\xi) = Ce^{-\xi^2}$, and hence $w(\xi) = A + C\int_0^\xi e^{-s^2}\,ds$, with A and C constants. Going back into the x and t variables, we have

$$u(x,t) = A + C\int_0^{\frac{x}{2\sqrt{t}}} e^{-s^2}\,ds, \qquad x < b(t), \quad t > 0.$$

We note that on the other side of the boundary, when $x \geq b(t)$, the solution is identically equal to zero, $u(x,t) = 0$ from condition (10.5.9). For the sake of simplicity, we shall use the *erf function*

$$\epsilon(\xi) = \frac{2}{\sqrt{\pi}}\int_0^\xi e^{-s^2}\,ds,$$

so we can write

$$u(x,t) = A + B\epsilon\left(\frac{x}{2\sqrt{t}}\right),$$

with $B = C\frac{\sqrt{\pi}}{2}$.

Using that $\epsilon(0) = 0$, then the initial condition (10.5.8) provides

$$1 = u(0, t) = A + B\epsilon(0) = A,$$

so $A = 1$. The continuity at the boundary condition (10.5.6) yields

$$u(b(t), t) = 1 + B\epsilon\left(\frac{b(t)}{2\sqrt{t}}\right) = 0,$$

which implies that

$$\epsilon\left(\frac{b(t)}{2\sqrt{t}}\right) = -\frac{1}{B}. \qquad (10.5.11)$$

has a constant value. Since $\epsilon(\cdot)$ is an increasing function, it follows that

$$\frac{b(t)}{2\sqrt{t}} = \epsilon^{-1}\left(-\frac{1}{B}\right) = \text{constant}.$$

Hence, there is a constant $c_0 > 0$ such that the free-boundary takes the explicit form

$$b(t) = c_0\sqrt{t}, \qquad t \geq 0. \qquad (10.5.12)$$

So far, the solution of the system (10.5.5 - 10.5.9) takes the form

$$u(x, t) = \begin{cases} 1 + B\epsilon\left(\frac{x}{2\sqrt{t}}\right), & \text{if } 0 < x < c_0\sqrt{t} \\ 0, & \text{if } x \geq c_0\sqrt{t}, \end{cases}$$

with constants B and c_0 subject to be found out.

First, we substitute relation (10.5.12) into formula (10.5.11) and obtain

$$1 + B\epsilon\left(\frac{c_0}{2}\right) = 0. \qquad (10.5.13)$$

A second relation between constants B and c_0 is obtained from the free-boundary condition (10.5.7). Using the previous formulas for the boundary and temperature, the equation

$$\frac{\partial u}{\partial x}(b(t), t) = -b'(t)$$

becomes

$$Be^{-c_0^2/4} = -c_0. \qquad (10.5.14)$$

Eliminating B from equations (10.5.13) and (10.5.14) yields the following equation for c_0

$$c_0 e^{c_0^2/4} \int_0^{c_0/2} e^{-s^2} \, ds = 1. \qquad (10.5.15)$$

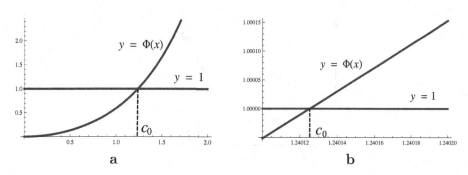

Figure 10.6: *The intersection point of the graphs of $y = \Phi(x)$ and $y = 1$:* **a.** *The rough picture;* **b.** *The zoom-in picture.*

This integral equation has a unique positive solution $c_0 > 0$. This follows from the fact that the right side function

$$\Phi(c_0) = c_0 e^{c_0^2/4} \int_0^{c_0/2} e^{-s^2} \, ds$$

is continuous and increasing, with $\Phi(0) = 0$ and $\lim_{c_0 \to \infty} \Phi(x) = \infty$. Therefore, there is only one value $c_0 > 0$ such that $\phi(c_0) = 1$, see Fig.10.6 a. The solution is obtained graphically, see Fig.10.6 b, and its value is given by

$$c_0 = 1.2401...$$

The value of B follows from (10.5.14)

$$B = -c_0 e^{c_0^2/4} = -1.82155...$$

In conclusion, the free-boundary for the one-phase Stefan problem for the semi-infinite rod is given by $b(t) = c_0 \sqrt{t}$ and the temperature in the pipe has the explicit form provided by the piece-wise function

$$u(x,t) = \begin{cases} 1 - c_0 e^{c_0^2/4} \epsilon\left(\frac{x}{2\sqrt{t}}\right), & \text{if } 0 < x < c_0\sqrt{t} \\ 0, & \text{if } x \geq c_0\sqrt{t}. \end{cases} \tag{10.5.16}$$

For more details regarding one and two-phase Stefan problems the reader is refered to Guenther and Lee [36].

Stochastic interpretation The solution (10.5.16) has been obtained by the methods of partial differential equations. There is a dual approach involving stochastic processes, which shall be presented in the following.

We shall start with the observation that water particles under the influence of heat will perform a diffusion in the pipe with a reflecting wall at $x = 0$. This

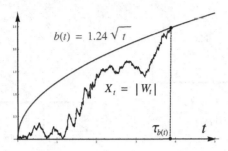

Figure 10.7: *The reflected Brownian motion* $X_t = |W_t|$ *reaches the free-boundary* $b(t)$ *at the random time* $\tau_{b(t)}$.

is modeled by a *reflected Brownian motion* process $X_t = |W_t|$. The process starts at the origin, $X_0 = 0$, and whenever reaches the wall at $x = 0$ it bounces back to the right side. These particles eventually reach the water-ice interface transferring energy to it, and hence moving it to the right. For each particle ω, the hitting time of the boundary $b(t) = c_0\sqrt{t}$ is a random variable $\tau_{b(t)}(\omega) > 0$, see Fig.10.7.

Let $u(x, t)$ be the probability density that a particle starting from $x = 0$ at time $t = 0$ arrives at x at time t, i.e.

$$P(X_t \in dx) = u(x, t)\, dx.$$

We also require that the density of particles at the origin is always constant, $u(0, t) = 1$.

Then $u(x, t)$ solves the Kolmogorov backwards equation

$$\frac{\partial u}{\partial t} - \frac{\partial^2 u}{\partial x^2} = 0, \qquad 0 < x < b(t).$$

Since the reflected Brownian motion is absorbed at the boundary, the probability density beyond that point is zero, i.e. $u(x, t) = 0$, for $x \geq b(t)$. The temperature in the pipe at the point x and time t is given by the density of particles in the proximity of x at time t, so $u(x, t)$ describe the temperature in the pipe.

American puts and calls on dividend-paying stocks are very similar with the one-phase Stefan problem studied previously. Understanding the Stefan problem first will make easier the comprehension of the pricing mechanisms for American options. However, because of their more complicated boundary conditions, both their free-boundary and value cannot be determined in closed form, and hence we shall work out an asymptotics formula for them.

American options can be approached from both PDEs and stochastic points of view, in a similar fashion as the Stefan problem. The dynamics

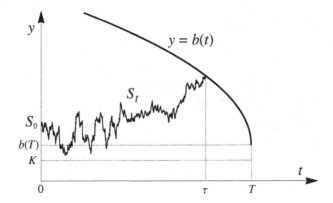

Figure 10.8: *The exercise time τ of a call is the first time when the stock price S_t rises above the value of the time-dependent boundary $b(t)$.*

of the free-boundary in this case cannot be obtained from an energy balance, but it is a consequence of the *no arbitrage opportunities* in the market. It can be inferred that the financial analog of the "conservation of energy" from Physics is the "lack of arbitrage opportunities in the market". Equivalently, the existence of arbitrage opportunities would correspond to breaking the law of conservation of energy, which even might happen sometimes, (see the case of tunneling effect in Quantum Mechanics) it does not last for too long.

10.6 Free-Boundary of a Call

We have seen that in the case of a perpetuity call it was useful to adopt the strategy of exercising the call when the stock reaches for the first time a barrier b, and then optimizing the value of the call over b. In the case of finite horizon calls on dividend-paying stocks we shall employ the same strategy, the difference being that the barrier in this case is time dependent, $b(t)$. The value of the call will be optimized over a family of curves rather than over a single parameter, which makes the problem more complex.

For the one-phase Stefan problem the particles diffuse under the heat influence in the pipe and reach at some random time the water-ice interface. In the case of the call we consider the Ito diffusion S_t, the stock price, that eventually hits the time dependent boundary $b(t)$ at a random time τ which is the exercise time of the call, see Fig.10.8.

The value of the call at time $t = 0$ is obtained by discounting the payoff $\max(S_\tau - K, 0) = \max(b(\tau) - K, 0) = b(\tau) - K$ with the random time τ and then take the average as

$$C(S, 0) = \widehat{\mathbb{E}}[e^{-r\tau}(b(\tau) - K)]. \tag{10.6.17}$$

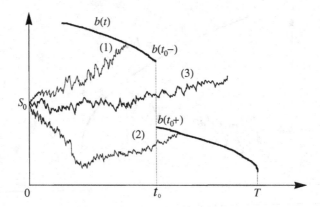

Figure 10.9: *The stock simulations (1) and (2) hit the barrier, while simulation (3) penetrate the barrier without intersecting it.*

The time-depending boundary $b(t)$ is a curve defined on $[0, T]$, with values larger than K and which maximizes the functional (10.6.17). Finding an explicit expression for $b(t)$ is difficult, but it can be shown that $b(t)$ satisfies a nonlinear Volterra integral equation of the second kind (see Peskir and Shiryaev [58]).

In the following we shall describe the shape and properties of the free-boundary $b(t)$ in a heuristic way.

1. *Relative relation with the stock:* $b(\tau) = S_\tau$, $b(t) > S_t$ for $t < \tau$.

The time-dependent boundary $b(t)$ denotes the level at which the holder prefers to exercise the call option. The exercise time is τ and for $t < \tau$ the stock S_t has the value less than the level of the boundary $b(t)$. At the exercise the stock and the boundary values coincide.

2. *Relation with the strike:* $b(t) > K$ for $0 \le t \le T$.

When a call is exercised the stock price has to exceed the strike K. At this time the value of the stock equals the value of the boundary, and hence the boundary has to be above the strike K.

3. *Monotonicity:* $b(t)$ is decreasing on $(0, T]$.

Let $0 < t_1 < t_2 < T$ be two time instances. Then the remaining lives of the option satisfy $T - t_2 < T - t_1$. Now, the shorter the remaining life the less hope we have that the stock will go up. Consequently, our strategy should be to exercise at lower levels of the stock when approaching the maturity of the option. Hence $b(t_1) > b(t_2)$, i.e. $b(t)$ is a decreasing function.

4. *Continuity:* $b(t)$ is continuous on $(0, T]$.

Assume that $b(t)$ is not continuous on $(0, T]$. It is known that a decreasing function has only jump-type discontinuities. Let t_0 one of these points, see

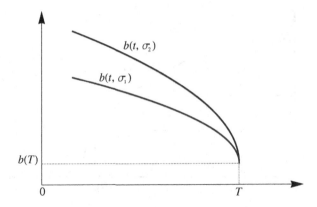

Figure 10.10: *If $\sigma_1 < \sigma_2$ then $b(t,\sigma_1) < b(t,\sigma_2)$, for $t < T$, and $b(T,\sigma_1) = b(T,\sigma_2)$.*

Fig.10.9. From the properties of the stock,

$$P\left(b(t_0+) < S_{t_0} < b(t_0-)\right) > 0,$$

i.e. there is a positive probability for the stock price to go through the barrier without intersecting it. In Fig.10.9 this possibility is represented by the simulation with label (3). In this case we lose option value since despite of the fact that the stock has a high value we do not exercise and hence end up with a zero value for the call. Therefore $b(t)$ must be a continuous curve.

5. *Convexity:* $b(t)$ is concave on $(0,T]$.

The concavity of the curve $b(t)$ is a consequence of the pessimistic attitude (driven by risk-aversion) of the option holder under the influence of a finite time horizon, feeling which accentuates as the option approaches maturity. Equivalently, between two equal time intervals Δt the holder is willing to change its preference more in the case of the latter interval, which is closer to expiration, see Fig.10.11. A similar mechanism is used by stores which decrease the price of a perishable item when the expiry date approaches. It is worth noting that $b(t)$ can be thought as a concave utility function that describes the option holder satisfaction for exercising the call.

6. *Dependence on σ:* The free boundary $b(t)$ increases with volatility, i.e. $b(t,\sigma_1) < b(t,\sigma_2)$ for $\sigma_1 < \sigma_2$, where $b(t,\sigma)$ denotes the free boundary corresponding to a stock with volatility σ, see Fig.10.10.

At time t, the remaining life of the option is equal to $T - t$. The larger the volatility of the stock, the larger the chance that the stock will rise to a higher level during this period. Consequently, the holder's exercise preference will be to a higher level if the volatility is higher. However, if $t \to T$, when the call approaches the maturity, we cannot distinguish much between stocks

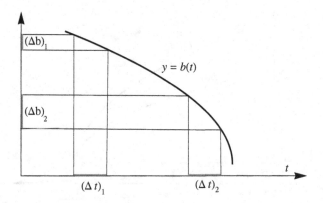

Figure 10.11: *For two equal time intervals,* $(\Delta t)_1 = (\Delta t)_2$, *the change in the preference is* $(\Delta b)_1 < (\Delta b)_2$.

with low or with large volatility, since there is no time left for the stock to rise. In this case the value of the boundary at maturity, $b(T)$, is independent of σ. Therefore, all boundary curves end at the same point $(T, b(T))$, see Fig.10.10.

7. *End point value:* $b(T)$ depends on r, δ and K but it is independent of σ, with the following variation table:

	r	δ	K	σ
$b(T)$	↗	↘	↗	constant

We study the effect of the interest rate on the boundary by considering two interest rates, with $r_1 < r_2$. In the latter case the holder would have the tendency of keeping the amount of money K in the bank at the risk-free rate r_2 for a longer period of time, which allows the stock more time to rise to a higher level and hence to increase the option value. Therefore a higher risk-free rate r allows for a higher exercise preference level $b(t)$, a fact that implies that $b(T)$ increases as r does so.

If we have two stocks with the continuous dividend rates δ_1 and δ_2, such that $\delta_1 < \delta_2$, then in the latter case the holder should prefer to exercise at a lower barrier than in the former case, since this choice gives him a longer time to gain dividend value from the stock.

To infer the variation of $b(T)$ with respect to strike K, we assume the case of two strikes, $K_1 < K_2$. The higher the strike, the larger the exercise preference level. This is also in agreement with the previous inequality $b(T) > K$.

The fact that $b(T)$ is independent of σ follows from point 6.

It can be proved mathematically that the actual terminal value of the free-boundary is given by

$$b(T) = \frac{r}{\delta}K,$$

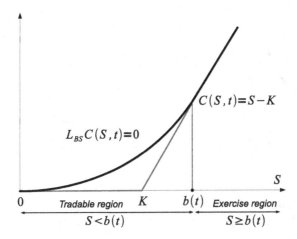

Figure 10.12: *The exercise and tradable regions of an American Call on a dividend-paying stock.*

see section 10.7.1. We notice that in the case of a non-dividend-paying stock we have $\delta = 0$. Then the value $b(T)$ becomes infinite and the entire boundary curve has only infinite values. In this case there is no preference in exercising the call at an earlier time, and the call is exercised at the maturity T and has a value equal to the value of a European call.

10.7 The Call as a Free Boundary Problem

The analogy with the one-phase Stefan problem continues. The temperature $u(x, t)$ in the pipe at point x and time t corresponds in the case of an American call to the price of the call $C(S, t)$, where the stock price is S at time t. In both problems there is a free-boundary, $b(t)$; in the case of the Stefan problem this corresponds to the water-ice interface, while in the case of the American call it represents the stock level at which the holder prefers to exercise the option.

For the Stefan problem the pipe region $0 < x < b(t)$ represents the water phase, whose temperature $u(x, t)$ changes according to the heat equation. For the American call the interval $0 < S < b(t)$ corresponds to the "tradable" region, where the call value $C(S, t)$ satisfies the Black-Scholes equation (which is also a heat-type equation after some transformations). For the Stefan problem the rest of the pipe, $b(t) \leq x$, is filled with ice at the constant temperature $u(x, t) = 0$. For the American call, the rest of the interval, $b(t) \leq S$, corresponds to the call value $S - K$, which is the price obtained by exercising the call at stock level S. All these features of the American call are shown in

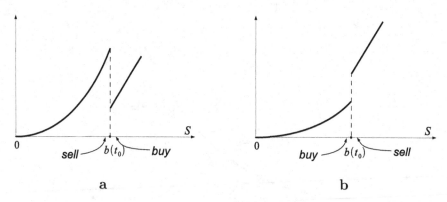

Figure 10.13: *Arbitrage opportunities:* **a.** *If* $C(b(t_0-), t_0-) > C(b(t_0+), t_0+)$ *then sell high, buy low;* **b.** *If* $C(b(t_0-), t_0+) < C(b(t_0+), t_0-)$ *then buy low, sell high.*

Fig.10.12.

The free-boundary problem satisfied by the American call is similar to the free-boundary Stefan problem (10.5.5-10.5.9). We shall show that the American call can be modeled by the following partial differential equation with free-boundary:

$$\mathcal{L}_{BS} C(S, t) = 0, \quad 0 < S < b(t) \tag{10.7.18}$$
$$C(b(t), t) = b(t) - K, \quad t > 0 \tag{10.7.19}$$
$$\frac{\partial C}{\partial S}(b(t), t) = 1, \quad t > 0 \tag{10.7.20}$$
$$C(0, t) = 0, \quad t > 0 \tag{10.7.21}$$
$$C(S, t) > S - K, \quad S < b(t), \tag{10.7.22}$$
$$C(S, t) = S - K, \quad S > b(t), \tag{10.7.23}$$

where

$$\mathcal{L}_{BS} = \frac{\partial}{\partial t} + (r - \delta) S \frac{\partial}{\partial S} + \frac{1}{2} \sigma^2 S^2 \frac{\partial^2}{\partial S^2} - r$$

is the Black-Scholes operator (8.11.23). The reasoning behind each of the previous equations and conditions is explained below.

Equation (10.7.18) states that the American call value $C(S, t)$ evolves according to the Black-Scholes equation for $0 < S < b(t)$, which is a consequence of the fact that the derivative is tradable as long as it is not exercised.

The boundary condition $C(0, t) = 0$ given by (10.7.21) says that the call value for a non-worthy stock vanishes. The fact that once the stock rises above the exercise level $b(t)$, i.e. if $S \geq b(t)$, the call value becomes after exercising equal to $S - K$ is described by (10.7.23).

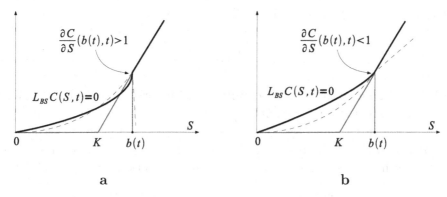

Figure 10.14: **a.** *The case* $\frac{\partial C}{\partial S}(b(t), t) > 1$; **b.** *The case* $\frac{\partial C}{\partial S}(b(t), t) < 1$.

The equation (10.7.19) assures the price continuity across the exercise boundary $b(t)$. This is a consequence of a non-arbitrage opportunities argument as follows. Assume there is a $t_0 > 0$ such that $C(S, t)$ is not continuous at $b(t_0)$. If there is a downwards jump, i.e. if $C(b(t_0-), t_0-) > C(b(t_0+), t_0+)$, then selling the call right before t_0 and buying it back right after t_0 will lead to an arbitrage opportunity, see Fig.10.13 a. The case of an upward jump is treated the same way, see Fig.10.13 b.

The inequality (10.7.22) assures that exercising the call leads to a value at least as high as holding the call.

Condition (10.7.20) states the continuity of the delta of the call at the free-boundary $b(t)$. This fact assures the smoothness of the call value at this point. This is a consequence of a *no arbitrage argument* and of an exercise stategy. It makes sense to recall that the analog relation for the Stefan problem was a consequence of the law of conservation of energy.

Figure Fig.10.14 presents the cases when the smooth-fit condition (10.7.20) does not hold.

Relation (10.7.20) will be proved by verifying a double inequality. First we shall introduce a few notations.

Let $x = b(t)$, and consider the stock S such that $K < S < b(t)$ and denote $\epsilon = x - S$. We shall evaluate the partial derivative to the left

$$\frac{\partial^- C}{\partial S}(b(t), t) = \lim_{S \to b(t)} \frac{C(b(t), t) - C(S, t)}{b(t) - S} = \lim_{\epsilon \to 0} \frac{C(x, t) - C(S, t)}{\epsilon}$$

in two different ways.

Using that $C(S, t) \geq S - K$ for $S < x$, we have the estimation

$$\frac{C(x, t) - C(S, t)}{\epsilon} \leq \frac{(x - K) - (S - K)}{\epsilon} = \frac{x - S}{\epsilon} = 1.$$

Then taking $\epsilon \to 0$ yields the inequality

$$\frac{\partial^- C}{\partial S}(b(t), t) \leq 1. \tag{10.7.24}$$

We can also arrive to the same inequality by the following heuristical argument. Assuming the converse inequality, $\frac{\partial C}{\partial S}(b(t), t) > 1$, holds true. Then using Fig.10.14 a we infer that $C(S, t) < S - K$ for values of S in the immediate left neighborhood of $b(t)$. This means the value of the call is smaller if exercised rather than if hold unexercised, which is a contradiction.

Next we shall show the opposite inequality of (10.7.24)

$$\frac{\partial^- C}{\partial S}(b(t), t) \geq 1. \tag{10.7.25}$$

It suffices to show that the slope is larger than or equal to 1, i.e.

$$\frac{C(x, t) - C(S, t)}{\epsilon} \geq 1$$

for $\epsilon > 0$ small enough. This is equivalent to the inequality

$$C(S, t) \leq C(x, t) - \epsilon. \tag{10.7.26}$$

Using that $C(x, t) = x - K$, $S = x - \epsilon$, and denoting by τ the first time when a stock with value S at time t rises above the curve $u \to b(u)$, then (10.7.26) becomes

$$\widehat{\mathbb{E}}_t[e^{-r\tau}(x - \epsilon - K)] \leq x - \epsilon - K,$$

which is equivalent to the obvious inequality $\widehat{\mathbb{E}}_t[e^{-r\tau}] \leq 1$. Hence, inequality (10.7.25) holds. The simultaneous inequalities (10.7.24) and (10.7.25) yield the condition

$$\frac{\partial^- C}{\partial S}(b(t), t) = 1.$$

Since for $S > x$ the payoff is $S - K$, the derivative on the right side is always equal to 1. Hence, the derivative exists at x and we are allowed to write the smooth-fit condition for the both-sided partial derivative as

$$\frac{\partial C}{\partial S}(b(t), t) = 1,$$

which states that the call value meets the payoff function smoothly.

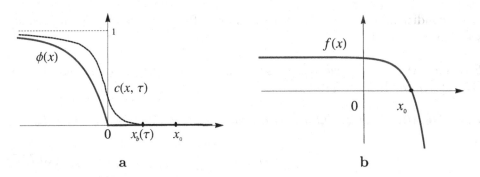

Figure 10.15: *Arbitrage opportunities:* **a.** *The free-boundary $x_b(\tau)$ lies between 0 and x_0;* **b.** *x_0 is the solution of $f(x) = 0$.*

10.7.1   Dynamics of the Free-Boundary

We start by observing that substitutions

$$S = Ke^x, \quad t = T - \tau/(\sigma^2/2), \quad C(S,t) = S - K + Kc(x,\tau) \qquad (10.7.27)$$

transform the system (10.7.18)-(10.7.23) into the following simpler one

$$\frac{\partial c}{\partial \tau} = \frac{\partial^2 c}{\partial x^2} + (k_2 - 1)\frac{\partial c}{\partial x} - k_1 c + f(x), \quad x < x_b(\tau) \qquad (10.7.28)$$

$$c\big(x_b(\tau), \tau\big) = 0, \qquad \tau > 0 \qquad (10.7.29)$$

$$\frac{\partial c}{\partial x}\big(x_b(\tau), \tau\big) = 0, \qquad \tau > 0 \qquad (10.7.30)$$

$$c(-\infty, \tau) = 1, \qquad \tau > 0 \qquad (10.7.31)$$

$$c(x, \tau) > \phi(x), \qquad x < x_b(\tau) \qquad (10.7.32)$$

$$c(x, \tau) = 0, \qquad x > x_b(\tau), \qquad (10.7.33)$$

where

$$k_1 = r/(\sigma^2/2), \quad k_2 = (r - \delta)/(\sigma^2/2) \qquad (10.7.34)$$

$$f(x) = (k_2 - k_1)e^x + k_1 = \frac{2}{\sigma^2}(r - \delta e^x) \qquad (10.7.35)$$

$$\phi(x) = \max(1 - e^x, 0) = \begin{cases} 1 - e^x, & \text{if } x < 0 \\ 0, & \text{if } x \geq 0 \end{cases} \qquad (10.7.36)$$

and $x_b(\tau)$ represents the new free-boundary, see Fig.10.15 a. We also note the initial condition $c(x, 0) = \phi(x)$.

Equation (10.7.28) is obtained from (10.7.18) by a procedure similar with the one applied in section 8.8 for solving the Black-Scholes equation. The other conditions follow easily and there are left as an exercise to the reader, see Exercise 10.7.1.

Exercise 10.7.1 *Use substitutions (10.7.27) to transform the system (10.7.18)-(10.7.23) into (10.7.28)-(10.7.33).*

Exercise 10.7.2 *Differentiate with respect to τ in (10.7.29) to show that*

$$\frac{\partial}{\partial \tau} c(x_b(\tau), \tau) = 0. \tag{10.7.37}$$

We note that the new boundary conditions (10.7.29)-(10.7.30) are homogeneous, which makes the problem easier to solve.

The region where the call is exercised, $\{S \leq b(t)\}$, becomes in this case $\{x < x_b(\tau)\}$, with $x_b(\tau) = \ln \frac{b(t)}{K}$. Using that $b(t)$ is decreasing and concave (i.e. $b'(t) < 0$, $b''(t) < 0$), the dynamics of the free-boundary $x_b(\tau)$ can be obtained now by differentiation

$$x_b'(\tau) = \frac{d}{dt}\left(\ln b(t)\right)\frac{dt}{d\tau} = -\frac{\sigma^2}{2}\frac{b'(t)}{b(t)} > 0 \tag{10.7.38}$$

$$x_b''(\tau) = \frac{\sigma^4}{4}\frac{b''(t)b(t) - b'(t)^2}{b(t)^2} < 0, \tag{10.7.39}$$

and hence, $x_b(\tau)$ is an increasing and concave function of τ.

Finding the value of $x_b(0)$ We shall provide heuristic arguments of the fact that $\lim_{\tau \to 0+} x_b(\tau) = x_0$, i.e. the free-boundary starts at x_0. Since the underlying stock pays dividends, the exercise of the call will happen before the maturity T. This implies that $C(S,t) = S - K$, for time t close to T. Then substitutions (10.7.27) imply that $c(x, \tau) = 0$, for τ close to 0. Consequently, we have

$$\frac{\partial c}{\partial x}(x, \tau) = 0, \qquad \frac{\partial^2 c}{\partial x^2}(x, \tau) = 0, \qquad 0 < \tau < \epsilon. \tag{10.7.40}$$

Making $x \to x_b(\tau)$ in (10.7.28) and using (10.7.40), continuity reasons imply

$$\frac{\partial}{\partial \tau} c(x_b(\tau), \tau) = f(x_b(\tau)), \qquad 0 < \tau < \epsilon.$$

Taking the side limit $\tau \to 0+$ yields

$$\frac{\partial}{\partial \tau} c(x_b(0+), 0+) = f(x_b(0+)). \tag{10.7.41}$$

Denote by x_0 the unique root of the equation $f(x) = 0$, see Fig.10.15 b. This is given by $x_0 = \ln \frac{r}{\delta}$. Under the assumption $r > \delta$, we have $x_0 > 0$.

Then taking $\tau \to 0+$ in relation (10.7.37) provides

$$\frac{\partial}{\partial \tau} c(x_b(0+), 0+) = 0. \tag{10.7.42}$$

Comparing (10.7.41) and (10.7.42) we infer

$$f(x_b(0+)) = 0.$$

Since x_0 is the unique root of the equation, it follows that $x_b(0+) = x_0$, i.e. the free-boundary starts at $\tau = 0$ from x_0. This can be used to find the value of the free-boundary $b(t)$ at $t = T$. Since $x_b(0) = \ln \frac{b(T)}{K}$, and $x_0 = \ln \frac{r}{\delta}$, it follows that $b(T) = \frac{rK}{\delta}$. It is worth noting that this value is independent of the volatility σ, and when $\delta \to 0+$ then $b(T) \to \infty$. Combining with the fact that $b(t)$ is decreasing, it follows that $b(t)$ is an infinite boundary, a fact that can be stated by saying that the call is not optimal to be exercised in the case of a non-divided-paying stock.

Exercise 10.7.3 (a) *Use relations (10.7.38) and (10.7.39) to compute $x_b'(0+)$ and $x_b''(0+)$.*
(b) *Sketch the graph of $x(\tau)$.*

Exercise 10.7.4 *In the previous proof we have assumed that $r > \delta$, which means $x_0 > 0$. What is the relation between this inequality and $b(T) > K$? Why is this assumption needed?*

10.7.2 Local Analysis near Maturity

In the absence of closed-form formulas for both the free-boundary $b(t)$ and call value $C(S, t)$, we are working out some approximations as $t \to T$. It suffices to find approximations just for the boundary $x_b(\tau)$ and $c(x, \tau)$ as $\tau \to 0+$. We shall do this following an idea from Wilmott et al. [69].

To simplify equation (10.7.28) we make the following assumptions:
(i) the heat source function $f(x)$ is replaced by its linear approximation about x_0:

$$f(x) \sim \underbrace{f(x_0)}_{=0} + f'(x_0)(x - x_0) = -k_1(x - x_0).$$

(ii) the free-boundary is of the form $x_b(\tau) = x_0 + c\sqrt{\tau}$, with the constant $c > 0$ subject to be determined.
(iii) the terms containing c and $\frac{\partial c}{\partial x}$ are removed, since at the boundary only the highest spatial derivative will make a difference.

Then the approximation system looks like in the following

$$\frac{\partial c}{\partial \tau} = \frac{\partial^2 c}{\partial x^2} - k_1(x - x_0), \quad x < x_b(\tau) \tag{10.7.43}$$

$$c(x_b(\tau), \tau) = 0, \qquad \tau > 0 \tag{10.7.44}$$

$$\frac{\partial c}{\partial x}(x_b(\tau), \tau) = 0, \qquad \tau > 0 \tag{10.7.45}$$

$$c(-\infty, \tau) = 1, \qquad \tau > 0 \tag{10.7.46}$$

$$c(x, \tau) > \phi(x), \qquad x < x_b(\tau) \tag{10.7.47}$$

$$c(x, \tau) = 0, \qquad x > x_b(\tau). \tag{10.7.48}$$

Let $\xi = \dfrac{x - x_0}{\sqrt{\tau}}$ and look for a solution of the type

$$c(x, \tau) = \tau^n v(\xi),$$

and determine n such that some homogeneity condition holds. Chain rule provides

$$\frac{\partial c}{\partial \tau} = n\tau^{n-1} v(\xi) - \frac{1}{2}\tau^{n-1}\xi v'(\xi)$$

$$\frac{\partial^2 c}{\partial x^2} = \tau^{n-1} v''(\xi),$$

and substituting into (10.7.43) yields

$$n\tau^{n-1} v(\xi) - \frac{1}{2}\tau^{n-1}\xi v'(\xi) = \tau^{n-1} v''(\xi) - k_1\xi\sqrt{\tau}.$$

Choosing $n = 3/2$ cancels the factors involving τ and provides the following equation in $v(\xi)$:

$$v''(\xi) + \frac{1}{2}\xi v'(\xi) - \frac{3}{2}v(\xi) = k_1\xi.$$

The solution of this equation is given by the sum between a particular solution, which is $v_p(\xi) = -k_1\xi$, and the general solution of the associated homogeneous equation (see Exercise 10.7.5)

$$v(\xi) = -k_1\xi + Av_1(\xi) + Bv_2(\xi). \tag{10.7.49}$$

Next we shall determine the value of A. Since $c(x, \tau) = \tau^{3/2} v(\xi)$, using $\xi = (x - x_0)/\sqrt{\tau}$, we obtain after a separation of variables

$$\frac{v(\xi)}{\xi^3} = \frac{c(x, \tau)}{(x - x_0)^3}.$$

Take the limit with $\xi \to -\infty$ on the left side and $x \to -\infty$ on the right side, and use condition (10.7.46) to obtain

$$\lim_{\xi \to -\infty} \frac{v(\xi)}{\xi^3} = 0. \tag{10.7.50}$$

On the other side, using (10.7.49) the limit becomes

$$
\begin{aligned}
\lim_{\xi \to -\infty} \frac{v(\xi)}{\xi^3} &= \lim_{\xi \to -\infty} \left(-\frac{k_1}{\xi^2} \right) + A \left(1 + \frac{6}{\xi^2} \right) \\
&\quad + B \lim_{\xi \to -\infty} \left(\frac{\xi^2 + 4}{\xi^3} e^{-\xi^2/4} + \frac{\xi^2 + 6\xi}{2\xi^3} \int_{-\infty}^{\xi} e^{-s^2/4} \, ds \right) \\
&= A. \tag{10.7.51}
\end{aligned}
$$

Comparing the limit values (10.7.50) and (10.7.51) yields $A = 0$. Therefore

$$v(\xi) = -k_1 \xi + B v_2(\xi). \tag{10.7.52}$$

Using assumption (ii), the boundary conditions (10.7.44)-(10.7.45) become

$$v(c) = 0, \qquad v'(c) = 0.$$

Using (10.7.52) these conditions yields a system of equations with unknowns c and B:

$$
\begin{aligned}
B v_2(c) &= k_1 c \\
B v_2'(c) &= k_1.
\end{aligned}
$$

Dividing the equations we eliminate B and k_1, obtaining the following equation for c

$$v_2(c) = c v'(c).$$

This equation can be written explicitly as

$$c^3 e^{c^2/4} \int_{-\infty}^{c} e^{-s^2/4} \, ds = 2(2 - c^2). \tag{10.7.53}$$

This equation has a unique positive solution $c > 0$. This follows from the analysis of the left and right sides variation. Let $\Phi_1(x) = x^3 e^{x^2/4} \int_{-\infty}^{x} e^{-s^2/4} \, ds$ and $\Phi_2(x) = 2(2 - x^2)$. Since

$\Phi_1(x)$ and $\Phi_2(x)$ are continuous;

$\Phi_1(x)$ is increasing and $\Phi_2(x)$ is decreasing on $(0, \infty)$;

$\Phi_1(0+) = 0 < \phi_2(0) = 4$;

$\Phi_1(\infty) = \infty > \phi_2(\infty) = -\infty$,

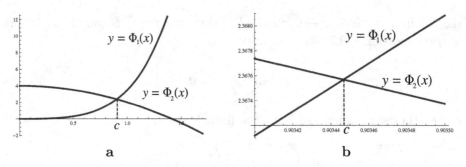

Figure 10.16: *The intersection point of the graphs of $y = \Phi_1(x)$ and $y = \Phi_2(x)$:* **a.** *The rough picture;* **b.** *The zoom-in picture.*

then the graphs of Φ_1 and Φ_2 have only one intersection point, see Fig.10.16 a. The solution is obtained graphically, see Fig.10.16 b, and its value is given by

$$c = 0.903446...$$

The value of the constant B is obtained now as $B = k_1 c / v_2(c) = 0.074718 k_1$.

Exercise 10.7.5 *Consider the homogeneous differential equation*

$$v''(\xi) + \frac{1}{2}\xi v'(\xi) - \frac{3}{2}v(\xi) = 0,$$

and let $v_1(\xi) = \xi^3 + 6\xi$ and

$$v_2(\xi) = (\xi^2 + 4)e^{-\xi^2/4} + \frac{1}{2}(\xi^2 + 6\xi)\int_{-\infty}^{\xi} e^{-s^2/4}\,ds.$$

(a) *Show that v_1 and v_2 satisfy the equation.*

(b) *Prove that v_1 and v_2 have a nonzero wronskian*

$$\begin{vmatrix} v_1 & v_2 \\ v_1' & v_2' \end{vmatrix} \neq 0.$$

(c) *Use (a) and (b) to obtain the general solution of the equation as*

$$v(\xi) = Av_1(\xi) + Bv_2(\xi),$$

with A and B real constants.

To conclude, the approximation of the free-boundary $x_b(\tau) = x_0 + c\sqrt{\tau}$ as $\tau \to 0+$ is

$$b(t) = \frac{rK}{\delta}\left(1 + c\sqrt{\sigma^2(T-t)/2}\right), \qquad t \to T.$$

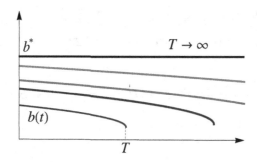

Figure 10.17: *If $T \to \infty$ then $b(t) \to b^*$.*

Reverting the variables, we obtain an asymptotics for the call price as $t \to T$:

$$
\begin{aligned}
C(S,t) &= S - K + Kc(x,\tau) \\
&= S - K + K\tau^{3/2}v(\xi) \\
&= S - K + K\tau^{3/2}\big(-k_1\xi + Bv_2(\xi)\big) \\
&= S - K + K\tau^{3/2}k_1\big(-\xi + 0.0747v_2(\xi)\big) \\
&= S - K + Kr\frac{\sigma}{\sqrt{2}}(T-t)^{3/2}\big(-\xi + 0.0747v_2(\xi)\big), \qquad (10.7.54)
\end{aligned}
$$

where

$$
\xi = \frac{x - x_0}{\sqrt{\tau}} = \frac{\ln\frac{S}{K} - \ln\frac{r}{\delta}}{\sqrt{\tau}} = -\frac{\ln\left(\frac{rK}{\delta S}\right)}{\sqrt{\sigma^2(T-t)/2}},
$$

and $v_2(\xi)$ is given by Exercise 10.7.5.

10.7.3   The Infinite Horizon Case

If $T \to \infty$, then the call becomes a perpetuity, and its value is not affected by the remaining life time of the option, $T - t$. The time independence of the perpetuity value can be written as

$$
C(S,t) = C(S).
$$

This dimensional reduction simplifies the problem considerably, closed form solutions being possible. The free-boundary becomes also time-invariant, i.e., $b(t) = b^*$, see Fig.10.17. The constant b^* is the value that maximizes the expression

$$
C(S) = \max_{b>K} \widehat{\mathbb{E}}[e^{-r\tau_b}(b - K)] = \widehat{\mathbb{E}}[e^{-r\tau_{b^*}}(b^* - K)],
$$

where τ_b denotes the first passage of time when the stock rises above level b.

The free-boundary problem (10.7.18)-(10.7.23) in this case can be written as

$$\frac{\sigma^2}{2} S^2 C''(S) + (r - \delta) S C'(S) - r C(S) = 0, \quad 0 < S < b^* \tag{10.7.55}$$

$$C(b^*) = b^* - K \tag{10.7.56}$$

$$C'(b^*) = 1, \tag{10.7.57}$$

$$C(0) = 0, \tag{10.7.58}$$

$$C(S) > S - K, \quad S < b^* \tag{10.7.59}$$

$$C(S) = S - K, \quad S > b^*. \tag{10.7.60}$$

This ordinary differential equation problem has two unknowns: the call value $C(S)$ and the barrier level b^*. Since (10.7.55) is an Euler equation, we are looking for a solution of the form $C(S) = S^h$. Substituting in (10.7.55) we arrive at the quadratic equation

$$h^2 - \left(1 - \frac{r - \delta}{\sigma^2/2}\right) h - \frac{r}{\sigma^2/2} = 0. \tag{10.7.61}$$

The equation (10.7.61) has two roots

$$h_1 = \frac{1}{2} - \frac{r - \delta}{\sigma^2} + \sqrt{\left(\frac{r - \delta}{\sigma^2} - \frac{1}{2}\right)^2 + \frac{2r}{\sigma^2}}$$

$$h_2 = \frac{1}{2} - \frac{r - \delta}{\sigma^2} - \sqrt{\left(\frac{r - \delta}{\sigma^2} - \frac{1}{2}\right)^2 + \frac{2r}{\sigma^2}}.$$

It is important to note that $h_2 < 0 < h_1$. (The fact that the roots have opposite signs is implied by the fact that the product $h_1 h_2 = -\frac{r}{\sigma^2/2}$ is negative). Then the general solution of the equation (10.7.55) can be written as a linear combination

$$C(S) = \alpha_1 S^{h_1} + \alpha_2 S^{h_2},$$

with α_1 and α_2 undetermined constants. Since $C(S)$ tends to zero as $S \to 0$, then condition (10.7.58) implies $\alpha_2 = 0$. Substituting $C(S) = \alpha_1 S^{h_1}$ in (10.7.56) and (10.7.57) we get two equations in the unknowns α_1 and b^*:

$$\alpha_1 (b^*)^{h_1} = b^* - K$$

$$\alpha_1 h_1 (b^*)^{h_1 - 1} = 1.$$

Dividing we obtain the linear equation $\dfrac{b^*}{h_1} = b^* - K$, which yields the optimal exercise boundary

$$b^* = K \frac{h_1}{h_1 - 1}.$$

The constant α_1 is obtained from the first equation as

$$\alpha_1 = \frac{b^* - K}{(b^*)^{h_1}}.$$

Substituting back into the solution we obtain

$$C(S) = \alpha_1 S^{h_1} = (b^* - K)\left(\frac{S}{b^*}\right)^{h_1} = \left(\frac{Kh_1}{h_1 - 1} - K\right)\frac{S^{h_1}}{\left(\frac{Kh_1}{h_1-1}\right)^{h_1}}$$

$$= \frac{K}{h_1 - 1}\left(\frac{h_1 - 1}{h_1}\frac{S}{K}\right)^{h_1},$$

where $0 < S < b^*$. This agrees with the value obtained in Exercise 10.1.9 using the optimization method. Using condition (10.7.60) we obtain the value of the call perpetuity on a dividend-paying stock as

$$C(S) = \begin{cases} \frac{K}{h_1-1}\left(\frac{h_1-1}{h_1}\frac{S}{K}\right)^{h_1}, & \text{if } S < b^* \\ S - K, & \text{if } S \geq b^*. \end{cases}$$

Exercise 10.7.6 (a) *Find the value of the optimal boundary b^* when the dividend yield $\delta \to 0$.*

(b) *What is the value of the call perpetuity in this case?*

Exercise 10.7.7 *Consider a perpetuity $V(S)$ that satisfies the free-boundary problem*

$$\frac{\sigma^2}{2}S^2 V''(S) + (r - \delta)SV'(S) - rV(S) = 0, \quad 0 < S < b^*$$
$$C(b^*) = 1$$
$$C'(b^*) = \beta$$
$$C(0) = 0$$
$$C(S) < 1, \qquad S < b^*$$
$$C(S) = 1, \qquad S > b^*,$$

with $\beta > 0$.
(a) *Find the value of the boundary b^*;*
(b) *Solve the free-boundary problem and find $V(S)$;*
(c) *What happens when $\beta \to 0$?*
(d) *State the financial significance of this problem.*

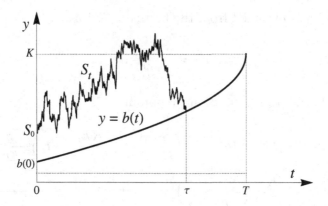

Figure 10.18: *The optimal exercising boundary $b(t)$ and the stopping time τ.*

10.8 American Put

This contract provides the holder the right to sell a stock for the given price K any time before or at the maturity T. In the case of an American put we assume the holder's strategy is based on the hope that the stock S_t goes low enough so it can obtain a large difference $K - S_t$. The preference level at which the holder exercises the contract is time dependent and constitutes the free-boundary $b(t)$, see Fig. 10.18.

Given the free-boundary $b(t)$, and the random time τ at which the stock goes under $b(t)$, the value of the put at time $t = 0$ is obtained by discounting the payoff $\max(K - S_\tau, 0) = \max(K - b(\tau), 0) = K - b(\tau)$ with the random time τ and then take the average as

$$P(S, 0) = \widehat{\mathbb{E}}[e^{-r\tau}(K - b(\tau))]. \tag{10.8.62}$$

The time-depending boundary $b(t)$ is a curve defined on $[0, T]$, with smaller values than K and which maximizes the functional (10.8.62).

10.8.1 Properties of the Free-Boundary of a Put

The shape and properties of the free-boundary $b(t)$ can be described employing similar heuristics used for the call, even if we shall not get into as much detail as in that case. The properties are as in the following:

1. *Relative relation with the stock:* $b(\tau) = S_\tau$, $b(t) < S_t$ for $t < \tau$.

Choosing a strategy under which the boundary is reached from above provides intrinsic value for the put.

2. *Relation with the strike:* $b(t) < K$ for $0 \le t < T$.

We prefer a strategy that chooses a boundary $b(t)$ such that $K - b(t)$ provides value for the put.

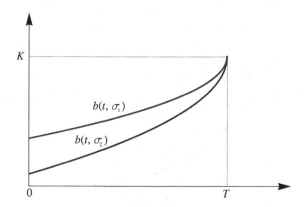

Figure 10.19: *The free-boundary for two volatilites: if $\sigma_1 < \sigma_2$ then $b(t,\sigma_1) > b(t,\sigma_2)$, for $t < T$, and $b(T,\sigma_1) = b(T,\sigma_2) = K$.*

3. *Monotonicity:* $b(t)$ is increasing on $(0,T]$.

If $0 < t_1 < t_2 < T$ are two time instances, then the remaining lives of the option satisfy $T - t_2 < T - t_1$. Now, the shorter the remaining life the less chance for the stock to go lower. Consequently, our strategy should be to exercise at higher levels of the stock when approaching the maturity of the option. Hence, $b(t_1) < b(t_2)$, i.e. $b(t)$ is an increasing function.

4. *Continuity:* $b(t)$ is continuous on $(0,T]$.

If $b(t)$ were not continuous on $(0,T]$, since it is decreasing, then it has only jump-type discontinuities. There is a positive probability for the stock price to go through the barrier without intersecting it, a fact that loses option value.

5. *Convexity:* $b(t)$ is convex on $(0,T]$.

6. *Dependence on σ:* The free boundary $b(t)$ decreases with volatility, i.e. $b(t,\sigma_1) > b(t,\sigma_2)$ for $\sigma_1 < \sigma_2$, where $b(t,\sigma)$ denotes the free-boundary corresponding to a stock with volatility σ, see Fig.10.19.

At time t, the remaining life of the option is equal to $T - t$. The larger the volatility of the stock, the larger the chance that the stock will get under a lower level during this period. Consequently, the holder's exercise preference will be to a lower level if the volatility is higher.

7. *Dependence on r:* The free boundary $b(t)$ decreases with respect to the free-interest rate.

The higher the interest rate r, the more incentive to get the cash amount K and deposit it in the bank sooner. This increases the put value.

8. *Dependence on δ:* The free boundary $b(t)$ increases with respect to the stock-dividend yield δ.

The larger the yield δ, the longer the stock is held for more profit. Hence, postponing the exercise increases the value of the put.

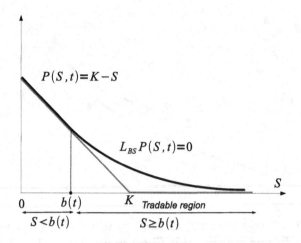

Figure 10.20: *The value of an American put as a function of S.*

9. *End point values:* $0 < b(0+) < K$ and $b(T) = K$.

The optimality of the exercising strategy requires that $0 < b(t) \le K$, so $0 < b(0+) < K$ and $b(T) \le K$. Assume, by contradiction, that $b(T) < K - \epsilon$, with $\epsilon > 0$. There is a positive probability that the stock, satisfying $S_t > b(t)$, will end up at maturity with a value S_T between $K - \epsilon$ and K, a case in which the put is not exercised, since the stock does not intersect the boundary $b(t)$. In this scenario we lose value for the put. Hence, the above inequality does not hold, and $b(T) = K$.

10.8.2   The Put as a Free-Boundary Problem

Similar arguments with the ones presented for the American call lead to the following free-boundary problem for the American put on a non-dividend-paying stock, with the unknown functions $P(S, t)$ and $b(t)$:

$$\mathcal{L}_{BS} P(S, t) = 0, \quad S > b(t) \tag{10.8.63}$$

$$P(b(t), t) = K - b(t), \quad t > 0 \tag{10.8.64}$$

$$\frac{\partial P}{\partial S}(b(t), t) = -1, \quad t > 0 \tag{10.8.65}$$

$$P(0, t) = K, \quad t > 0 \tag{10.8.66}$$

$$P(S, t) > K - S, \quad S > b(t), \tag{10.8.67}$$

$$P(S, t) = K - S, \quad S < b(t), \tag{10.8.68}$$

where

$$\mathcal{L}_{BS} = \frac{\partial}{\partial t} + rS \frac{\partial}{\partial S} + \frac{1}{2}\sigma^2 S^2 \frac{\partial^2}{\partial S^2} - r$$

is the Black-Scholes operator. The reasons behind each of the previous equations are similar with the case of an American call and there are left as an exercise for the reader.

Exercise 10.8.1 *Prove the smooth-fit condition (10.8.65) by showing the inequalities:*

(a) $\dfrac{\partial P}{\partial S}(b(t), t) \geq -1, \qquad t > 0;$

(b) $\dfrac{\partial P}{\partial S}(b(t), t) \leq -1, \qquad t > 0.$

The free-boundary $b(t)$ represents the stock level at which the holder prefers to exercise the option. The interval $S > b(t)$ corresponds to the "tradable" region, where the put value $P(S, t)$ satisfies the Black-Scholes equation and it is better off to hold the put. The rest of the interval, $S < b(t)$, represents the "exercise region" or the "stopping region", and corresponds to the put value $K - S$, which is the premium obtained by exercising the put at the stock level S. The smooth-fit condition (10.8.65) states that the delta of the put at the boundary has the value -1. Together with condition (10.8.64) they form two boundary smooth-pasting conditions imposed to avoid arbitrage opportunities. All these features are shown in Fig.10.20.

10.8.3 The Simplified Free-Boundary Problem

The substitutions

$$S = Ke^x, \quad t = T - \tau/(\sigma^2/2), \quad P(S, t) = K - S + Kp(x, \tau) \qquad (10.8.69)$$

transform the system (10.8.63)-(10.8.68) into the following simpler version:

$$\frac{\partial p}{\partial \tau} = \frac{\partial^2 p}{\partial x^2} + (k-1)\frac{\partial p}{\partial x} - kp + k(e^x - 1), \quad x > x_b(\tau) \quad (10.8.70)$$

$$p(x_b(\tau), \tau) = 0, \quad \tau > 0 \qquad (10.8.71)$$

$$\frac{\partial p}{\partial x}(x_b(\tau), \tau) = 0, \qquad \tau > 0 \qquad (10.8.72)$$

$$p(-\infty, \tau) = 0 \qquad (10.8.73)$$

$$p(x, \tau) > 0, \qquad x > x_b(\tau) \qquad (10.8.74)$$

$$p(x, \tau) = 0, \qquad x < x_b(\tau), \qquad (10.8.75)$$

where the new free-boundary is given by $x_b(\tau) = \ln(b(t)/K)$, with $k = \frac{2r}{\sigma^2}$. The computational details are left to the reader, see Exercise 10.8.4. Since $P(S, t) \sim 0$ as $S \to \infty$, then $p(x, \tau) \sim e^x - 1$ as $x \to \infty$, see Fig.10.21. The advantages of this formulation over the former consist in the homogeneous boundary conditions and in a more simple dynamics.

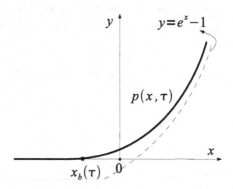

Figure 10.21: *The simplified boundary problem for $p(x, \tau)$ and $x_b(\tau)$.*

Exercise 10.8.2 *Provide a reason why $P(S, t) \sim 0$ as $S \to \infty$.*

Exercise 10.8.3 *Using the properties of the free-boundary $b(t)$ show the following:*

(a) $x_b(0+) = 0$ and $x_b(\tau) < 0$, $\tau > 0$;

(b) $x_b'(\tau) < 0$, i.e. the boundary $x_b(\tau)$ moves to the left as τ increases.

Exercise 10.8.4 *Show that substitutions (10.8.69) transform the system (10.8.63)-(10.8.68) into (10.8.70)-(10.8.75).*

The dynamics of the problem can be further simplified using the substitutions

$$\xi = x + (k-1)\tau, \qquad w(\xi, \tau) = e^{-\tau} p(x, \tau). \tag{10.8.76}$$

The resulting problem looks like in the following, see Exercise 10.8.5

$$\frac{\partial w}{\partial \tau} = \frac{\partial^2 w}{\partial \xi^2} - k(e^{\xi + \tau} - e^{k\tau}), \quad \xi > \xi_b(\tau) \tag{10.8.77}$$

$$w(\xi_b(\tau), \tau) = 0, \quad \tau > 0 \tag{10.8.78}$$

$$\frac{\partial w}{\partial x}(\xi_b(\tau), \tau) = 0, \quad \tau > 0 \tag{10.8.79}$$

$$w(-\infty, \tau) = 0 \tag{10.8.80}$$

$$w(\xi, \tau) > 0, \quad \xi > \xi_b(\tau) \tag{10.8.81}$$

$$w(\xi, \tau) = 0, \quad \xi < \xi_b(\tau), \tag{10.8.82}$$

where $\xi_b(\tau) = x_b(\tau) + (k-1)\tau$ is the new free-boundary.

Exercise 10.8.5 *Use substitutions (10.8.76) to transform the system (10.8.70)-(10.8.75) into (10.8.77)-(10.8.82).*

10.8.4   An Integral Equation for the Free-Boundary

The problem (10.8.77)-(10.8.82) can be solved explicitly using Duhamel's principle, see Appendix D.2. In order to apply this principle we need to extend the equation (10.8.77) to the positive half-axis. This idea was first used by Jamshidian [43] in the early 1990s.

First, we note that equation (10.8.77) holds in the domain $(\xi_b(\tau), +\infty)$. In order to construct the partial differential equation in the region $0 \le \xi < \infty$ that corresponds to the above free-boundary problem, we need to find the inhomogeneous term such that the equation holds for all values of ξ. Since (10.8.82) assures that $w(\xi, \tau) = 0$ for $\xi \le \xi_b(\tau)$, a direct substitution in the equation in the stopping region evaluates to

$$-\frac{\partial w}{\partial \tau} + \frac{\partial^2 w}{\partial \xi^2} - k(e^{\xi+\tau} - e^{k\tau}) = -k(e^{\xi+\tau} - e^{k\tau}).$$

Therefore, the inhomogeneous partial differential equation that holds for $0 \le \xi < \infty$ is

$$\frac{\partial w}{\partial \tau} = \frac{\partial^2 w}{\partial \xi^2} - k(e^{\xi+\tau} - e^{k\tau}) + k\mathbb{I}_{\{\xi<\xi_b(\tau)\}}(e^{\xi+\tau} - e^{k\tau}),$$

or, equivalently

$$\frac{\partial w}{\partial \tau} = \frac{\partial^2 w}{\partial \xi^2} - k(e^{\xi+\tau} - e^{k\tau})(1 - \mathbb{I}_{\{\xi<\xi_b(\tau)\}}), \tag{10.8.83}$$

where

$$\mathbb{I}_{\{\xi<\xi_b(\tau)\}}(\xi) = \begin{cases} 1, & \text{if } \xi \le \xi_b(\tau) \\ 0, & \text{if } \xi > \xi_b(\tau) \end{cases}$$

is the indicator function of the interval $(-\infty, \xi_b(\tau)]$.

Equation (10.8.83) represents the evolution of heat in a semi-infinite rod under the influence of a fancy heat source. In order to solve this equation, we need to find the initial condition, $w(\xi, 0)$, which corresponds to the initial heat distribution in the rod. Taking $\tau = 0$, we have

$$\begin{aligned} w(\xi, 0) &= p(x, 0) = \frac{1}{K}\left((K - S)^+ - (K - S)\right) \\ &= \left((1 - e^x)^+ - (1 - e^x)\right) = \psi(x) = \psi(\xi), \tag{10.8.84} \end{aligned}$$

with

$$\psi(x) = \begin{cases} 0, & \text{if } x < 0 \\ e^x - 1, & \text{if } x \ge 0. \end{cases}$$

The equation (10.8.83) together with condition (10.8.84) produce a inhomogeneous Cauchy problem for the heat operator

$$\frac{\partial w}{\partial \tau} - \frac{\partial^2 w}{\partial \xi^2} = -k(e^{\xi+\tau} - e^{k\tau})(1 - \mathbb{I}_{\{\xi < \xi_b(\tau)\}}), \qquad \xi \geq 0$$

$$w(\xi, 0) = \psi(\xi),$$

which can be solved using Duhamel's principle. Applying Proposition D.2.1 of Appendix D.2, we obtain the following solution

$$w(\xi, \tau) = \underbrace{\int G(\xi - y, \tau)\psi(y)\, dy}_{I_1} \tag{10.8.85}$$

$$- k \underbrace{\int_0^\tau \int G(\xi - y, \tau - s)(e^{y+s} - e^{ks})(1 - \mathbb{I}_{\{y < \xi_b(s)\}})\, dy\, ds}_{I_2},$$

where $G(\ ,\ )$ stands for the heat kernel

$$G(x, \tau) = \frac{1}{\sqrt{4\pi\tau}} e^{-\frac{x^2}{4\tau}}, \qquad \tau > 0.$$

The first integral of (10.8.85) can be computed as in the following

$$I_1 = \int G(\xi - y, \tau)\psi(y)\, dy = \frac{1}{\sqrt{4\pi\tau}} \int_0^\infty e^{-\frac{(y-\xi)^2}{4\tau}} (e^y - 1)\, dy$$

$$= \underbrace{\frac{1}{\sqrt{4\pi\tau}} \int_0^\infty e^{-\frac{(y-\xi)^2}{4\tau} + y}\, dy}_{A} - \underbrace{\frac{1}{\sqrt{4\pi\tau}} \int_0^\infty e^{-\frac{(y-\xi)^2}{4\tau}}\, dy}_{B}$$

Completing to a square and using the substitution $u = (y - \xi - 2\tau)/\sqrt{2\tau}$, the first of the above integrals becomes

$$A = \frac{1}{\sqrt{4\pi\tau}} \int_0^\infty e^{-\frac{(y-\xi)^2}{4\tau} + y}\, dy = \frac{1}{\sqrt{4\pi\tau}} e^{\xi+\tau} \int_0^\infty e^{-\frac{(y-\xi-2\tau)^2}{4\tau}}\, dy$$

$$= \frac{1}{\sqrt{2\pi}} e^{\xi+\tau} \int_{-(\xi+2\tau)/\sqrt{2\tau}}^\infty e^{-u^2/2}\, du = e^{\xi+\tau} N\left(\frac{\xi}{\sqrt{2\tau}} + \sqrt{2\tau}\right),$$

where $N(x) = \frac{1}{\sqrt{2\pi}} \int_{-\infty}^x e^{-u^2/2}\, du$.

Using the substitution $v = (y - \xi)/\sqrt{2\tau}$, the integral B can be evaluated as

$$B = \frac{1}{\sqrt{4\pi\tau}} \int_0^\infty e^{-\frac{(y-\xi)^2}{4\tau}}\, dy = \frac{1}{\sqrt{2\pi}} \int_{-\xi/\sqrt{2\tau}}^\infty e^{-v^2/2}\, dv$$

$$= N\left(\frac{\xi}{\sqrt{2\tau}}\right).$$

Putting the pieces together, we obtain

$$I_1 = A - B = e^{\xi+\tau} N\left(\frac{\xi}{\sqrt{2\tau}} + \sqrt{2\tau}\right) - N\left(\frac{\xi}{\sqrt{2\tau}}\right).$$

The second integral of (10.8.85) can be evaluated using four integrals as

$$I_2 = k(A_1 - A_2 - A_3 + A_4).$$

We shall compute each of these integrals in the following.

$$A_1 = \int_0^\tau \int G(\xi - y, \tau - s) e^y e^s \, dy ds = \int_0^\tau \int \frac{1}{\sqrt{4\pi(\tau - s)}} e^{-\frac{(y-\xi)^2}{4(\tau-s)}+y} \, dy ds$$

$$= \int_0^\tau e^s e^{\xi+\tau-s} \, ds = \tau e^{\xi+\tau}.$$

$$A_2 = \int_0^\tau \int G(\xi - y, \tau - s) e^{ks} \, dy ds = \int_0^\tau e^{ks} \underbrace{\int G(\xi - y, \tau - s) \, dy}_{=1} \, ds = \frac{e^{k\tau} - 1}{k}.$$

$$A_3 = \int_0^\tau \int G(\xi - y, \tau - s) e^{y+s} \, \mathbb{I}_{\{y<\xi_b(s)\}}(y) \, dy ds$$

$$= \int_0^\tau e^s \int_{-\infty}^{\xi_b(s)} \frac{1}{\sqrt{4\pi(\tau - s)}} e^{-\frac{(y-\xi)^2}{4(\tau-s)}+y} \, dy ds$$

$$= e^{\xi+\tau} \int_0^\tau N\left(\frac{\xi_b(s) - \xi}{\sqrt{2(\tau - s)}} - \sqrt{2(\tau - s)}\right) ds$$

$$= e^{\xi+\tau} \int_0^\tau N\left(\frac{\xi_b(\tau - u) - \xi}{\sqrt{2u}} - \sqrt{2u}\right) du.$$

$$A_4 = \int_0^\tau \int G(\xi - y, \tau - s) e^{ks} \, \mathbb{I}_{\{y<\xi_b(s)\}}(y) \, dy ds$$

$$= \int_0^\tau e^{ks} \int_{-\infty}^{\xi_b(s)} \frac{1}{\sqrt{4\pi(\tau - s)}} e^{-\frac{(y-\xi)^2}{4(\tau-s)}} \, dy ds$$

$$= \int_0^\tau e^{ks} N\left(\frac{\xi_b(s) - \xi}{\sqrt{2(\tau - s)}}\right) ds$$

$$= e^{k\tau} \int_0^\tau e^{-ku} N\left(\frac{\xi_b(\tau - u) - \xi}{\sqrt{2u}}\right) du.$$

Combining the pieces together we obtain the solution $w(\xi, \tau)$ in terms of the boundary $\xi_b(\tau)$ as follows

$$
\begin{aligned}
w(\xi, \tau) &= I_1 - I_2 \\
&= I_1 - k(A_1 - A_2 - A_3 + A_4) \\
&= e^{\xi + \tau} N\left(\frac{\xi}{\sqrt{2\tau}} + \sqrt{2\tau}\right) - N\left(\frac{\xi}{\sqrt{2\tau}}\right) - k\tau e^{\xi + \tau} + e^{k\tau} - 1 \\
&\quad + k e^{\xi + \tau} \int_0^\tau N\left(\frac{\xi_b(\tau - u) - \xi}{\sqrt{2u}} - \sqrt{2u}\right) du \\
&\quad - k e^{k\tau} \int_0^\tau e^{-ku} N\left(\frac{\xi_b(\tau - u) - \xi}{\sqrt{2u}}\right) du.
\end{aligned}
$$

After the factorization and using that $N(a) = 1 - N(-a)$, we obtain

$$
\begin{aligned}
w(\xi, \tau) &= e^{\xi + \tau}\left\{ N\left(\frac{\xi}{\sqrt{2\tau}} + \sqrt{2\tau}\right) - k\tau + k \int_0^\tau N\left(\frac{\xi_b(\tau - u) - \xi}{\sqrt{2u}} - \sqrt{2u}\right) du \right\} \\
&\quad - N\left(\frac{\xi}{\sqrt{2\tau}}\right) + e^{k\tau} - 1 - k e^{k\tau} \int_0^\tau e^{-ku} N\left(\frac{\xi_b(\tau - u) - \xi}{\sqrt{2u}}\right) du \\
&= e^{\xi + \tau}\left\{ N\left(\frac{\xi}{\sqrt{2\tau}} + \sqrt{2\tau}\right) - k \int_0^\tau N\left(-\frac{\xi_b(\tau - u) - \xi}{\sqrt{2u}} + \sqrt{2u}\right) du \right\} \\
&\quad - N\left(\frac{\xi}{\sqrt{2\tau}}\right) + k e^{k\tau} \int_0^\tau e^{-ku} N\left(-\frac{\xi_b(\tau - u) - \xi}{\sqrt{2u}}\right) du. \quad (10.8.86)
\end{aligned}
$$

From the solution uniqueness of the problem (10.8.77)-(10.8.82), the function (10.8.86) has the properties

$$
\begin{aligned}
w(\xi, \tau) &> 0, &\quad \xi &> \xi_b(\tau) \\
w(\xi, \tau) &= 0, &\quad \xi &\leq \xi_b(\tau).
\end{aligned}
$$

The equation satisfied by the free-boundary $\xi_b(\tau)$ is obtained substituting (10.8.86) into the boundary condition (10.8.77)

$$
\begin{aligned}
e^{\xi_b(\tau) + \tau}&\left\{ N\left(\frac{\xi_b(\tau)}{\sqrt{2\tau}} + \sqrt{2\tau}\right) - k \int_0^\tau N\left(-\frac{\xi_b(\tau - u) - \xi_b(\tau)}{\sqrt{2u}} + \sqrt{2u}\right) du \right\} \\
&= N\left(\frac{\xi_b(\tau)}{\sqrt{2\tau}}\right) - k e^{k\tau} \int_0^\tau e^{-ku} N\left(-\frac{\xi_b(\tau - u) - \xi_b(\tau)}{\sqrt{2u}}\right) du. \quad (10.8.87)
\end{aligned}
$$

Exercise 10.8.6 *Reverting the variables, show that*

(a) $k\tau = r(T - t)$;

(b) $e^{\xi_b(\tau) + \tau} = \frac{b(t)}{K} e^{r(T-t)}$;

(c) *Write the integral equation for the free-boundary $b(t)$ using (10.8.87).*

Exercise 10.8.7 *In order to perform an analysis near the boundary, we keep only the highest spatial derivative, assuming that $c(x, \tau) = 0$ and $\frac{\partial p}{\partial x} = 0$. Then the problem (10.8.70)-(10.8.75) becomes:*

$$\frac{\partial p}{\partial \tau} = \frac{\partial^2 p}{\partial x^2} + k(e^x - 1), \quad x > x_b(\tau)$$

$$p(-\infty, \tau) = 0$$

$$p(x, \tau) > 0, \qquad x > x_b(\tau)$$

$$p(x, \tau) = 0, \qquad x < x_b(\tau).$$

(a) Use Duhamel's principle to extend the problem to the interval $[0, \infty)$, and show that $p(x, \tau)$ satisfies the following initial value problem:

$$\frac{\partial p}{\partial \tau} = \frac{\partial^2 p}{\partial x^2} + k(e^x - 1)\mathbb{I}_{\{x > x_b(\tau)\}}, \quad x > 0$$

$$p(x, 0) = \psi(x).$$

(b) Show that the problem given at (a) has the following solution, which depends on the free-boundary x_b:

$$p(x, \tau) = e^{x+\tau} N\left(\frac{x}{\sqrt{2\tau}} + \sqrt{2\tau}\right) - N\left(\frac{x}{\sqrt{2\tau}}\right)$$

$$+ k \int_0^\tau e^{x+u} N\left(\frac{x - x_b(\tau - u)}{\sqrt{2u}} + \sqrt{2u}\right) du - k \int_0^\tau N\left(\frac{x - x_b(\tau - u)}{\sqrt{2u}}\right) du.$$

(c) Use the boundary condition $p(x_b(\tau), \tau) = 0$ to obtain an integral equation satisfied by the free-boundary x_b.

10.8.5   Approximation of the Boundary near Maturity

The integral equation (10.8.87) is far too complicated to be solved explicitly. However, we can find an approximation of the boundary, $\xi_b(\tau)$, as $\tau \to 0+$. Since the equation depends on the expression $\frac{\xi_b(\tau)}{\sqrt{2\tau}}$, it makes sense to look for an approximation of the form

$$\xi_b(\tau) = \xi_b(0) - c_0\sqrt{2\tau}, \qquad \tau \to 0+,$$

with $c_0 > 0$ constant. The minus sign was chosen to agree with the fact that $\xi_b(\tau)$ is decreasing. Since $\xi_b(0) = 0$, then $\frac{\xi_b(\tau)}{\sqrt{2\tau}} = -c_0$. Substituting in (10.8.87) and taking the limit $\tau \to 0+$ we obtain a tautology; hence, for τ small the solution is approximated by $\xi_b(\tau) = -c_0\sqrt{2\tau}$. This corresponds to the boundary

$$x_b(\tau) = \xi_b(\tau) - (k-1)\tau = -c_0\sqrt{2\tau} - \frac{k-1}{2}(2\tau), \qquad \tau \to 0+$$

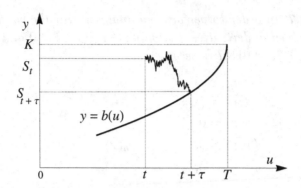

Figure 10.22: *The stopping time τ.*

which implies

$$b(t) = Ke^{x_b(\tau)} = Ke^{-c_0\sigma\sqrt{T-t}-(r-\frac{\sigma^2}{2})(T-t)}, \qquad t \to T. \qquad (10.8.88)$$

Using $e^{-x} = 1 - x + O(x^2)$ as $x \to 0$, it follows that the free-boundary of the put near maturity is given by

$$b(t) = K\left(1 - c_0\sigma\sqrt{T-t} - (r - \frac{\sigma^2}{2})(T-t)\right) + O(T-t)^2, \qquad t \to T. \quad (10.8.89)$$

The constant c_0 can be obtained substituting the solution (10.8.86) into the smooth-fit condition (10.8.79). We do not cover these calculations here, since they are pretty computationally involved. However, we shall provide in the following a stochastic interpretation of the constant c_0.

Let t be close to T and consider the stock starting at time t in the risk-neutral world

$$S_{t+u} = S_t e^{(r-\frac{\sigma^2}{2})u+\sigma W_u}.$$

Let $t + \tau$ denote the stopping time when the stock reaches the boundary

$$\tau = \inf\{u > 0; S_{t+u} \le b(t + u)\},$$

see Fig.10.22. Using the boundary formula (10.8.88), we can show by an algebraic manipulation that

$$S_{t+u} \le b(t+u) \iff c_0\sqrt{T-t-u} + W_u \le \frac{1}{\sigma}\ln\frac{K}{S_t} - \frac{1}{\sigma}\left(r - \frac{\sigma^2}{2}\right)(T-t).$$

Since $K > S_t$ and $T - t$ is small, the term on the right of the inequality is positive and does not depend on u. The stopping time $\tau = \tau(c_0)$ depends on the value of c_0, since

$$\tau = \inf\{u > 0; c_0\sqrt{T-t-u} + W_u \le x\}$$

where $x = \frac{1}{\sigma} \ln \frac{K}{S_t} - \frac{1}{\sigma}\left(r - \frac{\sigma^2}{2}\right)(T - t)$. The value of the positive constant c_0 should be chosen such that it maximizes the expression

$$\widehat{\mathbb{E}}_t[e^{-r\tau(c_0)}(K - S_{t+\tau(c_0)})] = \widehat{\mathbb{E}}_t[e^{-r\tau(c_0)}(K - b(t + \tau(c_0)))].$$

Using (10.8.89), the previous expression becomes

$$\widehat{\mathbb{E}}_t[e^{-r\tau(c_0)}(c_0\sigma\sqrt{T - t - \tau(c_0)} + \left(r - \frac{\sigma^2}{2}\right)(T - t - \tau(c_0))].$$

It is worth noting that the maximum value is just the put value $P(S_t, t)$.

In the next section we shall address the case of a put perpetuity and find its explicit valuation formula.

10.8.6   The Infinite Horizon Case

The put perpetuity is obtained taking $T \to \infty$. In this case the value of the option is not affected by the remaining life time of the option, $T - t$. This time independence provides the following dimensional reduction of variables for the value of the put perpetuity

$$P(S, t) = P(S).$$

In this case the free-boundary $b(t) = b^*$ is constant and the problem (10.8.63 - 10.8.68) becomes

$$\frac{\sigma^2}{2}S^2 P''(S) + rSP'(S) - rP(S) = 0, \quad S > b^* \tag{10.8.90}$$

$$P(b^*) = K - b^* \tag{10.8.91}$$

$$P'(b^*) = -1, \tag{10.8.92}$$

$$P(0) = K, \tag{10.8.93}$$

$$P(S) > K - S, \qquad S > b^* \tag{10.8.94}$$

$$P(S) = K - S, \qquad 0 < S < b^*. \tag{10.8.95}$$

This free-boundary problem has two unknowns: the put value $P(S)$ and the barrier level b^*. Equation (10.8.90) can be solved as an Euler equation, looking for a solutions of the form $P(S) = S^h$. Substituting in the equation we arrive at the quadratic equation

$$h^2 - \left(1 - \frac{r}{\sigma^2/2}\right)h - \frac{r}{\sigma^2/2} = 0, \tag{10.8.96}$$

which has two roots

$$h_1 = 1$$

$$h_2 = -\frac{2r}{\sigma^2}.$$

The general solution of the equation (10.8.90) is given as the linear combination

$$P(S) = \alpha_1 S + \alpha_2 S^{-2r/\sigma^2},$$

with α_1 and α_2 undetermined constants. Since $P(S) \le K$ (a put cannot value more than the strike price), taking $S \to \infty$ in the previous formula implies that $\alpha_1 = 0$. It follows that

$$P(S) = \alpha_2 S^{-2r/\sigma^2}.$$

Substituting in (10.8.91) and (10.8.92) we get a system of two equations in the unknowns α_2 and b^*:

$$\alpha_2 (b^*)^{-2r/\sigma^2} = K - b^*$$

$$\alpha_2 \left(-\frac{2r}{\sigma^2} \right) (b^*)^{-2r/\sigma^2 - 1} = -1.$$

Dividing, we eliminate α_2 and obtain an equation for b^* with the solution

$$b^* = \frac{K}{1 + \frac{\sigma^2}{2r}}.$$

Substituting back in the first of the equations yields

$$\alpha_2 = \frac{\sigma^2}{2r} \left(\frac{K}{1 + \frac{\sigma^2}{2r}} \right)^{1 + \frac{2r}{\sigma^2}}.$$

We conclude that the solution of the problem (10.8.90 - 10.8.95) is given by

$$P(S) = \begin{cases} \frac{\sigma^2}{2r} \left(\frac{K}{1 + \frac{\sigma^2}{2r}} \right)^{1 + \frac{2r}{\sigma^2}} S^{-2r/\sigma^2}, & \text{if} \quad S \ge \frac{K}{1 + \frac{\sigma^2}{2r}} \\[4mm] K - S, & \text{if} \quad 0 < S < \frac{K}{1 + \frac{\sigma^2}{2r}}. \end{cases}$$

This provides an explicit formula for the value of a put perpetuity when the underlying stock pays no dividends.

It is worth noting that this formula agrees with the expression (10.1.2) obtained using stochastic optimization, via the following algebraic calculation:

$$(K - b^*) \left(\frac{S}{b^*} \right)^{-2r/\sigma^2} = \left(K - \frac{K}{1 + \frac{\sigma^2}{2r}} \right) \left(\frac{S(1 + \frac{\sigma^2}{2r})}{K} \right)^{-2r/\sigma^2}$$

$$= \frac{\sigma^2}{2r} \frac{1}{1 + \frac{\sigma^2}{2r}} K^{1 + \frac{2r}{\sigma^2}} \frac{1}{\left(1 + \frac{\sigma^2}{2r} \right)^{2r/\sigma^2}} S^{-2r/\sigma^2}$$

$$= \frac{\sigma^2}{2r} \left(\frac{K}{1 + \frac{\sigma^2}{2r}} \right)^{1 + \frac{2r}{\sigma^2}} S^{-2r/\sigma^2}.$$

Exercise 10.8.8 (*a*) *Which American option is more expensive: a put perpetuity or a finitely lived put? Why?*

(*b*) *Prove that* $b^* < b(0)$, *where* $b(t)$ *is the free-boundary for a finitely lived American put.*

Exercise 10.8.9 *Use the method described in this section to find the value of a put perpetuity in the case of a dividend-paying stock with the continuous dividend yield rate* δ.

In the absence of closed-form expressions for the values of American options, some analytical approximations have been worked out, and the next section deals with some of them.

10.9   MacMillan-Barone-Adesi-Whaley Formula

This section presents an approximation formula for American calls and puts that is based on a technique initiated by MacMillan [51] and developed by Barone-Adesi and Whaley [5]-[6] in 1980s. The idea is to find a differential equation for the difference between the value of an American and a European option (called *early exercise premium*) and neglecting the small terms solve the equation explicitly. For the sake of simplicity, we shall do this for the call and put separately.

10.9.1   Analytic Approximation to a Call

Consider an American call on a dividend-paying stock, S, with yield δ, whose value is denoted by $C(S,t)$. Let $c(S,t)$ denote the associated European call, and consider the difference

$$v(S,t) = C(S,t) - c(S,t). \qquad (10.9.97)$$

Since $c(S,t)$ satisfies the Black-Scholes equation for any $S > 0$ and $C(S,t)$ satisfies the Black-Scholes equation in the tradable region, it follows by linearity that $v(S,t)$ satisfies

$$\frac{\partial v}{\partial t} + (r - \delta)S\frac{\partial v}{\partial S} + \frac{1}{2}\sigma^2 S^2 \frac{\partial^2 v}{\partial S^2} = rv, \qquad 0 < S < b(t), \qquad (10.9.98)$$

where $b(t)$ denotes the free-boundary at which the call is exercised. In the attempt of solving the equation for v and finding the boundary $b(t)$, we denote the time to expiration by $\tau = T - t$ and look for a solution of the form

$$v(S,t) = h(\tau)g(S,h), \qquad (10.9.99)$$

where the function $h(\tau)$ will be chosen in a convenient way later and the function g will satisfy a second order differential equation.

Chain and product rules provide

$$\frac{\partial v}{\partial t} = \frac{\partial v}{\partial \tau}\frac{\partial \tau}{\partial t} = -h'(\tau)\left(g + h\frac{\partial g}{\partial h}\right)$$

$$\frac{\partial v}{\partial S} = h\frac{\partial g}{\partial S}$$

$$\frac{\partial^2 v}{\partial S^2} = h\frac{\partial^2 g}{\partial S^2}.$$

Substituting in (10.9.98) we obtain the equation

$$-h'(\tau)\left(g + h\frac{\partial g}{\partial h}\right) + (r - \delta)Sh\frac{\partial g}{\partial S} + \frac{1}{2}\sigma^2 S^2 h\frac{\partial^2 g}{\partial S^2} - rhg = 0. \qquad (10.9.100)$$

Dividing by $h\sigma^2/2$ the equation becomes

$$S^2\frac{\partial^2 g}{\partial S^2} + k_2 S\frac{\partial g}{\partial S} - k_1 g - k_1\frac{h'(\tau)}{rh(\tau)}\left(g + h\frac{\partial g}{\partial h}\right) = 0, \qquad (10.9.101)$$

where

$$k_1 = \frac{2r}{\sigma^2}, \qquad k_2 = \frac{2(r - \delta)}{\sigma^2}.$$

Now, it is convenient to choose

$$h(\tau) = 1 - e^{-r\tau}.$$

Then the last two terms of (10.9.101) simplify as follows

$$-k_1 g - k_1\frac{h'(\tau)}{rh(\tau)}\left(g + h\frac{\partial g}{\partial h}\right) = -k_1 g - k_1\left(\frac{1}{h} - 1\right)\left(g + h\frac{\partial g}{\partial h}\right)$$

$$= -\frac{k_1}{h}g - k_1(1 - h)\frac{\partial g}{\partial h}.$$

Therefore, equation (10.9.101) becomes

$$S^2\frac{\partial^2 g}{\partial S^2} + k_2 S\frac{\partial g}{\partial S} - \frac{k_1}{h}g - k_1(1 - h)\frac{\partial g}{\partial h} = 0. \qquad (10.9.102)$$

Until now, no approximation has been made yet. Equation (10.9.102) can be simplified if the last term is dropped. The reason behind this is the following: when τ is large, then $h(\tau) \to 1$, and hence $k_1(1 - h)\frac{\partial g}{\partial h} \to 0$. When τ is small, then $h(\tau) \to 0$ and $\dfrac{\partial g}{\partial h} \to 0$. Therefore, the equation (10.9.102) can be approximated by the following second-order ordinary differential equation

$$S^2 \frac{\partial^2 g}{\partial S^2} + k_2 S \frac{\partial g}{\partial S} - \frac{k_1}{h} g = 0. \tag{10.9.103}$$

It is known that this equation has a solution that is a linear combination of solutions of the power type, S^p. Substituting $g = S^p$ yields a quadratic equation in p

$$p^2 + (k_2 - 1)p - \frac{k_1}{h} = 0$$

with roots

$$
\begin{aligned}
p_1 &= -(k_2 - 1 - \sqrt{(k_2 - 1)^2 + 4k_1/h})/2 & (10.9.104) \\
p_2 &= -(k_2 - 1 + \sqrt{(k_2 - 1)^2 + 4k_1/h})/2. & (10.9.105)
\end{aligned}
$$

Since $p_1 p_2 = -\dfrac{k_1}{h} g < 0$, the roots have opposite signs, with $p_1 < 0 < p_2$. The solution of (10.9.103) can now be written as

$$g(S, h) = a_1 S^{p_1} + a_2 S^{p_2}, \tag{10.9.106}$$

with a_1 and a_2 independent of S, but possibly dependent on τ.

It is worth noting that until now we did not use yet the fact that we are working with calls. All equations so far are still valid for the case of puts too.

Since a call becomes worthless when the stock tends to zero, then

$$\lim_{S \to 0} g(S, h) = 0.$$

Using equation (10.9.106), it follows that $a_1 = 0$ (otherwise the limit would be infinite). Then $g(S, h) = a_2 S^{p_2}$, and then substituting back into (10.9.114) and (11.4.70) yields

$$C(S, t) = c(S, t) + a_2 h S^{p_2}, \qquad 0 < S < b(t). \tag{10.9.107}$$

The unknowns a_2 and $b(t)$ will be determined from the free-boundary conditions

$$
\begin{aligned}
C(b(t), t) &= b(t) - K \\
\frac{\partial C}{\partial S}(b(t), t) &= 1.
\end{aligned}
$$

The first one writes as

$$b(t) - K = c(b(t), t) + a_2 h b(t)^{p_2}. \tag{10.9.108}$$

Using Exercise 10.9.1 (a), the second boundary condition becomes

$$1 = e^{-\delta(T-t)} N(d_1(b)) + a_2 h p_2 b(t)^{p_2-1}, \qquad (10.9.109)$$

with $b = b(t)$ and $h = h(\tau) = h(T-t)$. Solving for a_2 yields

$$a_2 = \frac{1 - e^{-\delta(T-t)} N(d_1(b))}{h p_2 b(t)^{p_2-1}},$$

and then substituting back into (10.9.108) provides the following equation for the free-boundary

$$b - K = c(b, t) + \left(1 - e^{-\delta(T-t)} N(d_1(b))\right) \frac{b}{p_2}. \qquad (10.9.110)$$

This non-linear equation has a solution b^* that can be obtained using a Newton-type method. With this value of b^* known, we then find a_2, and substituting it into (10.9.107), after simplification, we obtain

$$C(S, t) = c(S, t) + \frac{b^*}{p_2}\left(1 - e^{-\delta(T-t)} N(d_1(b^*))\right) \left(\frac{S}{b^*}\right)^{p_2}, \qquad 0 < S < b^*.$$

To conclude, an analytical approximation for the value of an American call is given by

$$C(S, t) = \begin{cases} c(S, t) + \dfrac{b^*}{p_2}\left(1 - e^{-\delta(T-t)} N(d_1(b^*))\right) \left(\dfrac{S}{b^*}\right)^{p_2}, & \text{if } 0 < S < b^* \\[2ex] S - K, & \text{if } S \geq b^*, \end{cases}$$

$$(10.9.111)$$

where p_2 is given by (10.9.105).

Exercise 10.9.1 *Consider a dividend-paying stock with yield δ.*
(a) Show that the delta of a European call is

$$\frac{\partial c}{\partial S}(S, t) = e^{-\delta(T-t)} N(d_1(S)).$$

(b) Show that the delta of a European put is

$$\frac{\partial p}{\partial S}(S, t) = -e^{-\delta(T-t)} N(-d_1(S)),$$

where

$$d_1(S) = \frac{\ln(S/K) + (r - \delta + \frac{\sigma^2}{2})(T-t)}{\sigma\sqrt{T-t}}.$$

Exercise 10.9.2 *When $\delta \to 0$, then $b \to \infty$. Find the expression of the formula (10.9.111) in this case.*

In the next section we shall provide an approximation for the value of an American put using the same basic procedure.

10.9.2 Analytic Approximation to a Put

Consider an American put on a dividend-paying stock, S, with yield δ, whose value is denoted by $P(S,t)$. Let $p(S,t)$ denote the associated European call, and consider the difference

$$u(S,t) = P(S,t) - p(S,t), \tag{10.9.112}$$

which satisfies the Black-Scholes equation

$$\frac{\partial u}{\partial t} + (r-\delta)S\frac{\partial u}{\partial S} + \frac{1}{2}\sigma^2 S^2 \frac{\partial^2 u}{\partial S^2} = ru, \qquad S > b(t), \tag{10.9.113}$$

where $b(t)$ denotes the free-boundary at which the put is exercised. Employing the same notations as in the previous section, assume

$$u(S,t) = h(\tau)g(S,h). \tag{10.9.114}$$

Similar computations lead to the following second-order ordinary differential equation

$$S^2\frac{\partial^2 g}{\partial S^2} + k_2 S\frac{\partial g}{\partial S} - \frac{k_1}{h}g = 0, \tag{10.9.115}$$

having the solution

$$g(S,h) = a_1 S^{p_1} + a_2 S^{p_2}, \tag{10.9.116}$$

with p_1 and p_2 given by (10.9.104) and (10.9.105) and satisfying $p_1 < 0 < p_2$. Since the difference $C(S,t) - c(S,t) \to 0$ as $S \to \infty$, then $\lim_{S\to\infty} g(S,h) = 0$. Since the first term of (10.9.116) also tends to zero, in order not to violate the limit condition we need to choose $a_2 = 0$. Therefore, $g(S,h) = a_1 S^{p_1}$, and hence

$$P(S,t) = p(S,t) + a_1 h S^{p_1}. \tag{10.9.117}$$

Using Exercise 10.9.1 (b), the conditions at the boundary

$$
\begin{aligned}
P(b(t),t) &= K - b(t)\\
\frac{\partial P}{\partial S}(b(t),t) &= -1
\end{aligned}
$$

can be written as

$$
\begin{aligned}
K - b(t) &= p(b(t),t) + a_1 h b(t)^{p_1}\\
-1 &= -e^{-\delta(T-t)}N(-d_1(b)) + a_1 h p_1 b(t)^{p_1-1}.
\end{aligned}
$$

Isolating a_1 from the second condition

$$a_1 = -\frac{1 - e^{-\delta(T-t)}N(-d_1(b))}{hp_1 b(t)^{p_1-1}} \tag{10.9.118}$$

and substituting in the first equation yields the following equation for the exercise boundary $b = b(t)$

$$K - b = p(b, t) - \left(1 - e^{-\delta(T-t)}N(-d_1(b))\right)\frac{b}{p_1},$$

which can be solved iteratively. Its solution, b^{**} is substituted into (10.9.118) to provide the value of the coefficient a_1. Therefore, substituting back into (10.9.117) yields

$$P(S, t) = p(S, t) - \frac{b^{**}}{p_1}\left(1 - e^{-\delta(T-t)}N(-d_1(b^{**}))\right)\left(\frac{S}{b^{**}}\right)^{p_1}, \qquad S > b^{**}.$$

Concluding, we arrived at the following approximation formula for the value of an American put

$$P(S, t) = \begin{cases} p(S, t) - \dfrac{b^{**}}{p_1}\left(1 - e^{-\delta(T-t)}N(-d_1(b^{**}))\right)\left(\dfrac{S}{b^{**}}\right)^{p_1}, & \text{if } S > b^{**} \\ K - S, & \text{if } S \leq b^{**}, \end{cases}$$

$$(10.9.119)$$

where p_2 is given by (10.9.105).

Part V

Stochastic Volatility and Return Models

Chapter 11

Heston Model

11.1 Heston Equation

This chapter is devoted to pricing derivatives provided the underlying asset has stochastic volatility. We shall price the options on the lines of Black-Scholes model, even if in this case some complexity arise. For more details, the interested reader is referred to the book [60].

11.1.1 Dynamics of the Model

The Heston model [39] assumes that the underlying stock price, S_t, follows a geometric Brownian motion, like in the case of Black-Scholes, but with the volatility allowed to follow a CIR process, while still correlated with the stock:

$$
\begin{aligned}
dS_t &= \mu S_t \, dt + \sqrt{v_t} S_t \, dW_t & (11.1.1) \\
dv_t &= k(\theta - v_t) dt + \sigma \sqrt{v_t} \, dZ_t. & (11.1.2)
\end{aligned}
$$

This system of stochastic differential equations depends on two coupled sources of randomness with correlation coefficient $cor(dW_t, dZ_t) = \rho$. Since the jumps of a Brownian motion are normally distributed, $dW_t, dZ_t \sim \mathcal{N}(0, dt)$, the correlation can be written using expectations as follows

$$
\begin{aligned}
cor(dW_t, dZ_t) &= \frac{cov(dW_t, dZ_t)}{\sqrt{Var(dW_t)}\sqrt{Var(dZ_t)}} = \frac{\mathbb{E}(dW_t \, dZ_t) - \mathbb{E}(dW_t)\mathbb{E}(dZ_t)}{\sqrt{dt}\sqrt{dt}} \\
&= \frac{\mathbb{E}(dW_t \, dZ_t)}{dt},
\end{aligned}
$$

so the correlation condition can be written equivalently as $\mathbb{E}(dW_t \, dZ_t) = \rho dt$. It has been noticed that volatility increases when stock prices fall, and volatility drops when the stock has a large upward move, a feature of stocks called

313

the *leverage effect*. To accommodate for the leverage effect, the stock prices and volatility are assumed to have negative correlation, i.e. $-1 < \rho < 0$.

The constant μ is the drift rate of the stock. The volatility of the stock, $\sqrt{v_t}$, is the square root of the variance v_t, which is described by the square root process (11.1.2). It is worth noting that this is the same as the process (3.3.13) used to model interest rates. The positive constant θ denotes the mean reverting level of variance, while $k > 0$ denotes the rate at which the variance v_t approaches the long-run level θ. The positive constant σ is called the *volatility of the variance* (or *volvol* in the financial jargon).

Remark 11.1.1 If assume for the time being that $\sigma = 0$, the variance equation (11.1.2) becomes the ordinary differential equation $dv_t = k(\theta - v_t)dt$, which is the same as the Newton's cooling law, where v_t denotes the temperature of a cooling body and θ is the environment temperature. Under this law the temperature reverts exponentially in the long run towards θ. Adding the random term $\sigma\sqrt{v_t}dZ_t$ gives a random kick to the temperature, keeping it still positive.

Following the computations of section 3.3.4, the mean and variance of v_t are given by

$$\mathbb{E}[v_t] = (v_0 - \theta)e^{-kt} + \theta \to \theta, \qquad t \to \infty$$

$$Var(v_t) = \frac{\sigma^2}{k}e^{-2kt}(e^{kt} - 1)\left(v_0 + \frac{\theta}{2}(e^{kt} - 1)\right) \to \frac{\theta\sigma^2}{2k}, \qquad t \to \infty,$$

where v_0 is the variance at $t = 0$, which is estimated from the market data.

Proposition 11.1.2 *The stock price, S_t, can be obtained in terms of the variance, v_t, as in the following*

$$S_t = S_0 e^{\mu t - \frac{1}{2}\int_0^t v_s\, ds + \int_0^t \sqrt{v_s}\, dW_s}. \qquad (11.1.3)$$

Proof: We shall show that (11.1.3) verifies the equation (11.1.1) using Ito's lemma. Consider the process $Y_t = \mu t - \frac{1}{2}\int_0^t v_s\, ds + \int_0^t \sqrt{v_s}\, dW_s$, so $S_t = S_0 e^{Y_t}$. Then

$$dY_t = (\mu - \frac{1}{2}v_t)dt + \sqrt{v_t}\, dW_t$$

$$(dY_t)^2 = v_t dt.$$

Substituting into Ito's lemma, we have

$$
\begin{aligned}
dS_t &= S_0 d(e^{Y_t}) = S_0 e^{Y_t} dY_t + \frac{1}{2} S_0 e^{Y_t} (dY_t)^2 \\
&= S_t dY_t + \frac{1}{2} S_t v_t \, dt \\
&= S_t (\mu - \frac{1}{2} v_t) dt + \sqrt{v_t} S_t \, dW_t + \frac{1}{2} S_t v_t \, dt \\
&= \mu S_t dt + \sqrt{v_t} S_t \, dW_t,
\end{aligned}
$$

and hence S_t verifies equation (11.1.1). ∎

Remark 11.1.3 If we make $v_t = $ constant in (11.1.3), we retrieve the geometric Brownian motion formula for the stock, as in the case of the Black-Scholes model.

Exercise 11.1.4 *Consider S_t given as in (11.1.3). Compute the following:*

(a) $\mathbb{E}[S_t]$;

(b) $\mathbb{E}[S_t | \mathcal{F}_s]$, $s < t$.

Next, for pricing purposes, we shall consider the stochastic system (11.1.1)-(11.1.2) in the risk-neutral world. According to Proposition 7.1.4, we can replace μ by the risk-free interest rate r in the case of a non-dividend-paying stock

$$
\begin{aligned}
dS_t &= r S_t \, dt + \sqrt{v_t} S_t \, d\widehat{W}_t & (11.1.4) \\
dv_t &= \widehat{k}(\widehat{\theta} - v_t) dt + \sigma \sqrt{v_t} \, d\widehat{Z}_t, & (11.1.5)
\end{aligned}
$$

with $cor(d\widehat{W}_t, d\widehat{Z}_t) = \rho$. The processes $\widehat{W}_t$ and $\widehat{Z}_t$ are Brownian motions with respect to the martingale measure Q (and Brownian motions with drift in the physical measure P). The new parameters $\widehat{k}$ and $\widehat{\theta}$ depend on the previous parameters k and θ. For the sake of simplicity, we shall remove the hats for the rest of the chapter. We notice that this action will suffice for our pricing purposes. However, this would be misleading for estimation purposes, since one should estimate k and θ (and not $\widehat{k}$ and $\widehat{\theta}$) from the physical world data. With these observations, the system (11.1.4)-(11.1.5) can be written in the following simplified form:

$$
\begin{aligned}
dS_t &= r S_t \, dt + \sqrt{v_t} S_t \, dW_t & (11.1.6) \\
dv_t &= k(\theta - v_t) dt + \sigma \sqrt{v_t} \, dZ_t. & (11.1.7)
\end{aligned}
$$

It is worth noting that in the case when the stock pays dividends at the continuous rate δ the equation (11.1.6) changes to

$$dS_t = (r - \delta)S_t \, dt + \sqrt{v_t} S_t \, dW_t.$$

For the sake of simplicity we shall assume $\delta = 0$, unless specified otherwise.

11.1.2   Setting up the Heston PDE

In this section we shall obtain a partial differential equation by eliminating the random terms introduced by the Brownian motions. The procedure is similar with the hedging argument followed in Black-Scholes model, the difference consisting of two risk sources and only one tradable asset, a fact that reflects the market incompleteness. In order to offset both risk sources, we need to use in addition to the underlying stock, S_t, an extra derivative, which will introduce a volatility risk premium not specified within the model. This can be considered as the compensation received for assuming the volatility risk, $\sigma \sqrt{v_t} \, dZ_t$.

Consider the hedging portfolio

$$\Pi = V + \alpha S + \beta U,$$

which consists in one unit of derivative V, α units of underlying stock S and β units of another derivative U. For the sake of simplicity we have dropped the time index. We notice that both derivatives are written on the same underlying stock and are affected by the same stochastic volatility, i.e. $V = V(t, S, v)$ and $U = U(t, S, v)$. The values of α and β will be determined later from the riskless condition on the portfolio Π. On the other side, since Π is considered self-financing (no outside inflow of capital allowed), then

$$d\Pi = dV + \alpha dS + \beta dU.$$

Each term in the right side can be computed using Ito's lemma. Using the following table of stochastic multiplication

$\cdot$	dt	dW_t	dZ_t
dt	0	0	0
dW_t	0	dt	ρdt
dZ_t	0	ρdt	dt

we have

$$
\begin{aligned}
(dS)^2 &= vS^2 \, dt \\
(dv)^2 &= \sigma^2 v \, dt \\
dS dv &= \rho \sigma S v \, dt.
\end{aligned}
$$

Using these relations, Ito's lemma applied to $V(t, S, v)$ provides

$$
\begin{aligned}
dV &= \frac{\partial V}{\partial t}dt + \frac{\partial V}{\partial S}dS + \frac{\partial V}{\partial v}dv \\
&\quad + \frac{1}{2}\frac{\partial^2 V}{\partial S^2}(dS)^2 + \frac{1}{2}\frac{\partial^2 V}{\partial v^2}(dv)^2 + \frac{\partial^2 V}{\partial S \partial v}dS dv \\
&= \left(\frac{\partial V}{\partial t} + \frac{1}{2}vS^2\frac{\partial^2 V}{\partial S^2} + \frac{1}{2}\sigma^2 v\frac{\partial^2 V}{\partial v^2} + \rho\sigma vS\frac{\partial^2 V}{\partial S \partial v}\right)dt + \frac{\partial V}{\partial S}dS + \frac{\partial V}{\partial v}dv.
\end{aligned}
$$

A similar relation holds for the derivative $U(t, S, v)$. Therefore

$$
\begin{aligned}
d\Pi &= dV + \alpha dS + \beta dU \\
&= \left(\frac{\partial V}{\partial t} + \frac{1}{2}vS^2\frac{\partial^2 V}{\partial S^2} + \frac{1}{2}\sigma^2 v\frac{\partial^2 V}{\partial v^2} + \rho\sigma vS\frac{\partial^2 V}{\partial S \partial v}\right)dt \\
&\quad + \beta\left(\frac{\partial U}{\partial t} + \frac{1}{2}vS^2\frac{\partial^2 U}{\partial S^2} + \frac{1}{2}\sigma^2 v\frac{\partial^2 U}{\partial v^2} + \rho\sigma vS\frac{\partial^2 U}{\partial S \partial v}\right)dt \\
&\quad + \left(\frac{\partial V}{\partial S} + \alpha + \beta\frac{\partial U}{\partial S}\right)dS + \left(\frac{\partial V}{\partial v} + \beta\frac{\partial U}{\partial v}\right)dv.
\end{aligned}
$$

Since the change $d\Pi$ should not depend on the movements in stock and volatility, the coefficients of dS and dv must be equal to zero. This yields a linear system in the hedge parameters α and β, having solutions

$$
\beta = -\frac{\partial V/\partial v}{\partial U/\partial v}, \qquad \alpha = -\frac{\partial V}{\partial S} - \beta\frac{\partial U}{\partial S}.
$$

It is worthy to note that β measures the ratio of vega sensitivities for the derivatives V and U, while α is a linear combination of their deltas. With this choice of parameters α and β we have

$$
\begin{aligned}
d\Pi &= dV + \alpha dS + \beta dU \\
&= \left(\frac{\partial V}{\partial t} + \frac{1}{2}vS^2\frac{\partial^2 V}{\partial S^2} + \frac{1}{2}\sigma^2 v\frac{\partial^2 V}{\partial v^2} + \rho\sigma vS\frac{\partial^2 V}{\partial S \partial v}\right)dt \\
&\quad + \beta\left(\frac{\partial U}{\partial t} + \frac{1}{2}vS^2\frac{\partial^2 U}{\partial S^2} + \frac{1}{2}\sigma^2 v\frac{\partial^2 U}{\partial v^2} + \rho\sigma vS\frac{\partial^2 U}{\partial S \partial v}\right)dt. \quad (11.1.8)
\end{aligned}
$$

Since the risk-less condition $d\Pi = r\Pi dt$ can be written as

$$
d\Pi = r(V + \alpha S + \beta U)dt, \qquad (11.1.9)
$$

equating the coefficients of dt in equations (11.1.8) and (11.1.9), and substituting for α and β, then a simple algebraic manipulation separates the terms involving V from the ones involving U:

$$\frac{\left(\frac{\partial V}{\partial t} + \frac{1}{2}vS^2\frac{\partial^2 V}{\partial S^2} + \frac{1}{2}\sigma^2 v\frac{\partial^2 V}{\partial v^2} + \rho\sigma v S\frac{\partial^2 V}{\partial S\partial v}\right) + rS\frac{\partial V}{\partial S} - rV}{\frac{\partial V}{\partial v}}$$

$$= \frac{\left(\frac{\partial U}{\partial t} + \frac{1}{2}vS^2\frac{\partial^2 U}{\partial S^2} + \frac{1}{2}\sigma^2 v\frac{\partial^2 U}{\partial v^2} + \rho\sigma v S\frac{\partial^2 U}{\partial S\partial v}\right) + rS\frac{\partial U}{\partial S} - rU}{\frac{\partial U}{\partial v}}.$$

It follows that there is a separation function, $f(t, S, v)$, independent of derivatives V and U, which equates both terms of the previous expression. Therefore, any derivative $V = V(t, S, v)$, with the underlying asset S and stochastic volatility v satisfies the partial differential equation

$$\boxed{\frac{\partial V}{\partial t} + \frac{1}{2}vS^2\frac{\partial^2 V}{\partial S^2} + \frac{1}{2}\sigma^2 v\frac{\partial^2 V}{\partial v^2} + \rho\sigma v S\frac{\partial^2 V}{\partial S\partial v} + rS\frac{\partial V}{\partial S} - f(t, S, v)\frac{\partial V}{\partial v} - rV = 0.}$$

$$(11.1.10)$$

Following Heston's anszats, the function f is specified as

$$f(t, S, v) = -k(\theta - v) + \lambda(t, S, v),$$

where $\lambda(t, S, v)$ is the price of volatility risk. It is computationally convenient and also economically sound[1] to assume that the price of volatility risk is proportional to the variance, $\lambda(t, S, v) = \lambda v$, with λ constant. Substituting this in the previous equation yields the Heston equation:

$$\frac{\partial V}{\partial t} + \frac{1}{2}vS^2\frac{\partial^2 V}{\partial S^2} + \frac{1}{2}\sigma^2 v\frac{\partial^2 V}{\partial v^2} + \rho\sigma v S\frac{\partial^2 V}{\partial S\partial v} + rS\frac{\partial V}{\partial S}$$

$$+ (k(\theta - v) - \lambda v)\frac{\partial V}{\partial v} - rV = 0. \qquad (11.1.11)$$

It is worth noting that removing the terms containing the derivatives with respect to the volatility v recovers the usual Black-Scholes equation.

11.1.3   Simplifying the Heston PDE

The Heston equation can be simplified by taking advantage of the homogeneity with respect to S. In order to do this, we employ the substitutions $x = \ln S$ and $W(t, x, v) = V(t, S, v)$. As a consequence of the chain rule, we obtain the

---

[1]This choice of the price of volatility risk is in agreement with Breeden's consumption model.

following relations

$$S\frac{\partial V}{\partial S} = \frac{\partial W}{\partial x}$$
$$S^2\frac{\partial^2 V}{\partial S^2} = \frac{\partial^2 W}{\partial x^2} - \frac{\partial W}{\partial x}$$
$$S\frac{\partial^2 V}{\partial S \partial v} = \frac{\partial^2 W}{\partial x \partial v},$$

which substituted into (11.4.57) yields the following simplified version of the Heston equation

$$\frac{\partial W}{\partial t} + \frac{1}{2}v\frac{\partial^2 W}{\partial x^2} + \frac{1}{2}\sigma^2 v\frac{\partial^2 W}{\partial v^2} + \rho\sigma v\frac{\partial^2 W}{\partial x \partial v} + \left(r - \frac{1}{2}v\right)\frac{\partial W}{\partial x}$$
$$+ (k(\theta - v) - \lambda v)\frac{\partial W}{\partial v} - rW = 0. \qquad (11.1.12)$$

This equation has the following extra features, which will play an important role in solving the equation later:

(*i*) The coefficients are independent of the variable x;

(*ii*) The coefficients of derivatives involving v are linear functions of v.

11.2 Boundary Conditions for European Call

Inspired by the price expression of a European call in the case of Black-Scholes model, which is given by

$$C(t, S) = SN(d_1) - Ke^{-r(T-t)}N(d_2),$$

we postulate for the Heston model the following similar call formula

$$C(t, S, v) = Sf_1 - Ke^{-r(T-t)}f_2, \qquad (11.2.13)$$

where $f_i = f(t, S, v)$ are functions subject to be found out. The idea is to set up the PDEs, find the boundary conditions satisfied by f_j, and solve them.

The boundary conditions for the call are given by

$$C(T, S, v) = \max(S - K, 0) \qquad (11.2.14)$$
$$C(t, 0, v) = 0 \qquad (11.2.15)$$
$$\frac{\partial C}{\partial S}(t, \infty, v) = 1 \qquad (11.2.16)$$
$$C(t, S, \infty) = S. \qquad (11.2.17)$$

Condition (11.2.14) comes from the definition of the call payoff. The fact that the call becomes worthless when the stock drops to zero is modeled by (11.2.15). Condition (11.2.16) states that the call price increases asymptotically with the stock price. Finally, condition (11.2.17) states that for large volatility the call price becomes the stock price.

We shall see in the following that all these conditions induce boundary conditions on the functions $f_j(t, S, v)$. Since

$$C(T, S, v) = S f_1(T, S, v) - K f_2(T, S, v),$$

then (11.2.14) infers the choice

$$f_j(T, S, v) = \begin{cases} 1, & \text{if } S > K \\ 0, & \text{otherwise.} \end{cases}$$

Condition (11.2.15) implies

$$C(t, 0, v) = -Ke^{-r(T-t)} f_2(t, 0, v) = 0,$$

so $f_2(t, 0, v) = 0$. Since

$$C(t, S, \infty) = S f_1(t, S, \infty) - Ke^{-r(T-t)} f_2(t, S, \infty) = S,$$

it makes sense to equate the coefficients and choose $f_1(t, S, \infty) = 1$ and $f_2(t, S, \infty) = 0$. Differentiating with respect to S

$$\frac{\partial C}{\partial S} = f_1(t, S, v) + S \frac{\partial f_1(t, S, v)}{\partial S} - Ke^{-r(T-t)} \frac{\partial f_2(t, S, v)}{\partial S}$$

and taking the limit $S \to \infty$, condition (11.2.16) yields

$$\frac{\partial f_1}{\partial S}(t, \infty, v) = 0$$

and

$$f_1(t, \infty, v) - Ke^{-r(T-t)} \frac{\partial f_2}{\partial S}(t, \infty, v) = 1.$$

It makes sense to choose $f_1(t, \infty, v) = 1$ and $\dfrac{\partial f_2}{\partial S}(t, \infty, v) = 0$.

To conclude, we list the boundary conditions satisfied by f_j as in the following

$$f_1(T, S, v) = f_2(T, S, v) = \begin{cases} 1, & \text{if } S > K \\ 0, & \text{otherwise.} \end{cases} \tag{11.2.18}$$

$$f_1(t, \infty, v) = 1, \quad f_2(t, 0, v) = 0 \tag{11.2.19}$$

$$f_1(t, S, \infty) = 1, \quad f_2(t, S, \infty) = 0 \tag{11.2.20}$$

$$\frac{\partial f_1}{\partial S}(t, \infty, v) = 0, \quad \frac{\partial f_2}{\partial S}(t, \infty, v) = 0. \tag{11.2.21}$$

Remark 11.2.1 When we chose the boundary conditions for f_j we also had some flexibility. However, we opted for the conditions that are in agreement with the Black-Scholes model. In the case when v is constant, we have $f_1 = N(d_1)$ and $f_2 = N(d_2)$, with

$$d_1 = \frac{\ln(S/K) + (r + v/2)(T - t)}{\sqrt{v(T - t)}}, \quad d_2 = \frac{\ln(S/K) + (r - v/2)(T - t)}{\sqrt{v(T - t)}}.$$

We shall verify that condition (11.2.18) is obtained as a limit case, and leave the others to be checked by the reader as an exercise. Using

$$\lim_{t \to T-} d_j = \begin{cases} +\infty, & \text{if } S > K \\ -\infty, & \text{if } S < K, \end{cases}$$

then

$$f_j(T, S, v) = \lim_{t \to T-} N(d_j) = \begin{cases} 1, & \text{if } S > K \\ 0, & \text{otherwise.} \end{cases}$$

The following exercise deals with the limits (11.2.19) and (11.2.21) in the case of Black-Scholes model.

Exercise 11.2.2 *Prove the following:*

(a) $\lim\limits_{S \to \infty} N(d_1) = 1$;

(b) $\lim\limits_{S \to 0} N(d_2) = 0$;

(c) $\lim\limits_{S \to \infty} \dfrac{\partial N(d_j)}{\partial S} = 0$, $j = 1, 2$.

We shall introducce new notations. With respect to the variable $x = \ln S$ the functions f_j are denoted by

$$\pi_j(t, x, v) = f_j(t, e^x, v).$$

It is not hard to see that conditions (11.2.18)-(11.2.21) can be rewritten in terms of π_j as in the following, see Exercise 11.2.3:

$$\pi_1(T, x, v) = \pi_2(T, x, v) = \begin{cases} 1, & \text{if } x > \ln K \\ 0, & \text{otherwise.} \end{cases} \tag{11.2.22}$$

$$\pi_1(t, \infty, v) = 1, \quad \pi_2(t, -\infty, v) = 0 \tag{11.2.23}$$

$$\pi_1(t, x, \infty) = 1, \quad \pi_2(t, x, \infty) = 0 \tag{11.2.24}$$

$$\lim_{x \to \infty} e^{-x} \frac{\partial \pi_j}{\partial x}(t, x, v) = 0, \quad j = 1, 2. \tag{11.2.25}$$

Exercise 11.2.3 *Show that conditions (11.2.22)-(11.2.25) are equivalent to conditions (11.2.18)-(11.2.21).*

11.2.1 PDEs for π_1 and π_2

The call formula (11.2.13) can be written in terms of x as the difference of two derivatives

$$
\begin{aligned}
W(t,x,v) = C(t,e^x,v) &= e^x \pi_1(t,x,v) - K e^{-r(T-t)} \pi_2(t,x,v) \\
&= W_1(t,x,v) - K W_2(t,x,v),
\end{aligned}
$$

where $W_1(t,x,v)$ and $W_2(t,x,v)$ correspond to an asset-or-nothing and all-or-nothing contracts, respectively. Consequently, as tradable derivatives, $W_1(t,x,v)$, and $W_2(t,x,v)$ satisfy the simplified Heston equation (11.1.12).

In order to obtain a PDE for π_1, we substitute $W_1(t,x,v) = e^x \pi_1(t,x,v)$ into (11.1.12). Differentiation rules yield

$$
\begin{aligned}
\frac{\partial W_1}{\partial x} &= e^x \pi_1 + e^x \frac{\partial \pi_1}{\partial x} \\
\frac{\partial^2 W_1}{\partial x^2} &= e^x \pi_1 + 2 e^x \frac{\partial \pi_1}{\partial x} + e^x \frac{\partial^2 \pi_1}{\partial x^2} \\
\frac{\partial^2 W_1}{\partial x \partial v} &= \frac{\partial}{\partial v} \frac{\partial W_1}{\partial x} = e^x \frac{\partial \pi_1}{\partial v} + e^x \frac{\partial^2 \pi_1}{\partial x \partial v}.
\end{aligned}
$$

After substituting into (11.1.12), dividing by e^x, canceling the terms containing π_1 and arranging over the derivatives of π_1, leads to the following PDE:

$$
\frac{\partial \pi_1}{\partial t} + \frac{1}{2} v \frac{\partial^2 \pi_1}{\partial x^2} + \frac{1}{2} \sigma^2 v \frac{\partial^2 \pi_1}{\partial v^2} + \rho \sigma v \frac{\partial^2 \pi_1}{\partial x \partial v} + \left(r + \frac{1}{2} v \right) \frac{\partial \pi_1}{\partial x}
$$

$$
+ \left[\rho \sigma v + k(\theta - v) - \lambda v \right] \frac{\partial \pi_1}{\partial v} = 0. \qquad (11.2.26)
$$

Equation (11.2.26) together with the final condition

$$
\pi_1(T,x,v) = \begin{cases} 1, & \text{if } x > \ln K \\ 0, & \text{otherwise} \end{cases} \qquad (11.2.27)
$$

and previous boundary conditions determine uniquely the function π_1.

In order to obtain a PDE for π_2, we substitute $W_2(t,x,v) = e^{-r(T-t)} \pi_2(t,x,v)$ into (11.1.12). Since

$$
\begin{aligned}
\frac{\partial W_2}{\partial t} &= r e^{-r(T-t)} \pi_2 + e^{-r(T-t)} \frac{\partial \pi_2}{\partial t} \\
\frac{\partial W_2}{\partial x} &= e^{-r(T-t)} \frac{\partial \pi_2}{\partial x} \\
\frac{\partial^2 W_2}{\partial x^2} &= e^{-r(T-t)} \frac{\partial^2 \pi_2}{\partial x^2} \\
\frac{\partial W_2}{\partial v} &= e^{-r(T-t)} \frac{\partial \pi_2}{\partial v},
\end{aligned}
$$

substituting into (11.1.12), simplifying by the factor $e^{-r(T-t)}$, and canceling the terms involving π_2 yields the following equation

$$\frac{\partial \pi_2}{\partial t} + \frac{1}{2}v\frac{\partial^2 \pi_2}{\partial x^2} + \frac{1}{2}\sigma^2 v\frac{\partial^2 \pi_2}{\partial v^2} + \rho\sigma v\frac{\partial^2 \pi_2}{\partial x \partial v} + (r - \frac{1}{2}v)\frac{\partial \pi_2}{\partial x}$$
$$+ [k(\theta - v) - \lambda v]\frac{\partial \pi_2}{\partial v} = 0. \qquad (11.2.28)$$

Equation (11.2.28) together with the final condition

$$\pi_2(T, x, v) = \begin{cases} 1, & \text{if } x > \ln K \\ 0, & \text{otherwise} \end{cases} \qquad (11.2.29)$$

and previous boundary conditions determine uniquely the function π_2.

Equations (11.2.26) and (11.2.28) are very similar, the only differences being a negative sign for $\frac{1}{2}v$ in the coefficient of $\frac{\partial \pi_2}{\partial x}$ and an extra term in the latter equation in the coefficient of $\frac{\partial \pi_2}{\partial v}$. Otherwise, both equations have coefficients independent of x and linear in v, a fact that will be useful when applying a Fourier transform in x.

11.2.2 Finding π_1 and π_2

Let $\tau = T - t$ be the time to expiration. If we denote $P_j(\tau, x, v) = \pi_j(T - \tau, x, v)$, then $P_1(\tau, x, v)$ verifies the forward equation

$$-\frac{\partial P_1}{\partial \tau} + \frac{1}{2}v\frac{\partial^2 P_1}{\partial x^2} + \frac{1}{2}\sigma^2 v\frac{\partial^2 P_1}{\partial v^2} + \rho\sigma v\frac{\partial^2 P_1}{\partial x \partial v} + (r + \frac{1}{2}v)\frac{\partial P_1}{\partial x}$$
$$+ [\rho\sigma v + k(\theta - v) - \lambda v]\frac{\partial P_1}{\partial v} = 0 \qquad (11.2.30)$$

with the initial condition given by the indicator function

$$P_1(0, x, v) = \mathbb{I}_{\{x > \ln K\}}(x) = \begin{cases} 1, & \text{if } x > \ln K \\ 0, & \text{otherwise.} \end{cases} \qquad (11.2.31)$$

We shall approach the problem using the heat kernel method introduced in section 8.1. Considering the differential operator

$$A_1 = \frac{1}{2}v\frac{\partial^2}{\partial x^2} + \frac{1}{2}\sigma^2 v\frac{\partial^2}{\partial v^2} + \rho\sigma v\frac{\partial^2}{\partial x \partial v} + (r + \frac{1}{2}v)\frac{\partial}{\partial x} + [\rho\sigma v + k(\theta - v) - \lambda v]\frac{\partial}{\partial v},$$

we obtain that P_1 is the solution of the Cauchy problem

$$\frac{\partial P_1}{\partial \tau} - A_1 P_1 = 0$$
$$P_1(0, x, v) = \mathbb{I}_{\{x > \ln K\}}(x).$$

Its solution can be obtained as a convolution between the initial condition and the heat kernel, $G(\tau, x, v)$, of the operator A_1 as

$$P_1(\tau, x, v) = \int_{-\infty}^{\infty} G(\tau, x - y, v) \mathbb{I}_{\{y > \ln K\}}(y) \, dy = \int_{\ln K}^{\infty} G(\tau, x - y, v) \, dy.$$

$$(11.2.32)$$

The heat kernel, $G(\tau, x, v)$, is the solution of the problem

$$\frac{\partial G}{\partial \tau} - A_1 G = 0 \qquad (11.2.33)$$

$$G(0+, x, v) = \delta(x), \qquad (11.2.34)$$

where $\delta(x)$ is the Dirac delta function and $0+$ denotes the side limit to the right of 0. In order to find $G(\tau, x, v)$ we shall Fourier transform the previous equation with respect to x. This proves to be a feasible method since the coefficients are independent of x.

Consider the partial Fourier transform of G with respect to the x-variable

$$\widehat{G}(\tau, \omega, v) = \mathcal{F}_x G(\tau, \omega, v) = \int_{-\infty}^{\infty} e^{-i\omega x} G(\tau, x, v) \, dx.$$

Elementary properties of the Fourier transform, see Appendix A.1, yield

$$\mathcal{F}_x\left(\frac{\partial G}{\partial x}\right) = -i\omega \widehat{G}$$

$$\mathcal{F}_x\left(\frac{\partial^2 G}{\partial x^2}\right) = -\omega^2 \widehat{G}$$

$$\mathcal{F}_x\left(\frac{\partial G}{\partial \tau}\right) = \frac{\partial \widehat{G}}{\partial \tau}$$

$$\mathcal{F}_x\left(\frac{\partial G}{\partial v}\right) = \frac{\partial \widehat{G}}{\partial v}$$

$$\mathcal{F}_x\left(\frac{\partial^2 G}{\partial v^2}\right) = \frac{\partial^2 \widehat{G}}{\partial v^2}$$

$$\mathcal{F}_x\left(\frac{\partial^2 G}{\partial x \partial v}\right) = -i\omega \frac{\partial \widehat{G}}{\partial v}$$

$$\mathcal{F}_x(\delta(x)) = 1.$$

Applying the partial Fourier transform $\mathcal{F}_x$ to equation (11.2.33) and using the previous properties, after some term arrangements, we get

$$\frac{\partial \widehat{G}}{\partial \tau} - \frac{1}{2}\sigma^2 v \frac{\partial^2 \widehat{G}}{\partial v^2} + [i\omega \rho \sigma v - \rho \sigma v - k(\theta - v) + \lambda v]\frac{\partial \widehat{G}}{\partial v} + \left(\frac{1}{2}\omega^2 v + i\omega r + \frac{1}{2}i\omega v\right)\widehat{G} = 0.$$

$$(11.2.35)$$

With notations

$$
\begin{aligned}
a &= -k\theta \\
b &= \rho\sigma(i\omega - 1) + k + \lambda \\
c &= i\omega r \\
d &= \frac{1}{2}(\omega^2 + i\omega)
\end{aligned}
$$

the equation (11.2.35) becomes

$$
\frac{\partial \widehat{G}}{\partial \tau} = \frac{1}{2}\sigma^2 v \frac{\partial^2 \widehat{G}}{\partial v^2} + (a + vb)\frac{\partial \widehat{G}}{\partial v} + (c + vd)\widehat{G}, \tag{11.2.36}
$$

which is a PDE with affine coefficients in v, with $v > 0$. Therefore, it makes sense to look for a solution in the form of an exponential of an affine expression in v

$$
\widehat{G}(\tau, \omega, v) = e^{C(\tau, \omega) + vD(\tau, \omega)}. \tag{11.2.37}
$$

The Fourier transform of the initial condition (11.2.34) becomes

$$
\widehat{G}(0+, \omega, v) = 1,
$$

which implies the initial conditions

$$
C(0, \omega) = D(0, \omega) = 0. \tag{11.2.38}
$$

Next we shall find a system of ordinary differential equations satisfied by the functions $C(\tau, \omega)$ and $D(\tau, \omega)$. In order to do this we substitute (11.2.37) into (11.2.36) and use that

$$
\begin{aligned}
\frac{\partial \widehat{G}}{\partial \tau} &= \widehat{G}\left(\frac{\partial C}{\partial \tau} + v\frac{\partial D}{\partial \tau}\right) \\
\frac{\partial \widehat{G}}{\partial v} &= \widehat{G}D \\
\frac{\partial^2 \widehat{G}}{\partial v^2} &= \widehat{G}D^2.
\end{aligned}
$$

After simplifying the factor $\widehat{G}$ and collecting terms, we obtain

$$
\left(\frac{\partial C}{\partial \tau} - aD - c\right) + v\left(\frac{\partial D}{\partial v} - \frac{1}{2}\sigma^2 D^2 - bD - d\right) = 0.
$$

Since the expressions in the parenthesis are independent of the variable v, it follows that

$$
\frac{\partial C}{\partial \tau} = aD + c \tag{11.2.39}
$$

$$
\frac{\partial D}{\partial v} = \frac{1}{2}\sigma^2 D^2 + bD + d, \tag{11.2.40}
$$

which is a system of equations satisfied by the functions $C(\tau, \omega)$ and $D(\tau, \omega)$, with the initial conditions (11.2.38). The first equation is an ordinary differential equation, which can be solved by direct integration, once $D(\tau, \omega)$ is known, while the latter is a Riccati equation in $D(\tau, \omega)$. We note that ω acts as a real parameter.

11.2.3   Solving the Riccati Equation

Equation (11.2.40) is a constant coefficients Riccati equation with homogeneous initial solution, which can be solved explicitly. Taking ω as a parameter, it makes sense to consider the simpler notation $\dot{D} = \dfrac{\partial D}{\partial \tau}$. The trick used to solve this equation is to look for a solution in the form

$$D(\tau, \omega) = \alpha \frac{\dot{u}}{u},$$

where u is a function of τ and α is a constant that will be determined later. In order to do this, we shall compute $\dot{D}$ in two different ways and then equate the results. First, we differentiate in the previous formula using the quotient rule

$$\dot{D} = \alpha \frac{\ddot{u}}{u} - \alpha \left(\frac{\dot{u}}{u} \right)^2 = \alpha \frac{\ddot{u}}{u} - \frac{1}{\alpha} D^2.$$

Then, we substitute the formula for D into equation (11.2.40)

$$\dot{D} = \frac{1}{2} \sigma^2 D^2 + b\alpha \frac{\dot{u}}{u} + d.$$

Equating the coefficients of D^2 from the last two equations yields

$$\alpha = -\frac{2}{\sigma^2}.$$

Equating the remaining coefficients, we obtain

$$\alpha \frac{\ddot{u}}{u} = b\alpha \frac{\dot{u}}{u} + d,$$

which can be easily transformed into the following second order ordinary differential equation with constant coefficients

$$\ddot{u} - b\dot{u} + \frac{\sigma^2}{2} du = 0. \tag{11.2.41}$$

In order to solve this equation we consider the associated characteristic equation

$$r^2 - br + \frac{\sigma^2}{2} d = 0,$$

which has the roots

$$r_{1,2} = \frac{b \pm \sqrt{b^2 - 2\sigma^2 d}}{2}.$$

Then the solution u of equation (11.2.41) is given by the linear combination

$$u(\tau) = M_1 e^{r_1 \tau} + M_2 e^{r_2 \tau},$$

with M_1 and M_2 independent of τ. The initial condition $D(0) = 0$ implies the following relation

$$r_1 M_1 + r_2 M_2 = 0,$$

or, equivalently

$$\frac{M_2}{M_1} = -\frac{r_1}{r_2}.$$

Then the solution D takes the form

$$D = \alpha \frac{\dot{u}}{u} = -\frac{2}{\sigma^2} \frac{r_1 M_1 e^{r_1 \tau} + r_2 M_2 e^{r_2 \tau}}{M_1 e^{r_1 \tau} + M_2 e^{r_2 \tau}} = -\frac{2}{\sigma^2} \frac{r_1 + r_2 \frac{M_2}{M_1} e^{(r_2 - r_1)\tau}}{1 + \frac{M_2}{M_1} e^{(r_2 - r_1)\tau}}$$

$$= -\frac{2r_1}{\sigma^2} \frac{1 - e^{(r_2 - r_1)\tau}}{1 - \frac{r_1}{r_2} e^{(r_2 - r_1)\tau}} = -\frac{2r_1}{\sigma^2} \frac{1 - e^{\tau\sqrt{b^2 - 2\sigma^2 d}}}{1 - \frac{r_1}{r_2} e^{\tau\sqrt{b^2 - 2\sigma^2 d}}},$$

which solves the Riccati equation.

11.2.4 Finding the Function $C(\tau, \omega)$

The function $C(\tau, \omega)$ can be obtained by direct integration from the equation (11.2.39) as

$$C(\tau, \omega) = C(0, \omega) + a \int_0^\tau D(u, \omega)\, du + c\tau$$

$$= -k\theta \int_0^\tau D(u, \omega)\, du + i\omega r\tau$$

$$= \frac{2k\theta r_1}{\sigma^2} \int_0^\tau \frac{1 - e^{(r_2 - r_1)t}}{1 - \frac{r_1}{r_2} e^{(r_2 - r_1)t}}\, dt + i\omega r\tau$$

$$= \frac{2k\theta r_1}{\sigma^2} \left\{ \tau + \frac{\frac{r_1}{r_2} - 1}{(r_2 - r_1)\frac{r_1}{r_2}} \ln\left(\frac{\frac{r_1}{r_2} e^{(r_2 - r_1)\tau} - 1}{\frac{r_1}{r_2} - 1} \right) \right\} + i\omega r\tau,$$

where we used the condition $C(0, \omega) = 0$ and Exercise 11.2.4.

Exercise 11.2.4 *Use the substitution $v = e^{at}$ to show that*

$$\int_0^\tau \frac{1 - e^{at}}{1 - b e^{at}}\, dt = \tau + \frac{b - 1}{ab} \ln\left(\frac{b e^{a\tau} - 1}{b - 1} \right).$$

11.2.5 The Final Formula

At this stage we have obtained explicit expressions for both functions $C(\tau, \omega)$ and $D(\tau, \omega)$. Then (11.2.37) provides a formula for the Fourier transform $\widehat{G}(\tau, \omega, v)$. The inverse Fourier transform supplies a formula for the heat kernel $G(\tau, x, v)$ as in the following

$$
G(\tau, x, v) = \frac{1}{2\pi} \int_{-\infty}^{\infty} e^{i\omega x} \widehat{G}(\tau, \omega, v) \, d\omega = \frac{1}{2\pi} \int_{-\infty}^{\infty} e^{C(\tau, \omega) + vD(\tau, \omega) + i\omega x} \, d\omega
$$

$$
= \frac{1}{2\pi} \int_{-\infty}^{\infty} \phi_1(\omega, x, v, \tau) \, d\omega,
$$

where for the sake of simplicity we have used the notation

$$
\phi_1(\omega, x, v, \tau) = e^{C(\tau, \omega) + vD(\tau, \omega) + i\omega x}.
$$

This formula implies

$$
G(\tau, x - y, v) = \frac{1}{2\pi} \int_{-\infty}^{\infty} \phi_1(\omega, x - y, v, \tau) \, d\omega = \frac{1}{2\pi} \int_{-\infty}^{\infty} \frac{\phi_1(\omega, x, v, \tau)}{e^{i\omega y}} \, d\omega.
$$

The term $P_1(\tau, x, v)$ is obtained by substituting the heat kernel into the integral formula (11.2.32)

$$
P_1(\tau, x, v) = \int_{\ln K}^{\infty} G(\tau, x - y, v) \, dy
$$

$$
= \frac{1}{2\pi} \int_{\ln K}^{\infty} \int_{-\infty}^{\infty} \frac{\phi_1(\omega, x, v, \tau)}{e^{i\omega y}} \, d\omega \, dy.
$$

Swapping the integrals, a fact allowed by Fubbini's theorem, we obtain

$$
P_1(\tau, x, v) = \frac{1}{2\pi} \int_{-\infty}^{\infty} \int_{\ln K}^{\infty} \frac{\phi_1(\omega, x, v, \tau)}{e^{i\omega y}} \, dy \, d\omega
$$

$$
= \frac{1}{2\pi} \int_{-\infty}^{\infty} \phi_1(\omega, x, v, \tau) \int_{\ln K}^{\infty} e^{-i\omega y} \, dy \, d\omega. \qquad (11.2.42)
$$

We shall compute first $\displaystyle \int_{\ln K}^{\infty} e^{-i\omega y} \, dy$, which is an integral that is not convergent in the orthodox way, in the sense that its value is not a regular function. In fact, this is nothing but the Fourier transform of the indicator function $\mathbb{I}_{\{x > \ln K\}}$.

Proposition 11.2.5 *We have*

$$
\int_{\ln K}^{\infty} e^{-i\omega y} \, dy = \pi \delta(\omega) + \frac{e^{-i\omega \ln K}}{i\omega},
$$

where δ is the Dirac delta function.

Proof: For simplicity reasons, denote $a = \ln K$. It suffices to compute the integral starting from 0, since a substitution transforms the integral as

$$I = \int_a^\infty e^{-i\omega y}\, dy = \int_a^\infty e^{-i\omega(a+x)}\, dx = e^{-i\omega a} \int_0^\infty e^{-i\omega x}\, dx.$$

The trick for computing the integral is to introduce a convergence factor and then apply the Fundamental Theorem of Calculus

$$\int_0^\infty e^{-i\omega x}\, dx = \lim_{\epsilon \to 0+} \int_0^\infty e^{-\epsilon x} e^{-i\omega x}\, dx = \lim_{\epsilon \to 0+} \int_0^\infty e^{-(\epsilon+i\omega)x}\, dx$$

$$= \lim_{\epsilon \to 0+} \frac{e^{-(\epsilon+i\omega)x}}{-(\epsilon+i\omega)}\bigg|_0^\infty$$

$$= \lim_{\epsilon \to 0+} \frac{1}{\epsilon+i\omega}\left(1 - \lim_{x \to \infty} e^{-(\epsilon+i\omega)x}\right)$$

$$= \lim_{\epsilon \to 0+} \frac{1}{\epsilon+i\omega}, \qquad (11.2.43)$$

since the limit in the parentheses vanishes. This fact is implied by the vanishing limit in the absolute value as $\lim_{x \to \infty}\left|e^{-(\epsilon+i\omega)x}\right| = \lim_{x \to \infty} e^{-\epsilon x} = 0$.

The next step is to show the following relation formula with the Dirac delta function[2]

$$\lim_{\epsilon \to 0+} \frac{1}{\epsilon+i\omega} = \pi\delta(\omega) + \frac{1}{i\omega}. \qquad (11.2.44)$$

Amplifying by the conjugate, we obtain

$$\frac{1}{\epsilon+i\omega} = \frac{\epsilon}{\epsilon^2 + \omega^2} - i\frac{\omega}{\epsilon^2 + \omega^2}. \qquad (11.2.45)$$

It is well known that the bell-shaped graph function

$$\phi_\epsilon(\omega) = \frac{1}{\pi}\frac{\epsilon}{\epsilon^2 + \omega^2},$$

which satisfies $\phi_\epsilon(x) > 0$ and $\int_{-\infty}^\infty \phi_\epsilon(\omega)\, d\omega = 1$, tends to the Dirac delta function as ϵ approaches zero:

$$\lim_{\epsilon \to 0+} \phi_\epsilon(\omega) = \delta(\omega).$$

---

[2]In the distributions theory this is known under the name of Sohotki's formula.

This translates as the limit $\lim_{\epsilon \to 0+} \dfrac{\epsilon}{\epsilon^2 + \omega^2} = \pi\delta(\omega)$. Then taking the limit in (11.2.45) yields

$$
\begin{aligned}
\lim_{\epsilon \to 0+} \frac{1}{\epsilon + i\omega} &= \lim_{\epsilon \to 0+} \frac{\epsilon}{\epsilon^2 + \omega^2} - i \lim_{\epsilon \to 0+} \frac{\omega}{\epsilon^2 + \omega^2} \\
&= \pi\delta(\omega) - \frac{i}{\omega} = \pi\delta(\omega) + \frac{1}{i\omega},
\end{aligned}
$$

which is formula (11.2.44). Then substituting in the integral formula (11.2.43), we obtain

$$
\int_0^\infty e^{-i\omega x} \, dx = \pi\delta(\omega) + \frac{1}{i\omega}.
$$

The initial integral is obtained now by a factor multiplication

$$
\begin{aligned}
I &= \int_a^\infty e^{-i\omega y} \, dy = \pi e^{-i\omega a} \delta(\omega) + e^{-i\omega a} \frac{1}{i\omega} \\
&= \pi\delta(\omega) + e^{-i\omega a} \frac{1}{i\omega},
\end{aligned}
$$

where we used formula (11.2.46). Reverting a into $\ln K$ yields the desired formula. ∎

Since the Dirac delta function is zero everywhere but at the origin, if this is multiplied by a function $f(x)$, the product depends only on the value of f at zero. Hence, we have

$$
f(x)\delta(x) = f(0)\delta(x). \tag{11.2.46}
$$

Integrating with respect to x, and using that the area under the "spike" δ is equal to 1, we have

$$
\int_{-\infty}^\infty f(x)\delta(x) \, dx = \int_{-\infty}^\infty f(0)\delta(x) \, dx = f(0) \int_{-\infty}^\infty \delta(x) \, dx = f(0).
$$

With this brief introduction into the calculus of delta function, we can continue the computation in formula (11.2.42):

$$
\begin{aligned}
P_1(\tau, x, v) &= \frac{1}{2\pi} \int_{-\infty}^\infty \phi_1(\omega, x, v, \tau) \int_{\ln K}^\infty e^{-i\omega y} \, dy \, d\omega \\
&= \frac{1}{2\pi} \int_{-\infty}^\infty \phi_1(\omega, x, v, \tau) \left(\pi\delta(\omega) + \frac{e^{-i\omega \ln K}}{i\omega} \right) d\omega \\
&= \frac{1}{2} \int_{-\infty}^\infty \phi_1(\omega, x, v, \tau)\delta(\omega) \, d\omega + \frac{1}{2\pi} \int_{-\infty}^\infty \phi_1(\omega, x, v, \tau) \frac{e^{-i\omega \ln K}}{i\omega} \, d\omega \\
&= \frac{1}{2} \phi_1(0, x, v, \tau) + \frac{1}{2\pi} \int_{-\infty}^\infty \frac{\phi_1(\omega, x, v, \tau) e^{-i\omega \ln K}}{i\omega} \, d\omega \\
&= \frac{1}{2} + \frac{1}{\pi} \int_0^\infty \mathrm{Re}\left\{ \frac{\phi_1(\omega, x, v, \tau) e^{-i\omega \ln K}}{i\omega} \right\} d\omega,
\end{aligned}
$$

where we used that $\phi_1(0, x, v, \tau) = 1$ and Exercise 11.2.6 part (c).

Exercise 11.2.6 *Let $A(x)$, $B(x)$ be real functions and $\Phi(x) = A(x) + iB(x)$ be a complex function of a real variable. Show the following:*

(a) *If $A(-x) = A(x)$, then $\displaystyle \int_{-\infty}^{\infty} A(x) \, dx = 2 \int_0^{\infty} A(x) \, dx$;*

(b) *If $B(-x) = -B(x)$, then $\displaystyle \int_{-\infty}^{\infty} A(x) \, dx = 0$;*

(c) *If $\overline{\Phi(x)} = \Phi(-x)$, then $\displaystyle \int_{-\infty}^{\infty} \Phi(x) \, dx = 2 \int_0^{\infty} Re\{\Phi(x)\} \, dx$, where $Re\{\Phi(x)\} = A(x)$ is the real part of the function Φ.*

(d) *Show that the function $\phi(\omega) = \dfrac{\phi_1(\omega, x, v, \tau)e^{-i\omega \ln K}}{i\omega}$ satisfies $\overline{\phi(\omega)} = \phi(-\omega)$.*

Similarly, we can obtain the expression for $P_2(\tau, x, v)$ given by

$$P_2(\tau, x, v) = \frac{1}{2} + \frac{1}{\pi} \int_0^{\infty} Re\left\{\frac{\phi_2(\omega, x, v, \tau)e^{-i\omega \ln K}}{i\omega}\right\} d\omega.$$

To show this, there is no need for a new explicit computation, the differences being minor. We shall point out these differences in the following.

Denoting $P_2(\tau, x, v) = \pi_2(T - \tau, x, v)$, then $P_2(\tau, x, v)$ verifies the forward equation

$$-\frac{\partial P_1}{\partial \tau} + \frac{1}{2}v\frac{\partial^2 P_1}{\partial x^2} + \frac{1}{2}\sigma^2 v\frac{\partial^2 P_1}{\partial v^2} + \rho\sigma v\frac{\partial^2 P_1}{\partial x \partial v} + \left(r - \frac{1}{2}v\right)\frac{\partial P_1}{\partial x}$$

$$+ [k(\theta - v) - \lambda v]\frac{\partial P_1}{\partial v} = 0, \qquad (11.2.47)$$

with the initial condition

$$P_2(0, x, v) = \mathbb{I}_{\{x > \ln K\}}(x) = \begin{cases} 1, & \text{if } x > \ln K \\ 0, & \text{otherwise.} \end{cases} \qquad (11.2.48)$$

In this case, the associated differential operator is

$$A_2 = \frac{1}{2}v\frac{\partial^2}{\partial x^2} + \frac{1}{2}\sigma^2 v\frac{\partial^2}{\partial v^2} + \rho\sigma v\frac{\partial^2}{\partial x \partial v} + \left(r - \frac{1}{2}v\right)\frac{\partial}{\partial x} + [k(\theta - v) - \lambda v]\frac{\partial}{\partial v},$$

where we notice some term adjustments in the coefficients of the last two terms. Then P_2 satisfies the Cauchy problem

$$\frac{\partial P_2}{\partial \tau} - A_2 P_2 = 0$$

$$P_2(0, x, v) = \mathbb{I}_{\{x > \ln K\}}(x).$$

The PDE with affine coefficients in v satisfied by the Fourier transform of the heat kernel is

$$\frac{\partial \widehat{G}}{\partial \tau} = \frac{1}{2}\sigma^2 v \frac{\partial^2 \widehat{G}}{\partial v^2} + (a + vb)\frac{\partial \widehat{G}}{\partial v} + (c + vd)\widehat{G}, \qquad (11.2.49)$$

with the coefficients given by

$$
\begin{aligned}
a &= -k\theta \\
b &= iw\rho\sigma + k + \lambda \\
c &= iwr \\
d &= \frac{1}{2}(\omega^2 - iw).
\end{aligned}
$$

We notice that coefficients a and c remain the same as in the case of P_1, while b has a term adjustment and d is replaced by its conjugate expression. The rest of the computation is identical. The function

$$\phi_2(\omega, x, v, \tau) = e^{C(\tau, \omega) + vD(\tau, \omega) + iwx},$$

is now computed in terms of the adjusted coefficients b and d. In fact, an algebraic computation shows the following dependence between ϕ_1 and ϕ_2:

$$\phi_2(\omega - i, x, v, \tau) = e^x e^{r\tau} \phi_1(\omega, x, v, \tau). \qquad (11.2.50)$$

11.2.6 Concluding Formula

Since finding the closed form expression for a European call option in the Heston model was long journey, we shall conclude in the following the results of the previous multi-section computation by reverting the variables into S and t. We obtain that the value of a European call at time t, with underlying stock, S, satisfying the stochastic volatility model (11.1.6)–(11.1.7) is given by

$$C(t, S, v) = S f_1(t, S, v) - Ke^{-r(T-t)} f_2(t, S, v), \qquad (11.2.51)$$

where

$$f_j(t, S, v) = \frac{1}{2} + \frac{1}{\pi}\int_0^\infty Re\left\{\frac{\psi_j(\omega, S, v, t)e^{-iw\ln K}}{iw}\right\} d\omega,$$

with

$$
\begin{aligned}
\psi_j(\omega, S, v, t) &= \phi_j(\omega, x, v, \tau) = e^{C_j(\tau, \omega) + vD_j(\tau, \omega) + iwx} \\
&= e^{C_j(T-t, \omega) + vD_j(T-t, \omega) + iw\ln S}, \qquad j = 1, 2
\end{aligned}
$$

the relation between the variables being given by $x = \ln S$ and $\tau = T - t$. The dependence between ψ_1 and ψ_2 follows from formula (11.2.50)

$$\psi_2(\omega - i, S, v, t) = Se^{r(T-t)}\psi_1(\omega, S, v, t). \qquad (11.2.52)$$

11.3 Delta of a Call

We shall compute in this section the sensitivity of the call price with respect to the stock. Even if this is an intricate computation, we shall end up with the coefficient of S from formula (11.2.51); this fact is similar to the formula of the delta for a call in the Black-Scholes model. The computation provided in this section gives an interpretation of the integral factor $f_1(t, S, v)$.

Differentiating by product rule in (11.2.51) we have

$$\Delta_E(t, S, v) = \frac{\partial C(t, S, v)}{\partial S} = f_1(t, S, v) + S\frac{\partial f_1(t, S, v)}{\partial S}$$

$$-Ke^{-r(T-t)}\frac{\partial f_2(t, S, v)}{\partial S}. \tag{11.3.53}$$

Using that operators $\frac{\partial}{\partial S}$ and Re commute, see Exercise 11.3.1, the derivative of f_1 can be evaluated as

$$\frac{\partial f_1(t, S, v)}{\partial S} = \frac{\partial}{\partial S}\left\{\frac{1}{2} + \frac{1}{\pi}\int_0^\infty Re\left\{\frac{\psi_1(\omega, S, v, t)e^{-i\omega \ln K}}{i\omega}\right\}d\omega\right\}$$

$$= \frac{1}{\pi}\int_0^\infty Re\left\{\frac{\partial \psi_1(\omega, S, v, t)}{\partial S}\frac{e^{-i\omega \ln K}}{i\omega}\right\}d\omega$$

$$= \frac{1}{\pi}\int_0^\infty Re\left\{\frac{i\omega}{S}\psi_1(\omega, S, v, t)\frac{e^{-i\omega \ln K}}{i\omega}\right\}d\omega$$

$$= \frac{1}{\pi}\frac{1}{S}\int_0^\infty Re\left\{\psi_1(\omega, S, v, t)e^{-i\omega \ln K}\right\}d\omega. \tag{11.3.54}$$

Using (11.2.52) the previous integral can be expressed in terms of ψ_2 and then transformed by the substitution $\eta = \omega - i$ as follows

$$\frac{\partial f_1(t, S, v)}{\partial S} = \frac{1}{\pi}\frac{1}{S}\int_0^\infty Re\left\{\frac{1}{S}e^{-r(T-t)}\psi_2(\omega - i, S, v, t)e^{-i\omega \ln K}\right\}d\omega$$

$$= \frac{1}{\pi}\frac{1}{S^2}Ke^{-r(T-t)}\int_0^\infty Re\left\{\psi_2(\eta, S, v, t)e^{-i\omega \ln K}\right\}d\eta. \tag{11.3.55}$$

A computation similar with the one done in (11.3.54) yields

$$\frac{\partial f_2(t, S, v)}{\partial S} = \frac{1}{\pi}\frac{1}{S}\int_0^\infty Re\left\{\psi_2(\omega, S, v, t)e^{-i\omega \ln K}\right\}d\omega. \tag{11.3.56}$$

Substituting (11.3.55) and (11.3.56) into (11.3.53) and noticing the cancelation of the last two terms, we obtain

$$
\begin{aligned}
\Delta_E(t, S, v) &= f_1(t, S, v) \\
&\quad + S\frac{1}{\pi}\frac{1}{S^2}Ke^{-r(T-t)} \int_0^\infty Re\Big\{\psi_2(\eta, S, v, t)e^{-i\omega \ln K}\Big\} d\eta \\
&\quad - Ke^{-r(T-t)}\frac{1}{\pi}\frac{1}{S} \int_0^\infty Re\Big\{\psi_2(\omega, S, v, t)e^{-i\omega \ln K}\Big\} d\omega \\
&= f_1(t, S, v).
\end{aligned}
$$

Hence, $f_1(t, S, v)$ represents the sensitivity of the call with respect to changes in the underlying stock, i.e. is the delta of the call.

Exercise 11.3.1 *Let* $\Phi : \mathbb{R} \to \mathbb{C}$ *be a real variable function with complex values. Show that*

$$
Re\left(\frac{d}{dx}\Phi(x)\right) = \frac{d}{dx}Re(\Phi(x)).
$$

Exercise 11.3.2 *Consider the portfolio* $\Pi = C - \alpha S$, *which consists in a long position in a call and* α *short positions in the underlying stock. Find* α *such that* Π *is hedged against changes in the stock* S.

Exercise 11.3.3 (Vega sensitivity) *(a) Show that*

$$
\frac{\partial f_j}{\partial v} = \frac{1}{\pi} \int_0^\infty Re\left(D_j(T - t, \omega)\psi_j(\omega, S, v, t)\frac{e^{-i\omega \ln K}}{i\omega}\right) d\omega.
$$

(b) Compute $\dfrac{\partial C(t, S, v)}{\partial v}$.

Exercise 11.3.4 *Use the put-call parity to compute the delta of a put, whose underlying asset satisfies a Heston model.*

11.3.1   Interpretation of f_2

It is known that if the stock follows a geometric Brownian motion given as in the Black-Scholes model, then $N(d_2)$ has the interpretation of a probability

$$
N(d_2) = P(S_T > K),
$$

where the probability is taken in the risk-neutral world, see section 6.9.

A similar probabilistic interpretation applies to the factor f_2

$$
f_2 = P(S_T > K),
$$

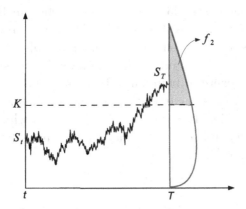

Figure 11.1: *f_2 is the fraction of times the terminal stock price S_T ends up above the strike K.*

the difference being that the stock is not a geometric Brownian motion any more, while still being an exponential process given by (11.1.3). The closed-form expression for f_2 is complex, involving improper integral evaluation. However, the previous interpretation provides a way of estimating f_2 from stock simulation as in the following. Make N simulations of the stock price S_t given by formula (11.1.3), and denote by n the number of times the ending value of the stock, S_T, ends up above the strike price K, see Fig.11.1. Then f_2 can be estimated by the empirical probability

$$f_2 = \frac{n}{N}.$$

In particular, a contract that pays \$1 at maturity, T, if the terminal stock value, S_T, exceeds the strike, K, is given by f_2.

Exercise 11.3.5 *Use the put-call parity to find the price of a European put for the Heston model.*

Exercise 11.3.6 *Find the value of a forward contract in the case of the Heston model.*

11.4 American Options

Pricing American options under stochastic volatility is similar to pricing American options in the Black-Scholes case. Likewise the constant volatility case, the price of the American call option for Heston model can be written as a sum between the corresponding European call and an early exercise premium term.

Jamshidian's approach (see [43]) can also be applied in this case, extending the equation for the option to the entire real line by introducing an indicator function and then solving it using Duhamel's principle.

However, there are also some important distinctions and complications related to the early exercise boundary. If in the constant volatility case the free-boundary $b = b(t)$ is a curve, then in the stochastic volatility case this becomes a two-dimensional free surface, $b = b(v, t)$, in which the early exercise value of the stock depends on both time horizon and volatility, see Lewis [50] and Detemple & Tian [24].

11.4.1　Calls

Let $C_A(t, S, v)$ denote the value of an American call at time t on a stock that pays dividends at a continuous rate δ. Then C_A satisfies the Heston equation (11.4.57) in the tradable region $0 < S < b(v, t)$, where $b(v, t)$ represents the free-boundary surface. Together with the boundary conditions, the problem looks like in the following:

$$\frac{\partial C_A}{\partial t} + \frac{1}{2}vS^2\frac{\partial^2 C_A}{\partial S^2} + \frac{1}{2}\sigma^2 v\frac{\partial^2 C_A}{\partial v^2} + \rho\sigma v S\frac{\partial^2 C_A}{\partial S\partial v} + (r - \delta)S\frac{\partial C_A}{\partial S}$$

$$+ (k(\theta - v) - \lambda v)\frac{\partial C_A}{\partial v} - rC_A = 0, \qquad 0 < S < b(v, t) \tag{11.4.57}$$

$$C_A(t, 0, v) = 0 \tag{11.4.58}$$

$$C_A\big(t, b(v, t), v\big) = b(v, t) - K \tag{11.4.59}$$

$$\lim_{S\to b(v,t)}\frac{\partial C_A}{\partial S}(t, S, v) = 1 \tag{11.4.60}$$

$$C_A(t, S, v) = S - K, \qquad S > b(v, t). \tag{11.4.61}$$

Condition (11.4.58) states that the value of the call vanishes when the stock price hits zero. The continuity of the solution is assured by the condition (11.4.59), while the smooth-fit condition is provided by (11.4.60). The last two conditions are imposed for avoiding arbitrage opportunities.

Differentiating implicitly with respect to v in (11.4.59)

$$\frac{\partial C_A}{\partial S}(t, b(v, t), v)\frac{\partial b}{\partial v}(v, t) + \frac{\partial C_A}{\partial v}(t, b(v, t), v) = \frac{\partial b}{\partial v}(v, t)$$

and then using (11.4.60) yields

$$\frac{\partial C_A}{\partial v}(t, b(v, t), v) = 0,$$

i.e. the call has zero vega sensitivity along the free-boundary surface.

We note that in the case of a dividend-paying stock at the continuous rate δ, the risk-free rate r in equation (11.4.58) is replaced by $r - \delta$.

The aforementioned free-boundary problem can be solved and a similar solution with the case of Black-Scholes model is obtained. The computational details can be found in Adolfsson et al. [2]:

Proposition 11.4.1 *Assume the stock pays dividends at a continuous rate δ. The price of an American call option, $C_A(t, S, v)$, in the Heston case is given by*

$$C_A(t, S, v) = C_E(t, S, v)$$
$$+ \int_0^{T-t} \int_0^\infty [\delta S e^{-\delta(T-t-\xi)} P_1^A(S, v, T - t - \xi, w, b(w, \xi))$$
$$- rK e^{-r(T-t-\xi)} P_2^A(S, v, T - t - \xi, w, b(w, \xi))] \, dw d\xi,$$

where $C_E(t, S, v)$ is the price of a European option given by (11.2.51) and

$$P_j^A(S, v, \tau - \xi, w, b(w, \xi)) = \frac{1}{2}$$
$$+ \frac{1}{\pi} \int_0^\infty Re\left(\frac{g_j(S, v, \tau - \xi, \phi, w)e^{-i\phi \ln b(w, \xi)}}{i\phi}\right) d\phi$$

for $j = 1, 2$, with

$$g_j(S, v, \tau - \xi, \phi, w) = e^{(\Theta_j - \Omega_j)(v - w + \alpha(\tau - \xi))/\sigma^2} e^{i\phi(r - \delta)(\tau - \xi)} e^{i\phi \ln S}$$
$$\times \frac{2\Omega_j e^{\Omega_j(\tau - \xi)}}{\sigma^2(e^{\Omega_j(\tau - \xi)} - 1)}\left(\frac{we^{\Omega_j(\tau - \xi)}}{v}\right)^{\frac{\alpha}{\sigma^2} - \frac{1}{2}}$$
$$\times \exp\left\{\frac{-2\Omega_j}{\sigma^2(e^{\Omega_j(\tau - \xi)} - 1)}(we^{\Omega_j(\tau - \xi)} + v)\right\}$$
$$\times I_{\frac{2\alpha}{\sigma^2} - 1}\left(\frac{4\Omega_j}{\sigma^2(e^{\Omega_j(\tau - \xi)} - 1)}(wve^{\Omega_j(\tau - \xi)})^{1/2}\right),$$

with notations

$$\Theta_1 = \Theta(i - \phi), \quad \Theta_2 = \Theta(-\phi), \quad \Theta(\phi) = k + \lambda + \rho\sigma i\phi, \quad \alpha = k\theta$$
$$\Omega_1 = \Omega(i - \phi), \quad \Omega_2 = \Omega(-\phi), \quad \Omega(\phi) = \sqrt{\Theta(\phi)^2 - \sigma^2(i\phi - \phi^2)},$$

$\tau = T - t$, and

$$I_k(z) = \sum_{n \geq 0} \frac{1}{\Gamma(n + k + 1)n!}\left(\frac{z}{2}\right)^{2n+k}$$

denoting the modified Bessel function of the first kind of order k.

The free-surface $b = b(v, \tau)$ satisfies the following two-dimensional Volterra integral equation

$$
b(v, \tau) - K = b(v, \tau)e^{-\delta\tau} f_1(b(v, \tau), v, \tau) - Ke^{-r\tau} f_2(b(v, \tau), v, \tau)
$$
$$
+ \int_0^{T-t} \int_0^\infty [\delta b(v, \tau)e^{-\delta(T-t-\xi)} P_1^A(b(v, \tau), v, T-t-\xi, w, b(w, \xi))
$$
$$
- rKe^{-r(T-t-\xi)} P_2^A(b(v, \tau), v, T-t-\xi, w, b(w, \xi))] \, dw d\xi.
$$

Solving the previous Volterra integral equation requires a great deal of computational sophistication. This is the reason why approximative solutions have been sought. One useful approach, based on the empirical analysis of Broadie et al. [14], considers that the logarithm of the boundary is approximated by a function that is linear in volatility

$$
\ln b(v, \tau) \approx b_0(\tau) + b_1(\tau)v,
$$

with the functions $b_0(\tau)$ and $b_1(\tau)$ satisfying the system

$$
C_A(b(v_0(\tau), \tau), v, \tau) = e^{b_0(\tau) + b_1(\tau)v_0(\tau)} - K
$$
$$
C_A(b(v_1(\tau), \tau), v, \tau) = e^{b_0(\tau) + b_1(\tau)v_1(\tau)} - K,
$$

where $v_0(\tau)$ and $v_1(\tau)$ are two values of the volatility that maximize the accuracy estimate. For instance, Tzavalis and Wang [65] suggest the choices

$$
v_0(\tau) = \mathbb{E}_T[v(0)], \qquad v_1(\tau) = \mathbb{E}_T[v(\tau)],
$$

where $\mathbb{E}_T[v(\tau)] = e^{-k(T-\tau)}(v(T)-\theta)+\theta$. However, when $v(T)$ is close to θ, the aforementioned values v_0 and v_1 are nearly identical, which creates difficulties in solving the previous system. To fix this problem, following Dufresne [29], the following volatility values were proposed

$$
v_0(\tau) = \mathbb{E}_T[v(\tau)] + \frac{\sigma}{|k|}\sqrt{\frac{k\theta}{2}}
$$
$$
v_1(\tau) = \mathbb{E}_T[v(\tau)] - \frac{\sigma}{|k|}\sqrt{\frac{k\theta}{2}}.
$$

For more details the reader is referred to Adolfsson et al. [2].

11.4.2   Call Perpetuity

If an American call has an infinite horizon, then its value, V, does not depend on the time to maturity, and hence, it is time independent, i.e. $V = V(S, v)$. In this case the free-boundary is one-dimensional, $b = b(v)$. Assuming that the volatility satisfies a dynamics given by the Heston model, the call perpetuity satisfies the following system:

$$\frac{1}{2}vS^2\frac{\partial^2 V}{\partial S^2} + \frac{1}{2}\sigma^2 v\frac{\partial^2 V}{\partial v^2} + \rho\sigma vS\frac{\partial^2 V}{\partial S\partial v} + rS\frac{\partial V}{\partial S}$$

$$+ (k(\theta - v) - \lambda v)\frac{\partial V}{\partial v} = rV, \qquad 0 < S < b(v) \qquad (11.4.62)$$

$$V(0, v) = 0 \qquad (11.4.63)$$

$$V(b(v), v) = b(v) - K \qquad (11.4.64)$$

$$\lim_{S \to b(v)} \frac{\partial V}{\partial S}(S, v) = 1 \qquad (11.4.65)$$

$$V(S, v) = S - K, \qquad S > b(v). \qquad (11.4.66)$$

We look for a solution with separable variables of the type $V(S, v) = S^n\phi(v)$, where n is real number. This form is suggested by the fact that the equation is of Euler-type in the variable S, with some multiplicative factors depending on v. Substituting into equation (11.4.62) and simplifying by S^n, yields the following ordinary differential equation

$$\frac{1}{2}\sigma^2 v\phi''(v) + [k\theta + (\rho\sigma n - k - \lambda)v]\phi'(v) + \left[n\left(r + \frac{n-1}{2}v\right) - r\right]\phi(v) = 0.$$

Dividing by $\frac{1}{2}\sigma^2$ we obtain a second order differential equation with linear coefficients

$$v\phi''(v) + (A + B_n v)\phi'(v) + (C_n + D_n v)\phi(v) = 0, \qquad (11.4.67)$$

where

$$A = \frac{2k\theta}{\sigma^2}, \quad B_n = \frac{2(\rho\sigma n - k - \lambda)}{\sigma^2}, \quad C_n = \frac{2r(n-1)}{\sigma^2}, \quad D_n = \frac{n(n-1)}{\sigma^2}.$$

The index n indicates the dependence on n. The equation (11.4.67) comes with two boundary conditions

$$\lim_{v \to 0} \phi(v) < \infty, \qquad \lim_{v \to \infty} \phi(v) = 1. \qquad (11.4.68)$$

The first condition states that the value of the call is finite in a world with zero volatility, while the latter is implied by the fact that for large volatility the value of the call becomes the stock price.

The solution of equation (11.4.67) can be written as a linear combination of Tricomi and Kummer confluent hypergeometric functions, see Appendix D.1, as

$$\phi(v) = e^{-\frac{1}{2}(B_n + \gamma_n)v}\left(\alpha_1 U(\xi_n, A; \gamma_n v) + \alpha_2 M(\xi_n, A; \gamma_n v)\right),$$

where

$$\gamma_n = \sqrt{B_n^2 - 4D_n}, \qquad \xi_n = \frac{AB_n - 2C_n + A\gamma_n}{2\gamma_n},$$

and α_1 and α_2 are constants. Since from the first condition of (11.4.68) we have $\lim_{v \to 0} \phi(v)$ finite, and since $U(a, b; x)$ is usually singular at $x = 0$, we need to take $\alpha_1 = 0$; therefore, the solution becomes

$$\phi(v) = e^{-\frac{1}{2}(B_n + \gamma_n)v} M(\xi_n, A; \gamma_n v).$$

The second condition of (11.4.68) implies that the values of n are such that

$$\lim_{v \to \infty} \frac{M(\xi_n, A; \gamma_n v)}{e^{\frac{1}{2}(B_n + \gamma_n)v}} = 1.$$

The above limit can be simplified using the following asymptotic formula for the Kummer function

$$M(a, b; z) \sim \Gamma(b) \left(\frac{e^z z^{a-b}}{\Gamma(a)} + \frac{(-z)^{-a}}{\Gamma(b-a)} \right), \qquad |z| \to \infty.$$

For the case when $a > 0$, just the first term matters. Using this formula with $a = \xi_n > 0$, $b = A$, and $z = \gamma_n v$, the aforementioned limit can be transformed as

$$
\begin{aligned}
1 &= \lim_{v \to \infty} \frac{M(\xi_n, A; \gamma_n v)}{e^{\frac{1}{2}(B_n + \gamma_n)v}} = \lim_{v \to \infty} \frac{\Gamma(A) e^{\gamma_n v}(\gamma_n v)^{\xi_n - A}}{\Gamma(\xi_n) e^{\frac{1}{2}(B_n + \gamma_n)v}} \\
&= \frac{\Gamma(A)}{\Gamma(\xi_n)} \lim_{v \to \infty} [e^{\frac{1}{2}(\gamma_n - B_n)v}(\gamma_n v)^{\xi_n - A}].
\end{aligned}
$$

An elementary Calculus exercise shows that a product between a non-degenerate exponential and a polynomial function is either zero or infinity (an iterative application of the L'Hospital formula provides that). Hence, the only possibility for the aforementioned limit to be finite is when both exponents vanish

$$\xi_n = A, \qquad \gamma_n = B_n. \tag{11.4.69}$$

In this case the previous limit has the value $\frac{\Gamma(A)}{\Gamma(\xi_n)} = 1$, which agrees with the requirement. The remaining problem is to find all numbers n for which conditions (11.4.69) hold, and then use the linearity of equation (11.4.62) to obtain a general solution as $V(S, v) = \sum_n C_n S^n \phi_n(v)$.

The former condition of (11.4.69) states that all ξ_n have the same value, namely

$$\xi_n = \frac{2k\theta}{\sigma^2}.$$

The latter condition of (11.4.69) can be written as $\sqrt{B_n^2 - 4D_n} = B_n$, a fact that is satisfied if and only if $D_n = 0$. Using the expression of $D_n = \frac{n(n-1)}{\sigma^2}$, it follows that $n = 0$ or $n = 1$. Hence,

$$
\begin{aligned}
V(S, v) &= C_0 S^0 \phi_0(v) + C_1 S \phi_1(v) \\
&= C_0 e^{-\frac{1}{2}(B_0 + \gamma_0)v} M(\xi_0, A; \gamma_0 v) + C_1 S e^{-\frac{1}{2}(B_1 + \gamma_1)v} M(\xi_1, A; \gamma_1 v) \\
&= C_0 e^{-\gamma_0 v} M(\xi_0, \xi_0; \gamma_0 v) + C_1 S e^{-\gamma_1 v} M(\xi_1, \xi_1; \gamma_1 v) \\
&= C_0 e^{-\gamma_0 v} e^{\gamma_0 v} + C_1 S e^{-\gamma_1 v} e^{\gamma_1 v} \\
&= C_0 + C_1 S,
\end{aligned}
$$

where we used the formula $M(a, a; z) = e^z$. The constants C_0 and C_1 are found from the boundary conditions. Condition (11.4.63) implies $C_0 = 0$ and from (11.4.65) we infer $C_1 = 1$, and hence $V(S, v) = S$. Substituting into the boundary condition (11.4.64) yields $b(v) = b(v) - K$, which is satisfied only for an infinite boundary, $b(v) = \infty$. This fact can be equivalently stated by saying that it is not optimal to exercise the call.

Exercise 11.4.2 *Using the integral representations provided in Appendix D.1, show that*

(a) $\dfrac{d}{dz} U(a, b; z) = -a U(a + 1, b + 1; z);$

(b) $\dfrac{d}{dz} M(a, b; z) = \dfrac{a}{b} M(a + 1, b + 1; z);$

(c) $M(a, b; 0) = 1;$

(d) $U(a, b; 0) = \dfrac{\Gamma(1 - b)}{\Gamma(1 - b + a)}$ *for $a > 0$ and $0 < b < 1$.*

11.4.3 Analytic Approximation to a Call

In this section we shall develop an analytic approximation for the price of a call when the underlying stock has a stochastic volatility given by the Heston model. We shall follow a method similar with the one of Baron-Adesi and Whaley used in the case of Black-Scholes model. Even if the method is not as accurate as a numerical method, we include it here for the sake of completeness and simplicity.

Consider an American call on a dividend-paying stock, S, with continuous yield rate δ, whose value is denoted by $C_A(t, S, v)$. Let $C_E(t, S, v)$ denote the associated European call, and consider the early exercise premium given by the difference

$$
V(t, S, v) = C_A(t, S, v) - C_E(t, S, v). \tag{11.4.70}
$$

Since $C_E(t, S, v)$ satisfies the Heston equation for any $S > 0$ and $C_A(t, S, v)$ satisfies the Heston equation in the tradable region, $0 < S < b(v, t)$, linearity reasons imply that $V(t, S, v)$ satisfies

$$\frac{\partial V}{\partial t} + \frac{1}{2}vS^2\frac{\partial^2 V}{\partial S^2} + \frac{1}{2}\sigma^2 v\frac{\partial^2 V}{\partial v^2} + \rho\sigma v S\frac{\partial^2 V}{\partial S\partial v} + (r - \delta)S\frac{\partial V}{\partial S}$$
$$+ (k(\theta - v) - \lambda v)\frac{\partial V}{\partial v} - rV = 0, \qquad 0 < S < b(v, t),$$

$$(11.4.71)$$

where $b = b(v, t)$ denotes the free-boundary surface at which the call is exercised. Our task is to find approximations for the early exercise premium $V(t, S, v)$ and the free-boundary $b(v, t)$.

In order to do this, we denote the time to expiration by $\tau = T - t$ and look for a solution in the following particular form

$$V(t, S, v) = h(\tau)g(S, h, v), \qquad (11.4.72)$$

with the function $h(\tau)$ conveniently chosen. Elementary rules of differentiation provide

$$\frac{\partial V}{\partial t} = -h'(\tau)\left(g + h\frac{\partial g}{\partial h}\right)$$
$$\frac{\partial V}{\partial S} = h(\tau)\frac{\partial g}{\partial S}$$
$$\frac{\partial^2 V}{\partial S^2} = h(\tau)\frac{\partial^2 g}{\partial S^2}$$
$$\frac{\partial V}{\partial v} = h(\tau)\frac{\partial g}{\partial v}$$
$$\frac{\partial^2 V}{\partial v^2} = h(\tau)\frac{\partial^2 g}{\partial v^2}$$
$$\frac{\partial^2 V}{\partial S\partial v} = h(\tau)\frac{\partial^2 g}{\partial S\partial v}.$$

Substituting into (11.4.71) and dividing by h, leads to the equation

$$-\frac{h'}{h}\left(g + h\frac{\partial g}{\partial h}\right) + (r - \delta)S\frac{\partial g}{\partial S} + \frac{1}{2}vS^2\frac{\partial^2 g}{\partial S^2} - rg$$
$$+ [k(\theta - v) - \lambda v]\frac{\partial g}{\partial v} + \frac{1}{2}\sigma^2 v\frac{\partial^2 g}{\partial v^2} + \rho\sigma v S\frac{\partial^2 g}{\partial S\partial v} = 0,$$

$$(11.4.73)$$

which holds in the tradable region $0 < S < b(v, T - \tau)$. We note that the four terms present in the first row of the previous equation are the same as in equation (10.9.100) that was addressed in the Black-Scholes model. Therefore, selecting

$$h(\tau) = 1 - e^{-r\tau},$$

and noting that $h' = r(1 - h)$, the first and the last of the terms simplify as follows

$$
\begin{aligned}
-\frac{h'}{h}\left(g + h\frac{\partial g}{\partial h}\right) - rg &= -\frac{h'}{h}g - h'\frac{\partial g}{\partial h} - rg = -\left(r + \frac{h'}{h}\right)g - h'\frac{\partial g}{\partial h} \\
&= -\left(r + \frac{r(1-h)}{h}\right)g - r(1-h)\frac{\partial g}{\partial h} \\
&= -\frac{r}{h}g - r(1-h)\frac{\partial g}{\partial h}.
\end{aligned}
$$

Assembling the pieces together, equation (11.4.73) takes the form

$$
\frac{1}{2}vS^2\frac{\partial^2 g}{\partial S^2} + (r - \delta)S\frac{\partial g}{\partial S} - \frac{r}{h}g - r(1-h)\frac{\partial g}{\partial h}
$$
$$
+ [k(\theta - v) - \lambda v]\frac{\partial g}{\partial v} + \frac{1}{2}\sigma^2 v\frac{\partial^2 g}{\partial v^2} + \rho\sigma vS\frac{\partial^2 g}{\partial S\partial v} = 0. \quad (11.4.74)
$$

Equation (11.4.74) can be simplified if some terms are neglected. Following the same argument as in the Barone-Adesi and Whaley, the term $r(1-h)\frac{\partial g}{\partial h}$ is small and hence it can be dropped (in fact, this is the only term involving a derivative in h). The resulting equation is the sum between an Euler-type equation with respect to S and a second order differential equation in v:

$$
\frac{1}{2}vS^2\frac{\partial^2 g}{\partial S^2} + (r - \delta)S\frac{\partial g}{\partial S} - \frac{r}{h}g
$$
$$
+ [k(\theta - v) - \lambda v]\frac{\partial g}{\partial v} + \frac{1}{2}\sigma^2 v\frac{\partial^2 g}{\partial v^2} + \rho\sigma vS\frac{\partial^2 g}{\partial S\partial v} = 0. \quad (11.4.75)
$$

It makes sense to look for solutions of the type

$$g(S, h, v) = S^p\phi(v), \quad (11.4.76)$$

with the function ϕ and exponent p depending on the parameter h. Substituting into (11.4.75) and canceling the factor S^p yields the following second order differential equation with linear coefficients

$$
\frac{1}{2}\sigma^2 v\phi''(v) + [k(\theta - v) - \lambda v + \rho\sigma pv]\phi'(v)
$$
$$
+ \left[\frac{1}{2}p(p - 1)v + (r - \delta)p - \frac{r}{h}\right]\phi(v) = 0. \quad (11.4.77)
$$

Since the early exercise premium vanishes when the stock reaches zero

$$\lim_{S \to 0} g(S, h) = \lim_{S \to 0} \frac{V(t, S, v)}{h(T-t)} = \lim_{S \to 0} \frac{C_A(t, S, v) - C_E(t, S, v)}{h(T-t)} = 0,$$

it follows that in formula (11.4.76) we have $p > 0$.

Assume for the time being that we are able to find all p_j and their corresponding eigenfunctions solutions ϕ_j for equation (11.4.77). Then we can write the solution as

$$V(t, S, v) = h(T-t) \sum_j C_j S^{p_j} \phi_j(v),$$

with C_j constants. We notice that p_j in formula (11.4.77) do not depend on the independent variables S and v.

We shall impose two boundary conditions with respect to v.
(i) When the volatility is constant, $v = v_0$, we expect to recover the early exercise premium for the Black-Scholes case given by (10.9.107). Since the functions $\phi_j(v)$ can be written as linear combinations of confluent hypergeometric functions (as we shall see shortly), they are continuous on $(0, \infty)$, and hence $\phi_j(v) < \infty$. Then

$$V(t, S, v_0) = h(T-t) \sum_j C_j S^{p_j} \phi_j(v_0), \qquad v_0 > 0.$$

Since the early exercise premium for the Black-Scholes case is given by (see formula (10.9.107))

$$a_2 h(T-t) S^{p_2},$$

with p_2 given by (10.9.105)

$$p_2 = -(k_2 - 1 + \sqrt{(k_2 - 1)^2 + 4k_1/h})/2,$$

by comparison it follows that the previous sum has only one non-zero coefficient, let's say $C_2 \neq 0$, and hence

$$V(t, S, v_0) = h(T-t) C_2 S^{p_2} \phi_2(v_0), \qquad v_0 > 0.$$

Considering v as a variable, this can be written as

$$V(t, S, v) = h(T-t) C_2 S^{p_2} \phi_2(v), \qquad v > 0. \tag{11.4.78}$$

For the sake of simplicity, in the formula of p_2 we consider k_1 and k_2 in terms of the average volatility, $\mathbb{E}[v_t] = (v_0 - \theta)e^{-kt} + \theta$, as

$$k_1 = \frac{2r}{(v_0 - \theta)e^{-kt} + \theta}, \qquad k_2 = \frac{2(r - \delta)}{(v_0 - \theta)e^{-kt} + \theta}.$$

(*ii*) When the volatility is large, $v \to \infty$, both prices of American and European calls becomes equal to the stock price S. Therefore, the early exercise premium in this limit case is equal to

$$\lim_{v\to\infty} V(t, S, v) = \lim_{v\to\infty} \left(C_A(t, S, v) - C_E(t, S, v) \right) = S - S = 0.$$

Comparing with (11.4.78) yields

$$\lim_{v\to\infty} \phi_2(v) = 0.$$

Equation (11.4.77) can be written as a second order ordinary differential equation with linear coefficients as

$$v\phi'' + (A + Bv)\phi' + (C + Dv)\phi = 0,$$

with

$$A = \frac{2k\theta}{\sigma^2}, \quad B = \frac{2}{\sigma^2}(\rho\sigma p - k - \lambda), \quad C = \frac{2}{\sigma^2}\left((r - \delta)p - \frac{r}{h}\right), \quad D = \frac{p(p-1)}{\sigma^2}.$$

The solution space is 2-dimensional and it is generated by the functions

$$\psi_1(v) = e^{-\frac{1}{2}(B+\gamma)v} U(\xi, A, \gamma v)$$

$$\psi_2(v) = e^{-\frac{1}{2}(B+\gamma)v} M(\xi, A, \gamma v),$$

with

$$\gamma = \sqrt{B^2 - 4D}, \qquad \xi = \frac{AB - 2C + A\gamma}{2\gamma},$$

where $U(\cdot)$ and $M(\cdot)$ denote the Tricomi and Kummer confluent hypergeometric functions, see Appendix D.1. The solution is given as a linear combination

$$\phi(v) = \alpha_1 \psi_1(v) + \alpha_2 \psi_2(v).$$

Since $\lim_{v\to 0} \phi(v)$ has to be finite, and $U(a, b, x)$ usually has a singularity at $x = 0$, we need to take $\alpha_1 = 0$, so the solution becomes

$$\phi(v) = \alpha_2 \psi_2(v) = \alpha_2 e^{-\frac{1}{2}(B+\gamma)v} M(\xi, A, \gamma v).$$

Therefore, the early exercise premium takes the form

$$V(t, S, v) = h(T - t)C_2 S^{p_2} e^{-\frac{1}{2}(B+\gamma)v} M(\xi, A, \gamma v), \qquad v > 0. \qquad (11.4.79)$$

The boundary conditions

$$C_A(t, b, v) = b - K$$
$$\frac{\partial C_A}{\partial S}(t, b, v) = 1$$

become

$$b - K = C_E(t, b, v) + h(T - t)C_2 b^{p_2} e^{-\frac{1}{2}(B+\gamma)v} M(\xi, A, \gamma v)$$
$$1 = \Delta_E(t, b, v) + p_2 h(T - t)C_2 b^{p_2 - 1} e^{-\frac{1}{2}(B+\gamma)v} M(\xi, A, \gamma v).$$

Eliminating C_2 yields the following equation for the free-boundary

$$b - K = C_E(t, b, v) + b(1 - \Delta_E(t, b, v))/p_2,$$

with the solution $b^* = b(t, v)$. Substituting into the early exercise risk premium we obtain

$$V(t, S, v) = S^{p_2} \frac{1 - \Delta_E(t, b^*, v)}{p_2 (b^*)^{p_2 - 1}} = \frac{1 - \Delta_E(t, b^*, v)}{p_2} \left(\frac{S}{b^*}\right)^{p_2} b^*.$$

Since the formula for the Delta of a call is given by $\Delta_E(t, S, v) = f_1(t, S, v)$, see section 11.3, the approximative formula for the price of an American call becomes

$$C_A(t, S, v) = \begin{cases} C_E(t, S, v) + \dfrac{1 - f_1(t, b^*, v)}{p_2} \left(\dfrac{S}{b^*}\right)^{p_2} b^*, & \text{if } 0 < S < b^* \\ \\ S - K, & \text{otherwise.} \end{cases}$$

The free-boundary $b^* = b(t, v)$ is obtained as the solution of the nonlinear equation by a Newton type method

$$b^* - K = C_E(t, b^*, v) + b^* \left(1 - f_1(t, b^*, v)\right)/p_2,$$

where C_E is given by formula (11.2.51).

Chapter 12

GARCH Model

This chapter develops a theory similar to the one done for the Heston model, with the difference that the underlying asset stochastic volatility follows this time a GARCH process.

12.1 GARCH(1,1) Differential Model

A popular model for estimating stochastic volatility is GARCH(1,1). This acronym stands for *Generalized Autoregressive Conditional Heteroskedasticity*, a concept introduced by Bollerslev [12] in mid 80's. In its differential form it assumes that the randomness of the variance process varies with the variance ν_t

$$\boxed{d\nu_t = a(\omega - \nu_t)dt + b\nu_t\, dW_t.} \tag{12.1.1}$$

Under this model the variance ν_t reverts to the long run average ω at the rate a, while the variance of ν_t approaches a non-zero constant. These statements are inferred more clearly from the following result:

Proposition 12.1.1 *(i) The expectation of the variance ν_t, which satisfies the model (12.1.1), is given by*

$$\mathbb{E}[\nu_t] = \omega + (\nu_0 - \omega)e^{-at}.$$

(ii) The second moment has the following expression

$$\mathbb{E}[\nu_t^2] = \frac{2a\omega^2}{2a - b^2} + \frac{2a\omega}{a - b^2}(\nu_0 - \omega)e^{-at}$$

$$+ \left(\nu_0^2 - \frac{2a\omega^2}{2a - b^2} - \frac{2a\omega}{a - b^2}(\nu_0 - \omega)\right)e^{-(2a - b^2)t}. \tag{12.1.2}$$

(iii) If $a > \frac{b^2}{2}$, then the long run limit of the variance $Var(\nu_t)$ is $\dfrac{\omega^2}{\frac{2a}{b^2} - 1}$.

Proof: (*i*) After integrating the stochastic differential equation (12.1.1) between 0 and t yields

$$\nu_t = \nu_0 + a \int_0^t (\omega - \nu_s)\, ds + b \int_0^t \nu_s\, dW_s.$$

Using that an Ito integral has zero mean, taking the expectation we obtain

$$\varphi(t) = \nu_0 + a \int_0^t (\omega - \varphi(s))\, ds,$$

where $\varphi(t) = \mathbb{E}[\nu_t]$. Differentiating, the previous integral equation becomes a linear differential equation

$$\varphi'(t) + a\varphi(t) = a\omega, \qquad \varphi(0) = \nu_0$$

with the solution

$$\varphi(t) \;=\; \omega + (\nu_0 - \omega)e^{-at}.$$

(*ii*) The stochastic process followed by ν_t^2 is obtained applying Ito's lemma

$$
\begin{aligned}
d(\nu_t^2) &= 2\nu_t\, d\nu_t + (\nu_t)^2 = 2\nu_t\, d\nu_t + b^2\nu_t^2 dt \\
&= \left(2a\omega\nu_t + (b^2 - 2a)\nu_t^2\right)dt + 2b\nu_t^2 dW_t.
\end{aligned}
$$

An integration provides

$$\nu_t^2 = \nu_0^2 + \int_0^t \left(2a\omega\nu_s + (b^2 - 2a)\nu_s^2\right)ds + 2b \int_0^t \nu_s^2\, dW_s.$$

Taking the expectation operator and using the properties of the Ito integral yields

$$\Phi(t) = \nu_0^2 + \int_0^t \left(2a\omega\varphi(s) + (b^2 - 2a)\Phi(s)\right)ds, \qquad (12.1.3)$$

where $\varphi(s) = \mathbb{E}[\nu_s]$ and $\Phi(s) = \mathbb{E}[\nu_s^2]$. The integral equation (12.1.3) can be solved after transforming it into the following differential equation by differentiation

$$\Phi'(t) + (2a - b^2)\Phi(t) = 2a\omega\varphi(t), \qquad \Phi(0) = \nu_0^2.$$

Multiplying by the integrating factor $e^{(2a-b^2)t}$, we obtain the exact equation

$$\left(e^{(2a-b^2)t}\Phi(t)\right)' = 2a\omega^2 e^{(2a-b^2)t} + 2a\omega(\nu_0 - \omega)e^{(a-b^2)t}.$$

Integrating and solving for $\Phi(t)$, yields

$$
\begin{aligned}
\Phi(t) \quad = \quad & \frac{2a\omega^2}{2a - b^2} + \frac{2a\omega}{a - b^2}(\nu_0 - \omega)e^{-at} \\
& + \left(\nu_0^2 - \frac{2a\omega^2}{2a - b^2} - \frac{2a\omega}{a - b^2}(\nu_0 - \omega)\right)e^{-(2a-b^2)t}.
\end{aligned}
$$

(*iii*) If $2a > b^2$, then $\displaystyle\lim_{t\to\infty} \Phi(t) = \frac{\omega^2}{1 - \frac{b^2}{2a}}$. In this case, the long run behavior of the variance is given by

$$
\lim_{t\to\infty} Var(\nu_t) = \lim_{t\to\infty}(\Phi(t) - \varphi(t)^2) = \frac{\omega^2}{1 - \frac{b^2}{2a}} - \omega^2 = \frac{\omega^2}{\frac{2a}{b^2} - 1}.
$$

∎

Remark 12.1.2 Theoretically, all the moments $\mathbb{E}[\nu_t^n]$ can be found by a similar procedure, but their expression becomes more complex as n increases.

The next result provides a closed form formula for ν_t that satisfies a GARCH(1,1) model.

Proposition 12.1.3 *The explicit formula for the variance ν_t satisfying the GARCH(1,1) model (12.1.1) is given by*

$$
\boxed{\nu_t = e^{-(\frac{1}{2}b^2 + a)t + bW_t}\left(\nu_0 + ab\int_0^t e^{(\frac{1}{2}b^2 + a)s - bW_s}\,ds\right).} \tag{12.1.4}
$$

Proof: We shall solve equation (12.1.1) by the method of integrating factors (see, for instance, section 8.8 of Calin [15]). Multiplying the equation by

$$
\rho_t = e^{\frac{1}{2}b^2 t - bW_t}
$$

yields the exact equation

$$
d(\rho_t \nu_t) = a\rho_t(b - \nu_t)dt.
$$

Using the substitution $Y_t = \rho_t \nu_t$, the equation becomes

$$
dY_t = (ab\rho_t - aY_t)dt,
$$

which can be regarded as a linear differential equation

$$
\frac{dY_t}{dt} + aY_t = ab\rho_t
$$

with initial condition $Y_0 = \nu_0$. Multiplying by e^{at} yields the exact equation

$$\left(e^{at}Y_t\right)' = ab\rho_t e^{at}.$$

Integrating, results

$$e^{at}Y_t = ab \int_0^t \rho_s e^{as}\, ds + Y_0.$$

Then solving for ρ_t yields

$$\nu_t = e^{-at}\rho_t^{-1}\left(ab \int_0^t \rho_s e^{as}\, ds + \nu_0\right),$$

which leads to (12.1.4) after replacing the value of ν_t. ∎

Remark 12.1.4 The transition density of the process ν_t is the heat kernel of the infinitesimal generator operator associated with the diffusion (12.1.1), see Appendix C.8

$$L_x = \frac{1}{2}b^2 x^2 \partial_x^2 + a(\omega - x)\partial_x, \qquad x > 0.$$

The substitution $x = e^y$ changes the operator into

$$L_y = \frac{1}{2}b^2 \partial_y^2 + \left(a\omega e^{-y} - a - \frac{1}{2}b^2\right)\partial_y,$$

whose explicit formula for the heat kernel is still difficult to obtain in closed form.

Remark 12.1.5 The discrete version of the diffusion model (12.1.1) takes the form

$$\nu_{k+1} - \nu_k = a(\omega - \nu_k)\Delta t + b\nu_k\,\Delta W_t,$$

with unit time step $\Delta t = 1$ and news term $\epsilon_k = \nu_k\,\Delta W_t$. The model becomes

$$\nu_{k+1} = a\omega + (1-a)\nu_k + b\epsilon_k, \tag{12.1.5}$$

which is a discrete GARCH(1,1) model. This represents the variance in the $(k+1)$th period, ν_{k+1}, as a weighted average of the long run average, ω, the variance for the kth period, ν_k, and the new information in the kth period, ϵ_k. Solving the recurrence (12.1.5) yields the variance as a weighted average of the previous news terms:

$$\nu_k = \omega + b\sum_{j\geq 1}(1-a)^{j-1}\epsilon_{k-j}.$$

12.2 Dynamics of the GARCH Model

The GARCH model assumes that the underlying stock price, S_t, follows a geometric Brownian motion, while the volatility, ν_t, satisfies a GARCH process:

$$dS_t = \mu S_t\, dt + \sqrt{\nu_t} S_t\, dW_t \qquad (12.2.6)$$
$$d\nu_t = a(\omega - \nu_t)dt + b\nu_t\, dZ_t, \qquad (12.2.7)$$

with $cor(dW_t, dZ_t) = \rho$. In order to allow for the leverage effect (see the discussion in section 11.1.1), the correlation coefficient ρ is negative.

The constant μ denotes the drift rate of the stock. The volatility of the stock, $\sqrt{\nu_t}$, is the square root of the variance ν_t, which is described by the GARCH(1,1) process (12.2.7). The mean reverting level of variance is denoted by the positive constant ω, while $a > 0$ denotes the rate at which the variance ν_t approaches the long-run level ω. The positive constant b is the volatility of the variance.

It is worth noting that the stock S_t in a GARCH model can be obtained explicitly in terms of parameters μ, ω, a, b, and Brownian motions W_t and Z_t. A similar application of Ito's lemma as in Proposition 11.1.2 provides

$$S_t = S_0 e^{\mu t - \frac{1}{2}\int_0^t \nu_s\, ds + \int_0^t \sqrt{\nu_s}\, dW_s},$$

where the variance ν_t is given by (12.1.4)

$$\nu_t = e^{-(\frac{1}{2}b^2 + a)t + bZ_t}\left(\nu_0 + ab\int_0^t e^{(\frac{1}{2}b^2 + a)s - bZ_s}\, ds\right).$$

For pricing purposes, we shall consider the stochastic system (12.2.6)-(12.2.7) in the risk-neutral world. This means that we can replace μ by the risk-free interest rate r in the case of a non-dividend-paying stock, and by $r - \delta$ in the case of a dividend-paying stock at the continuous rate δ. After re-scaling the parameters we may assume that in the risk-neutral world the stock satisfies the following system

$$dS_t = rS_t\, dt + \sqrt{\nu_t} S_t\, dW_t \qquad (12.2.8)$$
$$d\nu_t = a(\omega - \nu_t)dt + b\nu_t\, dZ_t, \qquad (12.2.9)$$

or

$$dS_t = (r - \delta)S_t\, dt + \sqrt{\nu_t} S_t\, dW_t \qquad (12.2.10)$$
$$d\nu_t = a(\omega - \nu_t)dt + b\nu_t\, dZ_t, \qquad (12.2.11)$$

depending whether the stock pays or does not pay dividends. For the sake of simplicity, for the rest of the section we shall assume $\delta = 0$, unless otherwise stated.

12.3 The GARCH PDE

In this section we shall follow a procedure similar with the one used in the Heston case to obtain a partial differential equation by eliminating the random terms introduced by the Brownian motions.

We start by applying Ito's lemma to a derivative $V = V(t, S, \nu)$. Using that

$$
\begin{aligned}
(dS)^2 &= \nu S^2 \, dt \\
(d\nu)^2 &= b^2 \nu^2 \, dt \\
dS d\nu &= \rho b S \nu^{3/2} \, dt,
\end{aligned}
$$

we obtain

$$
\begin{aligned}
dV &= \frac{\partial V}{\partial t} dt + \frac{\partial V}{\partial S} dS + \frac{\partial V}{\partial \nu} d\nu \\
&\quad + \frac{1}{2} \frac{\partial^2 V}{\partial S^2} (dS)^2 + \frac{1}{2} \frac{\partial^2 V}{\partial \nu^2} (d\nu)^2 + \frac{\partial^2 V}{\partial S \partial \nu} dS d\nu \\
&= \left(\frac{\partial V}{\partial t} + \frac{1}{2} \nu S^2 \frac{\partial^2 V}{\partial S^2} + \frac{1}{2} b^2 \nu^2 \frac{\partial^2 V}{\partial \nu^2} + \rho b \nu^{3/2} S \frac{\partial^2 V}{\partial S \partial \nu} \right) dt \\
&\quad + \frac{\partial V}{\partial S} dS + \frac{\partial V}{\partial \nu} d\nu.
\end{aligned}
$$

Since there are two noise sources, dW_t and dZ_t, we shall hedge the derivative $V = V(t, S, \nu)$ with the underlying stock S and another derivative $U = U(t, S, \nu)$, with the same underlying stock. We form the hedging portfolio

$$
\Pi = V + \alpha S + \beta U,
$$

which consists in one unit of derivative V, α units of underlying stock S and β units of another derivative U. The values of coefficients α and β will be determined later from the risk-less condition on the portfolio Π. Since Π is considered self-financing, we have

$$
\begin{aligned}
d\Pi &= dV + \alpha dS + \beta dU \\
&= \left(\frac{\partial V}{\partial t} + \frac{1}{2} \nu S^2 \frac{\partial^2 V}{\partial S^2} + \frac{1}{2} b^2 \nu^2 \frac{\partial^2 V}{\partial \nu^2} + \rho b \nu^{3/2} S \frac{\partial^2 V}{\partial S \partial \nu} \right) dt \\
&\quad + \beta \left(\frac{\partial U}{\partial t} + \frac{1}{2} \nu S^2 \frac{\partial^2 U}{\partial S^2} + \frac{1}{2} b^2 \nu^2 \frac{\partial^2 U}{\partial \nu^2} + \rho b \nu^{3/2} S \frac{\partial^2 U}{\partial S \partial \nu} \right) dt \\
&\quad + \left(\frac{\partial V}{\partial S} + \alpha + \beta \frac{\partial U}{\partial S} \right) dS + \left(\frac{\partial V}{\partial \nu} + \beta \frac{\partial U}{\partial \nu} \right) d\nu.
\end{aligned}
$$

In order to hedge the portfolio against movements in stock and volatility, we choose α and β such that the coefficients of dS and $d\nu$ cancel. Consequently, we obtain

$$
\beta = -\frac{\partial V/\partial \nu}{\partial U/\partial \nu}, \qquad \alpha = -\frac{\partial V}{\partial S} - \beta \frac{\partial U}{\partial S}.
$$

We notice that these expressions look the same as in the Heston case. With this parameters choice the change in the portfolio during time interval dt is

$$d\Pi = \left(\frac{\partial V}{\partial t} + \frac{1}{2}\nu S^2\frac{\partial^2 V}{\partial S^2} + \frac{1}{2}b^2\nu^2\frac{\partial^2 V}{\partial \nu^2} + \rho b\nu^{3/2}S\frac{\partial^2 V}{\partial S\partial \nu}\right)dt$$
$$+\beta\left(\frac{\partial U}{\partial t} + \frac{1}{2}\nu S^2\frac{\partial^2 U}{\partial S^2} + \frac{1}{2}b^2\nu^2\frac{\partial^2 U}{\partial \nu^2} + \rho b\nu^{3/2}S\frac{\partial^2 U}{\partial S\partial \nu}\right)dt.$$

Comparing with the risk-less condition

$$d\Pi = r\Pi dt = r(V + \alpha S + \beta U)dt, \qquad (12.3.12)$$

equating the coefficients of dt, substituting for α and β, and separating the terms involving V from the ones involving U, yields

$$\frac{\left(\frac{\partial V}{\partial t} + \frac{1}{2}\nu S^2\frac{\partial^2 V}{\partial S^2} + \frac{1}{2}b^2\nu^2\frac{\partial^2 V}{\partial \nu^2} + \rho b\nu^{3/2}S\frac{\partial^2 V}{\partial S\partial \nu}\right) + rS\frac{\partial V}{\partial S} - rV}{\frac{\partial V}{\partial \nu}}$$
$$= \frac{\left(\frac{\partial U}{\partial t} + \frac{1}{2}\nu S^2\frac{\partial^2 U}{\partial S^2} + \frac{1}{2}b^2\nu^2\frac{\partial^2 U}{\partial \nu^2} + \rho b\nu^{3/2}S\frac{\partial^2 U}{\partial S\partial \nu}\right) + rS\frac{\partial U}{\partial S} - rU}{\frac{\partial U}{\partial \nu}}.$$

Since the left term is a function of V and the right side depends on U, then both terms are equal to the same separation function $f(t, S, \nu)$, independent of derivatives V and U. Therefore, for any derivative $V(t, S, \nu)$ we have

$$\frac{\left(\frac{\partial V}{\partial t} + \frac{1}{2}\nu S^2\frac{\partial^2 V}{\partial S^2} + \frac{1}{2}b^2\nu^2\frac{\partial^2 V}{\partial \nu^2} + \rho b\nu^{3/2}S\frac{\partial^2 V}{\partial S\partial \nu}\right) + rS\frac{\partial V}{\partial S} - rV}{\frac{\partial V}{\partial \nu}} = f(t, S, \nu),$$

which can be brought to the following Black-Scholes type equation

$$\boxed{\frac{\partial V}{\partial t} + \frac{1}{2}\nu S^2\frac{\partial^2 V}{\partial S^2} + \frac{1}{2}b^2\nu^2\frac{\partial^2 V}{\partial \nu^2} + \rho b\nu^{3/2}S\frac{\partial^2 V}{\partial S\partial \nu} + rS\frac{\partial V}{\partial S} - f(t, S, \nu)\frac{\partial V}{\partial \nu} = rV.}$$

$$(12.3.13)$$

Equation (12.3.13) describes the dynamics of a derivative when the underlying stock satisfies the GARCH model (12.2.8)–(12.2.9).

12.4   Simplifying the PDE

Despite of some similarities between equation (12.3.13) and the Heston equation (11.1.10), there are a couple of important differences, which makes this model more complicated and harder to be tracked analytically:

- The coefficient of the third term, $\dfrac{\partial^2 V}{\partial \nu^2}$, in the GARCH case is quadratic in ν, while in the Heston model is just linear in ν.

- The coefficient of the mixed derivative, $\dfrac{\partial^2 V}{\partial S \partial \nu}$, in the GARCH case contains a fractional power, $\nu^{3/2}$, while in the Heston model is just linear in ν.

One way to go around the inconvenience raised by the second bullet is to assume $\rho = 0$, i.e. we require that random noises dW_t and dZ_t are not correlated. However, the statement of the first bullet turns out to be favorable, since makes the third term of the equation homogeneous in ν. Making the further assumption $f(t, S, \nu) = -\lambda \nu$, the equation becomes homogeneous in each of the variables S and ν:

$$\frac{\partial V}{\partial t} + \frac{1}{2} \nu S^2 \frac{\partial^2 V}{\partial S^2} + \frac{1}{2} b^2 \nu^2 \frac{\partial^2 V}{\partial \nu^2} + rS \frac{\partial V}{\partial S} + \lambda \nu \frac{\partial V}{\partial \nu} = rV. \qquad (12.4.14)$$

This homogeneity can be exploited using the change of variables $x = \ln S$ and $y = \ln \nu$. Denoting $W(t, x, y) = V(t, S, \nu)$, we have

$$S \frac{\partial V}{\partial S} = \frac{\partial W}{\partial x}$$

$$S^2 \frac{\partial^2 V}{\partial S^2} = \frac{\partial^2 W}{\partial x^2} - \frac{\partial W}{\partial x}$$

$$\nu \frac{\partial V}{\partial \nu} = \frac{\partial W}{\partial y}$$

$$\nu^2 \frac{\partial^2 V}{\partial \nu^2} = \frac{\partial^2 W}{\partial y^2} - \frac{\partial W}{\partial y}.$$

Consequently, the equation (12.4.14) transforms into a second order PDE

$$\frac{\partial W}{\partial t} + \frac{e^y}{2} \frac{\partial^2 W}{\partial x^2} + \frac{b^2}{2} \frac{\partial^2 W}{\partial y^2} + \left(r - \frac{e^y}{2}\right) \frac{\partial W}{\partial x} + \left(\lambda - \frac{b^2}{2}\right) \frac{\partial W}{\partial y} = rW. \quad (12.4.15)$$

12.5 Solving the Equation

Since the coefficients of equation (12.4.15) are independent of x, we shall apply a partial Fourier transform with respect to x, see Appendix A.1. Denoting

$$\widehat{W}(\omega) = \mathcal{F}_x W(\omega) = \int_{-\infty}^{\infty} e^{-i\omega x} W(t, x, y) \, dx,$$

and using that

$$\mathcal{F}_x \left(\frac{\partial W}{\partial x}\right) = -i\omega \widehat{W}, \qquad \mathcal{F}_x \left(\frac{\partial^2 W}{\partial x^2}\right) = -\omega^2 \widehat{W},$$

then equation (12.4.15) becomes

$$\frac{\partial \widehat{W}}{\partial t} + \alpha \frac{\partial^2 \widehat{W}}{\partial y^2} + \beta \frac{\partial \widehat{W}}{\partial y} = f(e^y)\widehat{W},$$

where $\alpha = \frac{1}{2}b^2$, $\beta = \lambda - \frac{1}{2}b^2$, and

$$f(u) = r + \frac{1}{2}\omega^2 u + i\omega(r - u/2).$$

The previous PDE can be further simplified if substitute

$$U(t, \omega, y) = e^{\gamma y}\widehat{W}(t, \omega, y), \qquad \gamma = \frac{\beta}{2\alpha}.$$

The resulting equation satisfied by U is

$$\frac{\partial U}{\partial t} + \alpha \frac{\partial^2 U}{\partial y^2} = \left(f(e^y) + \frac{\beta^2}{4\alpha}\right)U.$$

We can get rid of the coefficient α by substituting $s = y/\sqrt{\alpha}$. Then $R(t, \omega, s) = U(t, \omega, y)$ satisfies the following Schrödinger equation

$$\frac{\partial R}{\partial t} + \frac{\partial^2 R}{\partial s^2} = \Phi(s)R, \tag{12.5.16}$$

with exponential potential

$$\Phi(s) = \eta e^{\kappa s} + C,$$

where

$$\kappa = b/\sqrt{2}, \quad \eta = \frac{1}{2}\omega^2 - \frac{1}{2}i\omega, \quad C = r + i\omega r + \frac{1}{2}\left(\frac{\lambda}{b} - \frac{1}{2}\right)^2.$$

Looking for a solution with separable variables, $R(t, \omega, s) = \alpha(t)\psi(s)$, substituting in the equation (12.5.16) and dividing by $\alpha(t)\psi(s)$ yields

$$\frac{\alpha'(t)}{\alpha(t)} + \frac{\psi''(s)}{\psi(s)} = \Phi(s).$$

Therefore, there is a separation constant, E, such that

$$\alpha'(t) = -E\alpha(t)$$
$$\psi''(s) = E\Phi(s)\psi(s).$$

The second equation, together with boundary conditions $\lim_{s \to \pm\infty} \psi(s) = \psi_{\pm\infty}$, becomes a Sturm-Liouville boundary problem. Denote by E_n its eigenvalues and by ψ_n its normalized eigenfunctions, i.e. $\int_{\mathbb{R}} |\phi_n(s)|^2 \, ds = 1$. Then

$$R_n(t, s) = e^{-E_n t}\psi_n(s)$$

are solutions for the equation (12.5.16). Notice that for the sake of simplicity, the parameter ω has been dropped, and this policy will be applied as long as there are no confusions. We also make the extra assumption that $\{\psi_n\}$ form an orthonormal system, i.e.

$$\langle \psi_n, \psi_k \rangle = \int_{\mathbb{R}} \psi_n(s)\psi_k(s)\,ds = \delta_{nk}.$$

The system $\{\psi_n\}$ can always be orthonormalized by applying the well-known Gram-Schmidt process from linear algebra. Then a solution of equation (12.5.16) can be written as an infinite linear combination of R_n

$$R(t,s) = \sum_{n=0}^{\infty} c_n R_n(t,s) = \sum_{n=0}^{\infty} c_n e^{-E_n t}\psi_n(s). \tag{12.5.17}$$

The coefficients c_k are obtained by projecting the final condition $R_T(s) = R(T,s)$ onto the direction of ψ_k

$$\langle R_T(s), \psi_k \rangle = \sum_{n=0}^{\infty} c_n e^{-E_n T}\langle \psi_n, \psi_k \rangle = c_k e^{-E_k T},$$

and hence

$$c_k = e^{E_k T} \int_{-\infty}^{\infty} R_T(u)\psi_k(u)\,du.$$

Substituting back into (12.5.17) we can write the solution in the following integral form

$$\begin{aligned}
R(t,s) &= \sum_{n=0}^{\infty} e^{E_n(T-t)}\psi_n(s) \int_{-\infty}^{\infty} R_T(u)\psi_k(u)\,du \\
&= \int_{-\infty}^{\infty} \sum_{n=0}^{\infty} e^{E_n(T-t)}\psi_n(s)\psi_n(u)R_T(u)\,du \\
&= \int_{-\infty}^{\infty} H(u,s)R_T(u)\,du, \tag{12.5.18}
\end{aligned}$$

where $H(u,s) = \sum_{n=0}^{\infty} e^{E_n(T-t)}\psi_n(s)\psi_n(u)$ is the fundamental solution of the aforementioned Schrödinger operator.

It makes sense to emphasize that boundary values $\psi_{\pm\infty}$ determine the values of the eigenvalues sequence E_n as well as the expression of the eigenfunctions ψ_n. The values $\psi_{\pm\infty}$ depend on the derivative type. Furthermore, the final condition R_T can be expressed as

$$R_T(s) = U(T,\omega,y) = e^{\gamma y}\widehat{W}(T,\omega,y) = e^{\gamma y} \int_{-\infty}^{\infty} e^{-i\omega x}V_T(e^x,e^y)\,dx,$$

where $V_T(S, \nu) = V(T, S, \nu) = V(T, e^x, e^y)$ is the final condition for equation (12.4.14).

Next, we shall find a closed form expression for $V(t, S, \nu)$. Reverting variables, we have

$$\widehat{W}(t, \omega, y) = e^{-\gamma y} U(t, \omega, y) = e^{-\gamma y} R(t, \omega, s)$$

Applying the inverse Fourier transform, $\mathcal{F}_x^{-1}$, we obtain

$$
\begin{aligned}
W(t, x, y) &= \mathcal{F}_x^{-1}\left(e^{-\gamma y} R(t, \omega, s)\right) \\
&= \frac{1}{2\pi} \int_{-\infty}^{\infty} e^{i\omega x} e^{-\gamma y} R(t, \omega, s) \, d\omega.
\end{aligned}
$$

Therefore, using (12.5.18), we have

$$
\begin{aligned}
V(t, S, \nu) &= W(t, \ln S, \ln \nu) = \frac{\nu^{-\gamma}}{2\pi} \int_{-\infty}^{\infty} e^{i\omega x} R(t, \omega, s) \, d\omega \\
&= \frac{\nu^{-\gamma}}{2\pi} \iint_{\mathbb{R}^2} e^{i\omega x} H(u, s) R_T(\omega, u) \, d\omega du,
\end{aligned}
$$

with $s = \dfrac{\sqrt{2}}{b} \ln \nu$. This provides the solution of the simplified GARCH equation (12.4.14).

The case when the correlation coefficient $\rho \neq 0$ is more intricate and we shall not attempt it here.

Chapter 13

AR(1) Model

We include the AR(1) stochastic volatility model for pedagogical reasons, so that the reader can notice the similarities between the Heston, GARCH and AR(1) models as well as their main differences. The AR(1) model can be solved in a similar way as the Heston model.

13.1 AR(1)-Differential Model

A diffusion X_t satisfying the stochastic differential equation

$$dX_t = \alpha X_t dt + \sigma dW_t, \tag{13.1.1}$$

with $\alpha < 0$, constant, is called an Ornstein-Uhlenbeck process. This process comes up in diverse problems in physics, economics, and finance[1], but here we are interested in X_t from the statistical point of view. We shall make the point that X_t is a *differential auto-regressive AR(1) model*, which can be used to model volatility to some extent.

It is known that a (discrete) AR(1) model X_n satisfies the recurrence

$$X_{n+1} = \phi X_n + \sigma \epsilon_t,$$

where ϕ is a constant less than 1, and $\sigma \epsilon_t$ is a white noise controlled by the parameter σ. We note that the white noise term has the magnitude independent of the size of X_n (contrary to the case of GARCH model, where the noise is proportional to X_n). The equation can be written equivalently as

$$X_{n+1} - X_n = \alpha X_n + \sigma \epsilon_t,$$

---

[1]In section 3.2 it came out as a solution of the Langevin equation in modeling interest rates.

with $\alpha = \phi - 1 < 0$. In this case the time step is $\Delta t = 1$. If we consider the differential analog of this equation, i.e. the case when the time step $\Delta t \to 0$, we obtain

$$dX_t = \alpha X_t + \sigma \epsilon_t.$$

Replacing the white noise ϵ_t by increments of a Brownian motion, dW_t, we obtain the equation (13.1.1).

We have seen in section 3.2 (see equation (3.2.4)) that the solution X_t of the Langevin equation (13.1.1) is given by,

$$X_t = X_0 e^{\alpha t} + \sigma \int_0^t e^{\alpha(t-u)} dW_u. \tag{13.1.2}$$

Then X_t is the sum of two parts: a deterministic one, $X_0 e^{\alpha t}$, which is monotonically decreasing in time, and a stochastic part, given by the Ito integral $\sigma \int_0^t e^{\alpha(t-u)} dW_u$, which is normally distributed with the mean 0 and variance $\frac{\sigma^2}{2\alpha}(e^{2\alpha t} - 1)$. Consequently, X_t is normally distributed as

$$X_t \sim \mathcal{N}\left(X_0 e^{\alpha t}, \frac{\sigma^2}{2\alpha}(e^{2\alpha t} - 1)\right).$$

Estimates for parameters α and σ In order to use the model (13.1.1) for any practical purposes, we need to estimate its parameters first. Let

$$\widehat{\mu}(t) \sim \mathbb{E}[X_t], \qquad \widehat{\sigma^2}(t) \sim Var(X_t)$$

be estimates for the mean and variance. The solutions of the system

$$X_0 e^{\alpha t} = \widehat{\mu}(t)$$
$$\frac{\sigma^2}{2\alpha}(e^{2\alpha t} - 1) = \widehat{\sigma^2}(t)$$

are given by

$$\widehat{\alpha} = \frac{1}{t} \ln \frac{\widehat{\mu}(t)}{X_0}$$

$$\widehat{\sigma^2} = \frac{2\widehat{\alpha}\,\widehat{\sigma^2}(t)}{\dfrac{\widehat{\mu}(t)^2}{X_0^2} - 1}.$$

Then the values $\widehat{\alpha}$ and $\widehat{\sigma^2}$ are estimates for parameters α and σ^2 of the model. We make the remark that this model is appropriate if $\ln \widehat{\mu}(t)$ is almost linear in the variable t.

Prediction with $AR(1)$ The $AR(1)$-differential model can be used to make predictions of the variable X_t in the following way. Let

$$\widetilde{X}_{t+s} = \mathbb{E}[X_{t+s}|\mathcal{F}_t]$$

be the predicted value of the variable at time $t+s$, given the market information $\mathcal{F}_t$ (which includes knowledge about all values X_u, $u \le t$.) Using (13.1.2), the linearity of the expectation operator, and the martingale property of the Ito integral, we have

$$
\begin{aligned}
\widetilde{X}_{t+s} &= \mathbb{E}[X_{t+s}|\mathcal{F}_t] = \mathbb{E}\left[X_0 e^{\alpha(t+s)} + \sigma e^{\alpha(t+s)} \int_0^{t+s} e^{-\alpha u} dW_u \Big| \mathcal{F}_t\right] \\
&= X_0 e^{\alpha(t+s)} + \sigma e^{\alpha(t+s)} \mathbb{E}\left[\int_0^{t+s} e^{-\alpha u} dW_u \Big| \mathcal{F}_t\right] \\
&= X_0 e^{\alpha(t+s)} + \sigma e^{\alpha(t+s)} \int_0^t e^{-\alpha u} dW_u \\
&= X_0 e^{\alpha(t+s)} + e^{\alpha(t+s)}\left(e^{-\alpha t} X_t - X_0\right) \\
&= e^{\alpha s} X_t.
\end{aligned}
$$

Since $\alpha < 0$, the prediction $\widetilde{X}_{t+s}$ will tend exponentially to zero.

Exercise 13.1.1 *Consider the model*

$$dX_t = (\omega + \alpha X_t)dt + \sigma dW_t,$$

with ω and α parameters.

(a) Solve the equation;
(b) Find the mean and variance of X_t;
(c) Find the prediction $\widetilde{X}_{t+s} = \mathbb{E}[X_{t+s}|\mathcal{F}_t]$.
(d) Compute the limit $\lim_{s\to\infty} \widetilde{X}_{t+s}$, and provide an interpretation for the parameter ω.

13.2   The PDE

We shall assume that the underlying stock price, S_t, follows a geometric Brownian motion, like in the case of Black-Scholes, while the volatility satisfies an $AR(1)$-differential process:

$$
\begin{aligned}
dS_t &= \mu S_t\, dt + \sqrt{v_t} S_t\, dW_t & (13.2.3) \\
dv_t &= \alpha v_t dt + \sigma\, dZ_t, & (13.2.4)
\end{aligned}
$$

with $cor(dW_t, dZ_t) = \rho < 0$ and $\alpha < 0$. Similar with the case of Heston and GARCH(1,1) stochastic volatility models, this system of stochastic differential equations also depends on two correlated sources of randomness. As we have seen before, if the stock does not pay dividends, we consider for pricing purposes $\mu = r$. In the risk-neutral world the above system becomes

$$dS_t = rS_t\, dt + \sqrt{v_t}S_t\, dW_t \tag{13.2.5}$$

$$dv_t = \alpha v_t dt + \sigma\, dZ_t. \tag{13.2.6}$$

In the following we shall find the associated partial differential equation by the procedure introduced in the previous chapters. We should be brief, the reader being expected to fill in the details, see Exercise 13.2.1.

If $V = V(t, S, v)$ is a derivative, using

$$(dv)^2 = \sigma^2 dt, \qquad dSdv = \rho\sigma S\sqrt{v}dt,$$

an application of Ito's lemma yields

$$dV = \left(\frac{\partial V}{\partial t} + \frac{1}{2}vS^2\frac{\partial^2 V}{\partial S^2} + \frac{1}{2}\sigma^2\frac{\partial^2 V}{\partial v^2} + \rho\sigma S\sqrt{v}\frac{\partial V}{\partial^2 S\partial v}\right)dt + \frac{\partial V}{\partial S}dS + \frac{\partial V}{\partial v}dv.$$

Using a standard hedging procedure against changes in S and v of the portfolio $\Pi = V + aS + bU$, where U is another derivative with the same underlying assets, S, we arrive to the following partial differential equation

$$\frac{\partial V}{\partial t} + \frac{1}{2}vS^2\frac{\partial^2 V}{\partial S^2} + \frac{1}{2}\sigma^2\frac{\partial^2 V}{\partial v^2} + \rho\sigma S\sqrt{v}\frac{\partial V}{\partial^2 S\partial v} + rS\frac{\partial V}{\partial S} - f(t, S, v)\frac{\partial V}{\partial v} = rV.$$

The trouble given by the square root of v can be offset by taking $\rho = 0$, i.e. the noise sources are considered independent. Taking, for the sake of simplicity, $f = \lambda v$, we obtain the following simplified version of the equation

$$\frac{\partial V}{\partial t} + \frac{1}{2}vS^2\frac{\partial^2 V}{\partial S^2} + \frac{1}{2}\sigma^2\frac{\partial^2 V}{\partial v^2} + rS\frac{\partial V}{\partial S} - \lambda v\frac{\partial V}{\partial v} = rV. \tag{13.2.7}$$

Exercise 13.2.1 *Let $V(t, S, v)$ be a solution of equation (13.2.7).*

(a) Using the substitution $W(t, x, v) = V(t, S, v)$, with $x = \ln S$, find the partial differential equation satisfied by W;

(b) Let $\widehat{W}(t, v; \omega) = (\mathcal{F}_x W)(\omega)$. Find the equation satisfied by $\widehat{W}$;

(c) Using Feynman-Kac formula (see Theorem C.8.4 of Appendix C.8) write the solution of the equation obtained at (b) in the form of an expectation.

(d) Solve the equation given the final condition $V(T, S, v) = \delta(\ln S)$, where δ is the Dirac delta function.

Exercise 13.2.2 *Consider the mean-reverting AR(1) stochastic volatility model given by*

$$dS_t = rS_t\, dt + \sqrt{v_t} S_t\, dW_t$$
$$dv_t = \omega + \alpha v_t dt + \sigma\, dZ_t.$$

Find its associated stochastic differential equation.

Chapter 14

Stochastic Return Models

This chapter deals with option pricing in the case when the underlying asset has a stochastic rate of return. The reason for including this chapter is that the subject of this work was the topic of the 2013 Nobel Price in Economics, see Campbell [18] and Cochrane [22] for details. These conclusions were based on a long line of empirical research, which has documented the extent to which expected return on assets has predictive components. This research leads to an Ornstain-Uhlenbeck process for the expected return on stocks and bonds in continuous time. The reader is referred to Campbell et al. [19] for the explanation of how researchers estimate expected return models in continuous time.

14.1   Mean-reverting Ornstein-Uhlenbeck Process

Let k, θ and σ be three constants and consider the stochastic differential equation

$$dX_t = k(\theta - X_t)dt + \sigma dW_t. \tag{14.1.1}$$

We have got into this equation in section 3.3.2 when presenting Vasicek's model of spot rates. Recalling Proposition 3.3.1 with the present notations, we have:

Proposition 14.1.1 *The solution of equation* (14.1.1) *is given by*

$$X_t = \theta + (X_0 - \theta)e^{-kt} + \sigma \int_0^t e^{-k(t-s)} dW_s. \tag{14.1.2}$$

Furthermore, X_t is Gaussian distributed with the following mean and variance:

$$
\begin{aligned}
\mathbb{E}[X_t] &= \theta + (X_0 - \theta)e^{-kt} \\
Var(X_t) &= \frac{\sigma^2}{2k}(1 - e^{-2kt}).
\end{aligned}
$$

Since $\lim\limits_{t\to\infty} \mathbb{E}[X_t] = \theta$, the process is *mean-reverting*, i.e. in the long run the mean tends towards θ. The process (14.1.2) is called a *mean-reverting Ornstein-Uhlenbeck process* and will be used in the sequel as a model for the instantaneous rate of return of a stock.

Exercise 14.1.2 *A geometric Orstein-Uhlenbeck process is defined as the exponential process $Y_t = e^{X_t}$, where X_t is the mean-reverting Orstein-Uhlenbeck process given by (14.1.1). The process Y_t has been used in modeling volatility, since it requires it to stay positive.*

(a) Find the stochastic differential equation of Y_t;

(b) Find the mean $\mathbb{E}(Y_t)$ and the variance $Var(Y_t)$ of the process.

14.2   A Continuous VAR Process

The dynamics of a stock with a stochastic rate of return will be modeled using a continuous-time vector autoregressive (VAR) process. The model assumes that the underlying stock price, S_t, follows a geometric Brownian motion, while the instantaneous rate of return, μ_t, satisfies a mean reverting Ornstain-Uhlenbeck process:

$$dS_t = \mu_t S_t\, dt + \sigma_S S_t\, dW_t^S \tag{14.2.3}$$
$$d\mu_t = k(\theta - \mu_t)dt + \sigma_\mu\, dW_t^\mu, \tag{14.2.4}$$

with correlated sources of uncertainty, $cor(dW_t^S, dW_t^\mu) = \rho$, with $\rho < 0$, to accommodate for the leverage effect. The volatilities σ_S and σ_μ are considered constant. By Proposition 14.1.1, the second equation has the solution

$$\mu_t = \theta + (\mu_0 - \theta)e^{-kt} + \sigma_\mu \int_0^t e^{-k(t-s)}\, dW_s^\mu.$$

Substituting into (14.2.3) we obtain a stochastic differential equation with stochastic coefficients, which is difficult to approach directly. In order to continue the analysis, following Campbell et al. [19], we shall adopt the convenient transformation:

$$X_t = \ln S_t + (\tfrac{1}{2}\sigma_S^2 - \theta)t$$
$$Y_t = \mu_t - \theta.$$

Since the transformation is invertible, with the inverse given by

$$S_t = e^{X_t - (\sigma_S^2/2 - \theta)t}$$
$$\mu_t = X_t + \theta,$$

it suffices to find and study the processes X_t and Y_t. Applying Ito's lemma, see Appendix C.6, we have

$$
\begin{aligned}
dX_t &= d(\ln S_t) + \left(\frac{1}{2}\sigma_S^2 - \theta\right)dt \\
&= \frac{1}{S_t}dS_t - \frac{1}{2}\frac{1}{S_t^2}(dS_t)^2 + \left(\frac{1}{2}\sigma_S^2 - \theta\right)dt \\
&= \mu_t dt + \sigma_S\,W_t^S - \frac{1}{2}\sigma_S^2 dt + \left(\frac{1}{2}\sigma_S^2 - \theta\right)dt \\
&= (\mu_t - \theta)dt + \sigma_S dW_t^S \\
&= Y_t dt + \sigma_S dW_t^S
\end{aligned}
$$

and

$$
\begin{aligned}
dY_t &= d\mu_t = -k(\mu_t - \theta)dt + \sigma_\mu dW_t^\mu \\
&= -kY_t dt + \sigma_\mu dW_t^\mu.
\end{aligned}
$$

The last two formulas can be written in the following matrix form

$$
d\begin{pmatrix} X_t \\ Y_t \end{pmatrix} = \begin{pmatrix} 0 & 1 \\ 0 & -k \end{pmatrix}\begin{pmatrix} X_t \\ Y_t \end{pmatrix}dt + \begin{pmatrix} \sigma_S & 0 \\ 0 & \sigma_\mu \end{pmatrix}\begin{pmatrix} dW_t^S \\ dW_t^\mu \end{pmatrix}. \tag{14.2.5}
$$

For convenience reasons, we shall convert the correlated Brownian motions into independent Brownian motions. This is accomplished by the application of Cholesky decomposition

$$
dW_t^S = dW_t^1 \tag{14.2.6}
$$
$$
dW_t^\mu = \rho dW_t^1 + \sqrt{1-\rho^2}\,dW_t^2, \tag{14.2.7}
$$

with dW_t^1 and dW_t^2 uncorrelated sources of random noise. Substituting in (14.2.5) yields

$$
d\begin{pmatrix} X_t \\ Y_t \end{pmatrix} = \begin{pmatrix} 0 & 1 \\ 0 & -k \end{pmatrix}\begin{pmatrix} X_t \\ Y_t \end{pmatrix}dt + \begin{pmatrix} \sigma_S & 0 \\ \sigma_\mu\rho & \sigma_\mu\sqrt{1-\rho^2} \end{pmatrix}\begin{pmatrix} dW_t^1 \\ dW_t^2 \end{pmatrix}. \tag{14.2.8}
$$

Adopting the notations

$$
Z_t = \begin{pmatrix} X_t \\ Y_t \end{pmatrix}, \quad A = \begin{pmatrix} 0 & 1 \\ 0 & -k \end{pmatrix}, \quad C = \begin{pmatrix} \sigma_S & 0 \\ \sigma_\mu\rho & \sigma_\mu\sqrt{1-\rho^2} \end{pmatrix}, \quad dW_t = \begin{pmatrix} dW_t^1 \\ dW_t^2 \end{pmatrix},
$$

the previous equation can be written conveniently as the following linear stochastic differential equation in matrix form

$$
dZ_t = AZ_t dt + C dW_t. \tag{14.2.9}
$$

Next we shall solve this equation. Multiplying by the integrating factor e^{-At} we obtain the exact equation

$$d(e^{-At}Z_t) = e^{-At}C dW_t.$$

Integrating between 0 and t, and then multiplying by e^{At} yields the solution

$$Z_t = e^{At}Z_0 + e^{At}\int_0^t e^{-As}C\, dW_s,$$

where $Z_0 = (X_0, Y_0)^T = (\ln S_0, \mu_0 - \theta)^T$. The exponential can be interchanged with the integral, a fact that allows us to write

$$Z_t = e^{At}Z_0 + \int_0^t e^{A(t-s)}C\, dW_s. \tag{14.2.10}$$

The next step is to compute the exponential. We start by writing it as a power series

$$e^{At} = \sum_{n \geq 0} \frac{1}{n!}A^n t^n = \mathbb{I}_2 + A + \frac{1}{2}A^2 + \frac{1}{6}A^3 + \cdots,$$

and notice that we have inductively (see Exercise 14.2.1)

$$A^n = \begin{pmatrix} 0 & (-k)^{n-1} \\ 0 & (-k)^n \end{pmatrix}, \qquad n \geq 1. \tag{14.2.11}$$

Substituting the expression of A^n in the previous power series, we can manage to write each entry of the matrix in closed form

$$e^{At} = \mathbb{I}_2 + \sum_{n \geq 1}\frac{1}{n!}A^n t^n = \mathbb{I}_2 + \begin{pmatrix} 0 & \sum_{n \geq 1}\frac{1}{n!}(-k)^{n-1}t^n \\ 0 & \sum_{n \geq 1}\frac{1}{n!}(-k)^n t^n \end{pmatrix}$$

$$= \begin{pmatrix} 1 & -\frac{1}{k}\sum_{n \geq 1}\frac{1}{n!}(-k)^n t^n \\ 0 & 1+\sum_{n \geq 1}\frac{1}{n!}(-k)^n t^n \end{pmatrix} = \begin{pmatrix} 1 & \frac{1}{k}(1-e^{-kt}) \\ 0 & e^{-kt} \end{pmatrix}.$$

This expression needs to be substituted into (14.2.10). The components of the first term can be computed as

$$e^{At}Z_0 = \begin{pmatrix} 1 & \frac{1}{k}(1-e^{-kt}) \\ 0 & e^{-kt} \end{pmatrix}\begin{pmatrix} X_0 \\ Y_0 \end{pmatrix} = \begin{pmatrix} X_0 + \frac{1}{k}(1-e^{-kt})Y_0 \\ e^{-kt}Y_0 \end{pmatrix}$$

$$= \begin{pmatrix} \ln S_0 + \frac{1}{k}(1-e^{-kt})(\mu_0 - \theta) \\ e^{-kt}(\mu_0 - \theta) \end{pmatrix}.$$

A straightforward matrix multiplication shows that $e^{A(t-s)}C\,dW_s$ equals

$$
\begin{pmatrix}
\{\sigma_S + \sigma_\mu \frac{\rho}{k}(1 - e^{-k(t-s)})\}dW_s^1 + \{\frac{\sigma_\mu}{k}\sqrt{1-\rho^2}(1 - e^{-k(t-s)})\}dW_s^2 \\
\sigma_\mu \rho e^{-k(t-s)}\,dW_s^1 + \sigma_\mu \sqrt{1-\rho^2}e^{-k(t-s)}\,dW_s^2
\end{pmatrix}.
$$

Integrating each component we have

$$
\int_0^t e^{A(t-s)}C\,dW_s = \begin{pmatrix} u_1 \\ u_2 \end{pmatrix}
$$

with

$$
\begin{aligned}
u_1 &= \left(\sigma_S + \sigma_\mu \frac{\rho}{k}\right)W_t^1 - \sigma_\mu \frac{\rho}{k}e^{-kt}\int_0^t e^{ks}\,dW_s^1 \\
&\quad + \frac{\sigma_\mu}{k}\sqrt{1-\rho^2}\,W_t^2 - \frac{\sigma_\mu}{k}\sqrt{1-\rho^2}e^{-kt}\int_0^t e^{ks}\,dW_s^2 \\
u_2 &= \sigma_\mu \rho e^{-kt}\int_0^t e^{ks}\,dW_s^1 + \sigma_\mu \sqrt{1-\rho^2}e^{-kt}\int_0^t e^{ks}\,dW_s^2.
\end{aligned}
$$

The component u_1 is a linear combination of Brownian motions W_t^1 and W_t^2 and two Wiener integrals, while the component u_2 is a linear combination of the same Wiener integrals. Thus, u_1 and u_2 have a bivariate normal distribution.

Assembling pieces together into formula (14.2.10), we obtain the following components for Z_t:

$$
\begin{aligned}
X_t &= \ln S_0 + \frac{1}{k}(1 - e^{-kt})(\mu_0 - \theta) + u_1 \\
Y_t &= e^{-kt}(\mu_0 - \theta) + u_2.
\end{aligned}
$$

Using the inversion of the initial transformation we obtain the following formula for the stock

$$
\begin{aligned}
S_t &= e^{X_t - (\sigma_S^2/2 - \theta)t} \\
&= S_0 e^{\phi(t)}e^{\alpha_1 W_t^1 + \alpha_2 W_t^2}e^{\beta_1 \int_0^t e^{-k(t-s)}\,dW_s^1 + \beta_2 \int_0^t e^{-k(t-s)}\,dW_s^2} \quad (14.2.12)
\end{aligned}
$$

where

$$
\begin{aligned}
\phi(t) &= \frac{1}{k}(1 - e^{-kt})(\mu_0 - \theta) - \left(\frac{1}{2}\sigma_S^2 - \theta\right)t \\
\alpha_1 &= \sigma_S + \sigma_\mu \frac{\rho}{k} \\
\alpha_2 &= \frac{\sigma_\mu}{k}\sqrt{1-\rho^2} \\
\beta_1 &= -\sigma_\mu \frac{\rho}{k} \\
\beta_2 &= -\frac{\sigma_\mu}{k}\sqrt{1-\rho^2}.
\end{aligned}
$$

We note that formula (14.2.12) plays for the model (14.2.3)-(14.2.3) a role similar to the one played by the formula (11.1.3) for the Heston model (11.1.1)-(11.1.2). We also make the remark that formula (14.2.12) can be used successfully in Monte-Carlo simulations.

Exercise 14.2.1 *Consider the matrix* $A = \begin{pmatrix} 0 & 1 \\ 0 & -k \end{pmatrix}$.

(a) *Compute the powers* A^2, A^3 *and* A^4;

(b) *Inspired by the pattern obtained at* (a), *prove formula* (14.2.11).

Exercise 14.2.2 *Prove the existence of two Brownian motions* W_t^1 *and* W_t^2, *with uncorrelated jumps, which satisfy the system* (14.2.6)-(14.2.7).

14.3   Probability Density of S_t

It is desirable to find a formula for the density of the stock S_t, which is provided by formula (14.2.12). In order to accomplish this goal, it suffices to find the density of its log process

$$X_t = \ln S_t = \ln S_0 + \phi(t) + Z_t^1 + Z_t^2, \tag{14.3.13}$$

with

$$Z_t^1 = \alpha_1 W_t^1 + \beta_1 e^{-kt} \int_0^t e^{ks} \, dW_t^1 \tag{14.3.14}$$

$$Z_t^2 = \alpha_2 W_t^2 + \beta_2 e^{-kt} \int_0^t e^{ks} \, dW_t^2. \tag{14.3.15}$$

The processes Z_t^1 and Z_t^2 are independent, since they are Gaussian and are uncorrelated. So, it makes sense to study only the process

$$Z_t = \alpha W_t + \beta e^{-kt} \int_0^t e^{ks} \, dW_t,$$

which is a linear combination of two Gaussians

$$X_1 = W_t, \qquad X_2 = e^{-kt} \int_0^t e^{ks} \, dW_t,$$

with

$$X_1 \sim \mathcal{N}(0, t), \qquad X_2 \sim \mathcal{N}\left(0, \frac{1 - e^{-2kt}}{2k}\right).$$

Using the covariance property of Ito integrals, see Appendix C.6, we have

$$\mathbb{E}[X_1 X_2] = e^{-kt}\mathbb{E}\left[W_t \int_0^t e^{ks}\,dW_s\right] = e^{-kt}\mathbb{E}\left[\int_0^t dW_s \int_0^t e^{ks}\,dW_s\right]$$

$$= e^{-kt}\int_0^t e^{ks}\,ds = \frac{1-e^{-kt}}{k}.$$

Since $\mathbb{E}[X_1] = \mathbb{E}[X_2] = 0$, we obtain the covariance of X_1 and X_2

$$cov(X_1, X_2) = \mathbb{E}[X_1 X_2] - \mathbb{E}[X_1]\mathbb{E}[X_2] = \frac{1-e^{-kt}}{k}.$$

Then the correlation coefficient between X_1 and X_2 is given by

$$\rho = corr(X_1, X_2) = \frac{cov(X_1, X_2)}{\sqrt{Var(X_1)Var(X_2)}} = \sqrt{\frac{2}{kt}}\sqrt{\frac{1-e^{-kt}}{1+e^{-kt}}}.$$

The joint distribution of (X_1, X_2) is bivariate Gaussian and is obtained substituting

$$\mu_1 = \mathbb{E}[X_1] = 0$$
$$\mu_2 = \mathbb{E}[X_2] = 0$$
$$\sigma_1 = \sqrt{Var(X_1)} = \sqrt{t}$$
$$\sigma_2 = \sqrt{Var(X_2)} = \sqrt{\frac{1-e^{-kt}}{k}}$$

into the formula

$$f_{(X_1 X_2)}(x_1, x_2) = \frac{1}{2\pi\sigma_1\sigma_2\sqrt{1-\rho^2}}$$

$$\exp\left(-\frac{1}{1-\rho^2}\left[\frac{(x_1-\mu_1)^2}{2\sigma_1^2} - \rho\frac{(x_1-\mu_1)(x_2-\mu_2)}{\sigma_1\sigma_2} + \frac{(x_2-\mu_2)^2}{2\sigma_2^2}\right]\right),$$

which after a straightforward computation becomes

$$f_{X_1 X_2}(x_1, x_2) = \frac{1}{2\pi}\sqrt{\frac{k}{t(1-e^{-kt})}}\frac{1}{\sqrt{1-\rho^2}}$$

$$\exp\left(-\frac{1}{1-\rho^2}\left[\frac{x_1^2}{2t} - \frac{1}{t}\sqrt{\frac{2}{1+e^{-kt}}}x_1 x_2 + \frac{kx_2^2}{2(1-e^{-kt})}\right]\right), \quad (14.3.16)$$

with

$$1-\rho^2 = 1 - \frac{2}{kt}\frac{1-e^{-kt}}{1+e^{-kt}}.$$

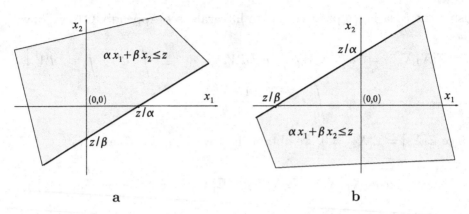

Figure 14.1: *The domain* $\{\alpha x_1 + \beta x_2 \leq z\}$ *in the cases:* **a.** $z > 0$ **b.** $z < 0$.

The cumulative distribution function of Z_t can now be found as

$$
\begin{aligned}
F_Z(z) &= P(Z_t \leq z) = P(\alpha X_1 + \beta X_2 \leq z) \\
&= \iint\limits_{\{\alpha x_1 + \beta x_2 \leq z\}} f_{X_1 X_2}(x_1, x_2)\, dx_1 dx_2.
\end{aligned} \tag{14.3.17}
$$

Before computing the integral, we notice that the previous formulas for α_i and β_i imply $\alpha > 0$ and $\beta < 0$. In fact, in order to have $\alpha_1 > 0$, we need to impose the slightly stronger condition $0 < \rho < -k\dfrac{\sigma_S}{\sigma_\mu}$. The line $\alpha x_1 + \beta x_2 = z$ has the x and the y-intercepts given respectively by z/α and z/β. The domain $\{\alpha x_1 + \beta x_2 \leq z\}$ is pictured in Fig.14.1 for both cases $z > 0$ and $z < 0$.

We shall approach only the case $z > 0$. The other case is similar and is left as an exercise to the reader. Applying Fubini's theorem to the integral (14.3.17) we obtain the double integral

$$
F_Z(z) = \int_{-\infty}^{z/\alpha} \int_{(z-\alpha x_1)/\beta}^{\infty} f_{X_1 X_2}(x_1, x_2)\, dx_2 dx_1.
$$

The density function of Z is obtained by differentiation

$$
f_Z(z) = F_Z'(z).
$$

In order to differentiate the right side we employ the following Leibnitz' formula of differentiating integrals

$$
\frac{d}{dz} \int_{a(z)}^{b(z)} f(z, x)\, dx = f\big(z, b(z)\big) b'(z) - f\big(z, a(z)\big) a'(z) + \int_{a(z)}^{b(z)} \frac{\partial}{\partial z} f(z, x)\, dx.
$$

$$\tag{14.3.18}$$

Proceeding with care, we have

$$f_Z(z) = \frac{1}{\alpha} \int_0^\infty f_{X_1 X_2}(z/\alpha, x_2)\, dx_2 - \frac{1}{\beta} \int_{-\infty}^{z/\alpha} f_{X_1 X_2}(x_1, (z - \alpha x_1)/\beta)\, dx_1.$$

$$(14.3.19)$$

Let

$$I_1 = I_1(z; \alpha, \beta) = \int_0^\infty f_{X_1 X_2}(z/\alpha, x_2)\, dx_2,$$

with $f_{X_1 X_2}$ given as in (14.3.16). This integral can be computed in closed form using the formula

$$\int_0^\infty e^{-a^2 y^2 + by}\, dx = \frac{\sqrt{\pi}}{2a} e^{\frac{b^2}{4a^2}} \left(1 + Erf\left(\frac{b}{2a}\right)\right).$$

$$(14.3.20)$$

Substituting $y = \dfrac{x_2}{\sqrt{1 - \rho^2}}$, and then taking out the factors under the integral we get

$$
\begin{aligned}
I_1 &= \frac{1}{2\pi\sigma_1\sigma_2\sqrt{1-\rho^2}} \int_0^\infty e^{-\frac{1}{1-\rho^2}\left[\frac{z^2}{2\alpha^2\sigma_1^2} - \frac{\rho z}{\alpha\sigma_1\sigma_2}x_2 + \frac{x_2^2}{2\sigma_2^2}\right]}\, dx_2 \\
&= \frac{1}{2\pi\sigma_1\sigma_2} e^{-\frac{z^2}{2\alpha^2\sigma_1^2(1-\rho^2)}} \int_0^\infty e^{-\frac{y^2}{2\sigma_2^2} + \frac{\rho}{\sqrt{1-\rho^2}}\frac{1}{\sigma_1\sigma_2}\frac{z}{\alpha}y}\, dy.
\end{aligned}
$$

Choosing

$$a^2 = \frac{1}{2\sigma_2^2}, \qquad b = \frac{\rho}{\sqrt{1-\rho^2}}\frac{1}{\sigma_1\sigma_2}\frac{z}{\alpha}$$

and then applying (14.3.20), some tedious algebra yields

$$I_1 = \frac{1}{2}\frac{1}{\sqrt{2\pi}\sigma_1} e^{-\frac{z^2}{2\alpha^2\sigma_1^2}} \left[1 + Erf\left(\frac{\rho}{\sqrt{1-\rho^2}}\frac{z}{\alpha\sigma_1\sqrt{2}}\right)\right].$$

Using the value $\sigma_1 = \sqrt{t}$ the expression becomes

$$I_1 = \frac{1}{2}\frac{1}{\sqrt{2\pi t}} e^{-\frac{z^2}{2\alpha^2 t}} \left[1 + Erf\left(\frac{\rho}{\sqrt{1-\rho^2}}\frac{z}{\alpha\sqrt{2t}}\right)\right].$$

We make the remark that if $\rho = 0$ then

$$I_1 = \frac{1}{2}\frac{1}{\sqrt{2\pi t}} e^{-\frac{z^2}{2\alpha^2 t}}.$$

Consider now the second integral

$$I_2 = I_2(z; \alpha, \beta) = \int_{-\infty}^{z/\alpha} f_{X_1 X_2}(x_1, (z - \alpha x_1)/\beta)\, dx_1.$$

This will be evaluated using the formula

$$\int_{-\infty}^{u} e^{-a^2 x^2 + bx}\, dx = \frac{\sqrt{\pi}}{2a} e^{\frac{b^2}{4a^2}} \left(1 - Erf\left(\frac{b}{2a} - au\right)\right). \tag{14.3.21}$$

Using the formula for $f_{X_1 X_2}$ and collecting the powers of x_1 in the exponent, the integral writes as

$$I_2 = \frac{1}{2\pi\sigma_1\sigma_2} \frac{1}{\sqrt{1-\rho^2}} \int_{-\infty}^{z/\alpha} e^{-\frac{1}{1-\rho^2}\left[\frac{x_1^2}{2\sigma_1^2} - \frac{\rho}{\sigma_1\sigma_2}x_1\frac{(z-\alpha x_1)}{\beta} + \frac{1}{2\sigma_2^2}\frac{(z-\alpha x_1)^2}{\beta^2}\right]} dx_1$$

$$= \frac{1}{2\pi\sigma_1\sigma_2} \frac{1}{\sqrt{1-\rho^2}} e^{-\frac{1}{1-\rho^2}\frac{z^2}{2\beta^2\sigma_2^2}} \int_{-\infty}^{z/\alpha} e^{-a^2 x_1^2 + bx_1}\, dx_1,$$

with

$$a^2 = \frac{1}{2\sigma_1^2(1-\rho^2)}\left(1 + \frac{\sigma_1}{\sigma_2}\frac{\alpha}{\beta}\right)^2$$

$$b = \frac{1}{1-\rho^2}\left(\frac{\rho}{\sigma_1} + \frac{\alpha}{\beta\sigma_2}\right)\frac{z}{\beta\sigma_2}.$$

At this point we apply formula (14.3.21) and after tedious algebra we arrive at

$$I_2 = \frac{\beta}{2\sqrt{2\pi}(\alpha\sigma_1 + \beta\sigma_2)} \exp\left[-\frac{1}{2\beta^2}\frac{1}{1-\rho^2}\frac{1-\sigma_1^2}{\sigma_2^2}\frac{\alpha\sigma_1 + \beta\rho\sigma_2}{\sigma_1(\alpha\sigma_1 + \beta\sigma_2)}z^2\right]$$

$$\times \left[1 - Erf\left(\frac{z}{\sqrt{2}\beta\sigma_2\sqrt{1-\rho^2}}\left(\frac{\alpha\sigma_1 + \beta\rho\sigma_2}{\alpha\sigma_1 + \beta\sigma_2} - \frac{\beta}{\alpha}\frac{\sigma_2}{\sigma_1} - 1\right)\right)\right].$$

Now, after I_1 and I_2 have been found, formula (14.3.19) provides the probability density of Z_t as

$$f_Z(z) = \frac{1}{\alpha}I_1 - \frac{1}{\beta}I_2.$$

Since the processes Z_t^1 and Z_t^2, which are given by (14.3.14)-(14.3.15), are independent, their joint probability density is given by the product of the individual densities as

$$f_{Z^1 Z^2}(z_1, z_2) = f_{Z^1}(z_1) f_{Z^2}(z_1)$$

$$= \left(\frac{1}{\alpha_1}I_1(z_1; \alpha_1, \beta_1) - \frac{1}{\beta_1}I_2(z_1; \alpha_1, \beta_1)\right)$$

$$\times \left(\frac{1}{\alpha_2}I_1(z_2; \alpha_2, \beta_2) - \frac{1}{\beta_2}I_2(z_2; \alpha_2, \beta_1)\right).$$

With this long preparation we are able now to compute the distribution of the log of the stock, $X_t = \ln S_t$. Using formula (14.3.13) the distribution function of X_t can be computed using integration as

$$
\begin{aligned}
F_X(x) &= P(X_t \le x) = P(Z_t^1 + Z_t^2 \le x - \ln S_0 - \phi(t)) \\
&= P(Z_t^1 + Z_t^2 \le \psi(t,x)) \\
&= \iint\limits_{\{z_1 + z_2 \le \psi(t,x)\}} f_{Z^1}(z_1) f_{Z^2}(z_2)\, dz_1 dz_2.
\end{aligned}
$$

The integration domain depends on the sign of $\psi(t,x)$. If $\psi(t,x) > 0$, then Fubini's theorem yields

$$
\begin{aligned}
F_X(x) &= \int_{-\infty}^{\psi(t,x)} \int_{-\infty}^{\psi(t,x)-z_1} f_{Z^1}(z_1) f_{Z^2}(z_2)\, dz_2 dz_1 \\
&= \int_{-\infty}^{\psi(t,x)} \left(f_{Z^1}(z_1) \int_{-\infty}^{\psi(t,x)-z_1} f_{Z^2}(z_2)\, dz_2 \right) dz_1.
\end{aligned}
$$

If $\psi(t,x) < 0$, we integrate on the complementary domain and subtract from 1 as

$$
\begin{aligned}
F_X(x) &= 1 - \int_{\psi(t,x)}^{\infty} \int_{\psi(t,x)-z_1}^{\infty} f_{Z^1}(z_1) f_{Z^2}(z_2)\, dz_2 dz_1 \\
&= 1 - \int_{\psi(t,x)}^{\infty} \left(f_{Z^1}(z_1) \int_{\psi(t,x)-z_1}^{\infty} f_{Z^2}(z_2)\, dz_2 \right) dz_1.
\end{aligned}
$$

The probability density of $X_t = \ln S_t$ can be obtained by differentiation, $f_X(x) = F'_X(x)$, using Leibnitz' formula (14.3.18). The computation details are left to the reader, see Exercise 14.3.2. It is not hard to find the law of S_t from the one of X_t, see Exercise 14.3.3.

As we have seen, there was a substantial effort to obtain the law of the stock S_t in the case when the rate of return follows an Ornstain-Uhlenbeck process. The bottom line is clear: despite of its complexity, the formula is analytically trackable and hence can be programmed. This is useful in simulations of the stock needed for Monte-Carlo evaluations of European options.

However, there is also another way to valuate options, by solving PDEs, and we shall deal with it in the next section.

Exercise 14.3.1 *Using Proposition C.6.3 of Appendix C.6 show that*

$$
Var\left(e^{-kt} \int_0^t e^{ks}\, dW_s\right) = \frac{1 - e^{-2kt}}{2k}.
$$

Exercise 14.3.2 *Apply Leibnitz' formula (14.3.18) to find the probability density of $X_t = \ln S_t$. Consider both cases $\psi > 0$ and $\psi < 0$.*

Exercise 14.3.3 *Show that the density of the stock, S_t, can be obtained from the density of its log, $X_t = \ln S_t$, by the following formula*

$$f_{S_t}(x) = \begin{cases} \frac{1}{x} f_X(\ln x), & \text{if } x > 0 \\ 0, & \text{otherwise.} \end{cases}$$

14.4   The PDE

The next step in the analysis is to produce a Heston type equation for this model. We proceed with a method similar with the one used in the case of Heston model, applied to the system of stochastic differential equations (14.2.3)-(14.2.4):

$$
\begin{aligned}
dS_t &= \mu_t S_t \, dt + \sigma_s S_t \, dW_t^S & (14.4.22) \\
d\mu_t &= k(\theta - \mu_t) dt + \sigma_\mu \, dW_t^\mu. & (14.4.23)
\end{aligned}
$$

We then consider a derivative, $V = V(t, S, \mu)$, whose value depends on time, t, stock value, S, and stock rate of return, μ. Using that

$$
\begin{aligned}
(dS)^2 &= \sigma_s^2 S^2 dt \\
(d\mu)^2 &= \sigma_\mu^2 dt \\
dS d\mu &= \rho \sigma_S \sigma_\mu S dt,
\end{aligned}
$$

then Ito's lemma applied to $V(t, S, v)$ provides

$$
\begin{aligned}
dV &= \frac{\partial V}{\partial t} dt + \frac{\partial V}{\partial S} dS + \frac{\partial V}{\partial \mu} d\mu \\
&\quad + \frac{1}{2} \frac{\partial^2 V}{\partial S^2} (dS)^2 + \frac{1}{2} \frac{\partial^2 V}{\partial \mu^2} (d\mu)^2 + \frac{\partial^2 V}{\partial S \partial \mu} dS d\mu \\
&= \left(\frac{\partial V}{\partial t} + \frac{1}{2} \sigma_s^2 S^2 \frac{\partial^2 V}{\partial S^2} + \frac{1}{2} \sigma_\mu^2 \frac{\partial^2 V}{\partial \mu^2} + \rho \sigma_S \sigma_\mu S \frac{\partial^2 V}{\partial S \partial \mu} \right) dt + \frac{\partial V}{\partial S} dS + \frac{\partial V}{\partial \mu} d\mu.
\end{aligned}
$$

There are two sources of uncertainty that need to be offset, which are due to changes in S and μ. The hedge will be done with the help of an extra derivative $U(t, S, v)$, by forming the portfolio

$$\Pi = V + \alpha S + \beta U.$$

Denoting for convenience

$$L(V) = \frac{\partial V}{\partial t} + \frac{1}{2} \sigma_s^2 S^2 \frac{\partial^2 V}{\partial S^2} + \frac{1}{2} \sigma_\mu^2 \frac{\partial^2 V}{\partial \mu^2} + \rho \sigma_S \sigma_\mu S \frac{\partial^2 V}{\partial S \partial \mu},$$

we obtain

$$
\begin{aligned}
d\Pi &= dV + \alpha dS + \beta dU \\
&= \Big(L(V) + \beta L(U)\Big)dt + \Big(\frac{\partial V}{\partial S} + \alpha + \beta\frac{\partial U}{\partial S}\Big)dS + \Big(\frac{\partial V}{\partial \mu} + \beta\frac{\partial U}{\partial \mu}\Big)d\mu.
\end{aligned}
$$

The values of the coefficients for which the portfolio becomes risk-less are

$$
\beta = -\frac{\partial V/\partial \mu}{\partial U/\partial \mu}, \qquad \alpha = -\frac{\partial V}{\partial S} - \beta\frac{\partial U}{\partial S},
$$

case in which

$$
d\Pi = \Big(L(V) + \beta L(U)\Big)dt.
$$

On the other side, in order to avoid arbitrage opportunities, we need to have

$$
d\Pi = r\Pi dt,
$$

where r denotes the risk-free rate, which is considered constant. Equating the right sides of the last two equations, we get

$$
L(V) + \beta L(U) = r(V + \alpha S + \beta U).
$$

Substituting for α and β, and separating the terms containing U from the ones depending on V, yields

$$
\frac{L(V) + rS\dfrac{\partial V}{\partial S} - rV}{\dfrac{\partial V}{\partial \mu}} = \frac{L(U) + rS\dfrac{\partial U}{\partial S} - rU}{\dfrac{\partial U}{\partial \mu}}.
$$

Since U and V are distinct and arbitrary derivatives, the previous quotient is the same for all derivatives. Hence, it must be equal to a function, $f(t, S, \mu)$, which depends on the market risk through the variables t, S and μ. Consequently, for any derivative V, we have

$$
\frac{L(V) + rS\dfrac{\partial V}{\partial S} - rV}{\dfrac{\partial V}{\partial \mu}} = f(t, S, \mu),
$$

or equivalently

$$
L(V) + rS\frac{\partial V}{\partial S} - f(t, S, \mu)\frac{\partial V}{\partial \mu} = rV.
$$

Substituting for $L(V)$ we obtain the following partial differential equation satisfied by any derivative:

$$\frac{\partial V}{\partial t} + \frac{1}{2}\sigma_S^2 S^2 \frac{\partial^2 V}{\partial S^2} + \frac{1}{2}\sigma_\mu^2 \frac{\partial^2 V}{\partial \mu^2} + \rho\sigma_S\sigma_\mu S \frac{\partial^2 V}{\partial S \partial \mu} + rS \frac{\partial V}{\partial S} - f(t, S, \mu)\frac{\partial V}{\partial \mu} = rV.$$

$$(14.4.24)$$

The previous equation is similar to Heston equation (11.1.10). Both equations have seven terms among which three are identical. However, there are also some differences, consisting in the form of the coefficients. In addition to this, we do not have available a "Heston's anszats" to tell us the form of the function $f(t, S, \mu)$. Since we plan to solve the equation, we shall choose a convenient form for $f(t, S, \mu)$ for which the equation can be solved. First, since we would like to exploit the homogeneity with respect to S, we require that $f = f(t, \mu)$ is independent of S.

Exercise 14.4.1 *Find all solutions of equation (14.4.24) of the type:*

(a) $V = V(t)$, time dependent;

(b) $V = V(S)$, stock dependent;

(c) $V = V(\mu)$, under the assumption $f(t, S, \mu) = -\lambda$, constant.

14.5 Simplifying the PDE

Similarly with the Heston model, we consider the substitution $x = \ln S$ and $W(t, x, \mu) = V(t, S, \mu)$. Then chain rule implies

$$S\frac{\partial V}{\partial S} = \frac{\partial W}{\partial x}$$

$$S^2 \frac{\partial^2 V}{\partial S^2} = \frac{\partial^2 W}{\partial x^2} - \frac{\partial W}{\partial x}$$

$$S\frac{\partial^2 V}{\partial S \partial \mu} = \frac{\partial^2 W}{\partial x \partial \mu}.$$

Substituting into (14.4.24) yields the following simplified version of the above equation:

$$\frac{\partial W}{\partial t} + \frac{1}{2}\sigma_S^2 \frac{\partial^2 W}{\partial x^2} + \frac{1}{2}\sigma_\mu^2 \frac{\partial^2 W}{\partial \mu^2} + \rho\sigma_S\sigma_\mu \frac{\partial^2 W}{\partial x \partial \mu} + \left(r - \frac{1}{2}\sigma_S^2\right)\frac{\partial W}{\partial x}$$

$$- f(t, \mu)\frac{\partial W}{\partial \mu} - rW = 0. \qquad (14.5.25)$$

The specific feature of this equation is that all coefficients, maybe with the exception of $f(t, \mu)$, are constant. This fact is not a surprise, since we have seen that the log of the stock, $X_t = \ln S_t$, given by formula (14.3.13) is bivariate Gaussian (since Z_t^1 and Z_t^2 are Gaussian). This implies that the equation (14.5.25), which depends on the infinitesimal generator of the bivariate Gaussian process X_t, is elliptic in variables (x, μ) with constant coefficients, see also Appendix C.8. This is the reason why in the following we shall price derivatives under the assumption that the risk $f(t, \mu) = -\lambda$ is constant.

For later convenience we employ new notations for coefficients

$$
\begin{aligned}
a_{11} &= \frac{1}{2}\sigma_S^2 & a_{12} = a_{21} &= \frac{1}{2}\rho\sigma_S\sigma_\mu \\
a_{22} &= \frac{1}{2}\sigma_\mu^2 & b = r - \frac{1}{2}\sigma_S^2 &= r - a_{11}.
\end{aligned}
$$

Then the equation writes neatly as in the following

$$
\frac{\partial W}{\partial t} + a_{11}\frac{\partial^2 W}{\partial x^2} + a_{22}\frac{\partial^2 W}{\partial \mu^2} + 2a_{12}\frac{\partial^2 W}{\partial x \partial \mu}
$$
$$
+ b\frac{\partial W}{\partial x} + \lambda\frac{\partial W}{\partial \mu} - rW = 0. \tag{14.5.26}
$$

The "ellipticity" specified before consists in the positivity[1] of the determinant made from the coefficients of the second order derivatives

$$
\begin{vmatrix} a_{11} & a_{12} \\ a_{21} & a_{22} \end{vmatrix} = a_{11}a_{22} - a_{12}a_{21} = \frac{1}{4}\sigma_S^2\sigma_\mu^2(1 - \rho^2) > 0, \tag{14.5.27}
$$

where σ_S and σ_μ are non-zero.

Remark 14.5.1 If consider the coefficients matrix $A = (a_{ij})$ and denote the gradient of W by $\nabla W = \left(\frac{\partial W}{\partial S}, \frac{\partial W}{\partial \mu}\right)$, then the previous equation can be written also as

$$
\frac{\partial W}{\partial t} + \langle A\nabla W, \nabla W \rangle + \langle (b, \lambda), \nabla W \rangle - rW = 0,
$$

where $\langle , \rangle$ denotes the usual scalar product.

---

[1] The case $\rho = -1$ is excluded from economic considerations, while $\rho = 1$ contradicts the leverage effect.

14.6   Solving the PDE

We start by noticing that equation (14.5.26) is a 2-dimensional version of equation (8.8.20). Hence, we shall use a similar reduction method to solve it. Consider the linear function

$$\varphi(t, x, \mu) = \alpha x + \beta t + \gamma \mu,$$

with the constant coefficients α, β and γ subject to be found later. Let

$$W(t, x, \mu) = e^{\varphi(t,x,\mu)} U(t, x, \mu).$$

Differentiation rules yield

$$
\begin{aligned}
\frac{\partial W}{\partial t} &= e^{\varphi}\left(\beta U + \frac{\partial U}{\partial t}\right) \\
\frac{\partial W}{\partial x} &= e^{\varphi}\left(\alpha U + \frac{\partial U}{\partial x}\right) \\
\frac{\partial W}{\partial \mu} &= e^{\varphi}\left(\gamma U + \frac{\partial U}{\partial \mu}\right) \\
\frac{\partial^2 W}{\partial x^2} &= e^{\varphi}\left(\alpha^2 U + 2\alpha\frac{\partial U}{\partial x} + \frac{\partial^2 U}{\partial x^2}\right) \\
\frac{\partial^2 W}{\partial \mu^2} &= e^{\varphi}\left(\gamma^2 U + 2\gamma\frac{\partial U}{\partial \mu} + \frac{\partial^2 U}{\partial \mu^2}\right) \\
\frac{\partial^2 W}{\partial x \partial \mu} &= e^{\varphi}\left(\alpha\gamma U + \alpha\frac{\partial U}{\partial \mu} + \gamma\frac{\partial U}{\partial x} + \frac{\partial^2 U}{\partial x \partial \mu}\right).
\end{aligned}
$$

Substituting in (14.5.26) and collecting similar terms, we obtain

$$
\begin{aligned}
\frac{\partial U}{\partial t} &+ a_{11}\frac{\partial^2 U}{\partial x^2} + a_{22}\frac{\partial^2 U}{\partial \mu^2} + 2a_{12}\frac{\partial^2 U}{\partial x \partial \mu} \\
&+ (2a_{11}\alpha + 2a_{12}\gamma + b)\frac{\partial U}{\partial x} \\
&+ (2a_{22}\gamma + 2a_{12}\alpha + \lambda)\frac{\partial U}{\partial \mu} \\
&+ (\beta + a_{11}\alpha^2 + a_{22}\gamma^2 + 2a_{12}\alpha\gamma + b\alpha + \lambda\gamma - r)U = 0.
\end{aligned}
$$

We choose the values of α, β and γ such that the coefficients of $\dfrac{\partial U}{\partial x}$, $\dfrac{\partial U}{\partial \mu}$, and U vanish. This implies that α and γ satisfy the linear system

$$
\begin{aligned}
a_{11}\alpha + a_{12}\gamma &= -b/2 \\
a_{12}\alpha + a_{22}\gamma &= -\lambda/2,
\end{aligned}
$$

which has a unique solution due to the ellipticity condition (14.5.27). Solving the system (for instance, by Cramer's rule) we obtain the solutions

$$\alpha = \frac{1}{\sigma_S^2(1-\rho^2)}\left(\rho\lambda\frac{\sigma_S}{\sigma_\mu} - b\right)$$

$$\gamma = \frac{1}{\sigma_\mu^2(1-\rho^2)}\left(\rho b\frac{\sigma_\mu}{\sigma_S} - \lambda\right).$$

The constant β is obtained from the vanishing condition on the coefficient of U:

$$\beta = r - a_{11}\alpha^2 - a_{22}\gamma^2 - 2a_{12}\alpha\gamma - b\alpha - \lambda\gamma.$$

The computation of β can be done in a more elegant way if we employ the matrix formalism and scalar products. Consequently, we can write

$$
\begin{aligned}
\beta &= r - \left\langle \begin{pmatrix} a_{11} & a_{12} \\ a_{21} & a_{22} \end{pmatrix}\begin{pmatrix} \alpha \\ \gamma \end{pmatrix}, \begin{pmatrix} \alpha \\ \gamma \end{pmatrix} \right\rangle - \left\langle \begin{pmatrix} b \\ \lambda \end{pmatrix}, \begin{pmatrix} \alpha \\ \gamma \end{pmatrix} \right\rangle \\
&= r - \left\langle \begin{pmatrix} -b/2 \\ -\lambda/2 \end{pmatrix}, \begin{pmatrix} \alpha \\ \gamma \end{pmatrix} \right\rangle - \left\langle \begin{pmatrix} b \\ \lambda \end{pmatrix}, \begin{pmatrix} \alpha \\ \gamma \end{pmatrix} \right\rangle \\
&= r - \frac{1}{2}\left\langle \begin{pmatrix} b \\ \lambda \end{pmatrix}, \begin{pmatrix} \alpha \\ \gamma \end{pmatrix} \right\rangle \\
&= r - \frac{1}{2}(\alpha b + \lambda\gamma).
\end{aligned}
\tag{14.6.28}
$$

Hence, choosing the aforementioned values for α, β and γ, the differential equation satisfied by U reduces to the following convenient form:

$$\frac{\partial U}{\partial t} + a_{11}\frac{\partial^2 U}{\partial x^2} + a_{22}\frac{\partial^2 U}{\partial\mu^2} + 2a_{12}\frac{\partial^2 U}{\partial x\partial\mu} = 0. \tag{14.6.29}$$

Substituting for the value of coefficients we have

$$\frac{\partial U}{\partial t} + \frac{1}{2}\sigma_S^2\frac{\partial^2 U}{\partial x^2} + \frac{1}{2}\sigma_\mu^2\frac{\partial^2 U}{\partial\mu^2} + \rho\sigma_S\sigma_\mu\frac{\partial^2 U}{\partial x\partial\mu} = 0. \tag{14.6.30}$$

Denote by T the maturity of the derivative and by $\tau = T - t$ the time to maturity. If substitute $\mathbf{U}(\tau, x, \mu) = U(t, x, \mu)$, then

$$\frac{\partial\mathbf{U}}{\partial\tau} = \frac{1}{2}\sigma_S^2\frac{\partial^2\mathbf{U}}{\partial x^2} + \frac{1}{2}\sigma_\mu^2\frac{\partial^2\mathbf{U}}{\partial\mu^2} + \rho\sigma_S\sigma_\mu\frac{\partial^2\mathbf{U}}{\partial x\partial\mu}. \tag{14.6.31}$$

If consider the second order differential operator

$$\mathcal{A} = \frac{1}{2}\sigma_S^2\frac{\partial^2}{\partial x^2} + \frac{1}{2}\sigma_\mu^2\frac{\partial^2}{\partial\mu^2} + \rho\sigma_S\sigma_\mu\frac{\partial^2}{\partial x\partial\mu},$$

the previous equation takes the simple form

$$\frac{\partial \mathbf{U}}{\partial \tau} = \mathcal{A}\mathbf{U},$$

which is a Kolmogorov backward equation, see Appendix C.8. In order to solve it we need to take the convolution between the heat kernel of operator $\mathcal{A}$ and the initial condition $\mathbf{U}_{|\tau=0}$, which depends on the specific derivative we want to price. Hence, the main concern is to obtain first the heat kernel for the operator $\mathcal{A}$. This means to obtain a function $p_\tau(x_0, \mu_0, x, \mu)$ which satisfies

$$\begin{aligned}
\frac{\partial p}{\partial t} &= \mathcal{A}p, \qquad t > 0 \\
\lim_{\tau \to 0+} p_\tau(x_0, \mu_0, x, \mu) &= \delta(x - x_0)\delta(\mu - \mu_0),
\end{aligned}$$

where $\delta(x)$ is Dirac's delta function. It is known that $p_\tau(x_0, \mu_0, x, \mu)$ is the transition density of the diffusion[2] whose generator is given by $\mathcal{A}$, see Appendix C.8. This 2-dimensional diffusion, (X_τ, M_τ), satisfies the stochastic equation

$$\begin{aligned}
dX_\tau &= \sigma_S dB_\tau^1 \\
dM_\tau &= \sigma_\mu dB_\tau^2,
\end{aligned}$$

with B_t^i correlated Brownian motions. Integrating yields

$$\begin{aligned}
X_\tau &= x_0 + \sigma_S B_\tau^1 \\
M_\tau &= \mu_0 + \sigma_\mu B_\tau^2.
\end{aligned}$$

Then the joint density of (X_τ, M_τ) is the aforementioned probability density

$$f_{X_\tau, M_\tau}(x, \mu) = p_\tau(x_0, \mu_0, x, \mu). \tag{14.6.32}$$

Since $\mathbb{E}[X_\tau] = x_0$, $Var(X_\tau) = \sigma_S^2 \tau$, $\mathbb{E}[M_\tau] = \mu_0$, and $Var(M_\tau) = \sigma_\mu^2 \tau$, then the formula for the bivariate normal distribution becomes

$$f_{X_\tau, M_\tau}(x, \mu) = \frac{1}{2\pi\tau\sigma_S\sigma_\mu\sqrt{1-\rho^2}}$$
$$\exp\left(-\frac{1}{1-\rho^2}\left[\frac{(x-x_0)^2}{2\tau\sigma_S^2} - \rho\frac{(x-x_0)(\mu-\mu_0)}{\tau\sigma_S\sigma_\mu} + \frac{(\mu-\mu_0)^2}{2\tau\sigma_\mu^2}\right]\right),$$

which via formula (14.6.32) produces the heat kernel for the operator $\mathcal{A}$.

We conclude the previous computation with the following result:

---

[2]Roughly, this means the probability density of a diffusion starting at (x_0, μ_0) and ending at (x, μ) at time t.

Proposition 14.6.1 *The solution of the Kolmogorov backward equation*

$$\frac{\partial \mathbf{U}}{\partial \tau} = \mathcal{A}\mathbf{U}$$
$$\mathbf{U}(0, x, \mu) = \Phi(x, \mu)$$

is given by

$$\mathbf{U}(\tau, x, \mu) = \iint p_\tau(x, \mu, x', \mu')\Phi(x', \mu')\, dx'd\mu',$$

where

$$p_\tau(x, \mu, x', \mu') = \frac{1}{2\pi\tau\sigma_S\sigma_\mu\sqrt{1-\rho^2}}$$

$$exp\left(-\frac{1}{1-\rho^2}\left[\frac{(x-x')^2}{2\tau\sigma_S^2} - \rho\frac{(x-x')(\mu-\mu')}{\tau\sigma_S\sigma_\mu} + \frac{(\mu-\mu')^2}{2\tau\sigma_\mu^2}\right]\right).$$

$$(14.6.33)$$

The initial condition $\Phi(x, \mu)$ depends on the initial derivative payoff $V(T, S, \mu)$. This can be seen if we follow backwards the previous transformations as

$$\begin{aligned} V(T, S, \mu) &= W(T, x, \mu) = e^{\varphi(T,x,\mu)}U(T, x, \mu) \\ &= e^{\varphi(T,x,\mu)}\mathbf{U}(0, x, \mu) = e^{\varphi(T,x,\mu)}\Phi(x, \mu). \end{aligned}$$

This implies the relation

$$\Phi(x, \mu) = e^{-\varphi(T,x,\mu)}V(T, e^x, \mu).$$

Consequently, we obtain the following result regarding the value of a given derivative.

Theorem 14.6.2 *The value of a European derivative with payoff* $V(T, S, \mu)$ *is given by*

$$\begin{aligned} V(t, S, \mu) &= e^{\varphi(t,\ln S,\mu)}\mathbf{U}(\tau, \ln S, \mu) \\ &= e^{\varphi(t,\ln S,\mu)}\iint p_\tau(\ln S, \mu, x', \mu')e^{-\varphi(T,x',\mu')}V(T, e^{x'}, \mu')\, dx'd\mu' \end{aligned}$$

$$(14.6.34)$$

with p_τ *given by (14.6.33), and* $\tau = T - t$.

We note that the factor in front of the integral is given by

$$e^{\varphi(t,\ln S,\mu)} = S^\alpha e^{\beta t + \gamma\mu}.$$

The possibility of obtaining closed form solutions for derivatives depends on the ability of computing explicitly the integral (14.6.34) in the case of diverse payoffs.

However, an explicit formula in terms of elementary functions cannot be found, but, as in the case of the Black-Scoles formula, the price can be written in terms of a distribution function of a well-known distribution, which in this case is the bivariate standard normal distribution:

$$M(a,b;\rho) = \frac{1}{2\pi\sqrt{1-\rho^2}} \int_\infty^a \int_{-\infty}^b e^{-\frac{1}{1-\rho^2}\left[\frac{x^2}{2}-\rho xy+\frac{y^2}{2}\right]} \, dx dy.$$

As a distribution function, this means

$$M(a,b;\rho) = P(X \le a, Y \le b),$$

i.e. the probability that the first variable is less than a and the second one is less than b, when the correlation coefficients of the variables is ρ, and $X \sim \mathcal{N}(0,1)$, $Y \sim \mathcal{N}(0,1)$. Numerical algorithms for the implementation of the function $M(a,b;\rho)$ are well-known, see for instance Drezner [28]. In some particular cases, when either a or b is infinite, the function $M(a,b,\rho)$ can be written in terms of the cumulative function of the standard normal distribution, $N(\cdot)$.

So far we have completely characterized from a theoretical point of view the pricing formulas for European derivatives. What it follows next is to try to work out the formula for a given particular option.

14.7 All-or-Nothing Option

Consider the European derivative with payoff

$$V(T,S,\mu) = \mathbb{I}_{[K,\infty)}(S) = \begin{cases} 1, & \text{if } S \ge K \\ 0, & \text{otherwise.} \end{cases}$$

For the sake of simplicity we shall use the notation $c = 1 - \rho^2$. Then formula (14.6.34) becomes

$$V(t,S,\mu) = S^\alpha e^{\beta t + \gamma\mu} \frac{1}{2\pi\tau\sigma_S\sigma_\mu\sqrt{c}}$$
$$\times \int_\mathbb{R} \int_{\ln K}^\infty \exp\left\{ -\frac{1}{c}\left[\frac{(\ln S - x')^2}{2\tau\sigma_S^2} - \rho\frac{(\ln S - x')(\mu - \mu')}{\tau\sigma_S\sigma_\mu} + \frac{(\mu - \mu')^2}{2\tau\sigma_\mu^2} \right] \right\}$$
$$\times e^{-(\alpha x' + \beta T + \gamma\mu')} dx' d\mu'.$$

Substituting

$$y = \frac{\ln S - x'}{\sqrt{\tau}\sigma_S}, \qquad z = \frac{\mu - \mu'}{\sqrt{\tau}\sigma_\mu}$$

after performing cancelations and applying Fubini's theorem, we get

$$V(t, S, \mu)$$
$$= \frac{1}{2\pi c} e^{-\beta(T-t)} \int_{\mathbb{R}} \int_{-\infty}^{\ln(S/K)/(\sqrt{\tau}\sigma_S)} e^{-\frac{1}{c}[\frac{y^2}{2} - \rho yz + \frac{z^2}{2}] + \alpha\sqrt{\tau}\sigma_S y + \gamma\sqrt{\tau}\sigma_\mu z} \, dy dz$$

$$= \frac{1}{2\pi c} e^{-\beta(T-t)} \int_{-\infty}^{\ln(S/K)/(\sqrt{\tau}\sigma_S)} e^{-\frac{y^2}{2c} + \alpha\sqrt{\tau}\sigma_S y} \left(\underbrace{\int_{\mathbb{R}} e^{-\frac{z^2}{2c} + (\gamma\sqrt{\tau}\sigma_\mu + \frac{\rho y}{c})z} \, dz}_{=J_1} \right) dy.$$

Next we shall compute the integral J_1 using the well-known formula of integration for Gaussians

$$\int_{\mathbb{R}} e^{-Az^2 + Bz} \, dz = \sqrt{\frac{\pi}{A}} e^{\frac{B^2}{4A}},$$

with $A = \frac{1}{2c}$ and $B = \gamma\sqrt{\tau}\sigma_\mu + \frac{\rho y}{c}$. We obtain $J_1 = \sqrt{2\pi c} e^{\frac{c}{2}(\gamma\sqrt{\tau}\sigma_\mu + \frac{\rho y}{c})^2}$. Substituting into the previous derivative formula yields

$$V(t, S, \mu) = \frac{1}{\sqrt{2\pi c}} e^{-\beta(T-t)} \underbrace{\int_{-\infty}^{\ln(S/K)/(\sqrt{\tau}\sigma_S)} e^{-\frac{y^2}{2c} + \alpha\sqrt{\tau}\sigma_S y + \frac{c}{2}(\gamma\sqrt{\tau}\sigma_\mu + \frac{\rho}{c}y)^2} \, dy}_{=J_2}$$

$$= \frac{1}{\sqrt{2\pi c}} e^{-\beta(T-t)} J_2. \tag{14.7.35}$$

We shall compute next the integral J_2. Ordering the exponent over the powers of y and then completing the square, we can reduce the integral to a standard normal distribution function

$$J_2 = e^{\frac{c}{2}\tau\gamma^2\sigma_\mu^2} \int_{-\infty}^{\ln(S/K)/(\sqrt{\tau}\sigma_S)} e^{-\frac{y^2}{2} + (\alpha\sigma_S + \gamma\rho\sigma_\mu)\sqrt{\tau}y} \, dy$$

$$= e^{\frac{c}{2}\tau\gamma^2\sigma_\mu^2} e^{\frac{1}{2}(\alpha\sigma_S + \gamma\rho\sigma_\mu)^2} \int_{-\infty}^{D_1} e^{\frac{-u^2}{2}} \, du$$

$$= \sqrt{2\pi} e^{\frac{c}{2}\tau\gamma^2\sigma_\mu^2} e^{\frac{1}{2}(\alpha\sigma_S + \gamma\rho\sigma_\mu)^2} N(D_2),$$

where

$$D_2 = \frac{\ln(S/K)}{\sqrt{\tau}\sigma_S} - (\alpha\sigma_S + \gamma\rho\sigma_\mu)\sqrt{\tau}, \tag{14.7.36}$$

and recall that $\tau = T - t$. Substituting the expression of J_2 back into formula (14.7.35) and reversing the variables, we get the value of the all-or-nothing option as

$$V(t, S, \mu) = \frac{1}{\sqrt{1 - \rho^2}} e^{\left(\frac{1-\rho^2}{2}\gamma^2\sigma_\mu^2 - \beta\right)(T-t)} N(D_2).$$

The value of the constant β can be computed using (14.6.28) in terms of the constants α and γ as follows

$$
\begin{aligned}
\beta &= r - \frac{1}{2}(\alpha b + \lambda\gamma) = r - \frac{1}{2}\left(\alpha(r - \frac{1}{2}\sigma_S^2) + \lambda\gamma\right) \\
&= r\left(1 - \frac{\alpha}{2}\right) + \frac{\alpha}{4}\sigma_S^2 - \frac{1}{2}\lambda\gamma.
\end{aligned}
$$

Remark 14.7.1 The previous formula simplifies even further if the following two additional assumptions are made:
- the source noises are independent, i.e. $\rho = 0$;
- the market risk is null, i.e. $\lambda = 0$.

In the case we have

$$\alpha = \frac{1}{2} - \frac{r}{\sigma_S^2}, \quad \gamma = 0, \quad \beta = r\left(1 - \frac{\alpha}{2}\right) + \frac{\alpha}{4}\sigma_S^2,$$

which implies the following value of D_2

$$D_2^* = \frac{\ln(S/K)}{\sigma_S\sqrt{T-t}} - \left(\frac{\sigma_S}{2} - \frac{r}{\sigma_S}\right)\sqrt{T-t}.$$

Consequently, value of the derivative becomes

$$V(t, S, \mu) = e^{-\beta(T-t)} N(D_2^*).$$

It is worth noting the similarity with the formula for a plan-vanilla all-or-nothing option given by (6.1.8)

$$V(t, S) = e^{-r(T-t)} N(d_2).$$

This is the reason why we denoted the expression (14.7.36) with a D_2.

Remark 14.7.2 We have solved the equation (14.6.31) by considering it as a Kolmogorov equation associated with a diffusion whose transition density was not hard to find. In the following we shall sketch a method which does not use any diffusions, but Fourier transforms. We shall apply both Fourier transforms in x and μ variables and denote

$$\Lambda(\tau; \omega, p) = \left(\mathcal{F}_x \mathcal{F}_\mu \mathbf{U}(\tau, x, \mu)\right)(\omega, p).$$

Applying both Fourier transforms to the equation (14.6.31)

$$\frac{\partial \mathbf{U}}{\partial \tau} = \frac{1}{2}\sigma_S^2 \frac{\partial^2 \mathbf{U}}{\partial x^2} + \frac{1}{2}\sigma_\mu^2 \frac{\partial^2 \mathbf{U}}{\partial \mu^2} + \rho \sigma_S \sigma_\mu \frac{\partial^2 \mathbf{U}}{\partial x \partial \mu},$$

we obtain the differential equation

$$\frac{\partial \Lambda}{\partial \tau} = -\frac{1}{2}(\sigma_S^2 \omega^2 + \sigma_\mu^2 p^2 + \rho \sigma_S \sigma_\mu \omega p)\Lambda,$$

with the solution

$$\Lambda(\tau; \omega, p) = e^{-\frac{1}{2}(\sigma_S^2 \omega^2 + \sigma_\mu^2 p^2 + \rho \sigma_S \sigma_\mu \omega p)\tau} \Lambda(0; \omega, p).$$

Inverting, we obtain

$$
\begin{aligned}
\mathbf{U}(\tau, x, \mu) &= \mathcal{F}_p^{-1} \mathcal{F}_\omega^{-1}\Big(\Lambda(\tau; \omega, p)\Big)(x, \mu) \\
&= \frac{1}{(2\pi)^2} \int_{\mathbb{R}} \int_{\mathbb{R}} e^{i\omega x} e^{ip\mu} \Lambda(\tau; \omega, p)\, d\omega dp \\
&= \frac{1}{(2\pi)^2} \int_{\mathbb{R}} \int_{\mathbb{R}} e^{i(\omega x + p\mu)} e^{-\frac{1}{2}(\sigma_S^2 \omega^2 + \sigma_\mu^2 p^2 + \rho \sigma_S \sigma_\mu \omega p)\tau} \Lambda(0; \omega, p)\, d\omega dp.
\end{aligned}
$$

Using the definition of the double Fourier transform

$$\Lambda(0; \omega, p) = \int_{\mathbb{R}} \int_{\mathbb{R}} e^{-i\omega x'} e^{-ip\mu'} \mathbf{U}(0, x', \mu')\, dx' d\mu'$$

the previous formula becomes

$$\mathbf{U}(\tau, x, \mu) = \frac{1}{(2\pi)^2} \int_{\mathbb{R}^4} e^{i[\omega(x-x') + p(\mu - \mu')] - \frac{1}{2}(\sigma_S^2 \omega^2 + \sigma_\mu^2 p^2 + \rho \sigma_S \sigma_\mu \omega p)\tau} \mathbf{U}(0; x', \mu')\, dv$$

where $dv = dx' d\mu' d\omega dp$. Because of its complex analysis character, this formula is harder to use than the one presented in the section and which was solely based on the heat kernel.

Exercise 14.7.3 (Forward contract) *Find the price of a derivative with the payoff $V(T, S, \mu) = S - K$, where K is a constant.*

Exercise 14.7.4 (Asset-or-nothing option) *Using a similar computation as in the last section find the price of a derivative with the payoff*

$$V(T, S, \mu) = \begin{cases} S, & \text{if, } S \geq K \\ 0, & \text{otherwise.} \end{cases}$$

Conclusions

We shall conclude with a few final remarks regarding the experience gained through our journey into the mathematical modeling of financial markets wonderland. The book starts with the most simple and classical models of assets and derivatives and ends with some of the most sophisticated, but extremely popular models today, such as Heston volatility model or VAR stochastic rate of return model.

We have seen that there are two main methods for computing the price of a traded contract. The first one, credited to Harrison and Kreps, is using the risk-neutral method, by which the derivative price is obtained by discounting at the risk-free rate the conditional expectation of the payoff with respect to the present market information. The second method involves solving a partial differential equation of first order in time and second order in the underlying asset, which is generically called the Black-Scholes equation. Even if the methods are equivalent, each of them have some advantages and disadvantages. For instance, the increasing popularity of the risk-neutral valuation is that it can be computed by a Monte-Carlo simulation if the derivative is European. Its advantage over the numerical analysis of PDEs is that it can accomodate high dimensionality. The disadvantage is that it takes a long time and it is not as easy to implement for American options.

The main idea and effort of this book is dedicated to the explicit computation of the price of European and American derivatives by both methods, the risk-neutral valuation and solving partial differential equations.

Pricing by the risk-neutral method requires computation skills involving conditional expectations. We have used this method to compute prices for the most familiar options as well as to evaluate bonds and swap prices in the case when the underlying interest rate is stochastic.

Solving the Black-Scholes partial differential equation for European options is done by reducing it to a simple forward heat equation, whose solution can be obtained by a convolution with the well known heat kernel. In the case of American options we obtain a free-boundary value problem, which cannot be solved explicitly, unless the time to maturity is infinite. However, we can obtain implicit pricing formulas as well as approximations of prices for calls and puts near the maturity or near the boundary. Even if in low dimensions the partial differential equation can be solved by the machinery of numerical analysis, the idea of the book is pushing to obtain closed form formulas as far as we could.

The first and most popular model to model assets is Samuelson's model, by which the asset's price is given by a geometric Brownian motion, which depends on two parameters, the drift rate, μ, and the volatility, σ. Using this model, Black and Scholes were able to price calls and puts in early 1970s. The work was awarded the 1997 Nobel Price, including R. Merton. Even if the Black-Scholes model is still used today, there are some problems this model cannot accommodate, the most popular being the variability of the volatility with respect to strike price and time to maturity. The fix to this problem was done by considering stochastic volatility models, the most famous being the Heston model introduced in 1994. The popularity of this model stems in the fact that the prices of options can be represented in closed form formulas, similar with the ones proposed by Black and Scholes. Heston model depends on two market sources of noise, eventually correlated, and introduces a market price of risk which was postulated to be linear in volatility and independent of the stock. Heston partial differential equation can then be solved by exploiting its homogeneity in the asset price and then use the fact that its coefficients are linear in volatility and independent of the asset, a fact that can be pursued by applying a Fourier transform. The equation thus obtained can be solved and the price of the option can be retrieved by an inverse Fourier transform.

There is a long line of empirical research regarding the expected return on assets. This work, which was the topic of the 2013 Nobel Price in Economics, leads to an Orstein-Uhlenbeck process for the expected return on stocks in continuous time. Given the importance of this new direction, the book dedicates the last chapter to finding a formula for the European options whose underlying asset verifies a VAR process. Even if the construction of this model is similar with the Heston model, the partial differential can be brought to a constant coefficients equation. The associated diffusion of the differential operator is easy to find and its transition probability is bivariate Gaussian, a fact that provides the heat kernel of the equation. A convolution of the final condition with the heat kernel provides the derivative price. The study of derivatives with stochastic expected return on assets is still a subject of ongoing research. The interested reader is referred to the books [33] and [34] for a more complete and updated exposition.

Chapter 15

Hints and Solutions

Here we provide hints and solutions to selected exercises. The interested reader should be able to derive the solutions to the rest based on the experience gained from the chapter examples and other similar solved exercises.

Chapter 1

Exercise 1.6.1 Just (f) and (g) are not stopping times, since they require information about the future.

Exercise 1.10.1 (a) $M(t) = M_0 + N_t$; (b) $M'(t) = rM(t)$.

Exercise 1.10.2 (a) $M(t) = M_0 + N_t - \widetilde{N}_t$; (b) $M'(t) = (r - \lambda)M(t)$.

Chapter 2

Exercise 2.1.3 (a) $a = 0.4934$, $b = -1.9181$, and $c = 2.8141$.

Exercise 2.2.3 Apply the definition of expectation and obtain

$$
\begin{aligned}
\mathbb{E}[X_t] &= \int_{-\infty}^{\infty} |x| \frac{1}{\sqrt{2\pi t}} e^{-\frac{x^2}{2t}}\, dx = \int_{0}^{\infty} 2x \frac{1}{\sqrt{2\pi t}} e^{-\frac{x^2}{2t}}\, dx \\
&= \frac{1}{\sqrt{2\pi t}} \int_{0}^{\infty} e^{-\frac{y}{2t}}\, dy = \sqrt{2t/\pi}.
\end{aligned}
$$

Exercise 2.2.4 The goodness of fit function

$$
F(b, \sigma) = \sigma^2 \sum t_j + 2\sigma \sqrt{\frac{2}{\pi}} \sum (b - y_j)\sqrt{t_j} + \sum (b - y_j)^2
$$

is quadratic in both variables. The equations $\partial_b F = 0$, $\partial_\sigma F = 0$ form a linear system in b and σ with an unique solution.

Exercise 2.2.6 (*b*) Integrating in (*a*) we get

$$W_t^4 = 4 \int_0^t W_s^3 \, dW_s + 6 \int_0^t W_s^2 \, ds.$$

Apply the expectation and obtain

$$\mathbb{E}[W_t^4] = 4\mathbb{E}\left[\int_0^t W_s^3 \, dW_s \right] + 6 \int_0^t \mathbb{E}[W_s^2] \, ds = 6 \int_0^t s \, ds = 3t^2.$$

(*c*) The goodness of fit is $F(\sigma) = 3\sigma^2 \sum t_j^2 - 2\sigma \sum y_j t_j + \sum y_j^2$, with the minimum reached for $\sigma = \dfrac{1}{3} \dfrac{\sum y_j t_j}{\sum t_j^2}$.

Exercise 2.4.2 (*a*) $p_t(u) = \dfrac{2}{\sigma\sqrt{2\pi t}} e^{-\frac{u^2}{2\sigma^2 t}}, \ t \geq 0.$ (*b*) $\sigma = \sqrt{\dfrac{1}{n} \sum \dfrac{y_j^2}{t_j}}.$

Chapter 3

Exercise 3.3.4 (*a*) Since $r_t \sim N(\mu, s^2)$, with $\mu = b + (r_0 - b)e^{-at}$ and $s^2 = \frac{\sigma^2}{2a}(1 - e^{-2at})$. Then

$$P(r_t < 0) = \int_{-\infty}^0 \frac{1}{\sqrt{2\pi}s} e^{-\frac{(x-\mu)^2}{2s^2}} \, dx = \frac{1}{\sqrt{2\pi}} \int_{-\infty}^{-\mu/s} e^{-v^2/2} \, dv = N(-\mu/s),$$

where by computation

$$-\frac{\mu}{s} = -\frac{1}{\sigma}\sqrt{\frac{2a}{e^{2at}-1}} \left(r_0 + b(e^{at} - 1) \right).$$

(*b*) Since

$$\lim_{t\to\infty} \frac{\mu}{s} = \frac{\sqrt{2a}}{\sigma} \lim_{t\to\infty} \frac{r_0 + b(e^{at} - 1)}{\sqrt{e^{2at} - 1}} = \frac{b\sqrt{2a}}{\sigma},$$

then

$$\lim_{t\to\infty} P(r_t < 0) = \lim_{t\to\infty} N(-\mu/s) = N(-\frac{b\sqrt{2a}}{\sigma}).$$

It is worth noting that the previous probability is less than 0.5.
(*c*) The rate of change is

$$\frac{d}{dt} P(r_t < 0) = -\frac{1}{\sqrt{2\pi}} e^{-\frac{\mu^2}{2s^2}} \frac{d}{ds}\left(\frac{\mu}{s} \right)$$

$$= -\frac{1}{\sqrt{2\pi}} e^{-\frac{\mu^2}{2s^2}} \frac{ae^{2at}[b(e^{2at} - 1)(e^{-at} - 1) - r_0]}{(e^{2at} - 1)^{3/2}}.$$

Exercise 3.3.5 By Ito's formula

$$d(r_t^n) = \left(na(b - r_t)r_t^{n-1} + \frac{1}{2}n(n-1)\sigma^2 r_t^{n-1}\right)dt + n\sigma r_t^{n-\frac{1}{2}}dW_t.$$

Integrating between 0 and t and taking the expectation yields

$$\mu_n(t) = r_0^n + \int_0^t \left[nab\mu_{n-1}(s) - na\mu_n(s) + \frac{n(n-1)}{2}\sigma^2\mu_{n-1}(s)\right]ds.$$

Differentiating yields

$$\mu_n'(t) + na\mu_n(t) = \left(nab + \frac{n(n-1)}{2}\sigma^2\right)\mu_{n-1}(t).$$

Multiplying by the integrating factor e^{nat} yields the exact equation

$$[e^{nat}\mu_n(t)]' = e^{nat}\left(nab + \frac{n(n-1)}{2}\sigma^2\right)\mu_{n-1}(t).$$

Integrating yields the following recursive formula for moments

$$\mu_n(t) = r_0^n e^{-nat} + \left(nab + \frac{n(n-1)}{2}\sigma^2\right)\int_0^t e^{-na(t-s)}\mu_{n-1}(s)\,ds.$$

Exercise 3.5.1 (a) The spot rate r_t follows the process

$$d(\ln r_t) = \theta(t)dt + \sigma dW_t.$$

Integrating yields

$$\ln r_t = \ln r_0 + \int_0^t \theta(u)\,du + \sigma W_t,$$

which is normally distributed. (b) Then $r_t = r_0 e^{\int_0^t \theta(u)\,du + \sigma W_t}$ is log-normally distributed. (c) The mean and variance are

$$\mathbb{E}[r_t] = r_0 e^{\int_0^t \theta(u)\,du} e^{\frac{1}{2}\sigma^2 t}, \quad Var(r_t) = r_0^2 e^{2\int_0^t \theta(u)\,du} e^{\sigma^2 t}(e^{\sigma^2 t} - 1).$$

Exercise 3.5.2 Substitute $u_t = \ln r_t$ and obtain the linear equation

$$du_t + a(t)u_t dt = \theta(t)dt + \sigma(t)dW_t.$$

The equation can be solved multiplying by the integrating factor $e^{\int_0^t a(s)\,ds}$.

Chapter 4

Exercise 4.5.1 (*a*) Swapping the expectation with the integral we have

$$\mathbb{E}[X_t] = \int_0^t e^{-as}\mathbb{E}[\beta_s]\,ds = 0,$$

since β_t is a Wiener integral satisfying the normality condition $\beta_t \sim \mathcal{N}(0, \frac{e^{2at}-1}{2a})$.
(*b*) The product rule yields

$$
\begin{aligned}
d(X_s\beta_s) &= \beta_s\,dX_s + X_s\,d\beta_s + \underbrace{dX_s\,d\beta_s}_{=0} \\
&= e^{-as}\beta_s^2\,ds + X_s e^{as}\,dW_s.
\end{aligned}
$$

Integrating we obtain

$$X_t\beta_t = \int_0^t e^{-as}\beta_s^2\,ds + \int_0^t X_s e^{as}\,dW_s.$$

Taking the expectation and using that a Wiener integral has a zero expectation, we get

$$\mathbb{E}[X_t\beta_t] = \int_0^t e^{-as}\mathbb{E}[\beta_s^2]\,ds = \int_0^t e^{-as}\frac{e^{2at}-1}{2a}\,ds = \frac{1}{a^2}\left(\frac{e^{at}+e^{-at}}{2}-1\right).$$

(*c*) Applying Ito's lemma

$$d(X_t^2) = 2X_t\,dX_t + (dX_t)^2 = 2X_t e^{-at}\beta_t\,dt$$

and then integrating and applying part (*b*) yields

$$
\begin{aligned}
\mathbb{E}[X_t^2] &= 2\int_0^t e^{-as}\mathbb{E}[X_s\beta_s]\,ds = \frac{2}{a^2}\int_0^t\left(\frac{1+e^{-2as}}{2}-e^{-as}\right)ds \\
&= \frac{1}{2a^2}\left(t+\frac{1-e^{-2at}}{2a}+\frac{2}{a}(1-e^{-at})\right).
\end{aligned}
$$

(*d*) Using a stochastic variant of Fubini's theorem we can interchange the Riemannian and the Wiener integrals as follows

$$
\begin{aligned}
X_t &= \int_0^t e^{-as}\beta_s\,ds = \int_0^t e^{-as}\int_0^s e^{a\tau}\,dW_\tau\,ds \\
&= \int_0^s e^{a\tau}\int_0^t e^{-as}\,ds\,dW_\tau = \int_0^t e^{-as}\,ds\int_0^s e^{a\tau}\,dW_\tau \\
&= \frac{1}{a}(1-e^{at})\int_0^s e^{a\tau}\,dW_\tau,
\end{aligned}
$$

which implies that X_t is normally distributed with mean 0 and variance $\mathbb{E}[X_t^2]$ computed in part (c). Part (d) follows from the formula

$$\mathbb{E}[e^{\sigma X_t}] = e^{(\sigma^2/2)Var(X_t)}.$$

Exercise 4.5.2 We obtain

$$\theta(t) = \partial_t f(0, t) + \sigma^2 t,$$

which recovers the expression for the Ho-Lee model.

Exercise 4.7.1 Apply the same method used in the Application 4.7. The price of the bond is: $P(t, T) = e^{-r_t(T-t)} \mathbb{E}[e^{-\sigma \int_0^{T-t} N_s \, ds}]$.

Exercise 4.9.2 The price of an infinitely lived bond at time t is given by $\lim_{T \to \infty} P(t, T)$. Using $\lim_{T \to \infty} B(t, T) = \dfrac{1}{a}$, and

$$\lim_{T \to \infty} A(t, T) = \begin{cases} +\infty, & \text{if } b < \sigma^2/(2a) \\ 1, & \text{if } b > \sigma^2/(2a) \\ (\frac{1}{a} + t)(b - \frac{\sigma^2}{2a^2}) - \frac{\sigma^2}{4a^3}, & \text{if } b = \sigma^2/(2a), \end{cases}$$

we can get the price of the bond in all three cases.

Exercise 4.12.3 (a) If S_n denotes the time to the nth jump of the Poisson process N_t, then we have

$$\begin{aligned} d[(1+\sigma)^{N_t}] &= \begin{cases} 0, & \text{if } t \neq S_n \\ (1+\sigma)^{n+1} - (1+\sigma)^n = \sigma(1+\sigma)^n, & \text{if } t = S_n \end{cases} \\ &= \sigma(1+\sigma)^{N_t} \, dN_t. \end{aligned}$$

(b) The solution is $X_t = (1+\sigma)^{N_t}$.

Exercise 4.12.4 The equation can be written as

$$dr_t + \sigma \lambda r_t dt = \sigma r_t \, dN_t.$$

Multiplying by the integrating factor $e^{\sigma \lambda t}$ yields

$$d(e^{\sigma \lambda t} r_t) = \sigma e^{\sigma \lambda t} r_t \, dN_t.$$

Denote by $X_t = e^{\sigma \lambda t} r_t / r_0$. Then X_t satisfies

$$dX_t = \sigma X_t \, dN_t, \qquad X_0 = 1.$$

Applying Exercise 4.12.3 and solving for r_t yields

$$r_t = r_0 e^{-\sigma \lambda t} \sigma (1+\sigma)^{N_t}.$$

Chapter 5

Exercise 5.1.1 Dividing the equations

$$S_t = S_0 e^{(\mu - \frac{\sigma^2}{2})t + \sigma W_t}$$
$$S_u = S_0 e^{(\mu - \frac{\sigma^2}{2})u + \sigma W_u}$$

yields

$$S_t = S_u e^{(\mu - \frac{\sigma^2}{2})(t-u) + \sigma(W_t - W_u)}.$$

Taking the predictable part out and dropping the independent condition we obtain

$$
\begin{aligned}
\mathbb{E}[S_t | \mathcal{F}_u] &= S_u e^{(\mu - \frac{\sigma^2}{2})(t-u)} \mathbb{E}[e^{\sigma(W_t - W_u)} | \mathcal{F}_u] \\
&= S_u e^{(\mu - \frac{\sigma^2}{2})(t-u)} \mathbb{E}[e^{\sigma(W_t - W_u)}] \\
&= S_u e^{(\mu - \frac{\sigma^2}{2})(t-u)} \mathbb{E}[e^{\sigma W_{t-u}}] \\
&= S_u e^{(\mu - \frac{\sigma^2}{2})(t-u)} e^{\frac{1}{2}\sigma^2(t-u)} \\
&= S_u e^{\mu(t-u)}.
\end{aligned}
$$

Similarly we obtain

$$
\begin{aligned}
\mathbb{E}[S_t^2 | \mathcal{F}_u] &= S_u^2 e^{2(\mu - \frac{\sigma^2}{2})(t-u)} \mathbb{E}[e^{2\sigma(W_t - W_u)} | \mathcal{F}_u] \\
&= S_u^2 e^{2\mu(t-u)} e^{\sigma^2(t-u)}.
\end{aligned}
$$

Then

$$
\begin{aligned}
Var(S_t | \mathcal{F}_u) &= \mathbb{E}[S_t^2 | \mathcal{F}_u] - \mathbb{E}[S_t | \mathcal{F}_u]^2 \\
&= S_u^2 e^{2\mu(t-u)} e^{\sigma^2(t-u)} - S_u^2 e^{2\mu(t-u)} \\
&= S_u^2 e^{2\mu(t-u)} \left(e^{\sigma^2(t-u)} - 1 \right).
\end{aligned}
$$

When $s = t$ we get $\mathbb{E}[S_t | \mathcal{F}_t] = S_t$ and $Var(S_t | \mathcal{F}_t) = 0$.

Exercise 5.1.2 By Ito's formula

$$
\begin{aligned}
d(\ln S_t) &= \frac{1}{S_t} dS_t - \frac{1}{2}\frac{1}{S_t^2}(dS_t)^2 \\
&= (\mu - \frac{\sigma^2}{2})dt + \sigma dW_t,
\end{aligned}
$$

so $\ln S_t = \ln S_0 + (\mu - \frac{\sigma^2}{2})t + \sigma W_t$, and hence $\ln S_t$ is normally distributed with $\mathbb{E}[\ln S_t] = \ln S_0 + (\mu - \frac{\sigma^2}{2})t$ and $Var(\ln S_t) = \sigma^2 t$.

Exercise 5.1.4 (a) $d\left(\frac{1}{S_t}\right) = (\sigma^2 - \mu)\frac{1}{S_t}dt - \frac{\sigma}{S_t}dW_t$;
(b) $d(S_t^n) = n(\mu + \frac{n-1}{2}\sigma^2)dt + n\sigma S_t^n dW_t$;
(c) We have

$$
\begin{aligned}
d\left((S_t - 1)^2\right) &= d(S_t^2) - 2dS_t = 2S_t dS_t + (dS_t)^2 - 2dS_t \\
&= \left((2\mu + \sigma^2)S_t^2 - 2\mu S_t\right)dt + 2\sigma S_t(S_t - 1)dW_t.
\end{aligned}
$$

Exercise 5.1.5 (b) Since $S_t = S_0 e^{(\mu - \frac{\sigma^2}{2})t + \sigma W_t}$, then $S_t^n = S_0^n e^{(n\mu - \frac{\sigma^2}{2})t + n\sigma W_t}$
and hence

$$
\begin{aligned}
\mathbb{E}[S_t^n] &= S_0^n e^{(n\mu - \frac{\sigma^2}{2})t}\mathbb{E}[e^{n\sigma W_t}] \\
&= S_0^n e^{(n\mu - \frac{\sigma^2}{2})t}e^{\frac{1}{2}n^2\sigma^2 t} \\
&= S_0^n e^{(n\mu + \frac{n(n-1)}{2}\sigma^2)t}.
\end{aligned}
$$

(a) Let $n = 2$ in (b) and obtain $\mathbb{E}[S_t^2] = S_0 e^{(2\mu + \sigma^2)t}$.

Exercise 5.1.6 (a) Let $X_t = S_t W_t$. The product rule yields

$$
\begin{aligned}
d(X_t) &= W_t dS_t + S_t dW_t + dS_t dW_t \\
&= W_t(\mu S_t dt + \sigma S_t dW_t) + S_t dW_t + (\mu S_t dt + \sigma S_t dW_t)dW_t \\
&= S_t(\mu W_t + \sigma)dt + (1 + \sigma W_t)S_t dW_t.
\end{aligned}
$$

Integrating

$$
X_s = \int_0^s S_t(\mu W_t + \sigma)dt + \int_0^s (1 + \sigma W_t)S_t dW_t.
$$

Let $f(s) = \mathbb{E}[X_s]$. Then

$$
f(s) = \int_0^s (\mu f(t) + \sigma S_0 e^{\mu t})\,dt.
$$

Differentiating yields

$$
f'(s) = \mu f(s) + \sigma S_0 e^{\mu s}, \qquad f(0) = 0,
$$

with the solution $f(s) = \sigma S_0 s e^{\mu s}$. Hence $\mathbb{E}[W_t S_t] = \sigma S_0 t e^{\mu t}$.
(b) $cov(S_t, W_t) = \mathbb{E}[S_t W_t] - \mathbb{E}[S_t]\mathbb{E}[W_t] = \mathbb{E}[S_t W_t] = \sigma S_0 t e^{\mu t}$. Then

$$
\begin{aligned}
cor(S_t, W_t) &= \frac{cov(S_t, W_t)}{\sigma_{S_t}\sigma_{W_t}} = \frac{\sigma S_0 t e^{\mu t}}{S_0 e^{\mu t}\sqrt{e^{\sigma^2 t} - 1}\sqrt{t}} \\
&= \sigma\sqrt{\frac{t}{e^{\sigma^2 t} - 1}} \to 0, \quad t \to \infty.
\end{aligned}
$$

Exercise 5.2.4 From the definition of the correlation and properties of Brownian motions, we have

$$\rho = cor(dW_1, dW_2) = \frac{cov(dW_1, dW_2)}{\sqrt{Var(dW_1)}\sqrt{Var(dW_1)}}$$

$$= \frac{\mathbb{E}[dW_1\,dW_2] - \mathbb{E}[dW_1]\mathbb{E}[dW_2]}{\sqrt{\mathbb{E}[dW_1^2]}\sqrt{\mathbb{E}[dW_2^2]}} = \frac{\mathbb{E}[dW_1\,dW_2]}{dt}.$$

Exercise 5.2.7 It follows from $cor(S_1, S_2)^2 \le 1$.

Exercise 5.5.3 $d = S_d/S_0 = 0.5$, $u = S_u/S_0 = 1.5$, $\gamma = -6.5$, $p = 0.98$.

Exercise 5.3.8 Let T_a be the time S_t reaches level a for the first time. Then

$$F(a) = P(\overline{S}_t < a) = P(T_a > t)$$

$$= \ln\frac{a}{S_0}\int_t^\infty \frac{1}{\sqrt{2\pi\sigma^3\tau^3}}e^{-\frac{(\ln\frac{a}{S_0} - \alpha\sigma\tau)^2}{2\tau\sigma^3}}\,d\tau,$$

where $\alpha = \mu - \frac{1}{2}\sigma^2$.

Exercise 5.3.9 (a) $P(T_a < T) = \ln\frac{a}{S_0}\int_0^T \frac{1}{\sqrt{2\pi\sigma^3\tau^3}}e^{-(\ln\frac{a}{S_0} - \alpha\sigma\tau)^2/(2\tau\sigma^3)}\,d\tau.$

(b) $\int_{T_1}^{T_2} p(\tau)\,d\tau.$

Exercise 5.8.4 $\mathbb{E}[A_t] = S_0\left(1 + \frac{\mu t}{2}\right) + O(t^2)$, $Var(A_t) = \frac{S_0}{3}\sigma^2 t + O(t^2)$.

Exercise 5.8.7 (a) Since $\ln G_t = \frac{1}{t}\int_0^t \ln S_u\,du$, using the product rule yields

$$d(\ln G_t) = d\left(\frac{1}{t}\right)\int_0^t \ln S_u\,du + \frac{1}{t}d\left(\int_0^t \ln S_u\,du\right)$$

$$= -\frac{1}{t^2}\left(\int_0^t \ln S_u\,du\right)dt + \frac{1}{t}\ln S_t\,dt$$

$$= \frac{1}{t}(\ln S_t - \ln G_t)dt.$$

(b) Let $X_t = \ln G_t$. Then $G_t = e^{X_t}$ and Ito's formula yields

$$dG_t = de^{X_t} = e^{X_t}dX_t + \frac{1}{2}e^{X_t}(dX_t)^2 = e^{X_t}dX_t$$

$$= G_t d(\ln G_t) = \frac{G_t}{t}(\ln S_t - \ln G_t)dt.$$

Exercise 5.8.8 Using the product rule we have

$$d\left(\frac{H_t}{t}\right) = d\left(\frac{1}{t}\right)H_t + \frac{1}{t}dH_t$$

$$= -\frac{1}{t^2}H_t dt + \frac{1}{t^2}H_t\left(1 - \frac{H_t}{S_t}\right)dt$$

$$= -\frac{1}{t^2}\frac{H_t^2}{S_t}dt,$$

so, if $dt > 0$, then $d\left(\frac{H_t}{t}\right) < 0$, and hence $\frac{H_t}{t}$ is decreasing. For the second part, try to apply l'Hospital rule.

Exercise 5.8.9 By continuity, the inequality (5.8.29) is preserved under the limit.

Exercise 5.8.10 (a) Use that $\frac{1}{n}\sum S_{t_k}^\alpha = \frac{1}{t}\sum S_{t_k}^\alpha \frac{t}{n} \to \frac{1}{t}\int_0^t S_u^\alpha \, du$. (b) Let $I_t = \int_0^t S_u^\alpha \, du$. Then

$$d\left(\frac{1}{t}I_t\right) = d\left(\frac{1}{t}\right)I_t + \frac{1}{t}dI_t = \frac{1}{t}\left(S_t^\alpha - \frac{I_t}{t}\right)dt.$$

Let $X_t = A_t^\alpha$. Then by Ito's formula we get

$$
\begin{aligned}
dX_t &= \alpha\left(\frac{1}{t}I_t\right)^{\alpha-1}d\left(\frac{1}{t}I_t\right) + \frac{1}{2}\alpha(\alpha-1)\left(\frac{1}{t}I_t\right)^{\alpha-2}\left(d\left(\frac{1}{t}I_t\right)\right)^2 \\
&= \alpha\left(\frac{1}{t}I_t\right)^{\alpha-1}d\left(\frac{1}{t}I_t\right) = \alpha\left(\frac{1}{t}I_t\right)^{\alpha-1}\frac{1}{t}\left(S_t^\alpha - \frac{I_t}{t}\right)dt \\
&= \frac{\alpha}{t}\left(S_t(A_t^\alpha)^{1-1/\alpha} - A_t^\alpha\right)dt.
\end{aligned}
$$

(c) If $\alpha = 1$ we get the continuous arithmetic mean, $A_t^1 = \frac{1}{t}\int_0^t S_u \, du$. If $\alpha = -1$ we obtain the continuous harmonic mean, $A_t^{-1} = \frac{t}{\int_0^t \frac{1}{S_u}\, du}$.

Exercise 5.8.11 (a) By the product rule we have

$$
\begin{aligned}
dX_t &= d\left(\frac{1}{t}\right)\int_0^t S_u dW_u + \frac{1}{t}d\left(\int_0^t S_u \, dW_u\right) \\
&= -\frac{1}{t^2}\left(\int_0^t S_u \, dW_u\right)dt + \frac{1}{t}S_t dW_t \\
&= -\frac{1}{t}X_t dt + \frac{1}{t}S_t dW_t.
\end{aligned}
$$

Using the properties of Ito integrals, we have

$$
\begin{aligned}
\mathbb{E}[X_t] &= \frac{1}{t}\mathbb{E}\left[\int_0^t S_u \, dW_u\right] = 0 \\
Var(X_t) &= \mathbb{E}[X_t^2] - \mathbb{E}[X_t]^2 = \mathbb{E}[X_t^2] \\
&= \frac{1}{t^2}\mathbb{E}\left[\left(\int_0^t S_u \, dW_u\right)\left(\int_0^t S_u \, dW_u\right)\right] \\
&= \frac{1}{t^2}\int_0^t \mathbb{E}[S_u^2]\, du = \frac{1}{t^2}\int_0^t S_0 e^{(2\mu+\sigma^2)u}\, du \\
&= \frac{S_0^2}{t^2}\frac{e^{(2\mu+\sigma^2)t} - 1}{2\mu + \sigma^2}.
\end{aligned}
$$

(b) The stochastic differential equation of the stock price can be written in the form

$$\sigma S_t dW_t = dS_t - \mu S_t dt.$$

Integrating between 0 and t yields

$$\sigma \int_0^t S_u \, dW_u = S_t - S_0 - \mu \int_0^t S_u \, du.$$

Dividing by t yields the desired relation.

Exercise 5.9.5 Using independence $\mathbb{E}[S_t] = S_0 e^{\mu t}\mathbb{E}[(1+\rho)^{N_t}]$. Then use

$$\mathbb{E}[(1+\rho)^{N_t}] \;=\; \sum_{n \geq 0} \mathbb{E}[(1+\rho)^n | N_t = n] P(N_t = n) = \sum_{n \geq 0}(1+\rho)^n \frac{(\lambda t)^n}{n!} = e^{\lambda(1+\rho)t}.$$

Exercise 5.9.6 $\mathbb{E}[\ln S_t] = \ln S_0 \left(\mu - \lambda\rho - \frac{\sigma^2}{2} + \lambda\ln(\rho+1)\right)t.$

Exercise 5.9.7 $\mathbb{E}[S_t | \mathcal{F}_u] = S_u e^{(\mu - \lambda\rho - \frac{\sigma^2}{2})(t-u)}(1+\rho)^{N_t - u}.$

Exercise 5.9.8 Recall that the jump ratios are $Y_j = S_{t_j}/S_{t_j -}$, where the jumps occur at time instances $0 < t_1 < t_2 < \cdots < t_n$, where $n = N_t$. The stock price at maturity is given by the formula

$$S_t = S_0 e^{(\mu - \lambda\rho - \frac{1}{2}\sigma^2)t + \sigma W_t} \prod_{j=1}^{N_t} Y_j,$$

where $\rho = \mathbb{E}[Y_j] - 1$ is the expected jump size, and $Y_1, \cdots, Y_n$ are considered independent among themselves and also with respect to W_t. Conditioning over $N_t = n$ yields

$$\mathbb{E}\Big[\prod_{j=1}^{N_t} Y_j\Big] \;=\; \sum_{n \geq 0} \mathbb{E}\Big[\prod_{j=1}^{N_t} Y_j | N_t = n\Big] P(N_t = n)$$

$$= \sum_{n \geq 0}\prod_{j=1}^{n} \mathbb{E}[Y_j] P(N_t = n)$$

$$= \sum_{n \geq 0}(\rho + 1)^n \frac{(\lambda t)^n}{n!} e^{-\lambda t} = e^{\lambda t(\rho+1) - \lambda t}.$$

Since W_t is independent of N_t and Y_j we have

$$\mathbb{E}[S_t] \;=\; S_0 e^{(\mu - \lambda\rho - \frac{\sigma^2}{2})t}\mathbb{E}[e^{\sigma W_t}]\mathbb{E}\Big[\prod_{j=0}^{N_t} Y_j\Big] = S_0 e^{\mu t}.$$

Chapter 6

Exercise 6.1.1 Since $\ln S_T \sim N\left(\ln S_t + (\mu - \frac{\sigma^2}{2}(T-t)), \sigma^2(T-t)\right)$, we have

$$
\begin{aligned}
\widehat{\mathbb{E}}_t[f_T] &= \mathbb{E}\left[(\ln S_T)^2 | \mathcal{F}_t, \mu = r\right] \\
&= Var\left(\ln S_T | \mathcal{F}_t, \mu = r\right) + \mathbb{E}[\ln S_T | \mathcal{F}_t, \mu = r]^2 \\
&= \sigma^2(T-t) + \left[\ln S_t + (r - \frac{\sigma^2}{2})(T-t)\right]^2,
\end{aligned}
$$

so

$$
f_t = e^{-r(T-t)}\left[\sigma^2(T-t) + \left[\ln S_t + (r - \frac{\sigma^2}{2})(T-t)\right]^2\right].
$$

Exercise 6.1.3 Let $n = 2$.

$$
\begin{aligned}
f_t &= e^{-r(T-t)}\widehat{\mathbb{E}}_t[(S_T - K)^2] \\
&= e^{-r(T-t)}\widehat{\mathbb{E}}_t[S_T^2] + e^{-r(T-t)}\left(K^2 - 2K\widehat{\mathbb{E}}_t[S_T]\right) \\
&= S_t^2 e^{(r+\sigma^2)(T-t)} + e^{-r(T-t)}K^2 - 2KS_t.
\end{aligned}
$$

Let $n = 3$. Then

$$
\begin{aligned}
f_t &= e^{-r(T-t)}\widehat{\mathbb{E}}_t[(S_T - K)^3] \\
&= e^{-r(T-t)}\widehat{\mathbb{E}}_t[S_T^3 - 3KS_T^2 + 3K^2 S_T - K^3] \\
&= e^{-r(T-t)}\widehat{\mathbb{E}}_t[S_T^3] - 3Ke^{-r(T-t)}\widehat{\mathbb{E}}_t[S_T^2] + e^{-r(T-t)}(3K^2\widehat{\mathbb{E}}_t[S_T] - K^3) \\
&= e^{2(r+3\sigma^2/2)(T-t)}S_t^3 - 3Ke^{(r+\sigma^2)(T-t)}S_t^2 + 3K^2 S_t - e^{-r(T-t)}K^3.
\end{aligned}
$$

Exercise 6.1.5 (a) Substitute $\lim_{S_t \to \infty} N(d_1) = \lim_{S_t \to \infty} N(d_2) = 1$ and $\lim_{S_t \to \infty} \frac{K}{S_t} = 0$ in

$$
\frac{c(t)}{S_t} = N(d_1) - \frac{K}{S_t}e^{-r(T-t)}N(d_2).
$$

(b) Use $\lim_{S_t \to 0} N(d_1) = \lim_{S_t \to 0} N(d_2) = 0$. (c) It comes from an analysis similar to the one used at (a).

Exercise 6.1.6 Differentiating in $c(t) = S_t N(d_1) - Ke^{-r(T-t)}N(d_2)$ yields

$$
\begin{aligned}
\frac{dc(t)}{dS_t} &= N(d_1) + S_t\frac{dN(d_1)}{dS_t} - Ke^{-r(T-t)}\frac{dN(d_2)}{dS_t} \\
&= N(d_1) + S_t N'(d_1)\frac{\partial d_1}{\partial S_t} - Ke^{-r(T-t)}N'(d_2)\frac{\partial d_2}{\partial S_t} \\
&= N(d_1) + S_t\frac{1}{\sqrt{2\pi}}e^{-d_1^2/2}\frac{1}{\sigma\sqrt{T-t}}\frac{1}{S_t} - Ke^{-r(T-t)}\frac{1}{\sqrt{2\pi}}e^{-d_2^2/2}\frac{1}{\sigma\sqrt{T-t}}\frac{1}{S_t} \\
&= N(d_1) + \frac{1}{\sqrt{2\pi}}\frac{1}{\sigma\sqrt{T-t}}\frac{1}{S_t}e^{-d_2^2/2}\left[S_t e^{(d_2^2-d_1^2)/2} - Ke^{-r(T-t)}\right].
\end{aligned}
$$

It suffices to show

$$S_t e^{(d_2^2 - d_1^2)/2} - K e^{-r(T-t)} = 0.$$

Since

$$d_2^2 - d_1^2 = (d_2 - d_1)(d_2 + d_1) = -\sigma\sqrt{T-t}\frac{2\ln(S_t/K) + 2r(T-t)}{\sigma\sqrt{T-t}}$$

$$= -2\Big(\ln S_t - \ln K + r(T-t)\Big),$$

then we have

$$e^{(d_2^2 - d_1^2)/2} = e^{-\ln S_t + \ln K - r(T-t)} = e^{\ln \frac{1}{S_t}} e^{\ln K} e^{-r(T-t)}$$

$$= \frac{1}{S_t} K e^{-r(T-t)}.$$

Therefore

$$S_t e^{(d_2^2 - d_1^2)/2} - K e^{-r(T-t)} = S_t \frac{1}{S_t} K e^{-r(T-t)} - K e^{-r(T-t)} = 0,$$

which shows that $\frac{dc(t)}{dS_t} = N(d_1)$.

Exercise 6.1.7 $\ln S_T^n$ is normally distributed with

$$\ln S_T^n = n \ln S_T \sim N\Big(n \ln S_t + n\big(\mu - \frac{\sigma^2}{2}(T-t)\big), n^2 \sigma^2 (T-t)\Big).$$

Redo the computation of Proposition 6.1.4 in this case.

Exercise 6.1.8 The payoff is

$$f_T = \begin{cases} 1, & \text{if } \ln K_1 \leq X_T \leq \ln K2 \\ 0, & \text{otherwise.} \end{cases}$$

where $X_T \sim N\Big(\ln S_t + (\mu - \frac{\sigma^2}{2})(T-t), \sigma^2(T-t)\Big)$.

$$\widehat{\mathbb{E}}_t[f_T] = E[f_T | \mathcal{F}_t, \mu = r] = \int_{-\infty}^{\infty} f_T(x) p(x)\, dx$$

$$= \int_{\ln K_1}^{\ln K_2} \frac{1}{\sigma\sqrt{2\pi}\sqrt{T-t}} e^{-\frac{[x - \ln S_t - (r - \frac{\sigma^2}{2})(T-t)]^2}{2\sigma^2(T-t)}}\, dx$$

$$= \int_{-d_2(K_1)}^{-d_2(K_2)} \frac{1}{\sqrt{2\pi}} e^{-y^2/2}\, dy = \int_{d_2(K_2)}^{d_2(K_1)} \frac{1}{\sqrt{2\pi}} e^{-y^2/2}\, dy$$

$$= N\big(d_2(K_1)\big) - N\big(d_2(K_2)\big),$$

where

$$d_2(K_1) = \frac{\ln S_t - \ln K_1 + (r - \frac{\sigma^2}{2})(T - t)}{\sigma\sqrt{T - t}}$$

$$d_2(K_2) = \frac{\ln S_t - \ln K_2 + (r - \frac{\sigma^2}{2})(T - t)}{\sigma\sqrt{T - t}}.$$

Hence the value of a box-contract at time t is $f_t = e^{-r(T-t)}[N(d_2(K_1)) - N(d_2(K_2))]$.

Exercise 6.1.9 The payoff is

$$f_T = \begin{cases} e^{X_T}, & \text{if } X_T > \ln K \\ 0, & \text{otherwise.} \end{cases}$$

$$f_t = e^{-r(T-t)}\widehat{\mathbb{E}}_t[f_T] = e^{-r(T-t)} \int_{\ln K}^{\infty} e^x p(x)\, dx.$$

The computation is similar with the one of the integral I_2 from Proposition 6.1.4.

Exercise 6.1.10 (*a*) The payoff is

$$f_T = \begin{cases} e^{nX_T}, & \text{if } X_T > \ln K \\ 0, & \text{otherwise.} \end{cases}$$

$$\begin{aligned}
\widehat{\mathbb{E}}_t[f_T] &= \int_{\ln K}^{\infty} e^{nx} p(x)\, dx \\
&= \frac{1}{\sigma\sqrt{2\pi(T-t)}} \int_{\ln K}^{\infty} e^{nx} e^{-\frac{1}{2}\frac{[x - \ln S_t - (r - \sigma^2/2)(T-t)]^2}{\sigma^2(T-t)}}\, dx \\
&= \frac{1}{\sqrt{2\pi}} \int_{-d_2}^{\infty} e^{n\sigma\sqrt{T-t}\,y} e^{\ln S_t^n} e^{n(r - \frac{\sigma^2}{2})(T-t)} e^{-\frac{1}{2}y^2}\, dy \\
&= \frac{1}{\sqrt{2\pi}} S_t^n e^{n(r - \frac{\sigma^2}{2})(T-t)} \int_{-d_2}^{\infty} e^{-\frac{1}{2}y^2 + n\sigma\sqrt{T-t}\,y}\, dy \\
&= \frac{1}{\sqrt{2\pi}} S_t^n e^{n(r - \frac{\sigma^2}{2})(T-t)} e^{\frac{1}{2}n^2\sigma^2(T-t)} \int_{-d_2}^{\infty} e^{-\frac{1}{2}(y - n\sigma\sqrt{T-t})^2}\, dy \\
&= \frac{1}{\sqrt{2\pi}} S_t^n e^{(nr + n(n-1)\frac{\sigma^2}{2})(T-t)} \int_{-d_2 - n\sigma\sqrt{T-t}}^{\infty} e^{-\frac{z^2}{2}}\, dz \\
&= S_t^n e^{(nr + n(n-1)\frac{\sigma^2}{2})(T-t)} N(d_2 + n\sigma\sqrt{T - t}).
\end{aligned}$$

The value of the contract at time t is

$$f_t = e^{-r(T-t)}\widehat{\mathbb{E}}_t[f_T] = S_t^n e^{(n-1)(r + \frac{n\sigma^2}{2})(T-t)} N(d_2 + n\sigma\sqrt{T - t}).$$

(b) Using $g_t = S_t^n e^{(n-1)(r+\frac{n\sigma^2}{2})(T-t)}$, we get $f_t = g_t N(d_2 + n\sigma\sqrt{T-t})$.

(c) In the particular case $n = 1$, we have $N(d_2 + \sigma\sqrt{T-t}) = N(d_1)$ and the value takes the form $f_t = S_t N(d_1)$.

Exercise 6.2.1 Choose $c_n = 1/n!$ in the general contract formula.

Exercise 6.2.2 Similar computation with the one done for the call option.

Exercise 6.3.1 Write the payoff as a difference of two puts, one with strike price K_1 and the other with strike price K_2. Then apply the superposition principle.

Exercise 6.3.2 Write the payoff as $f_T = c_1 + c_3 - 2c_2$, where c_i is a call with strike price K_i. Apply the superposition principle.

Exercise 6.3.4 (a) By inspection. (b) Write the payoff as the sum between a call with strike price K_2 and a put with strike price K_1, and then apply the superposition principle. (c) The strangle is cheaper.

Exercise 6.3.5 The payoff can be written as a sum between the payoffs of a (K_3, K_4)-bull spread and a (K_1, K_2)-bear spread. Apply the superposition principle.

Exercise 6.7.2 By computation.

Exercise 6.7.3 The risk-neutral valuation yields

$$
\begin{aligned}
f_t &= e^{-r(T-t)}\widehat{\mathbb{E}}_t[S_T - A_T] = e^{-r(T-t)}\widehat{\mathbb{E}}_t[S_T] - e^{-r(T-t)}\widehat{\mathbb{E}}_t[A_T] \\
&= S_t - e^{-r(T-t)}\frac{t}{T}A_t - \frac{1}{rT}S_t\left(1 - e^{-r(T-t)}\right) \\
&= S_t\left(1 - \frac{1}{rT} + \frac{1}{rT}e^{-r(T-t)}\right) - e^{-r(T-t)}\frac{t}{T}A_t.
\end{aligned}
$$

Exercise 6.7.5 We have

$$
\begin{aligned}
\lim_{t\to 0} f_t &= S_0 e^{-rT+(r-\frac{\sigma^2}{2})\frac{T}{2}+\frac{\sigma^2 T}{6}} - e^{-rT}K \\
&= S_0 e^{-\frac{1}{2}(r+\frac{\sigma^2}{6})T} - e^{-rT}K.
\end{aligned}
$$

Exercise 6.7.6 Since $G_T < A_T$, then $\widehat{\mathbb{E}}_t[G_T] < \widehat{\mathbb{E}}_t[A_T]$ and hence

$$
e^{-r(T-t)}\widehat{\mathbb{E}}_t[G_T - K] < e^{-r(T-t)}\widehat{\mathbb{E}}_t[A_T - K],
$$

so the forward contract on the geometric average is cheaper.

Exercise 6.7.7 Dividing

$$
\begin{aligned}
G_t &= S_0 e^{(\mu-\frac{\sigma^2}{2})\frac{t}{2}}e^{\frac{\sigma}{t}\int_0^t W_u\,du} \\
G_T &= S_0 e^{(\mu-\frac{\sigma^2}{2})\frac{T}{2}}e^{\frac{\sigma}{T}\int_0^t W_u\,du}
\end{aligned}
$$

yields

$$\frac{G_T}{G_t} = e^{(\mu-\frac{\sigma^2}{2})\frac{T-t}{2}} e^{\frac{\sigma}{T}\int_0^T W_u\,du - \frac{\sigma}{t}\int_0^t W_u\,du}.$$

An algebraic manipulation leads to

$$G_T = G_t e^{(\mu-\frac{\sigma^2}{2})\frac{T-t}{2}} e^{\sigma(\frac{1}{T}-\frac{1}{t})\int_0^t W_u\,du} e^{\frac{\sigma}{T}\int_t^T W_u\,du}.$$

Exercise 6.8.6 Let $m = \ln\frac{m_1^2}{\sqrt{m_2}} - \ln T$ and $s^2 = \ln\frac{m_2}{m_1^2}$. Then $\ln A_T \sim N(m, s^2)$. We have

$$
\begin{aligned}
\widehat{\mathbb{E}}_t[f_T] &= \int_{\ln K}^{\infty} \frac{1}{\sqrt{2\pi}s} e^{-\frac{(x-m)^2}{2s^2}}\,dx = \int_{\frac{\ln K-m}{s}}^{\infty} \frac{1}{\sqrt{2\pi}} e^{-y^2/2}\,dy \\
&= 1 - N\left(\frac{\ln K - m}{s}\right) = N\left(\frac{m - \ln K}{s}\right) = N(\check{d}_2).
\end{aligned}
$$

Discounting we get $h_0 = e^{-rT} N(\check{d}_2)$.

Exercise 6.8.7 Consider the notations $m = \ln\frac{m_1^2}{\sqrt{m_2}} - \ln T$ and $s^2 = \ln\frac{m_2}{m_1^2}$, so $\ln A_T \sim N(m, s^2)$. We have

$$
\begin{aligned}
\widehat{\mathbb{E}}_t[f_T] &= \int_{\ln K}^{\infty} e^x \frac{1}{\sqrt{2\pi}s} e^{-\frac{(x-m)^2}{2s^2}}\,dx = \frac{1}{\sqrt{2\pi}s} \int_{\ln K}^{\infty} e^{-\frac{1}{2}\left(\frac{x-m-s^2}{s}\right)^2} e^{m+s^2/2}\,dx \\
&= \frac{1}{\sqrt{2\pi}} e^{m+s^2/2} \int_{(\ln K-m-s^2)/s}^{\infty} e^{-z^2/2}\,dz \\
&= e^{m+s^2/2} N\left((m+s^2-\ln K)/s\right).
\end{aligned}
$$

A computation provides

$$e^{m+s^2/2} = e^m e^{s^2/2} = \frac{m_1}{T} = \frac{S_0(e^{rT}-1)}{rT},$$

$$(m+s^2-\ln K)/s = \check{d}_2 + \sqrt{\ln\frac{m_2}{m_1^2}} = \check{d}_1.$$

Substituting we get $\widehat{\mathbb{E}}_t[f_T] = \frac{S_0(e^{rT}-1)}{rT} N(\check{d}_1)$. After discounting we obtain the desired formula.

Exercise 6.8.9 (a) It follows from the superposition principle and Exercises 6.8.6 and 6.8.7. (b) Redo the computations.

Exercise 6.13.2 Use Exercise 5.4.5. We have

$$
\begin{aligned}
V_0 &= \omega S_0[N(\omega\tilde{d}_5) - \frac{\sigma^2}{2r} N(-\omega\tilde{d}_5)] \\
&\quad - \omega \tilde{S}_t e^{-rT}[N(\omega\tilde{d}_6) - \frac{\sigma^2}{2r}\left(\frac{S_t}{\tilde{S}_t}\right)^{1-\frac{2r}{\sigma^2}} N(-\omega\tilde{d}_8)]
\end{aligned}
$$

where

$$
\begin{aligned}
\tilde{d}_5 &= [\ln(S_t/\tilde{S}_t) + (r + \sigma^2/2)T]/(\sigma\sqrt{T}) \\
\tilde{d}_6 &= \tilde{d}_5 - \sigma\sqrt{T} \\
\tilde{d}_7 &= [-\ln(S_t/\tilde{S}_t) + (r + \sigma^2/2)T]/(\sigma\sqrt{T}) \\
\tilde{d}_8 &= \tilde{d}_7 - \sigma\sqrt{T}.
\end{aligned}
$$

The value for a look-back call is obtained for $\tilde{S}_t = \underline{S}_t$ and $\omega = 1$. The value for a look-back put is obtained for $\tilde{S}_t = \bar{S}_t$ and $\omega = -1$.

Chapter 7

Exercise 7.1.2 (a) Using that $S_t = S_u e^{(\mu - \frac{1}{2}\sigma^2)(t-u) + \sigma(W_t - W_u)}$, for $u < t$, then

$$
\ln S_t \sim \mathcal{N}\left((\mu - \frac{1}{2}\sigma^2)(t-u), \sigma^2(t-u) \right),
$$

i.e. the density function of $Y_t = \ln S_t$, given $\mathcal{F}_u$, is

$$
f(y|\mathcal{F}_u) = \frac{1}{\sqrt{2\pi(t-u)}\,\sigma} e^{-\frac{[x - \ln S_u - (\mu - \sigma^2/2)(t-u)]^2}{2\sigma^2(t-u)}}.
$$

Then $S_t = e^{Y_t}$ is log-normally distributed with the density

$$
p(x|\mathcal{F}_u) = \frac{1}{x\sqrt{2\pi(t-u)}\,\sigma} e^{-\frac{[\ln\frac{x}{S_u} - (\mu - \sigma^2/2)(t-u)]^2}{2\sigma^2(t-u)}}.
$$

(b) Take $u = 0$. Then $W_t = W_t - 0 = W_t - W_0$ is a jump independent of $\mathcal{F}_0$. Then we may drop the condition in the conditional expectation.

Exercise 7.1.3 (a) The expected gains are $\mu_A = (\$2 - \$1)/2 = \$0.5$ for option A and $\mu_B = (\$20,000 - \$10,000)/2 = \$5,000$ for option B. Even if we expect to make much more money in the case of option B, most players will not choose it, since there is a lot of risk involved. The risk is measured by the standard deviation which in case B is much larger than in the case A. (b) I would choose game A since I cannot afford to loose $\$10,000$. (c) I am not risk-neutral in my decision, since my decision is affected by the risk involved.

Exercise 7.1.5 (a) This is a computation using conditional expectations:

$$\mathbb{E}^P[e^{-rt}S_t|\mathcal{F}_u] = \mathbb{E}^Q[e^{-rt}S_t e^{\frac{1}{2}(\frac{\mu-r}{\sigma})^2 T + \lambda W_T}|\mathcal{F}_u]$$

$$= e^{-rt}e^{\frac{1}{2}(\frac{\mu-r}{\sigma})^2 T} e^{\lambda W_u} \mathbb{E}^Q[S_t e^{\lambda(W_T - W_u)}|\mathcal{F}_u]$$

$$= e^{-rt}e^{\frac{1}{2}(\frac{\mu-r}{\sigma})^2 T} e^{\lambda W_u} S_u e^{(\mu-\frac{\sigma^2}{2})(t-u)} \mathbb{E}^Q[e^{\sigma(W_t - W_u)}e^{\lambda(W_T - W_u)}|\mathcal{F}_u]$$

$$= e^{-rt}e^{\frac{1}{2}(\frac{\mu-r}{\sigma})^2 T} e^{\lambda W_u} S_u e^{(\mu-\frac{\sigma^2}{2})(t-u)} \mathbb{E}^Q[e^{\sigma(W_t - W_u)}] \mathbb{E}^Q[e^{\lambda(W_T - W_u)}]$$

$$= e^{-rt}e^{\frac{1}{2}(\frac{\mu-r}{\sigma})^2 T} e^{\lambda W_u} S_u e^{(\mu-\frac{\sigma^2}{2})(t-u)} e^{\frac{1}{2}\sigma^2(t-u)} e^{\frac{1}{2}\sigma^2(T-u)}$$

$$= e^{-ru}S_u e^{\frac{1}{2}(\frac{\mu-r}{\sigma})^2 T + \lambda W_u} e^{(\mu-r)(t-u)} e^{\frac{1}{2}\lambda^2(T-u)}$$

$$\neq e^{-ru}S_u,$$

and hence not a martingale. (b) Similar computation.

Exercise 7.1.6 Note that $g(S_t) = f(\lambda t + W_t)$, with $f(x) = g(S_0 e^{\sigma x})$. Then
(a) $\mathbb{E}[g(S_t)] = \mathbb{E}[f(\lambda t + W_t)] = e^{-\frac{\lambda^2 t}{2}}\mathbb{E}[f(W_t)e^{\lambda W_t}] = e^{-\frac{\lambda^2 t}{2}}\mathbb{E}[g(S_0 e^{\sigma W_t})e^{\lambda W_t}]$;
(b) $\mathbb{E}[g(S_t)e^{-\lambda W_t}] = \mathbb{E}[f(\lambda t + W_t)e^{-\lambda W_t}] = e^{\frac{\lambda^2 t}{2}}\mathbb{E}[f(W_t)] = e^{\frac{\lambda^2 t}{2}}\mathbb{E}[g(S_0 e^{\sigma W_t})]$.

Exercise 7.1.7 (a) The left side evaluates as $\mathbb{E}[g(S_t)] = \mathbb{E}[\ln S_t] = \mathbb{E}[\ln S_0 + (\mu - \frac{\sigma^2}{2})t + \sigma W_t] = \ln S_0 + \sigma \lambda t$, while the right side is computed as

$$e^{-\lambda^2 t/2}\mathbb{E}[\ln(S_0 e^{\sigma W_t})e^{\lambda W_t}] = e^{-\lambda^2 t/2}\ln S_0 \mathbb{E}[e^{\lambda W_t}] + \sigma e^{-\lambda^2 t/2}\mathbb{E}[W_t e^{\lambda W_t}]$$

$$= \ln S_0 + \sigma e^{-\lambda^2 t/2}\mathbb{E}[W_t e^{\lambda W_t}].$$

Equating the left and the right sides and solving for $\mathbb{E}[W_t e^{\lambda W_t}]$ yields $\mathbb{E}[W_t e^{\lambda W_t}] = \lambda t e^{\lambda^2 t/2}$; then take $\lambda = 1$.
(b) The left side is $\mathbb{E}[g(S_t)] = \mathbb{E}[S_t]$ while the right side can be evaluated as
$\mathbb{E}[S_t] = S_0 e^{-\lambda^2 t/2}e^{(\sigma+\lambda)^2 t/2} = S_0 e^{\frac{t}{2}[(\sigma+\lambda)^2 - \lambda^2]} = S_0 e^{\mu t}$.

Exercise 7.3.2 The portfolio is $V(t) = \theta_1 S_1 + \theta_2 S_2 = \theta_1 + \theta_2 W_t$. We need to show $dV(t) = \theta_1 dS_1 + \theta_2 dS_2$. Since $dS_1 = 0$ and $dS_2 = dW_t$, it suffices to show just that $dV(t) = 2W_t dW_t$. In order to do this, we compute $dV(t)$ using Ito's lemma

$$dV(t) = d\theta_1 + d\theta_2\, W_t + \theta_2\, dW_t + d\theta_2\, dW_t$$

$$= -dt - 2W_t dW_t - dt + 2dW_t\, W_t + 2W_t dW_t + 2dW_t dW_t$$

$$= 2W_t dW_t,$$

where we used that $d(W_t^2) = 2W - t dW_t + dt$.

Exercise 7.4.3 Solve the linear equation by cross multiplication $\frac{10-r}{15} = \frac{12-r}{25}$. The result is $r = 7\%$.

Exercise 7.5.2 Standard computation using conditional expectations.

Exercise 7.5.3 (a) We have

$$\widehat{\mathbb{E}}\left[\int_0^t S_u du | \mathcal{F}_s\right] = \widehat{\mathbb{E}}\left[\int_0^s S_u du | \mathcal{F}_s\right] + \widehat{\mathbb{E}}\left[\int_s^t S_u du | \mathcal{F}_s\right]$$

$$= \widehat{\mathbb{E}}\left[\int_0^s S_u du | \mathcal{F}_s\right] + \widehat{\mathbb{E}}\left[\int_s^t S_s e^{(r-\sigma^2/2)(u-s)+\sigma(W_u-W_s)} du | \mathcal{F}_s\right]$$

$$= \int_0^s S_u du + \int_s^t S_s e^{(r-\sigma^2/2)(u-s)} e^{\frac{1}{2}\sigma^2(u-s)} du$$

$$= \int_0^s S_u du + S_0 \frac{e^{r(t-s)} - 1}{r}.$$

(c) Using that Ito integrals have mean zero, $\widehat{\mathbb{E}}\left[\int_s^t S_u dW_u\right] = 0$, we have

$$\widehat{\mathbb{E}}\left[\int_0^t S_u du | \mathcal{F}_s\right] = \widehat{\mathbb{E}}\left[\int_0^s S_u dW_u + \int_s^t S_u dW_u | \mathcal{F}_s\right]$$

$$= \int_0^s S_u dW_u + \widehat{\mathbb{E}}\left[\int_s^t S_u dW_u\right]$$

$$= \int_0^s S_u dW_u.$$

(e) Use the decomposition

$$\widehat{\mathbb{E}}\left[\left(\int_0^t S_u du\right)^2 | \mathcal{F}_s\right] = \left(\int_0^s S_u du\right)^2 + 2\left(\int_0^s S_u du\right)\widehat{\mathbb{E}}\left[\int_s^t S_u du | \mathcal{F}_s\right]$$

$$+ \widehat{\mathbb{E}}\left[\left(\int_s^t S_u du\right)^2 | \mathcal{F}_s\right],$$

and part (a). The last integral results from a conditional variance.

Exercise 7.5.4 (a) $\widehat{\mathbb{E}}[e^{-r(T-t)} T S_T | \mathcal{F}_t] = T S_t$;

(b) Using part (b) of Exercise 7.5.3 we have

$$\mathbb{E}[e^{-r(T-t)} \int_0^T S_u du | \mathcal{F}_s] = e^{-r(T-t)}\left\{\int_0^s S_u du + \frac{S_0}{r}(e^{r(T-t)} - 1)\right\}$$

$$= e^{-r(T-t)} \int_0^s S_u du + \frac{S_0}{r}\left(e^{r(t-s)} - 1\right).$$

Chapter 8

Exercise 8.6.1 Show that $F = \ln S$ does not satisfy Black-Scholes equation.

Exercise 8.6.3 Straightforward verification.

Exercise 8.6.4 Just verify that $F(t) = e^{-r(T-t)}K$ satisfies Black-Scholes equation.

Exercise 8.6.5 (a)-(b) Straightforward skilful computation; (c) It follows from (a)-(b) by superposition; (d) asset-or-nothing, all-or-nothing and call options.

Exercise 8.11.3 Consequence of properties of derivatives.

Exercise 8.11.4 If follows from the linearity property of the scalar product $(\,,\,)$.

Exercise 8.11.5 (a) $C^{2,1}$; (b) 0.

Exercise 8.11.6 The equation $(x, g) = 0$ becomes $g(x,t) + \frac{\sigma^2}{r}x + \partial_x g(x,t) = 0$. Multiply by x^{r/σ^2} to get an exact equation, which after integration yields $g = x^{-r/\sigma^2}\psi(t)$.

Exercise 8.11.7 It follows from Proposition 8.11.1 and non-negativity property of $(\,,\,)$.

Exercise 8.11.8 Since $\mathcal{L}_{BS}(f) = f'(t)$, $\mathcal{L}_{BS}(g) = 0$, and $(f, g) = rfg$, then Proposition 8.11.1 provides $g(x)f'(t) + rf(t)g(x) = 0$. Divide by $g(x)$ and solve the linear equation in $f(t)$.

Exercise 8.11.9 (a) Use that $\mathcal{L}_{BS}(f) = f'(t) - rf(t)$; (b) Using Proposition 8.11.1 we obtain a Euler equation $xg'(x) + \frac{\sigma^2}{2r}x^2g''(x) - g(x) = 0$. Solve it by substituting $g(x) = x^m$.

Exercise 8.12.1 (a) We have

$$
\begin{aligned}
d\Pi &= (\mu_1 S_1 - \alpha\mu_2 S_2 - \beta\mu_3 S_3)dt + (S_1\sigma_{11} - \alpha S_2\sigma_{21} - \beta S_3\sigma_{31})dW_1 \\
&\quad + (S_1\sigma_{12} - \alpha S_2\sigma_{22} - \beta S_3\sigma_{32})dW_2.
\end{aligned}
$$

Equating the coefficients of dW_1 and dW_2 to zero yields the linear system

$$
\begin{aligned}
\alpha S_2\sigma_{21} + \beta S_3\sigma_{31} &= S_1\sigma_{11} \\
\alpha S_2\sigma_{22} + \beta S_3\sigma_{32} &= S_1\sigma_{12}
\end{aligned}
$$

with solutions

$$
\alpha = \frac{S_1}{S_2}\frac{\Sigma_2}{\Sigma_1}, \quad \beta = \frac{S_1}{S_3}\frac{\Sigma_3}{\Sigma_1}.
$$

(b) Comparing the coefficients of dt in the two expressions of $d\Pi$ we have

$$
\mu_1 S_1 - \alpha\mu_2 S_2 - \beta\mu_3 S_3 = rS_1 - \alpha rS_2 - r\beta S_3.
$$

This is equivalent to

$$
(\mu_1 - r)S_1 - \alpha(\mu_2 - r)S_2 - \beta(\mu_3 - r)S_3 = 0.
$$

Substituting the expressions for α and β and simplifying by S_1 yields

$$(\mu_1 - r) - \frac{\Sigma_2}{\Sigma_1}(\mu_2 - r) - \frac{\Sigma_3}{\Sigma_1}(\mu_3 - r) = 0.$$

Dividing by $\Sigma_2\Sigma_3$ yields

$$\frac{\mu_1 - r}{\Sigma_2\Sigma_3} = \frac{\mu_2 - r}{\Sigma_1\Sigma_3} + \frac{\mu_3 - r}{\Sigma_1\Sigma_2}.$$

Exercise 8.12.2 (a) $\dfrac{\partial P}{\partial t} + \mu r\dfrac{\partial P}{\partial r} + \dfrac{1}{2}\sigma^2 r^2\dfrac{\partial^2 P}{\partial r^2} = rP$; (b) Use the separation method and solve the differential equations for $A(t, T)$ and $B(t, T)$.

Exercise 8.12.3 (a) $\dfrac{\partial P}{\partial t} + a\dfrac{\partial P}{\partial r} + \dfrac{1}{2}\sigma^2\dfrac{\partial^2 P}{\partial r^2} = rP$; (b) Use the separation method.

Exercise 8.13.1 Use the swap formula (8.13.40).

Exercise 8.13.2 The balance equation during time interval Δt is

$$\Delta V(t) = (\Delta V)_{in} - (\Delta V)_{out} = rV(t)\Delta t + \rho\Delta t - r^*\Delta t.$$

Taking $\Delta t \to 0$ yields $V'(t) = rV + \rho - r^*$, $V(T) = 0$. Solving it as a linear equation to get $e^{-rt}V(t) = (\rho - r^*)\frac{e^{-rT}-1}{r} - (\rho - r^*)\frac{e^{-rt}-1}{r}$. Taking $t = 0$ implies $V(0) = \frac{r^*-\rho}{r}(1 - e^{-rT})$.

Exercise 8.13.3 (a) The price is zero, since the parties do not change any principal; (b) Elementary change of variable.

Exercise 8.13.4 (a), (b) obvious; (c) Making $r(s) = 0$ yields $D(t, s) = 1$, and hence $V(t) = Z\int_t^T(-r^*)ds = -r^*Z(T - t)$.

Exercise 8.13.6 Easy quantitative reasoning.

Chapter 9

Exercise 9.1.2 This is a Calculus exercise. Substituting $x = t/T$ and $g(t) = S_t$, we have $S_{ave}^{(n)} = (n + 1)\int_0^1 x^n g(Tx)dx = (n + 1)\int_0^{1-\epsilon} x^n g(Tx)dx + (n + 1)\int_{1-\epsilon}^1 x^n g(Tx)dx$. The first integral tends to zero

$$(n + 1)\int_0^{1-\epsilon} x^n g(Tx)dx \le (n + 1)\max_{[0,T]} g\int_0^{1-\epsilon} x^n dx = (\max_{[0,T]} g)(1 - \epsilon)^{n+1} \to 0$$

as $n \to \infty$. The second integral tends to $g(T)$:

$$(n+1) \int_{1-\epsilon}^{1} x^n g(Tx) dx = \int_{1-\epsilon}^{1} (x^{n+1})' g(Tx) dx$$

$$= g(T) - (1-\epsilon)^{n+1} g(T(1-\epsilon)) - \int_{1-\epsilon}^{1} x^{n+1} \frac{d}{dx} g(Tx) \, dx.$$

The middle term and the last integral tend to zero:
$(1-\epsilon)^{n+1} g(T(1-\epsilon)) \to 0$, as $n \to \infty$,

$$\int_{1-\epsilon}^{1} x^{n+1} \frac{d}{dx} g(Tx) \, dx \leq \int_{1-\epsilon}^{1} \frac{d}{dx} g(Tx) \, dx = g(T) - g(T(1-\epsilon)) \to g(T) - g(T) = 0.$$

Hence $S_{ave}^{(n)} \to g(T) = S_T$, as $n \to \infty$. Note that $\dfrac{n+1}{T^{n+1}} t^n dt$ is a probability measure on $[0, T]$ which tends to a point measure sitting at $t = T$, and hence the integral picks just the value of the stock at T.

Exercise 9.1.4 (a) Integrating in the equation yields $X_t = S_0 + \int_0^t \frac{f(u)}{g(u)}(S_u - X_u) \, ds$. Then apply the expectation operator and get $x(t) = S_0 + \int_0^t \frac{f(u)}{g(u)}(S_0 e^{\mu u} - x(u)) \, du$, with $x(0) = S_0$. Differentiate in t and use the Fundamental Theorem of Calculus; (b) Solve it as a linear ordinary equation.

Exercise 9.1.5 (a) Apply Ito using $(dX_t)^2 = 0$ and get

$$d(X_t^2) = 2X_t dX_t = 2X_t \frac{f(t)}{g(t)}(S_t - X_t) dt.$$

(b) Integrate between 0 and t and take the expectation

$$y(t) = X_0^2 + 2 \int_0^t \frac{f(u)}{g(u)} \left(\mathbb{E}[X_u S_u] - y_u \right) du.$$

Differentiate and get $y'(t) = 2\frac{f(t)}{g(t)}(\mathbb{E}[X_t S_t] - y_t)$, $y(0) = X_0^2$.
(c) $\mathbb{E}[X_t S_t]$ is obtained by integrating and then taking the expectation in Ito's formula

$$d(X_t S_t) = X_t dS_t + dX_t S_t + dX_t dS_t$$

and then solving a differential equation.

Exercise 9.4.1 Assume the limits exist, but there are not zero. Then L'Hospital's rule applied twice provides

$$\lim_{x \to 0} \frac{f(x)}{\ln x} = \lim_{x \to 0} x f'(x) = -\lim_{x \to 0} x^2 f''(x) = c \neq 0.$$

Therefore, $f(x) \sim c \ln x$, as $x \to 0$, i.e. $f(x)$ is unbounded at $x = 0$, contradiction.

Exercise 9.4.2 Similar computation as in the case of $\mathbb{E}[R]$.

Exercise 9.4.3 Interchanging expectation with the integral, we have

$$\mathbb{E}[R_t] = \int_0^t e^{(\mu - \frac{1}{2}\sigma^2)(u-t)} \mathbb{E}[e^{\sigma(W_u - W_t)}] f(u) \, du,$$

and then use the stationarity property $\mathbb{E}[e^{-\sigma(W_t - W_u)}] = e^{\frac{1}{2}\sigma^2(t-u)}$.

Exercise 9.4.5 (a) Let $G(t, R) = a(t)R^2 + b(t)R + c(t)$. The boundary condition implies $a(T) = 1$, $b(T) = 0$, $c(T) = 0$. Substituting $G(t, R)$ into the equation and collecting the powers of R, and equating them to zero provides three ordinary differential equations: $a'(t) = (2r - \sigma^2)a(t)$, $b'(t) - rb(t) = -2a(t)$, and $c'(t) = -b(t)$. Solve them using the previous boundary conditions.

(b) Since the payoff is $V_T = \frac{I_T^2}{S_T} = S_T R_T^2$, with $R_T = \frac{I_T}{S_T}$, it makes sense to look for a solution of type $V_t = SG(t, R)$, where $G(t, R)$ satisfies the partial differential equation given by (a) with condition $G(T, R_T) = R_T^2$. Then $V_t = S(a(t)R^2 + b(t)R + c(t))$, with $a(t)$, $b(t)$, and $c(t)$ computed at (a).

Exercise 9.5.2 Apply Proposition 9.5.1 for $S_{ave} = A_t$, $\rho = \frac{1}{t}$, $g(t) = t$ and obtain $F_t = \alpha(t)S_t + \beta(t)A_t$, with $\alpha(t) = 1 - \frac{1 - e^{-r(T-t)}}{rt}$ and $\beta(t) = -\frac{t}{T} e^{-r(T-t)}$.

Exercise 9.5.3 Apply Proposition 9.5.1.

Exercise 9.5.4 If $n = 1$, the payoff is $V_T = I_T = S_T R_T$, so we look for a solution $V_t = S_t G(t, R_t)$. Then determine the equation of $G(t, R)$ with the condition $G(T, R_T) = R_T$. Similar for $n = 2$, choose $V_t = S_t^2 G(t, R_t)$.

Exercise 9.6.1 Apply formula (9.6.18). (a) Using $g(t) = t$ and $\rho(u) = 1/u$.

Exercise 9.6.2 Apply formula (9.6.20).

Chapter 10

Exercise 10.1.3 The event $\{S_t$ hits $b\}$ is the same as $\{\mu t + \sigma W_t$ hits $x\}$, where $\mu = r - \frac{\sigma^2}{2} > 0$ and $x = \ln \frac{b}{S} > 0$. Use the recurrence property of the Brownian motion with positive drift to reach any positive barrier with probability 1.

Exercise 10.1.5 We have $\tau_b = \inf\{\mu t + \sigma W_t \geq x\}$ and use part (b) of Appendix C.7.

Exercise 10.1.6 (a) Use the same idea as in Remark 10.1.4. (b) – (d) Just computation.

Exercise 10.1.8 Similar computation to Remark 10.1.4.

Exercise 10.1.9 Use Exercise 10.1.6 and the method presented in the section.

Exercise 10.1.10 (a) The probability $P(S_t < b)$ can be written as

$$
= P\Big(\big(r - \frac{\sigma^2}{2}\big)t + \sigma W_t < \ln \frac{b}{S_0}\Big) = P\Big(\sigma W_t < \ln \frac{b}{S_0} - \big(r - \frac{\sigma^2}{2}\big)t\Big)
$$

$$
= P\big(\sigma W_t < -\lambda\big) = P\big(\sigma W_t > \lambda\big) \le e^{-\frac{\lambda^2}{2\sigma^2 t}} = e^{-\frac{1}{2\sigma^2 t}[\ln(S_0/b) + (r - \sigma^2/2)t]^2},
$$

where we used an upper bound of the probability of the tail of a normally distributed variable; (b) Obvious.

Exercise 10.1.11 Similar to Exercise 10.1.9.

Exercise 10.2.3 Follow the standard technique.

Exercise 10.3.1 We have $f(b) = \widehat{\mathbb{E}}[e^{-rT_b}S_{T_b}] = b\widehat{\mathbb{E}}[e^{-rT_b}] = b\dfrac{S_0}{b} = S_0$.

Exercise 10.3.3 Use $(K - S_t)^2 = S_t^2 - 2KS_t + K^2$ and write the contract as a sum of three contracts, then optimize over b.

Exercise 10.4.3 Consider a contract with the payoff $S - K$, which can be exercised at any time t before or at maturity T. Its value at $t = 0$ is given by $V = \sup_{0 \le \tau \le T} \widehat{\mathbb{E}}[e^{-r\tau}(S_\tau - K)]$, where τ is a stopping time. Since $M_t = e^{-rt}S_t$ is a martingale, by the Optional Stopping Theorem, see Theorem 1.8.1, we have $\widehat{\mathbb{E}}[M_\tau] = M_0 = S_0$, for any stopping time $\tau \le T$. Therefore,

$$
\begin{aligned}
V &= \sup_{0 \le \tau \le T} \Big\{ \widehat{\mathbb{E}}[e^{-r\tau}S_\tau] - K\widehat{\mathbb{E}}[e^{-r\tau}] \Big\} = \sup_{0 \le \tau \le T} \Big\{ \widehat{\mathbb{E}}[M_\tau] - K\widehat{\mathbb{E}}[e^{-r\tau}] \Big\} \\
&= S_0 - K \inf_{0 \le \tau \le T} \widehat{\mathbb{E}}[e^{-r\tau}] = S_0 - Ke^{-rt},
\end{aligned}
$$

since $\widehat{\mathbb{E}}[e^{-r\tau}]$ reaches the smallest value for $\tau = T$. Hence, an American future contract is not optimal to be exercised before maturity T; its value is equal to the value of a regular future contract.

Exercise 10.4.4 Use the usual technique and watch for conditions on K.

Exercise 10.7.1 Elementary computation involving changes of variables.

Exercise 10.7.2 Let $f(\tau) = c\big(x_b(\tau), \tau\big)$. Then chain rule yields $f'(\tau) = \frac{\partial c}{\partial x}\big(x_b(\tau), \tau\big)x_b'(\tau) + \frac{\partial c}{\partial \tau}\big(x_b(\tau), \tau\big)$. Then use the boundary condition $\frac{\partial c}{\partial x}\big(x_b(\tau), \tau\big) = 0$ and the fact that $f(\tau) = 0$.

Exercise 10.7.3 $x_b'(0+) = -\dfrac{\sigma^2}{2}\dfrac{b'(T)}{b(T)} = -\dfrac{\delta\sigma^2}{2rK}b'(T)$.

Exercise 10.7.5 (a) One way is just to substitute into the equation and verify that $v_1(\xi)$ and $v_2(\xi)$ are solutions. A smarter way (applicable in the case when the solutions are unknown) is to look for a polynomial solution of degree 3. Using the method of undetermined coefficients we obtain $v_1(\xi) = \xi^3 + 6\xi$. Then look for the second solution, $v_2(\xi) = u(\xi)v_1(\xi)$, using the reduction method; (b) The reduction method always provides solutions with non-zero wronskian; (c) Use the fact that the space of solutions of a second order differential equation is 2-dimensional.

Exercise 10.7.6 (a) $b^* = \infty$; (b) It is not optimal to exercise the call perpetuity early.

Exercise 10.7.7 Solve it as an Euler equation, looking for solutions as powers of S. Then consider a linear combination of solutions and determine the coefficients from the boundary conditions. Substitute into the boundary conditions to obtain an equation for b^*; (d) all-or-nothing bet.

Exercise 10.8.1 (a) Assume the contrary: $\frac{\partial P}{\partial S}(b(t), t) < -1$. Then when S increases over $b(t)$, the put $P(S, t)$ drops below the payoff (which has slope -1), which leads to an arbitrage opportunity; (b) This is more delicate. Assuming $\frac{\partial P}{\partial S}(b(t), t) > -1$ implies that the option is misvalued: the option value near $S = b(t)$ can be increased by choosing a lower value of $b(t)$. Or, use a similar technique applied to calls in section 10.7.

Exercise 10.8.4 Standard change of variables technique.

Exercise 10.8.6 Elementary reversion of variables.

Exercise 10.8.7 It follows closely the computation idea of section 10.8.4.

Exercise 10.8.8 (a) A put perpetuity provides more freedom for exercise than a finitely lived put, so it must be more expensive; (b) Assume $b^* \geq b(0)$ and arrive to a contradiction.

Exercise 10.8.9 Equation (10.8.90) is replaced by $\frac{\sigma^2}{2} S^2 P''(S) + (r-\delta)SP'(S) - rP(S) = 0$, $S > b^*$. Its solutions are linear combinations of S^{h_1} and S^{h_2}, with h_1 and h_2 solutions of a quadratic equation. Determine the coefficients from boundary conditions.

Exercise 10.9.1 (a) This is just a computation using chain rule; (b) Use put-call parity and part (a).

Chapter 11

Exercise 11.1.4 (a) Integrate from 0 to t in (11.1.1) to get $S_t = S_0 + \mu \int_0^t S_u du + \int_0^t \sqrt{v_u}\, dW_u$ and take the expectation $\mathbb{E}[S_t] = S_0 + \mu \int_0^t \mathbb{E}[S_u]\, du$. Denote $f(t) = \mathbb{E}[S_t]$. Then $f'(t) = \mu f(t)$, with $f(0) = S_0$. The solution is $f(t) = S_0 e^{\mu t}$.

(b) Similarly, integrate and take $\mathbb{E}[\cdot|\mathcal{F}_s]$ in (11.1.1): $\mathbb{E}[S_t|\mathcal{F}_s] = S_s + \mu \int_s^t \mathbb{E}[S_u|\mathcal{F}_s] \, du$, where we used that dW_u is independent of $\mathcal{F}_s$ for $s < u < t$. We obtain $\mathbb{E}[S_t|\mathcal{F}_s] = S_s e^{\mu(t-s)}$.

Exercise 11.2.2 (a) $\lim\limits_{S\to\infty} N(d_1) = N\left(\lim\limits_{S\to\infty} d_1\right) = N(+\infty) = 1$;

(b) $\lim\limits_{S\to0} N(d_2) = N\left(\lim\limits_{S\to0} d_2\right) = N(-\infty) = 0$;

(c) $\lim\limits_{S\to\infty} \dfrac{\partial N(d_j)}{\partial S} = \lim\limits_{S\to\infty} N'(d_j) \dfrac{\partial d_j}{\partial S} = \lim\limits_{S\to\infty} \dfrac{1}{\sqrt{2\pi}} e^{-d_j^2/2} \dfrac{1}{S\sqrt{v(T-t)}} = 0.$

Exercise 11.2.3 Elementary change of variables.

Exercise 11.2.4 $\int_0^\tau \frac{1-e^{at}}{1-be^{at}} dt = \frac{1}{a} \int_1^{e^{a\tau}} \frac{1-v}{v(1-bv)} dv = \frac{1}{a} \int_1^{e^{a\tau}} \frac{1}{v} dv + \frac{1-b}{a} \int_1^{e^{a\tau}} \frac{1}{bv-1} dv$

$= \tau + \frac{1-b}{ab} \int_{b-1}^{be^{a\tau}-1} \frac{1}{z} dz = \tau + \frac{1-b}{ab} \left(\ln(be^{a\tau} - 1) - \ln(b - 1) \right).$

Exercise 11.2.6 (a) property of even functions; (b) property of odd functions; (c) Equating $\overline{\Phi(x)} = A(x) - iB(x)$ against $\Phi(-x) = A(-x) + iB(-x)$ yields $A(-x) = A(x)$ and $B(-x) = -B(x)$. Applying (a) and (b) we get $\int_\mathbb{R} \Phi(x) dx = \int_\mathbb{R} A(x) dx + i \int_\mathbb{R} B(x) dx = 2 \int_0^\infty A(x) dx = 2 \int_0^\infty Re\{\Phi(x)\} dx.$
(d) Using that $\overline{\phi_1(\omega)} = e^{\overline{C(\tau,\omega)} + v\overline{D(\tau,\omega)} - i\omega x} = e^{C(\tau,-\omega) + vD(\tau,-\omega) - i\omega x} = \phi_1(-\omega)$ it easily follows that $\overline{\phi(\omega)} = \phi(-\omega).$

Exercise 11.3.1 $Re\left(\frac{d}{dx}\Phi(x)\right) = Re\left(A'(x) + iB'(x)\right) = \frac{d}{dx} Re(\Phi(x)).$

Exercise 11.3.2 The delta $f_1(t, S, v).$

Exercise 11.3.3 Just a computation.

Exercise 11.3.6 Look for a solution of the Heston PDE (11.4.57) such that $V(T, S_T) = S_T - K$. Obtain $V(t, S) = S - Ke^{-r(T-t)}.$

11.4.2 Use the integral representations given in Appendix D.1.

Chapter 13

Exercise 13.2.1 (a) We obtain a PDE with coefficients linear functions of v

$$\frac{\partial W}{\partial t} + \frac{1}{2} v \frac{\partial^2 W}{\partial x^2} + \left(r - \frac{1}{2}v\right) \frac{\partial W}{\partial x} + \frac{1}{2} \sigma^2 \frac{\partial^2 W}{\partial v^2} - \lambda v \frac{\partial W}{\partial v} = rW.$$

(b) If $U(v) = r + \frac{1}{2}\omega^2 v + i\omega\left(r - \frac{v}{2}\right)$, then $\widehat{W}$ satisfies

$$\frac{\partial \widehat{W}}{\partial t} + \frac{1}{2} \sigma^2 \frac{\partial^2 \widehat{W}}{\partial v^2} - \lambda v \frac{\partial \widehat{W}}{\partial v} = U(v)\widehat{W}.$$

(c) Let $\tau = T - t$ and $\Lambda(\tau, v; \omega) = \widehat{W}(t, v; \omega)$. Then $\dfrac{\partial \Lambda}{\partial \tau} = \mathcal{A}\Lambda - U\Lambda$, where $\mathcal{A} = \dfrac{1}{2}\sigma^2 \dfrac{\partial^2}{\partial v^2} - \lambda v \dfrac{\partial}{\partial v}$. The associated diffusion to $\mathcal{A}$ is $dX_t = -\lambda X_t + \sigma dW_t$, with the solution $X_t = x + \int_0^t e^{-\lambda(t-s)} dW_s$. We have

$$\int_0^t U(X_s) ds = (r + i\omega r)t + \frac{1}{2}(\omega^2 - i\omega)xt + \frac{1}{2}(\omega^2 - i\omega)\int_0^t e^{-\lambda s}\left[\int_0^s e^{\lambda u} dW_u\right] ds$$

Feynman-Kac formula provides the solution

$$\begin{aligned}
\Lambda(\tau, v; \omega) &= \mathbb{E}\left[\Lambda(0, X_t; \omega)e^{-\int_0^t U(X_u)du} | X_0 = x\right] \\
&= e^{(r+i\omega r)t} e^{\frac{1}{2}(\omega^2 - i\omega)xt} \mathbb{E}\left[\Lambda(0, X_t; \omega)e^{-\frac{1}{2}(\omega^2 - i\omega)\int_0^t e^{-\lambda s}\left(\int_0^s e^{\lambda u} dW_u\right) ds}\right].
\end{aligned}$$

(d) If $V(T, S, v) = \delta(\ln S)$, then $W(T, x, v) = \delta(x)$, and then $\widehat{W}(T, v; \omega) = \hat{\delta}(x) = 1$. Hence, substituting $\Lambda(0, X_t; \omega) = 1$ in the previous equation yields

$$\Lambda(\tau, v; \omega) = e^{(r+i\omega r)t} e^{\frac{1}{2}(\omega^2 - i\omega)xt} \mathbb{E}\left[e^{-\frac{1}{2}(\omega^2 - i\omega)\int_0^t e^{-\lambda s}\left(\int_0^s e^{\lambda u} dW_u\right) ds}\right].$$

The expectation can be computed using $\mathbb{E}[e^{cZ_t}] = e^{\frac{1}{2}c^2 Var(Z_t)}$, with $Z_t = \int_0^t X_s ds \sim \mathcal{N}\left(0, \frac{1 - e^{-2\lambda t}}{2\lambda}\right)$.

Chapter 14

Exercise 14.1.2 (a) Apply Ito's lemma and get

$$dY_t = (k(\theta - \ln Y_t) + \frac{1}{2}\sigma^2)Y_t dt + \sigma Y_t dW_t.$$

(b) Use that $Y_t = e^{X_t}$ has a log-normal distribution. The mean and variance of X_t are given in the same section.

Exercise 14.2.2 Write the system (14.2.6)-(14.2.7) in a matrix form, and then invert it to obtain

$$\begin{aligned}
dW_t^1 &= dW_t^S \\
dW_t^2 &= -\frac{\rho}{\sqrt{1 - \rho^2}}dW_t^S + \frac{1}{\sqrt{1 - \rho^2}}dW_t^\mu.
\end{aligned}$$

Then use the bilinearity of covariance to show that $cov(dW_t^1, dW_t^2) = 0$.

Exercise 14.3.1 $Var(e^{-kt}\int_0^t e^{ks} dW_s) = e^{-2kt}\int_0^t e^{2ks} ds = \frac{1 - e^{-2kt}}{2k}$.

Exercise 14.3.3 $F_S(x) = P(S_t \leq x) = P(\ln S_t \leq \ln x) = P(X_t \leq \ln x) = F_X(\ln x)$, $x > 0$. Differentiate with respect to x and obtain $f_S(x) = F_S'(x) = F_X'(x)\frac{1}{x} = \frac{1}{x}f_X(\ln x)$, for $x > 0$. If $x \leq 0$, then $F_S(x) = 0$.

Exercise 14.4.1 (a) $V(t) = V_T e^{-r(T-t)}$; (b) $V(S) = \alpha_1 S^{n_1} + \alpha_2 S^{n_2}$, with $\alpha_j \in \mathbb{R}$ and $n_j^2 + \left(\frac{2r}{\sigma_S^2} - 1\right)n_j - \frac{2r}{\sigma_S^2} = 0$; (c) $V(\mu) = \alpha_1 e^{\rho_1 \mu} + \alpha_2 e^{\rho_2 \mu}$ with $\rho_{1,2} = \frac{-\lambda \pm \sqrt{\lambda^2 + 2\sigma_\mu^2 r}}{\sigma_\mu^2}$.

Exercise 14.7.3 Use superposition: one term is given by the integral

$$S^\alpha e^{\beta t + \gamma u} \iint p_T(\ln S, \mu, x', \mu') e^{-\varphi(T, x', \mu')} e^{x'} \, dx' d\mu',$$

and the other by $-Ke^{-r(T-t)}$.

Appendix A

Useful Transforms

A.1 The Fourier Transform

The Fourier transform of a function $f(x)$ is defined by

$$\mathcal{F}(f)(\omega) = \int_{\mathbb{R}} e^{-i\omega x} f(x)\, dx.$$

If the function $y = f(x)$ is seen as the amplitude of a signal in terms of time, then $\mathcal{F}(f)(\omega)$ represents how much amplitude is contained in the signal at frequency ω. We note that the domain of $f(x)$ is the entire real line. If f is a function fast decreasing to infinity, in the sense that

$$\lim_{|x| \to \infty} |x^n f^{(k)}(x)| < \infty,$$

then its Fourier transform $\widehat{f}(\omega) = \mathcal{F}(f)(\omega)$ is also fast decreasing at infinity.

The compatibility between the Fourier transform and differentiation is given by the following set of formulas:

$$\widehat{f'}(\omega) = i\omega\widehat{f}(\omega), \qquad \widehat{f''}(\omega) = -\omega^2 \widehat{f}(\omega), \qquad \widehat{f^{(k)}}(\omega) = (i\omega)^k \widehat{f}(\omega).$$

$$\big(\widehat{f}(\omega)\big)' = -\widehat{ixf(x)}(\omega), \quad \big(\widehat{f}(\omega)\big)'' = -\widehat{x^2 f(x)}(\omega), \quad \big(\widehat{f}(\omega)\big)^{(k)} = (-\widehat{ix)^k f(x)}(\omega).$$

These properties make the Fourier transform useful when solving a differential equation; the Fourier transform changes the differential equation into an algebraic equation, which eventually can be solved by classical methods. Applying an inverse Fourier transform to the solution yields the solution of the initial differential equation.

The Fourier transform is also compatible with convolution

$$\widehat{f * g}(\omega) = \widehat{f}(\omega)\widehat{g}(\omega),$$

419

where the convolution is defined by

$$(f * g)(x) = \int_{\mathbb{R}} f(x - y)g(y)\, dy.$$

The Fourier transform is invertible, with the inverse given by the formula

$$(\mathcal{F}^{-1}g)(x) = \frac{1}{2\pi} \int_{\mathbb{R}} e^{ix\omega} g(\omega)\, d\omega.$$

An equivalent statement for the inversion formula in terms of the Fourier transform is

$$(\mathcal{F}^{-1}g)(x) = \frac{1}{2\pi}\widehat{g}(-x).$$

The following energy identity is due to Parseval:

$$\int_{\mathbb{R}} \widehat{f}(\omega)^2\, d\omega = 2\pi \int_{\mathbb{R}} f(x)^2\, dx, \qquad \forall f \in L^1(\mathbb{R}) \cap L^2(\mathbb{R}).$$

The left term represents the energy of the signal in frequency domain, while the right terms is the energy in the time domain.

The relation with probability theory If X is a random variable with the probability density function f, i.e.

$$P(X < x) = \int_{-\infty}^{x} f(u)\, du,$$

then the *characteristic function* of X is defined by

$$\varphi_X(\omega) = \mathbb{E}[e^{-iX}] = \int_{\mathbb{R}} e^{-i\omega x} f(x)\, dx,$$

and hence $\widehat{f}(\omega) = \varphi_X(\omega)$.

Let X and Y be two independent random variables, with probability densities f_X and f_Y. Then the statement

$$\varphi_{X+Y} = \varphi_X \varphi_Y$$

in terms of Fourier transform writes as $\widehat{f_{X+Y}} = \widehat{f_X}\, \widehat{f_Y}$, and is based on the fact that $f_{X+Y} = f_X * f_Y$.

Table of usual Fourier transforms Consider $a > 0$. Then we have:

$f(x)$	$\hat{f}(\omega) = \mathcal{F}(f)(\omega)$
1	$2\pi\delta$
x^k	$2\pi i^k \delta^{(k)}$
δ	1
δ_a	$e^{-ia\omega}$
$\cos(ax)$	$\pi(\delta_a + \delta_{-a})$
$\sin(ax)$	$-i\pi(\delta_a - \delta_{-a})$
$\frac{1}{a^2+x^2}$	$\frac{\pi}{a}e^{-a|\omega|}$
$e^{-a|x|}$	$\frac{2a}{a^2+\omega^2}$
e^{-ax^2}	$\sqrt{\frac{\pi}{a}}e^{-\frac{\omega^2}{4a}}$

A.2   The Laplace Transform

Consider a function $f(x)$ defined for $x \geq 0$, which does not increase to infinity faster than an exponential. The integral

$$\mathcal{L}\{f(t)\}(s) = \int_0^t e^{-sx} f(x)\, dx$$

is called the *Laplace transform* of f and it is customarily denoted by $F(s)$.

1. The Laplace transform is used to solve differential equations, since it transforms derivatives into powers of s:

$$\begin{aligned}
\mathcal{L}\{f'(t)\}(s) &= sF(s) - f(0) \\
\mathcal{L}\{f''(t)\}(s) &= s^2 F(s) - sf(0) - f'(0).
\end{aligned}$$

The algebraic equation thus obtained is solved and then applying the inverse Laplace transform we obtain the solution $f(t)$. The inverse Laplace transform for the usual functions is well tabulated, even if there is a closed form formula involving a complex integral, which we do not use in this book.

2. The relation with the probability theory is made through the moment generating function. If a random variable X takes positive values, its moment generating function is given by

$$\mu_X(t) = \mathbb{E}[e^{tX}] = \int_0^\infty e^{tx} f(x)\, dx,$$

where $f(x)$ is the probability density of X, i.e.

$$P(X < u) = \int_0^u f(x)\, dx.$$

Therefore the Laplace transform of a probability density function is the reflected moment generating function of the associated random variable

$$\mathcal{L}(f)(t) = \mu_X(-t).$$

3. The Laplace transform can also be interpreted as the present value of a perpetuity. If

$$0 = t_0 < t_1 < t_2 < \cdots < t_n < \infty,$$

is an increasing unbounded sequence, with $\Delta t = t_{k+1} - t_k$, then the present value of an infinite sequence of cash inflows, $f(t_k)\Delta t$, paid at time instances t_k, is given by the sum of the discounted values

$$PV = \sum_{k\geq 0} e^{-rt_k} f(t_k)\Delta t,$$

where r denotes the risk-free interest rate. We note that the term $f(t_k)$ represents the average payment over the period $[t_k, t_{k+1}]$. When taking $\Delta t \to 0$, the above sum becomes an improper integral

$$PV = \lim_{\Delta t \to 0} \sum_{k\geq 0} e^{-rt_k} f(t_k)\Delta t = \int_0^\infty e^{-rt} f(t)\, dt = \mathcal{L}(f)(r),$$

which is the Laplace transform of the payment density, $f(t)$, evaluated at r.

Appendix B

Probability Concepts

B.1 Events and Probability

The set of parts of the set Ω, is the set of all subsets

$$2^\Omega = \{A; A \subset \Omega\}.$$

It can be shown that for Ω finite, with $|\Omega| = n$ elements, then the number of elements of 2^Ω is 2^n. The set of parts verifies the following properties:

1. It contains the empty set $\emptyset$;

2. If it contains a set A, then it also contains its complement $\bar{A} = \Omega \backslash A$;

3. It is closed with regard to unions, i.e., if $A_1, A_2, \ldots$ is a sequence of sets, then their union $A_1 \cup A_2 \cup \cdots$ also belongs to 2^Ω.

A σ-field is any subset $\mathcal{F}$ of 2^Ω that satisfies the previous three properties. The sets belonging to $\mathcal{F}$ are called *events*. Hence, the complement of an event, or the union of events is also an event. We say that an event occurs if the outcome of the experiment is an element of that subset.

The chance of occurrence of an event is measured by a probability function $P : \mathcal{F} \to [0, 1]$ which assigns a number between 0 and 1 with each event. This function satisfies the following two properties:

1. $P(\Omega) = 1$;

2. For any mutually disjoint events $A_1, A_2, \cdots \in \mathcal{F}$,

$$P(A_1 \cup A_2 \cup \cdots) = P(A_1) + P(A_2) + \cdots.$$

The triplet $(\Omega, \mathcal{F}, P)$ is called a *probability space*, and is the main setup in which the probability theory works.

B.2   Stochastic Processes

A *stochastic process* on the probability space $(\Omega, \mathcal{F}, P)$ is a family of random variables X_t parameterized by $t \in \mathbf{T}$, where $\mathbf{T} \subset \mathbb{R}_+$. If $\mathbf{T}$ is an interval we say that X_t is a stochastic process in *continuous time*. If $\mathbf{T} = \{1, 2, 3, \dots\}$ we shall say that X_t is a stochastic process in *discrete time*. The latter case describes a sequence of random variables.

The evolution in time of a given state of the market $\omega \in \Omega$ given by the function $t \longmapsto X_t(\omega)$ is called a *path* or *realization* of X_t. The study of stochastic processes using computer simulations is based on retrieving information about the process X_t given a large number of its realizations.

Next we shall structure the information field $\mathcal{F}$ with an order relation parametrized by the time t. Consider that all the information accumulated until time t is contained in the σ-field $\mathcal{F}_t$. This means that $\mathcal{F}_t$ contains the information containing events that have already occurred until time t, and which did not. Since the information is growing in time, we have

$$\mathcal{F}_s \subset \mathcal{F}_t \subset \mathcal{F}$$

for any $s, t \in \mathbf{T}$ with $s \leq t$. The family $\mathcal{F}_t$ is called a *filtration*.

A stochastic process X_t is called *adapted* to the filtration $\mathcal{F}_t$ if X_t is $\mathcal{F}_t$-measurable, for any $t \in \mathbf{T}$. This means that the information at time t determines the value of the random variable X_t.

B.3   Expectation

A random variable $X : \Omega \to \mathbb{R}$ is called *integrable* if

$$\int_\Omega |X(\omega)| \, dP(\omega) = \int_\mathbb{R} |x| p(x) \, dx < \infty,$$

where $p(x)$ denotes the probability density function of X. The *expectation* of an integrable random variable X is defined by

$$\mathbb{E}[X] = \int_\Omega X(\omega) \, dP(\omega) = \int_\mathbb{R} x \, p(x) \, dx.$$

Customarily, the expectation of X is denoted by μ and is called the *mean*.

The expectation operator $\mathbb{E}$ is linear. For any integrable random variables X and Y the following holds:

1. $\mathbb{E}[cX] = c\mathbb{E}[X], \qquad \forall c \in \mathbb{R}$;
2. $\mathbb{E}[X + Y] = \mathbb{E}[X] + \mathbb{E}[Y]$.

3. If X and Y are independent integrable random variables, then

$$\mathbb{E}[XY] = \mathbb{E}[X]\mathbb{E}[Y].$$

The *covariance* of two random variables is defined by

$$cov(X, Y) = \mathbb{E}[XY] - \mathbb{E}[X]\mathbb{E}[Y] = \mathbb{E}(X - \mu_X)\mathbb{E}(Y - \mu_Y).$$

The variance of X is given by $Var(X) = cov(X, X)$. If X and Y are independent, then $cov(X, Y) = 0$. It is worth to note that the converse is not necessarily true (however, the converse holds true if both X and Y are assumed normally distributed). The *correlation* coefficient between X and Y is defined by

$$\rho = cor(X, Y) = \frac{cov(X, Y)}{\sqrt{Var(X)Var(Y)}},$$

and has the property that $-1 \leq \rho \leq 1$.

B.4   Conditional Expectations

Let X be a random variable on the probability space $(\Omega, \mathcal{F}, P)$, and $\mathcal{G}$ be a σ-field contained in $\mathcal{F}$.

The random variable that makes a prediction for X based on the information $\mathcal{G}$ is denoted by $\mathbb{E}[X|\mathcal{G}]$, and is called the *conditional expectation* of X given $\mathcal{G}$. This is defined as the random variable $Y = \mathbb{E}[X|\mathcal{G}]$, which is the best approximation of X in the least squares sense, i.e.

$$\mathbb{E}[(X - Y)^2] \leq \mathbb{E}[(X - Z)^2], \tag{B.4.1}$$

for any $\mathcal{G}$-measurable random variable Z.

Proposition B.4.1 *Let X and Y be two random variables on the probability space $(\Omega, \mathcal{F}, P)$. We have*
1. *Linearity:*

$$\mathbb{E}[aX + bY|\mathcal{G}] = a\mathbb{E}[X|\mathcal{G}] + b\mathbb{E}[Y|\mathcal{G}], \qquad \forall a, b \in \mathbb{R};$$

2. *Factoring out the measurable part:*

$$\mathbb{E}[XY|\mathcal{G}] = X\mathbb{E}[Y|\mathcal{G}]$$

if X is $\mathcal{G}$-measurable. In particular, $\mathbb{E}[X|\mathcal{G}] = X$.
3. *Tower property ("the coarser information wins"):*

$$\mathbb{E}[\mathbb{E}[X|\mathcal{G}]|\mathcal{H}] = \mathbb{E}[\mathbb{E}[X|\mathcal{H}]|\mathcal{G}] = \mathbb{E}[X|\mathcal{H}], \text{ if } \mathcal{H} \subset \mathcal{G}.$$

4. *Positivity:*

$$\mathbb{E}[X|\mathcal{G}] \geq 0, \, if \, X \geq 0.$$

5. *Expectation of a constant is a constant:*

$$\mathbb{E}[c|\mathcal{G}] = c.$$

6. *An independent condition drops out:*

$$\mathbb{E}[X|\mathcal{G}] = \mathbb{E}[X],$$

if X is independent of $\mathcal{G}$.

B.5   Martingales

We shall define next an important type of stochastic process, which was introduced by Lévy in 1934.

Definition B.5.1 *A process X_t, $t \in \mathbf{T}$, is called a martingale with respect to the filtration $\mathcal{F}_t$ if*
1. *X_t is integrable for each $t \in \mathbf{T}$;*
2. *X_t is adapted to the filtration $\mathcal{F}_t$;*
3. *$X_s = \mathbb{E}[X_t|\mathcal{F}_s]$, $\forall s < t$.*

The first condition states that the unconditional forecast is finite $\mathbb{E}[|X_t|] = \int_\Omega |X_t| \, dP < \infty$. Condition 2 says that the value X_t is known, given the information set $\mathcal{F}_t$. This can be also stated by saying that X_t is $\mathcal{F}_t$-measurable. The third relation asserts that the best forecast of unobserved future values is the last observation on X_t.

Example B.5.2 1. If X is an integrable random variable on $(\Omega, \mathcal{F}, P)$, and $\mathcal{F}_t$ is a filtration. Than $X_t = \mathbb{E}[X|\mathcal{F}_t]$ is a martingale.

2. Let X_t and Y_t be martingales with respect to the filtration $\mathcal{F}_t$. Then for any $a, b, c \in \mathbb{R}$ the process $Z_t = aX_t + bY_t + c$ is an $\mathcal{F}_t$-martingale.

3. Two processes X_t and Y_t are called conditionally uncorrelated, given $\mathcal{F}_t$, if

$$\mathbb{E}[(X_t - X_s)(Y_t - Y_s)|\mathcal{F}_s] = 0, \qquad \forall 0 \leq s < t < \infty.$$

Let X_t and Y_t be martingale processes. Then the process $Z_t = X_t Y_t$ is a martingale if and only if X_t and Y_t are conditionally uncorrelated. (Assume that X_t, Y_t and Z_t are integrable).

In the following, if X_t is a stochastic process, the minimum amount of information resulted from knowing the process X_s until time t is denoted by $\mathcal{F}_t = \sigma(X_s; s \leq t)$. This is the σ-algebra generated by the events $\{\omega; X_s(\omega) \in (a, b)\}$, for any real numbers $a < b$ and $s \leq t$.

In the case of a discrete process, the minimum amount of information resulted from knowing the process X_k until time n is $\mathcal{F}_n = \sigma(X_k; k \leq n)$, the σ-algebra generated by the events $\{\omega; X_k(\omega) \in (a, b)\}$, for any real numbers $a < b$ and $k \leq n$.

B.6   Submartingales and their Properties

A stochastic process X_t on the probability space $(\Omega, \mathcal{F}, P)$ is called *submartingale* with respect to the filtration $\mathcal{F}_t$ if:

(a) $\int_\Omega |X_t| \, dP < \infty$ (X_t integrable);

(b) X_t is known if $\mathcal{F}_t$ is given (X_t is adaptable to $\mathcal{F}_t$);

(c) $\mathbb{E}[X_{t+s}|\mathcal{F}_t] \geq X_t, \forall t, s \geq 0$ (future predictions exceed the present value).

Here are a few examples of submartingales:
1. The process $X_t = \mu t + \sigma W_t$, with $\mu > 0$.
2. The square of the Brownian motion, W_t^2.
3. The exponential of a Brownian motion e^{W_t}.

The following result supplies examples of submatingales starting from martingales or submartingales.

Proposition B.6.1 (a) *If X_t is a martingale and ϕ a convex function such that $\phi(X_t)$ is integrable, then the process $Y_t = \phi(X_t)$ is a submartingale.*

(b) *If X_t is a submartingale and ϕ an increasing convex function such that $\phi(X_t)$ is integrable, then the process $Y_t = \phi(X_t)$ is a submartingale.*

(c) *If X_t is a martingale and $f(t)$ is an increasing, integrable function, then $Y_t = X_t + f(t)$ is a submartingale.*

Corollary B.6.2 (a) *Let X_t be a right continuous martingale. Then X_t^2, $|X_t|$, e^{X_t} are submartingales.*

(b) *Let $\mu > 0$. Then $e^{\mu t + \sigma W_t}$ is a submartingale.*

The following result provides important inequalities involving submartingales, see for instance Doob [27].

Proposition B.6.3 (Doob's Submartingale Inequality) (a) *Let X_t be a non-negative submartingale. Then*

$$P(\sup_{s \leq t} X_s \geq x) \leq \frac{\mathbb{E}[X_t]}{x}, \qquad \forall x > 0.$$

(b) *If X_t is a right continuous submartingale, then for any $x > 0$*

$$P(\sup_{s \leq t} X_t \geq x) \leq \frac{\mathbb{E}[X_t^+]}{x},$$

where $X_t^+ = \max\{X_t, 0\}$.

B.7   Jensen's Inequality

A function φ is called *convex* if its graph looks like the cross-section of a bowl. Algebraically, this can be written as

$$\varphi(\lambda x_1 + (1 - \lambda)x_2) \leq \lambda\varphi(x_1) + (1 - \lambda)\varphi(x_2), \qquad \lambda \in (0, 1).$$

In particular,

$$\varphi\left(\frac{x_1 + x_2}{2}\right) \leq \frac{\varphi(x_1) + \varphi(x_2)}{2}.$$

If φ is differentiable, then the convexity is equivalent with the condition $\varphi''(x) \geq 0$. A function is called *concave* if its graph looks like the cross section of an upside-down bowl. It can be described algebraically by reverting the above inequalities.

Theorem B.7.1 (Jensen's inequality) *Let $\varphi : \mathbb{R} \to \mathbb{R}$ be a convex function and let X be an integrable random variable on the probability space $(\Omega, \mathcal{F}, P)$. If $\varphi(X)$ is integrable, then*

$$\varphi(\mathbb{E}[X]) \leq \mathbb{E}[\varphi(X)].$$

We shall present next a couple of applications. If $\varphi(x) = x^2$, then we obtain

$$\mathbb{E}[X]^2 \leq \mathbb{E}[X^2].$$

This implies that $Var(X) = \mathbb{E}[X^2] - \mathbb{E}[X]^2 \geq 0$. If $\varphi(x) = e^x$, then

$$e^{\mathbb{E}[X]} \leq \mathbb{E}[e^X].$$

If the distribution of X is symmetric, then the distribution of $\varphi(X)$ is skewed, with $\varphi(\mathbb{E}[X]) < \mathbb{E}[\varphi(X)]$.

It is worth noting that the inequality is reversed for φ concave. An application of Jensen's inequality in this case is the fact that a risk-averse investor, whose utility function is concave, has an expected utility associated with a gamble less than the utility from receiving the expected value of the gamble:

$$\mathbb{E}[U(X)] \leq U(\mathbb{E}[X]).$$

Appendix C

Elements of Stochastic Calculus

C.1   The Poisson Process

A *Poisson process* describes the number of occurrences of a certain event before time t. A few examples are: the number of shocks in the stock market from the beginning of the year until time t; the number of cars arriving at a gas station until time t; the number of phone calls received on a certain day until time t; the number of emails received on a certain day until time t, etc.

The definition of a Poisson process is stated below.

Definition C.1.1 *A Poisson process is a stochastic process N_t, $t \geq 0$, which satisfies*

1. The process starts at the origin, $N_0 = 0$;
2. N_t has independent increments;
3. The process N_t is right continuous in t, with left hand limits;
4. The increments $N_t - N_s$, with $0 < s < t$, have a Poisson distribution with parameter $\lambda(t - s)$, i.e.

$$P(N_t - N_s = k) = \frac{\lambda^k (t - s)^k}{k!} e^{-\lambda(t-s)}.$$

It is worth noting that condition *4* in the previous definition can be replaced by the following two conditions:

$$P(N_t - N_s = 1) \;=\; \lambda(t - s) + o(t - s) \tag{C.1.1}$$
$$P(N_t - N_s \geq 2) \;=\; o(t - s), \tag{C.1.2}$$

where $o(h)$ denotes a quantity such that $\lim_{h \to 0} o(h)/h = 0$. Then the probability that a jump of size 1 occurs in the infinitesimal interval dt is equal to

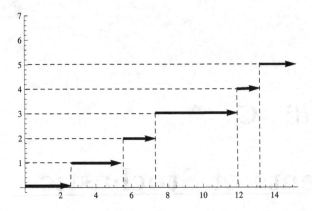

Figure C.1: *The Poisson process N_t.*

λdt, and the probability that at least 2 events occur in the same small interval is zero. This implies that the random variable dN_t may take only two values, 0 and 1, and hence satisfies

$$P(dN_t = 1) = \lambda\, dt \tag{C.1.3}$$
$$P(dN_t = 0) = 1 - \lambda\, dt. \tag{C.1.4}$$

Its graph looks like an increasing stair-type function with unit jumps, see Fig. C.1.

The mean and variance of increments of a Poisson process are given by

$$\mathbb{E}[N_t - N_s] = \lambda(t - s), \qquad Var[N_t - N_s] = \lambda(t - s).$$

In particular, the random variable N_t is Poisson distributed with $\mathbb{E}[N_t] = \lambda t$ and $Var[N_t] = \lambda t$. The parameter λ is called the *rate* of the process. This means that the events occur at the constant rate λ, with $\lambda > 0$.

As a consequence of the fact that the increments of N_t are independent, we have that if $0 \le s \le t$, then

1. $cov(N_s, N_t) = \lambda s$;
2. $cor(N_s, N_t) = \sqrt{\dfrac{s}{t}}.$

It is worth noting that the Poisson process N_t is not a martingale. However, subtracting the mean from a Poisson process, we obtain the $\mathcal{F}_t$-martingale $M_t = N_t - \lambda t$, which is called the *compensated Poisson process.*

The moment generating function of the random variable N_t is

$$m_{N_t}(x) = e^{\lambda t(e^x - 1)}.$$

The expressions for the first few moments are

$$
\begin{aligned}
\mathbb{E}[N_t] &= \lambda t \\
\mathbb{E}[N_t^2] &= \lambda^2 t^2 + \lambda t \\
\mathbb{E}[N_t^3] &= \lambda^3 t^3 + 3\lambda^2 t^2 + \lambda t \\
\mathbb{E}[N_t^4] &= \lambda^4 t^4 + 6\lambda^3 t^3 + 7\lambda^2 t^2 + \lambda t.
\end{aligned}
$$

And the first few central moments are given by

$$
\begin{aligned}
\mathbb{E}[N_t - \lambda t] &= 0 \\
\mathbb{E}[(N_t - \lambda t)^2] &= \lambda t \\
\mathbb{E}[(N_t - \lambda t)^3] &= \lambda t \\
\mathbb{E}[(N_t - \lambda t)^4] &= 3\lambda^2 t^2 + \lambda t.
\end{aligned}
$$

Interarrival times The path $t \to N_t$ is a step function that exhibits unit jumps at random times. Let T_1 be the random variable which describes the time of the 1st jump. Let T_2 be the time between the 1st jump and the second one. In general, denote by T_n the time elapsed between the $(n-1)$th and nth jumps. The random variables T_n are called *interarrival times*.

The random variables T_n are independent and exponentially distributed with mean $\mathbb{E}[T_n] = 1/\lambda$.

Waiting times The random variable $S_n = T_1 + T_2 + \cdots + T_n$ is called the *waiting time until the nth jump*. The event $\{S_n \le t\}$ means that there are n jumps that occurred before or at time t, i.e. there are at least n events that happened up to time t; the event is equal to $\{N_t \ge n\}$. The distribution function of S_n is given by

$$
F_{S_n}(t) = P(S_n \le t) = P(N_t \ge n) = e^{-\lambda t} \sum_{k=n}^{\infty} \frac{(\lambda t)^k}{k!}.
$$

Differentiating we obtain the density function of the waiting time S_n

$$
f_{S_n}(t) = \frac{t^{n-1} e^{-\lambda t}}{(1/\lambda)^n \Gamma(n)},
$$

i.e. S_n has a gamma distribution with parameters $\alpha = n$ and $\beta = 1/\lambda$. We have

$$
\mathbb{E}[S_n] = \frac{n}{\lambda}, \quad Var[S_n] = \frac{n}{\lambda^2}.
$$

The relation $\lim_{n \to \infty} \mathbb{E}[S_n] = \infty$ states that the expectation of the waiting time is unbounded as $n \to \infty$.

C.2   The Brownian Motion

Definition C.2.1 *A Brownian motion process is a stochastic process W_t, $t \geq 0$, which satisfies:*

1. *The process starts at the origin, $W_0 = 0$;*
2. *W_t has independent increments;*
3. *The process W_t is continuous in t;*
4. *The increments $W_t - W_s$ are normally distributed with mean zero and variance $|t - s|$,*

$$W_t - W_s \sim \mathcal{N}(0, |t - s|).$$

Condition 4 states that the increments of a Brownian motion are stationary, i.e. the distribution of $W_t - W_s$ depends only on the time interval $t - s$.

Even if W_t is continuous, it is nowhere differentiable. From condition 4 W_t is normally distributed with mean $\mathbb{E}[W_t] = 0$ and variance $Var[W_t] = t$. Its second moment is $\mathbb{E}[W_t^2] = t$.

A Brownian motion process W_t is a martingale with respect to the information set $\mathcal{F}_t = \sigma(W_s; s \leq t)$.

A process with similar properties as the Brownian motion was introduced by Wiener.

Definition C.2.2 *A Wiener process W_t is a process adapted to a filtration $\mathcal{F}_t$ such that*

1. *The process starts at the origin, $W_0 = 0$;*
2. *W_t is an $\mathcal{F}_t$-martingale with $\mathbb{E}[W_t^2] < \infty$ for all $t \geq 0$ and*

$$\mathbb{E}[(W_t - W_s)^2] = t - s, \qquad s \leq t;$$

3. *The process W_t is continuous in t.*

It can be shown that a Brownian process W_t is a Winer process. The converse also holds true.

Theorem C.2.3 (Lévy) *A Wiener process is a Brownian motion process.*

Sometimes the infinitesimal notation dW_t is used to denote the infinitesimal increment of a Wiener process in the time interval dt. In this case $dW_t \sim \mathcal{N}(0, dt)$, $\mathbb{E}[dW_t] = 0$, and $\mathbb{E}[(dW_t)^2] = dt$.

The Brownian motion is a Markov process (i.e. it is memoryless): The conditional distribution of W_{t+s}, given the present W_t and the past W_u, $0 \leq u < t$, depends only on the present.

The density function of the Brownian motion W_t is

$$\phi_t(x) = \frac{1}{\sqrt{2\pi t}} e^{-\frac{x^2}{2t}}.$$

Then its distribution function is given in the following

$$F_t(x) = P(W_t \le x) = \frac{1}{\sqrt{2\pi t}} \int_{-\infty}^{x} e^{-\frac{u^2}{2t}} \, du.$$

The probability that W_t is between the values a and b is given by

$$P(a \le W_t \le b) = \frac{1}{\sqrt{2\pi t}} \int_{a}^{b} e^{-\frac{u^2}{2t}} \, du, \qquad a < b.$$

The values of a Brownian motion are correlated as follows. Let $0 \le s \le t$. Then

1. $cov(W_s, W_t) = s$;
2. $cor(W_s, W_t) = \sqrt{\dfrac{s}{t}}.$

Integrated Brownian Motion The stochastic process

$$Z_t = \int_{0}^{t} W_s \, ds, \qquad t \ge 0$$

is called *integrated Brownian motion.*

Proposition C.2.4 *The integrated Brownian motion Z_t has a normal distribution with mean 0 and variance $t^3/3$.*

Brownian Motion with Drift The process $Y_t = \mu t + W_t$, $t \ge 0$, is called *Brownian motion with drift.* The process Y_t tends to drift off at a rate μ. It starts at $Y_0 = 0$ and it is a Gaussian process with mean $\mathbb{E}[Y_t] = \mu t$ and variance $Var[Y_t] = t$.

Multiplication Table The compatibility among dt, dW_t and dN_t is given by the following stochastic multiplication table:

	dt	dN_t	dW_t
dt	0	0	0
dN_t	0	dN_t	0
dW_t	0	0	dt

C.3   Exponential Process

Let $u : [0, T] \to \mathbb{R}$ be a continuous function. Then

$$M_t = e^{\int_0^t u(s)\, dW_s - \frac{1}{2}\int_0^t u^2(s)\, ds}$$

is an $\mathcal{F}_t$-martingale for $0 \le t \le T$. The condition that $u(s)$ is continuous on $[0, T]$ can be relaxed by asking only

$$u \in L^2[0, T] = \{u : [0, T] \to \mathbb{R}; \text{measurable and } \int_0^t |u(s)|^2\, ds < \infty\}.$$

It is worth noting that the conclusion still holds if the function $u(s)$ is replaced by a stochastic process $u(t, \omega)$ satisfying Novikov's condition

$$\mathbb{E}\left[e^{\frac{1}{2}\int_0^T u^2(s,\omega)\, ds}\right] < \infty.$$

Definition C.3.1 *Let $u \in L^2[0, T]$ be a deterministic function. Then the stochastic process*

$$M_t = e^{\int_0^t u(s)\, dW_s - \frac{1}{2}\int_0^t u^2(s)\, ds}$$

is called the exponential process induced by u.

Particular cases

1. Let $u(s) = \sigma$, constant, then $M_t = e^{\sigma W_t - \frac{\sigma^2}{2}t}$ is an $\mathcal{F}_t$-martingale.
2. Let $u(s) = s$. Integrating in $d(tW_t) = t\, dW_t - W_t\, dt$ yields

$$\int_0^t s\, dW_s = tW_t - \int_0^t W_s\, ds.$$

Let $Z_t = \int_0^t W_s\, ds$ be the integrated Brownian motion. Then

$$
\begin{aligned}
M_t &= e^{\int_0^t s\, dW_s - \frac{1}{2}\int_0^t s^2\, ds} \\
&= e^{tW_t - \frac{t^3}{6} - Z_t}
\end{aligned}
$$

is an $\mathcal{F}_t$-martingale.

C.4   Ito's Lemma

Ito's lemma is an analog of chain rule for stochastic calculus. We shall explain it by a heuristic analogy with deterministic calculus. Denote by f a twice continuously differentiable function of a real variable x. Let x_0 be fixed and consider the changes $\Delta x = x - x_0$ and $\Delta f(x) = f(x) - f(x_0)$. We write the following second order Taylor approximation

$$\Delta f(x) = f'(x)\Delta x + \frac{1}{2}f''(x)(\Delta x)^2 + O(\Delta x)^3.$$

When x is infinitesimally close to x_0, we replace Δx by the differential dx and obtain

$$df(x) = f'(x)dx + \frac{1}{2}f''(x)(dx)^2 + O(dx)^3. \tag{C.4.5}$$

In the elementary Calculus, all terms involving terms of equal or higher order to dx^2 are neglected; then the aforementioned formula becomes

$$df(x) = f'(x)dx.$$

Now, if we consider $x = x(t)$ to be a differentiable function of t, substituting into the previous formula we obtain the differential form of the well known chain rule

$$df\big(x(t)\big) = f'\big(x(t)\big)dx(t) = f'\big(x(t)\big)x'(t)\,dt.$$

A similar formula, involving an extra term, holds in the stochastic environment. In this case the deterministic function $x(t)$ is replaced by a stochastic process X_t. The composition between the differentiable function f and the process X_t is a process denoted by $F_t = f(X_t)$.

Neglecting the increment powers higher than or equal to $(dX_t)^3$, the expression (C.4.5) becomes

$$dF_t = f'(X_t)dX_t + \frac{1}{2}f''(X_t)\left(dX_t\right)^2. \tag{C.4.6}$$

In the computation of dX_t we may take into the account stochastic relations such as $dW_t^2 = dt$, or $dt\,dW_t = 0$.

Theorem C.4.1 (Ito's formula) *Let X_t be a stochastic process satisfying*

$$dX_t = b_t dt + \sigma_t dW_t,$$

with $b_t(\omega)$ and $\sigma_t(\omega)$ measurable processes. Let $F_t = f(X_t)$, with f twice continuously differentiable. Then

$$dF_t = \left[b_t f'(X_t) + \frac{\sigma_t^2}{2}f''(X_t)\right]dt + \sigma_t f'(X_t)\,dW_t. \tag{C.4.7}$$

Ito's formula can also be written under the following equivalent integral form

$$F_t = F_0 + \int_0^t \left(b_s f'(X_s) + \frac{1}{2} \sigma_s^2 f''(X_s) \right) ds + \int_0^t \sigma_s f'(X_s) \, dW_s.$$

In the particular case, when $X_t = W_t$ and $F_t = f(W_t)$, then

$$dF_t = \frac{1}{2} f''(W_t) dt + f'(W_t) \, dW_t.$$

For instance, the following formulas hold:

$$
\begin{aligned}
d(W_t^2) &= 2W_t \, dW_t + dt \\
d(W_t^3) &= 3W_t^2 \, dW_t + 3W_t dt \\
d(e^{W_t}) &= e^{W_t} dW_t + \frac{1}{2} e^{W_t} \, dt \\
d(\sin W_t) &= \cos W_t \, dW_t - \frac{1}{2} \sin W_t \, dt.
\end{aligned}
$$

Ito diffusion Consider the process X_t given by

$$dX_t = b(X_t, t)dt + \sigma(X_t, t)dW_t. \tag{C.4.8}$$

A process $X_t = (X_t^i) \in \mathbb{R}^n$ satisfying this relation is called an *Ito diffusion* in $\mathbb{R}^n$. Equation (C.4.8) models the position of a small particle that moves under the influence of a drift force $b(X_t, t)$, and is subject to random deviations. This situation occurs in the physical world when a particle suspended in a moving liquid is subject to random molecular bombardments. The amount $\frac{1}{2} \sigma \sigma^T$ is called the *diffusion coefficient* and describes the diffusion of the particle.

Ito's multidimensional formula If the process F_t depends on several Ito diffusions, say $F_t = f(t, X_t, Y_t)$, then the following formula holds

$$
\begin{aligned}
dF_t &= \frac{\partial f}{\partial t}(t, X_t, Y_t)dt + \frac{\partial f}{\partial x}(t, X_t, Y_t)dX_t + \frac{\partial f}{\partial y}(t, X_t, Y_t)dY_t \\
&+ \frac{1}{2} \frac{\partial^2 f}{\partial x^2}(t, X_t, Y_t)(dX_t)^2 + \frac{1}{2} \frac{\partial^2 f}{\partial y^2}(t, X_t, Y_t)(dY_t)^2 \\
&+ \frac{\partial^2 f}{\partial x \partial y}(t, X_t, Y_t)dX_t \, dY_t.
\end{aligned}
$$

C.5   Girsanov Theorem

The main usage of Girsanov's theorem is the reduction of drift. This result states that a Brownian motion with drift can be viewed as a regular Brownian motion under a certain change of the probability measure.

Theorem C.5.1 *Let W_t be a Brownian motion on the probability space $(\Omega, \mathcal{F}, P)$. Then the process*

$$X_t = \lambda t + W_t, \qquad 0 \le t \le T$$

becomes a Brownian motion on the probability space $(\Omega, \mathcal{F}, Q)$, where

$$dQ = e^{-\frac{1}{2}\lambda^2 T - \lambda W_T} dP.$$

A useful application of Girsanov theorem are the following reduction of drift formulas:

$$(i) \qquad \mathbb{E}[f(\lambda t + W_t)] = e^{-\frac{\lambda^2 t}{2}} \mathbb{E}[f(W_t)e^{\lambda W_t}]$$

$$(ii) \qquad \mathbb{E}[f(W_t)] = e^{-\frac{\lambda^2 t}{2}} \mathbb{E}[f(\lambda t + W_t)e^{-\lambda W_t}],$$

with f measurable function.

Girsanov's Theorem works also in the more general setup when the drift rate is not constant:

Theorem C.5.2 (General Girsanov's Theorem) *Let $u \in L^2[0,T]$ be a deterministic function. Then the process*

$$X_t = \int_0^t u(s)\, ds + W_t, \qquad 0 \le t \le T$$

is a Brownian motion with respect to the probability measure Q given by

$$dQ = e^{-\int_0^T u(s)\, dW_s - \frac{1}{2}\int_0^T u(s)^2\, ds} dP.$$

For a gentle introduction into this subject the interested reader can consult Baxter and Rennie [8], or Neftici [56].

C.6   Ito Integral

Let $F_t = f(W_t, t)$ be a nonanticipating process (i.e. a process which does not depend on future increments of W_t), with $a \le t \le b$. Divide the interval $[a, b]$ into n subintervals using the partition points

$$a = t_0 < t_1 < \cdots < t_{n-1} < t_n = b,$$

and consider the partial sums

$$S_n = \sum_{i=0}^{n-1} F_{t_i}(W_{t_{i+1}} - W_{t_i}).$$

We emphasize that the intermediate points are the left endpoints of each interval, and this is the way they should always be chosen.

The *Ito integral* is the mean square limit of the partial sums S_n

$$\text{ms-}\lim_{n \to \infty} S_n = \int_a^b F_t \, dW_t,$$

provided the limit exists. The previous convergence means

$$\lim_{n \to \infty} \mathbb{E}\left[\left(S_n - \int_a^b F_t \, dW_t\right)^2\right] = 0.$$

It can be shown that the choice of partition does not influence the value of the Ito integral. This is the reason why, for practical purposes, it suffices to assume the intervals equidistant, i.e.

$$t_{i+1} - t_i = \frac{(b-a)}{n}, \qquad i = 0, 1, \cdots, n-1.$$

The following properties of the Ito integral are similar with those of the Riemannian integral.

Proposition C.6.1 *Let $f(W_t, t)$, $g(W_t, t)$ be non-anticipating processes and $c \in \mathbb{R}$. Then we have*

1. *Additivity:*

$$\int_0^T [f(W_t, t) + g(W_t, t)] \, dW_t = \int_0^T f(W_t, t) \, dW_t + \int_0^T g(W_t, t) \, dW_t.$$

2. *Homogeneity:*

$$\int_0^T cf(W_t, t) \, dW_t = c \int_0^T f(W_t, t) \, dW_t.$$

3. *Partition property:*

$$\int_0^T f(W_t, t) \, dW_t = \int_0^u f(W_t, t) \, dW_t + \int_u^T f(W_t, t) \, dW_t, \qquad \forall 0 < u < T.$$

As a random variable, the Ito integral satisfies the following properties:

Proposition C.6.2 *We have*
1. *Zero mean:*

$$\mathbb{E}\left[\int_a^b f(W_t, t) \, dW_t\right] = 0.$$

2. *Isometry:*

$$\mathbb{E}\left[\left(\int_a^b f(W_t, t)\, dW_t\right)^2\right] = \mathbb{E}\left[\int_a^b f(W_t, t)^2\, dt\right].$$

3. *Covariance:*

$$\mathbb{E}\left[\left(\int_a^b f(W_t, t)\, dW_t\right)\left(\int_a^b g(W_t, t)\, dW_t\right)\right] = \mathbb{E}\left[\int_a^b f(W_t, t)g(W_t, t)\, dt\right].$$

Another important property of Ito integrals states that if $X_t = \int_0^t f(W_s, s)\, dW_s$, with $E\left[\int_0^\infty f^2(W_s, s)\, ds\right] < \infty$, then X_t is a continuous $\mathcal{F}_t$-martingale.

Wiener integral The *Wiener integral* is a particular case of the Ito integral, which is obtained by replacing the non-anticipating stochastic process $f(W_t, t)$ by the deterministic function $f(t)$. The Wiener integral $\int_a^b f(t)\, dW_t$ is the mean square limit of the partial sums

$$S_n = \sum_{i=0}^{n-1} f(t_i)(W_{t_{i+1}} - W_{t_i}).$$

In addition to all the properties of Ito integrals, Wiener integrals satisfy the following stronger result:

Proposition C.6.3 *The Wiener integral $I(f) = \int_a^b f(t)\, dW_t$ is a normal random variable with mean 0 and variance*

$$Var[I(f)] = \int_a^b f(t)^2\, dt.$$

C.7 Brownian Motion with Drift

1. Assume $\mu, x > 0$. Let τ be the time when the process

$$X_t = \mu t + \sigma W_t$$

hits x for the first time.

(*a*) We have

$$\mathbb{E}[e^{-s\tau}] = e^{\frac{1}{\sigma^2}(\mu - \sqrt{2s\sigma^2 + \mu^2})x}.$$

(*b*) The density function of τ is given by

$$p(\tau) = \frac{x}{\sigma\sqrt{2\pi}\tau^{3/2}} e^{-\frac{(x-\mu\tau)^2}{2\tau\sigma^2}}.$$

(c) The mean of τ is
$$\mathbb{E}[\tau] = \frac{x}{\mu}.$$

2. Assume $\mu, x > 0$. Let τ be the time when the process
$$X_t = \mu t + \sigma W_t$$

hits $-x$ for the first time.

(a) We have
$$\mathbb{E}[e^{-s\tau}] = e^{-\frac{1}{\sigma^2}(\mu + \sqrt{2s\sigma^2 + \mu^2})x}.$$

(b) The density function of τ is given by
$$p(\tau) = \frac{x}{\sigma\sqrt{2\pi}\tau^{3/2}} e^{-\frac{(x+\mu\tau)^2}{2\tau\sigma^2}}.$$

(c) The mean of τ is
$$\mathbb{E}[\tau] = \frac{x}{\mu} e^{-\frac{2\mu x}{\sigma^2}}.$$

For the proof, see Propositions 4.6.4, 4.6.5 and 4.6.6. in Calin [15].

C.8   The Generator of an Ito Diffusion

Let $(X_t)_{t \geq 0}$ be a stochastic process with $X_0 = x$. The rate of change of a function f, which depends smoothly on X_t, is given by
$$\mathcal{A}f(x) = \lim_{t \searrow 0} \frac{\mathbb{E}^x[f(X_t)] - f(x)}{t},$$

where f is at least of class C^2, with compact support, i.e. $f : \mathbb{R}^n \to \mathbb{R}$, $f \in C_0^2(\mathbb{R}^n)$. Here $\mathbb{E}^x$ stands for the expectation operator given the initial condition $X_0 = x$, i.e.,
$$\mathbb{E}^x[f(X_t)] = \mathbb{E}[f(X_t)|X_0 = x] = \int_{\mathbb{R}^n} f(y) p_t(x, y) \, dy,$$

where $p_t(x, y) = p(x, y; t, 0)$ is the transition density of X_t, given $X_0 = x$ (the initial value X_0 is a deterministic value x).

The operator $\mathcal{A}$ describes infinitesimally the rate of change of the function f along the process X_t. It is a second order partial differential operator and is called the *infinitesimal generator* of the stochastic process X_t.

In the case when X_t is an Ito diffusion, the generator $\mathcal{A}$ takes a distinguished form.

Theorem C.8.1 *Consider the Ito diffusion*

$$dX_t = b(X_t)dt + \sigma(X_t)dW(t), \qquad t \geq 0, X_0 = x, \qquad (C.8.9)$$

where $W(t) = (W_1(t), \ldots, W_m(t))$ *is an m-dimensional Brownian motion, with* $b : \mathbb{R}^n \to \mathbb{R}^n$ *and* $\sigma : \mathbb{R}^n \to \mathbb{R}^{n \times m}$ *measurable functions. The generator of the Ito diffusion (C.8.9) is given by*

$$\mathcal{A} = \frac{1}{2} \sum_{i,j} (\sigma\sigma^T)_{ij} \frac{\partial^2}{\partial x_i \partial x_j} + \sum_k b_k \frac{\partial}{\partial x_k}. \qquad (C.8.10)$$

The matrix σ is called *dispersion* and the product $\sigma\sigma^T$ is called *diffusion* matrix. These names are related with their physical significance.

Particular cases:
(*i*) If $X_t = W_t$ is a one-dimensional Brownian motion, then $\mathcal{A} = \frac{1}{2}\frac{d^2}{dx^2}$.
(*ii*) If $X_t = \mu t + \sigma W_t$ is a one-dimensional Brownian motion with drift, then the infinitesimal generator is $\mathcal{A} = \frac{1}{2}\sigma^2 \frac{d^2}{dx^2} + \mu \frac{d}{dx}$.
(*iii*) The infinitesimal generator for the geometric Brownian motion $dX_t = \mu X_t dt + \sigma X_t dW_t$ is given by $\mathcal{A} = \frac{1}{2}\sigma^2 x^2 \frac{d^2}{dx^2} + \mu x \frac{d}{dx}$.
(*iv*) If X_t and Y_t are two one-dimensional independent Ito diffusions with infinitesimal generators $\mathcal{A}_X$ and $\mathcal{A}_Y$, and denote $Z_t = (X_t, Y_t)$, then the infinitesimal generator of Z_t is given by $\mathcal{A}_Z = \mathcal{A}_X + \mathcal{A}_Y$.

The relation between the infinitesimal generator and the expectation operator is given by the following formula:

Theorem C.8.2 (Dynkin's formula) *Let* $f \in C_0^2(\mathbb{R}^n)$, *and* X_t *be an Ito diffusion starting at* x. *If* τ *is a stopping time with* $\mathbb{E}[\tau] < \infty$, *then*

$$\mathbb{E}^x[f(X_\tau)] = f(x) + \mathbb{E}^x\left[\int_0^\tau \mathcal{A}f(X_s)\,ds\right], \qquad (C.8.11)$$

where A is the infinitesimal generator of X_t.

We note the use of the conditional expectation notation

$$\mathbb{E}^x[f(X_\tau)] = \mathbb{E}[f(X_\tau)|X_0 = x].$$

An application of Dynkin's formula is given in the following:

Theorem C.8.3 (Kolmogorov's backward equation) *For any* $f \in C_0^2(\mathbb{R}^n)$ *the function* $v(t, x) = \mathbb{E}^x[f(X_t)]$ *satisfies the following Cauchy's problem*

$$\frac{\partial v}{\partial t} = \mathcal{A}v, \qquad t > 0$$
$$v(0, x) = f(x),$$

where $\mathcal{A}$ *denotes the generator of the Ito's diffusion (C.8.9).*

Using $\mathbb{E}^x[f(X_t)] = \int_{\mathbb{R}^n} f(y)p_t(x,y)\, dy$, where $p_t(x,y) = p(x,y;t,0)$ is the transition density of X_t, given $X_0 = x$, it follows that the solution can also be written as

$$v(t,x) = \int_{\mathbb{R}^n} f(y)p_t(x,y)\, dy.$$

In fact, the probability density $p_t(0,y)$ is the heat kernel of the infinitesimal generator operator A, i.e.:

$$\frac{\partial p}{\partial t} = Ap, \qquad t > 0$$
$$\lim_{t \to 0+} p_t(0,y) = \delta(y),$$

where $\delta(x)$ is Dirac's delta function.

A generalization of the Kolmogorov's backward equation in the case when a potential function is included, was done by Feynman and Kac, see [45].

Theorem C.8.4 (Feynman-Kac) *Let X_t be an Ito diffusion satisfying the stochastic differential equation (C.8.1), and A be its generator. Let $f \in C_0^2(\mathbb{R}^n)$ and $U(x)$ be a lower-bounded continuous function on $\mathbb{R}^n$. Then the function*

$$v(t,x) = \mathbb{E}[f(X_t)e^{-\int_0^t U(X_s)\, ds}|X_0 = x], \qquad t \geq 0, x \in \mathbb{R}^n$$

satisfies the initial value problem

$$\frac{\partial v}{\partial t} = Av - Uv, \qquad t > 0$$
$$v(0,x) = f(x).$$

We note that the Kolmogorov's backward equation is recovered for the case $U = 0$.

Retrieving Black-Scholes In the following we show how the Black-Scholes equation and its solution can be obtained from the Feynman-Kac formula. The stock price satisfies in the risk-neutral world the Ito diffusion

$$dS = rS dt + \sigma S dW_t,$$

whose generator is

$$A = \frac{1}{2}\sigma^2 S^2 \frac{\partial^2}{\partial S^2} + \mu S \frac{\partial}{\partial S}.$$

Consider the constant potential $U = r$, where r denotes the risk-free rate. Also consider the time to maturity $\tau = T - t$. Then the Feynman-Kac equation becomes

$$\frac{\partial v}{\partial \tau} + \mathcal{A}v = rv, \quad 0 < \tau < T$$
$$v(T, S) = f(S).$$

This is nothing but the Black-Scholes equation for a derivative $v = v(\tau, S)$ with payoff $f(S)$. The derivative price is given by the Feynman-Kac formula, which in the new variables takes the form

$$v(\tau, S) = \mathbb{E}[f(S_t)e^{-r(T-\tau)}|X_\tau = S] = e^{-r(T-\tau)}\widehat{\mathbb{E}}_t[f(S_t)].$$

Hence, we retrieved the risk-neutral valuation formula.

Appendix D

Series and Equations

D.1 Confluent Hypergeometric Equation

The ordinary differential equation

$$z\frac{d^2w}{dz^2} + (b - z)\frac{dw}{dz} - aw = 0$$

has a regular singular point at $z = 0$. Then its solution space is 2-dimensional and any solution is a linear combination of two linear independent solutions $M(a, b; z)$ and $U(a, b; z)$, i.e

$$w(z) = c_1 M(a, b; z) + c_2 U(a, b; z), \qquad c_1, c_2 \in \mathbb{R}.$$

The first solution is the Kummer's confluent hypergeometric function

$$M(a, b; z) = {}_1F_1(a, b; z) = \sum_{n=0}^{\infty} \frac{(a)^n}{(b)^n} \frac{z^n}{n!},$$

with $a^{(0)} = 1$ and $a^{(n)} = a(a + 1)(a + 2) \cdots (a + n - 1)$. We note the relation with the binomial coefficient

$$\frac{(a)^n}{n!} = \binom{a+n-1}{n}.$$

Consequently, when $a = b$, we obtain $M(a, a; z) = e^z$.

The second solution is given by Tricomi's function

$$U(a, b; z) = \frac{\Gamma(1 - b)}{\Gamma(a - b + 1)} M(a, b; z) + \frac{\Gamma(b - 1)}{\Gamma(a)} z^{1-b} M(a - b + 1, 2 - b, z).$$

The integral representations of these two functions are given by

$$M(a, b; z) = \frac{\Gamma(b)}{\Gamma(a)\Gamma(b-a)} \int_0^1 e^{zu} u^{a-1}(1-u)^{b-a-1}\, du$$

$$U(a, b; z) = \frac{1}{\Gamma(a)} \int_0^\infty e^{-zt} t^{a-1}(1+t)^{b-a-1}\, dt,$$

with a, b real and positive. More formulas of this type can be found in Erdélyi [30]. It is worth noting that for $a > 0$ and $0 < b < 1$ the initial values are given by

$$M(a, b; 0) = 1$$

$$U(a, b; 0) = \frac{\Gamma(1-b)}{\Gamma(1-b+a)}.$$

D.2 Duhamel's Principle

This section presents a method for solving non-homogeneous Cauchy problems of the type

$$x' - Ax = b \tag{D.2.1}$$

$$x(0) = x_0, \tag{D.2.2}$$

where b is a function of t and A denotes a linear operator. Using linearity, the previous problem can be written as the sum of the following two problems:

$$x' - Ax = 0 \qquad\qquad\qquad x' - Ax = b$$

$$x(0) = x_0 \qquad\qquad\qquad x(0) = 0.$$

Let $x_1(t)$ and $x_2(t)$ be the solutions of the above problems, respectively. We can easily verify that $x_1(t) = e^{At}x_0$. To find the solution $x_2(t)$ we apply the Laplace transform to the second problem

$$sX(s) - x_2(0) - AX(s) = B(s),$$

where $X(s) = \mathcal{L}(x)(s)$, $B(s) = \mathcal{L}(b)(s)$, and $x_2(0) = 0$. Solving for $X(s)$, we have

$$X(s) = \frac{1}{s-A} B(s) = \mathcal{L}(e^{At})(s)\mathcal{L}(b)(s).$$

The convolution property of Laplace transforms yields

$$x_2(t) = \mathcal{L}^{-1}(X(s))(t) = e^{At} * b(t) = \int_0^t e^{A(t-\tau)} b(\tau)\, d\tau.$$

Duhamel's principle states that x_2 can be obtained from x_1 as follows. First, we associate the homogeneous Cauchy problem

$$x' - Ax = 0$$
$$x(0) = b(\tau),$$

with solution $y(t, \tau) = e^{At}b(\tau)$. Then in the light of the previous integral formula, we infer that

$$x_2(t) = \int_0^t y(t - \tau, \tau) \, d\tau.$$

Then the solution of the initial problem (D.2.1)-(D.2.1) is obtained by super-position

$$x(t) = x_1(t) + x_2(t)$$
$$= e^{At}x_0 + \int_0^t e^{A(t-\tau)}b(\tau) \, d\tau. \tag{D.2.3}$$

The case when A is a differential operator is of distinguished importance. In this case the expression $e^{At}x_0$ can be computed as an integral involving the heat kernel $G(\ ,\ )$ of the operator A as

$$e^{At}x_0(x, t) = \int G(x - y, t)x_0(y) \, dy.$$

We shall present Duhamel's formula for the case of the differential operator $A = \frac{d^2}{dx^2}$, which plays a central role for the material of this book.

Proposition D.2.1 *The Cauchy problem*

$$\frac{\partial}{\partial t}u(x, t) - \frac{\partial^2}{\partial x^2}u(x, t) = F(x, t)$$
$$u(x, 0) = f(x) \tag{D.2.4}$$

has the solution

$$u(x, t) = \int G(x - y, t)f(y) \, dy$$
$$+ \int_0^t \int G(x - y, t - \tau)F(y, \tau) \, dy d\tau, \tag{D.2.5}$$

where $G(\ ,\ )$ stands for the heat kernel

$$G(x, \tau) = \frac{1}{\sqrt{4\pi\tau}}e^{-\frac{x^2}{4\tau}}, \qquad \tau > 0.$$

It is worth noting that formula (D.2.5) provides the temperature in an infi-nite rod, given the initial temperature distribution $f(x)$ and continuous time-depending heat source density $F(x, t)$.

Bibliography

[1] M. Abramovitz and I. Stegun. *Handbook of Mathematical Functions with Formulas, Graphs, and Mathematical Tables.* New York: Dover, 1965.

[2] T. Adolfsson, C. Chiarella, A. Ziogas, and J. Ziveyi. *Representation and Numerical Approximation of American Option Prices under Heston Stochastic Volatility Dynamics.* Quantitative Finance Research Centre, University of Technology Sydney, research paper, no. 327, 2013.

[3] V. I. Arnold. *Ordinary Differential Equations.* MIT Press, Cambridge, MA, London, 1973.

[4] I. G. Avramidi. *Heat Kernel Method and Its Applications.* Birkhäuser, Heidelberg, 2015.

[5] G. Barone-Adesi and R. E. Whaley. *Efficient Analytic Approximation of American Option Values.* The Journal of Finance 42, 1987, pp. 301–320.

[6] G. Barone-Adesi and R. E. Whaley. *On the Valuation of American Put Options on Dividend-Paying Stocks.* Advances in Futures and Options Research 3, 1988, pp. 1–13.

[7] D. S. Bates. *Pricing Options Under Jump-Diffusion Processes.* The Wharton School, 1988.

[8] M. Baxter and A. Renie. *Financial Calculus.* Cambridge University Press, 1996.

[9] J. Bertoin. *Lévy Processes.* Cambridge University Press, 121, 1996.

[10] F. Black, E. Derman, and W. Toy. *A One-Factor Model of Interest Rates and Its Application to Treasury Bond Options.* Financial Analysts Journal, Jan.-Febr., 1990, pp. 33–39.

[11] F. Black and P. Karasinski. *Bond and option pricing when short rates are lognormal.* Financial Analysts Journal, 1991, pp. 52–59.

[12] T. Bollerslev. *Generalized Autoregressive Conditional Heteroskedasticity.* Journal of Econometrics, 31, 1986, pp. 307–327.

[13] L. Bouaziz, E. Briys, and M. Crouhy. *The pricing of forward-starting Asian options.* Journal of Banking & Finance, 18, issue 5, 1994, pp. 823–839.

[14] M. Broadie, J. Detemple, E. Ghysels, and O. Torres. *American Options with Stochastic Dividends and Volatility: A Nonparametric Investigation.* Journal of Econometrics, 94, 2000, pp. 53–92.

[15] O. Calin. *An Informal Introduction to Stochastic Calculus with Applications.* World Scientific, Singapore, 2015.

[16] O. Calin and F. Al-Azemi. *Asian Options with Harmonic Average.* Appl. Math. Inf. Sci. 9 (6), 2015, pp. 2803–2811.

[17] O. Calin and C. Udriste. *Geometric Modeling in Probability and Statistics.* Springer, New York, 2014.

[18] J. Y. Campbell. *Empirical asset pricing: Eugene Fama, Lars Peter Hansen and Robert Shiller.* working paper, Harvard University, 2014.

[19] J. Y. Campbell, G. Chacko, J. Rodriguez, and L. M. Viceira. *Strategic asset allocation in a continuous-time VAR model.* Journal of Economic Dynamics and Control 28, no. 11, 2004, pp. 2195–2214.

[20] R. Carmona. *Statistical Analysis of Financial Data in R, 2nd ed.* Springer, New York, 2014.

[21] E. Çinlar. *Probability and Stochastics.* Springer, New York, 2011.

[22] J. H. Cochrane. *Gene Fama's Nobel Prize.* working paper, University of Chicago, 2013.

[23] J. C. Cox, J. E. Ingersoll, and S. A. Ross. *A Theory of the term Structure of Interest Rates.* Econometrica, 53, 1985, pp. 385–407.

[24] J. Detemple and W. Tian. *The Valuation of American Options for a Class of Difussion Processes.* Management Science **48**(7), 2002, pp. 917–937.

[25] P. Diaconis, S. Holmes, and R. Montgomery. *Dynamical Bias in the Coin Toss.* SIAM Review, 49, no. 2, 2007, pp. 211–235.

[26] C. Donati-Martin, R. Ghomrasni, and M. Yor. *On certain Markov processes attached to exponential functionals of Brownian motion; application to Asian options.* Revista Matematica Iberoamericana 17, pp. 179–193, 2001.

[27] J. L. Doob. *Stochastic Processes*. John Wiley and Sons, 1953.

[28] Z. Drezner. *Computation of the Bivariate Normal Integral*. Mathematics of Computation, 32, 1978, pp. 277–79.

[29] D. Dufresne. *The Integrated Square Root Process*. Centre for Actuarial Studies, Department of Economics, University of Melbourne, research paper, no. 90, 2001.

[30] A. Erdélyi. *Higher Transcendental Functions, vols. I, II, III*. McGraw-Hill, Inc. Springer, 1953.

[31] A. Etheridge. *A Course in Financial Calculus*. Cambridge University Press, Cambridge, 2008.

[32] A. Eydeland and H. Geman. *Asian options revisited: inverting the Laplace transform*. RISK Magazine, March, 1995.

[33] J. P. Fouque, G. Papanicolaou, and R. Sircar. *Derivatives in Financial Markets with Stochastic Volatility*. Cambridge University Press, London, 2000.

[34] J. P. Fouque, G. Papanicolaou, R. Sircar, and K. Sølna. *Multiscale Stochastic Volatility for Equity, Interest Rate, and Credit Derivatives*. Cambridge University Press, London, 2011.

[35] H. Geman and M. Yor. *Bessel processes, Asian options, and perpetuities*. Mathematical Finance, 1993, 3, pp. 349–375.

[36] R. B. Guenther and J. W. Lee. *Partial Differential Equations of Mathematical Physics and Integral Equations*. Prentice-Hall, London, 1988.

[37] A. Gulisashvili. *Analytically Tractable Stochastic Stock Price Models*. Springer-Verlag, Berlin, Heidelberg, 2012.

[38] D. Health, R. Jarrow, and A. Morton. *Bond pricing and the Term Structure of Interest Rates: a new methodology for contingent claims valuation*. Econometrica, 60, 1, 1992, pp. 77–105.

[39] S. L. Heston. *A Closed-Form Solution for Options with Stochastic Volatility with Applications to Bonds and Currency Options*. Review of Financial Studies, 6 (2), 1993, pp. 327–343.

[40] T. S. Y. Ho and S. B. Lee. *Term Structure Movements and Pricing Interest Rate Contingent Claims*. Journal of Finance, 41, 1986, pp. 1011–1029.

[41] J. Hull and A. White. *Pricing Interest Rates Derivative Securities*. Review of Financial Studies, 3, 4, 1990, pp. 573–592.

[42] J. C. Hull. *Options, Futures, and other Derivatives*. 3rd edition, Prentice-Hall, 1996.

[43] F. Jamshidian. *An Analysis of American Options*. Review of Futures Markets, 11, 1992, pp. 72–80.

[44] P. E. Jones. *Option Arbitrage and Strategy with Large Price Changes*. Journal of Financial Economics, 13, no. 1, 1984, pp. 91–113.

[45] M. Kac. *On distributions of certain Wiener functionals*. Trans. Am. Math. Soc., 65, 1949, pp. 1–13.

[46] I. Karatzas and S. E. Shreve. *Brownian Motion and Stochastic Calculus, 2nd ed.* Springer-Verlag, 1991.

[47] J. Keller. *The probability of heads*. Amer. Math. Monthly, 93, 1986, pp. 191–197.

[48] A. G. Z. Kemna and A. C. F. Vorst. *A pricing method for options based on average asset values*. Journal of Banking and Finance, 14, 1990, pp. 113–129.

[49] M. Kijima. *Stochastic Processes with Applications to Finance*. Chapman and Hall, Boca Raton, 2003.

[50] A. L. Lewis. *Option Valuation under Stochastic Volatility*. Finance Press, California, 2000.

[51] L. V. MacMillan. *Analytic Approximation for the American Put Option*. Advances in Futures and Options Research 1, 1986, pp. 119–139.

[52] B. G. Malkiel and E. Fama. *Efficient Capital Markets: a Review of Theory and Empirical Work*. The Journal of Finance, 25, issue 2, 1970, pp. 383–417.

[53] R. L. McDonald. *Derivatives Markets, 2nd ed.* Addison Wesley, Boston, 2006.

[54] R. C. Merton. *Option Pricing with Discontinuous Returns*. Journal of Financial Economics, 3, 1976, pp. 125–144.

[55] R. C. Merton. *Theory of Rational Option Pricing*. The Bell Journal of Economics and Management Science, Vol. 4, No. 1, (Spring, 1973), pp. 141–183.

[56] S. Neftici. *Mathematics of Financial Derivatives*. Academic Press, 1996.

[57] B. Øksendal. *Stochastic Differential Equations, An Introduction with Applications, 6th ed.* Springer-Verlag Berlin Heidelberg New-York, 2003.

[58] G. Peskir and A. Shiryaev. *Optimal Stopping and Free-Boundary Problems*. Birkhäuser Verlag, Basel, 2006.

[59] R. Rendleman and B. Bartter. *The Pricing of Options on Debt Securities*. Journal of Financial and Quantitative Analysis, 15, 1980, pp. 11–24.

[60] D. Fabrice Rouah. *Heston Model and Its Extensions in Matlab and C#*. John Wiley, Somerset, 2013.

[61] P. A. Samuelson. *Proof that Properly Anticipated Prices Fluctuate Randomly*. Industrial Management Review, 6, 1965, pp. 41–49.

[62] A. N. Shiryaev. *Probability*. Springer, New York, 1996.

[63] R. S. Tsay. *Analysis of Financial Time Series*. John Wiley & Sons, New York, 2002.

[64] S. M. Turnbull and L. M. Wakeman. *A quick algorithm for pricing European average options*. Journal of Financial and Quantitative Analysis, 26, 1991, pp. 377–389.

[65] E. Tzavalis and S. Wang. *Pricing American Options under Stochastic Volatility: A New Method using Chebyshev Polynomials to Approximate the Early Exercise Boundary*. Department of Economics, Queen Mary, University of London, Working paper no. 488, 2003.

[66] O. A. Vasicek. *An Equilibrium Characterization of the Term Structure*. Journal of Financial Economics, 5, 1977, pp. 177–188.

[67] J. Vecer. *Asian options on the harmonic average*. Quantitative Finance, 14 (8), 2014, pp. 1315–1322, DOI:10.1080/14697688.2013.847281.

[68] T. Vorst. *Prices and hedge rations of average exchange options*. International Review of Financial Analysis, 1, issue 3, 1992, pp. 179–193.

[69] P. Wilmott, J. Dewynne, and S. Howison. *Option Pricing - Mathematical Models and Computation*. Oxford Financial Press, 1993.

[70] M. Yor. *On some exponential functionals of Brownian motion*. Adv. Appl. Prob., 24, 1992, pp. 509–531.

Index

Printed in the United States
By Bookmasters